Oxidative Stress in Applied Basic Research and Clinical Practice

Series Editor

Narasimham Parinandi, Dept Internal Med, Pulmonatry Division, Ohio State University, Columbus, USA

The current definition of oxidative stress (OS) is an abnormal level of reactive oxygen-nitrogen species leading to cellular injury, which can be prevented by antioxidant (AOX) intake providing a therapeutic advantage. When this delicate balance is perturbed, homeostasis is compromised and disease may develop. Extensive research has shown a strong correlation with these events. Springer has been a leader in publishing books related to oxidative stress since 1998. The goal of the Oxidative Stress in Basic Research and Clinical Practice series is to present data showing relationships between OS and biomarkers, as well as results from clinical trials illustrating the efficacy of AOX supplementation. Each volume covers a medical specialty, or major disease known to be associated with an imbalance between free radical production and AOX defense. The expectation of the series is to critically evaluate the potential of AOXs in therapeutic intervention using simple and sophisticated methodology that could be adapted to point-of-care testing for diagnosis, intervention strategies and prognosis. This approach of combining methodology, technology and epidemiology will benefit medical practitioners and may subsequently improve health care and reduce medical cost. Editors anticipate updating their respective volumes as frequently as warranted by emerging new studies.

Nidhi Srivastava • Shubhini A. Saraf
Editors

Pharmacological Strategies for Prevention and Treatment of Oxidative Stress

Humana Press

Editors
Nidhi Srivastava
NIPER-Raebareli
Lucknow, Uttar Pradesh, India

Shubhini A. Saraf
Department of Pharmaceutics
NIPER-Raebareli
Lucknow, Uttar Pradesh, India

ISSN 2197-7224 ISSN 2197-7232 (electronic)
Oxidative Stress in Applied Basic Research and Clinical Practice
ISBN 978-3-032-19836-5 ISBN 978-3-032-19837-2 (eBook)
https://doi.org/10.1007/978-3-032-19837-2

This Humana imprint is published by the registered company Springer Nature Switzerland AG
The registered company address is: Gewerbestrasse 11, 6330 Cham, Switzerland

Foreword

It is my greatest pleasure to write this foreword for the book entitled *Pharmacological Strategies for Prevention and Treatment of Oxidative Stress*, edited by Prof. Shubhini A. Saraf and Dr. Nidhi Srivastava. The book is a collection of unique and up-to-date chapters on current and advancing scientific research in the field of strategies for pharmacological prevention of oxidative stress. The basic biology and biochemistry of reactive oxygen species generation, leading to oxidative stress, and the pharmacological attenuation and protection of oxidative stress, along with their implications in debilitating diseases and health, have been extensively covered in the chapters by experts in these fields of research. As the mechanistic paradigms of antioxidant actions in normal physiology and pathophysiology are rapidly evolving, this book will be a leader showing lucid directions in the arena of pharmacological intervention of chronic and life-threatening disease states (e.g., neurological diseases, psychiatric and behavioral disorders, neurodegenerative diseases, cancer, diabetes, cardiovascular diseases) by antioxidant treatments in humans encountering oxidative stress. Thus, this book will play a crucial role in the advancement of pharmacological interventions for chronic diseases faced by people on a global scale.

It has been clearly established that oxidative stress is caused by endogenous metabolic dysregulation through pathophysiological, toxicological, and genetic factors; however, the implementation of antioxidant interventions through appropriate pharmacological treatments has been complicated due to the complexity of the cascades of oxidant generation in cells and the targets of the chosen pharmacological antioxidant molecules. Each system in every disease is unique, contributing to a distinct scenario and mechanism of oxidative stress induction, as well as a diverse range of cellular, biochemical, and molecular targets. This complexity creates a tangled network of signaling events that are intervened by pharmacological antioxidants, specifically and effectively, in a particular disease state. We are at a stage of untangling the complexity of oxidative stress and antioxidant interventions in various disease states, aiming to achieve an effective and targeted treatment of oxidative stress to combat oxidant-driven pathophysiological states.

At this juncture, combating oxidative stress-induced/mediated human diseases through the pharmacological intervention of oxidative stress, utilizing effective antioxidants, is highly promising for accomplishing the pharmacological treatment of oxidant-mediated diseases. Each chapter in this book is unique, and the book will serve not only as a textbook in the classroom but also as excellent reference material for investigators in the field of oxidative stress and pharmacological intervention of oxidant injury in pathophysiological states.

I immensely enjoyed reading every chapter written by the authors. It is also promising that this book will have a significant impact on the advancement of the field of biomedical oxidative stress and the pharmacological intervention of oxidant-mediated diseases. I congratulate both editors for their indefatigable tenacity and commitment to assembling exceptional chapters contributed by expert scientists. I wholeheartedly thank all the authors and editors for their outstanding contributions in bringing this unique book to fruition, which will aid in the advancement of the treatment of oxidative stress-induced/mediated human diseases.

Department of Internal Medicine,
Dorothy M. Davis Heart & Lung
Research Institute, The Ohio State
University Wexner Medical Center &
College of Medicine, Columbus, OH,
USA

Narasimham L. Parinandi B.Sc. (Hons),
M.Sc., Ph.D.

Preface

This book enumerates the significance of oxidants and antioxidants in human health and disease. It explains how the excessive production of reactive oxygen species (ROS) leads to oxidative stress, a condition that plays a significant role in the development of various disorders. The book also discusses methods for preventing and treating these conditions.

Knowledge is shared in the chapters by leading experts from around the world. This book provides a clear understanding of how oxidative stress affects the body at the molecular level. It explores both the problems caused by oxidative stress and the potential of antioxidants as therapeutic agents. Readers will gain knowledge ranging from basic concepts to advanced research related to oxidative stress and its impact on human health.

The chapters cover various aspects of oxidative stress in health and disease.

- The first chapter gives an overview of oxidants and oxidative stress in the human body.
- The subsequent chapters describe different types of oxidants and antioxidants and how they are linked to human diseases.
- One chapter explains the role of *lipid peroxidation* in oxidative stress.
- Another discusses the *signaling pathways* that become active when ROS balance is disturbed, as well as the body's detoxification systems.
- Further chapters describe how oxidative stress activates *transcription factors* that trigger inflammation, and how this relates to disease mechanisms.
- The book also examines how oxidative stress can disturb the *gut–brain axis*, weaken the blood-brain and intestinal barriers, and contribute to disorders such as *cancer, diabetes, neurodegenerative diseases,* and *depression*.

Specific sections describe:

- How ROS promote cancer growth through cellular signaling.
- The role of oxidative stress in diabetes and its contribution to disease progression.

- How oxidative stress impairs brain function and how antioxidants can help restore it.
- The involvement of oxidative stress in diseases such as *ischemia-reperfusion injury, Alzheimer's, Parkinson's, Huntington's, multiple sclerosis,* and *malaria* (*Plasmodium falciparum*).

The book also examines the application of phytochemical antioxidants in preventing and treating diseases, as well as the challenges and strategies for mitigating oxidative stress. It reviews current and emerging therapeutic approaches and diagnostic tests used to measure oxidative stress. Additionally, it discusses *structure–activity relationships* used to study antioxidant effectiveness, *clinical applications*, and *patents* related to oxidative stress therapies.

Overall, this book serves as a comprehensive resource for scientists and healthcare professionals seeking to understand or initiate research on oxidative stress. It is helpful for researchers in both basic and clinical sciences, as well as for practitioners interested in pharmacological and nutraceutical treatments. By bringing together knowledge on oxidants, antioxidants, and their biological roles, this book provides an integrated view of oxidative stress and its contribution to human disease, helping guide future research and clinical applications.

The inspiration behind this book has been Dr. Narasimham Parinandi, and we thought it fit to get the foreword written by him. We also acknowledge the efforts put in by Reetika Tandon, while compiling the book. We are thankful to our family for their continuous support and inspiration to complete this task. We also acknowledge our publisher (Springer) for their interest in this book. We hope that the efforts put in by all the authors will be appreciated and that this book will prove to be a valuable resource.

Lucknow, Uttar Pradesh, India
November 2025

Nidhi Srivastava
Shubhini A. Saraf

Contents

Chapter 1
Introduction: Oxidants and Oxidative Stress in the Human Body

Shivali Tiwari, Olibha Xalxo, Kush Kumar Nayak, and Anju Meshram

Abbreviations

4-HNE	4-hydroxy-2-nonenal
AD	Alzheimer's disease
CAT	Catalase
EGFR	Epidermal growth factor receptor
GPx	Glutathione peroxidase
GPx1	Glutathione peroxidase-1
GSH	Reduced glutathione
GSSG	Oxidized glutathione
H_2O_2	Hydrogen peroxide
Keap1	Kelch-like ECH-associated protein 1
MDA	Malondialdehyde
NO	Nitric oxide
NOX	NADPH oxidase
Nrf2	Nuclear factor erythroid 2-related factor 2
$O_2{\bullet}^-$	Superoxide anion radical
$ONOO^-$	Peroxynitrite
•OH	Hydroxyl radical
PD	Parkinson's disease
RNS	Reactive nitrogen species
ROS	Reactive oxygen species

S. Tiwari · O. Xalxo · A. Meshram (✉)
Amity Institute of Biotechnology, Amity University Chhattisgarh, Raipur, Chhattisgarh, India

K. K. Nayak
School of Studies in Biotechnology, Shaheed Mahendra Karma Vishwavidyalaya, Jagdalpur, Chhattisgarh, India

N. Srivastava, S. A. Saraf (eds.), *Pharmacological Strategies for Prevention and Treatment of Oxidative Stress*, Oxidative Stress in Applied Basic Research and Clinical Practice, https://doi.org/10.1007/978-3-032-19837-2_1

SOD	Superoxide dismutase
SOD1	Cu/Zn- superoxide dismutase
SOD2	Mn-superoxide dismutase
SOD3	EC-superoxide dismutase
TBARS	Thiobarbituric acid reactive substances

1.1 Introduction

The phrase "oxidative stress" was initially used by Helmut Sies to describe an imbalance that could harm biological systems between the generation of oxidants and antioxidant defenses. Since then, theories of oxidative stress in pathology have given way to redox signaling in physiology (Flohe 2020; Sies et al. 2017) in the subject of redox biology. Numerous diseases, including cancer, Alzheimer's, atherosclerosis, and chronic obstructive pulmonary disease (COPD), have been linked to oxidative stress. These findings have elucidated several pathways by which oxidants cause cellular damage (Valko et al. 2007). Nonetheless, the degree to which oxidative stress contributes to the pathophysiology of illnesses varies widely, meaning that in certain conditions, boosting antioxidant defense may not be as beneficial. The chemistry of processes involving so-called reactive species formed from oxygen and nitrogen is what is known as oxidative stress. Knowing which of these species damages macromolecules helps to justify the development of more effective antioxidant defense treatments. (Forman and Zhang 2021).

The term "oxidative stress" refers to an oxidative imbalance caused by a failure to detoxify reactive products, which are created when cells produce reactive oxygen species (ROS) during metabolism. The formation of free radicals from biological oxidation leads to various consequences, such as disruption of intracellular protein structure and functions, damage to membranes from polyunsaturated fatty acid peroxidation with lipids, modifications to nucleic acid bases and chromosomes (DNA single-strand and double-strand breaks, DNA and protein cross-links), and oxidative damage to cellular structures and components like polysaccharide depolymerization and carbohydrate degeneration (Kiran et al. 2023).

Antioxidants are the mechanism that lessens or fixes the damage caused by free radicals while shielding the body from their damaging effects. It is well recognized that the pathophysiology of numerous diseases, including cancer, atherosclerosis, malaria, metabolic syndrome, Alzheimer's disease, rheumatoid arthritis, neurological diseases, and preeclampsia, is directly correlated with the mechanisms resulting from the rise in ROS brought on by oxidative stress (Kiran et al. 2023).

Maintaining redox equilibrium, regulating critical transcription factors, and sustaining several critical physiological processes all depend on low to intermediate levels of ROS and RNS (Jomova et al. 2023). On the other hand, excessive ROS production causes redox equilibrium to be upset, which then causes oxidative stress and ROS-mediated damage to all significant biomolecules, such as proteins, DNA, and membranes (Liguori et al. 2018).

A class of proteins called antioxidant enzymes, also known as metalloproteins, catalyze the conversion of reactive oxygen species (ROS) and their metabolites into more stable, generally less dangerous species. An essential defense mechanism against ROS-induced oxidative stress, which damages every component of the cell, is represented by antioxidant enzymes (Mates et al. 1999). Certain enzymes, including glutathione peroxidase, catalase, and superoxide dismutase, work together to defend the body against reactive oxygen species (ROS).

1.2 Superoxide Dismutase (SOD)

Irwin Fridovich made the discovery of SOD in 1967 (McCord and Fridovich 1969). Humans contain three primary types of superoxide dismutase: EC-SOD (SOD3), Mn-SOD (SOD2), and Cu, Zn-SOD (SOD1) (Zelko et al. 2002). Cu, Zn-SOD (SOD1) is a homodimer that comprises ions related to copper and zinc. SOD is one of the most powerful enzymes and is found in the cytoplasm. Based on the reaction data, this enzyme catalyzes the dismutation of superoxide anion radical very efficiently using a "Ping Pong" mechanism at remarkably high reaction rates (Zelko et al. 2002)

$$2H^+ + 2O_2^{\bullet -} \rightarrow H_2O_2 + O_2\left(\text{fast; K} \sim 2 \times 10^9\ M^{-1}S^{-1}\right)$$

Different SODs have varying affinities for negatively charged anions like fluorides and azides. Cuprizone, a copper-specific chelator, and EDTA are typical copper chelators that can also inhibit SODs (Zhu et al. 2014). The homotetrameric enzyme Mn-SOD (SOD2) has Mn in its active core. Situated in the mitochondria, it makes up around 15% of all superoxide dismutase. The Mn ion is cycling from a more stable Mn (III) oxidation state to a less stable Mn (II) oxidation state during the dismutation reaction. Numerous factors, such as inflammatory cytokines, changes in redox equilibrium, ionizing radiation, and even cigarette smoke, can increase the production of the Mn-SOD gene. The main extracellular SOD that is mostly located in the extracellular matrix of tissues is extracellular superoxide dismutase (EC-SOD, SOD3). Primarily expressed in the lungs is SOD3. Similar to SOD1 and SOD2, SOD3 is also involved in the detoxification of ROS by the catalytic dismutation of superoxide radical anion. EC-SOD has affinity towards heparin and regulates levels of NO·EC-SOD and glutathione peroxidase represent an important first line of defense against oxidants (Rahman et al. 2006).

1.3 Catalase

One of the most significant antioxidant enzymes found in aerobic organisms is catalase. It has been extracted from prokaryotic and eukaryotic creatures alike. Two molecules of hydrogen peroxide are broken down by catalase into two molecules of water and one molecule of oxygen (Nandi et al. 2019).

$$2H_2O_2 \rightarrow 2H_2O + O_2$$

The reaction occurs in two stages: the first is the reduction of the first H_2O_2 molecule, which is followed by the formation of oxyferryl species (FeIVO); the second is the reduction of the FeIVO complex to O_2, H_2O molecules, and free enzyme by the second H_2O_2 molecule (Nandi et al. 2019).

$$\text{Catalase}[\text{Fe (III)}(\text{Porphyrin})] + H_2O_2 \rightarrow \text{Catalase }[\text{Fe (IV} - \text{O(Porphyrin)}] + H_2O_2$$

and

$$\text{Catalase }[\text{Fe (IV} - \text{O(Porphyrin)}] + H_2O_2 \rightarrow \text{Catalase (Fe(III)}(\text{Porphyrin})] + H_2O + O_2$$

With conditions linked to oxidative stress, like those involving inflammation, mutagenesis, and the suppression of apoptosis, catalase is crucial (Sandstrom and Buttke 1993). Numerous illnesses, including neurological conditions like Parkinson's, bipolar disorder, diabetes I and II, hypertension, and insulin resistance, as well as metabolic conditions including cancer, anemia, and asthma, are associated with catalase deficiency. Catalase may be used to treat disorders linked to oxidative stress.

1.4 Glutathione Peroxidase

A family of enzymes known as glutathione peroxidase (GPx) has an active site, including a selenocysteine residue and peroxidase activity. Lipid hydroperoxides are reduced to alcohol and hydrogen peroxide to water by GPx (Lubos et al. 2011). Hydrogen peroxide is the substrate for glutathione peroxidase-1 (GPx-1), the most prevalent isoenzyme in the GPx family. Signal transduction, thiol redox homeostasis, and mitochondrial function all depend on the physiological level of hydrogen peroxide, which is maintained by GPx-1 in the intracellular environment (Brigelius-Flohe and Maiorino 2013). The reaction describes the overall outcome of peroxide inactivation (hydrogen peroxide):

$$\text{ROOH} + 2\text{GSH} \rightarrow \text{ROH} + \text{GSSG} + H_2O$$

where glutathione is oxidized (GSSG) and reduced (GSH). The activity of GPx-1 can be inhibited by a variety of ROS and RNS, including superoxide radicals, peroxynitrite, and nitric oxide, which is produced, for instance, by inducible NO synthase. The intracellular redox state and the ratio of GSH to GSSG, as well as the concentration of reduced glutathione GSH, are the most important factors influencing the degree of inactivation. GPX-1 inhibits carcinogenic potential and stops DNA alterations from occurring. If oxidants are depleted too much, reductive stress can be induced. Reductive stress is characterized by a deterioration in physiological

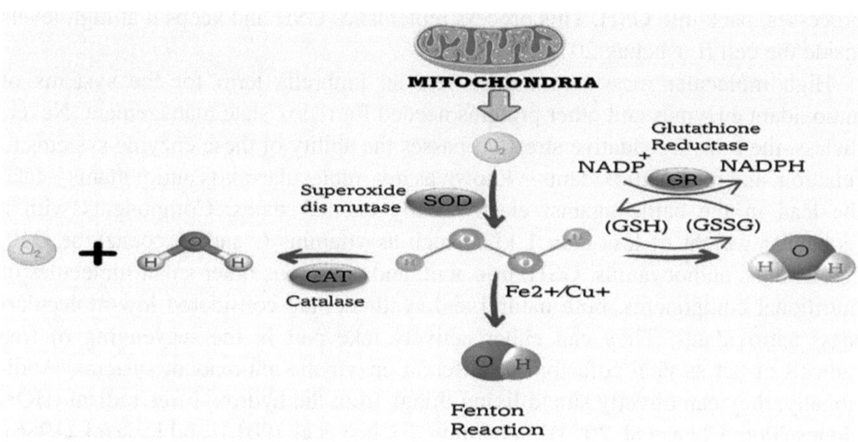

Fig. 1.1 The action of antioxidant enzymes. As a team, glutathione peroxidase, catalase, and superoxide dismutase convert superoxide radical anions into hydrogen peroxide, maintaining physiological levels. The formation of hydrogen radicals can be catalyzed by redox metals such as Cu and Fe. Glutathione peroxidase is essential for cellular redox defense because it transforms hydrogen peroxide into 2H2O, GSSG, and GSH

processes, including cell proliferation, inhibited mitochondrial activity, the emergence of cardiomyopathies, and other diseases (Rajasekaran et al. 2007). The reciprocal interactions of superoxide dismutase, catalase, and glutathione peroxidase are shown in Fig. 1.1 (Weaver and Skouta 2022).

1.5 The Antioxidant System: Cellular Defense Mechanism Against Oxidative Stress

Mechanisms to remove these oxidants are essential because of the detrimental damage to cells that excessive oxidant generation causes. For cells to remain healthy, these systems must function properly. Dismutase (SOD) is a biological enzyme system that catalyzes the dismutation of $O_2^{\bullet-}$. It was first found by McCord and Fridovich (1988). Superoxide anions are primarily generated by cyclooxygenases and lipoxygenases. The primary sources of superoxide anions are NADPH oxidases (NOXs), xanthine oxidase, lipoxygenase, and cyclooxygenases. SOD catalyses the dismutation of the anions into H_2O_2. Another enzymatic antioxidant system is catalases, which could dismutase H_2O_2 into O_2 and H_2O. A wide family of peroxidase enzymes and other enzymatic antioxidant mechanisms in the cell are also capable of reducing H_2O_2. H_2O_2 and other hydroperoxides can be reduced by peroxidases, the most famous of which being the GPx. With the aid of reduced glutathione (GSH), GPx can lower H_2O_2 or lipid peroxides. Glutathione reductase converts oxidized glutathione disulphide (GSSG), which is a result of GPx processes, back into GSH. This process replenishes GSH and keeps it at high levels inside the cell (Lushchak 2014).

High molecular mass antioxidants are an umbrella term for the systems of antioxidant enzymes and other proteins needed for redox state management. Nevertheless, the body's oxidative stress surpasses the ability of these enzyme systems to function, and other antioxidants—known as low-molecular-mass antioxidants—take the lead in the battle against electronically excited states. Components with a molecular weight of less than 1 kDa, such as vitamins C and E, coenzyme Q10, carotenoids, anthocyanins, GSH, uric acid, and numerous other small molecules or nutritional components, both natural and synthetic, are considered low-molecular-mass antioxidants. They can either actively take part in the scavenging of free radicals or act as vital cofactors in different enzymatic antioxidant systems. Additionally, they can directly shield living things from the hydroxyl free radical (HO•) (Jones 2006; Chen et al. 2013). According to Chen et al. (2013) and Dziezak (1986), low molecular weight antioxidants can generally function as (1) free-radical terminators, (2) reducing agents or oxygen scavengers, and (3) chelating agents that form stable complexes with prooxidant metal ions. The three main ways that antioxidants work to combat oxidative stress are through repair, interception, and prevention (Sies 1993).

1.6 Reactive Oxygen Species (ROS): Bridging Physiology and Pathology

Unpaired electrons in atomic orbitals are found in free radicals, which are autonomous chemical entities. These radicals are quite erratic and flammable. They conduct the roles of oxidants and reductants by either giving or receiving an electron (Lobo et al. 2010). An essential component of regular cellular processes, such as fertilization, ovulation, arachidonic acid metabolism, phagocytosis, and the mitochondrial respiratory chain, is the production of extremely reactive ROS. Pathological problems lead to a multiplication of their generation by many times. Numerous pathogenic stimuli have been shown to generate oxygen-free radicals during recovery stages, which have been shown to reach cerebral tissue (Snezhkina et al. 2019). Nitrogen oxide (NO), alkoxy (-OR), peroxyl (ROO), hydroxyl (OH), hydrogen peroxide (H_2O_2), and superoxide ($O_2^{\bullet-}$) are examples of reactive species. Superoxide radicals are mostly produced by the electron transport chains in mitochondria and microsomes. To shield the cell from oxidative damage, all other components of the mitochondrial respiratory chain transfer electrons directly to oxygen, except for cytochrome oxidase, which retains reduced forms of oxygen intermediates at their active sites. The auto-oxidation of semiquinones may result in the production of superoxide anions on the internal membrane of mitochondria. Dismutation to H_2O_2 is a primary process of superoxide anions produced by mitochondria. The reactive alkoxyl and hydroxyl free radicals quickly target the main macromolecules in cells

(Strzelak et al. 2018). Free radicals (ROS) cause lipid peroxidation, DNA fragmentation, cell death, DNA damage, protein alteration, and membrane damage, leading to various illnesses such as ischemia reperfusion injury, vascular endothelium, deep injuries, organ dysfunction, shock, inflammation, sepsis, diabetic retinopathy, cancer, cognitive dysfunction, cataract, and heart disease. Iron buildup in neurodegenerative diseases like amyotrophic lateral sclerosis, spastic paraplegia, and multiple sclerosis may be linked to gliosis or alter the blood-brain barrier's integrity, causing inflammatory reactions and altered tissue vascularization (Stephenson et al. 2014).

1.7 Protecting the Cell: Antioxidant Defense Against Cellular Component Damage

Reduction scavenging mechanisms typically counteract the excess of different oxidant molecules produced by normal metabolic activities, which might result in an oxidative state. Nevertheless, oxidative distress develops when this oxidative state varies much above the ideal threshold. Excessive oxidation can result in oxidative damage to a variety of biomolecules, including proteins, lipids, carbohydrates, and nucleic acids.

Oxidative surroundings negatively affect both DNA and RNA at the nucleic acid level. In fact, oxidative stress increases the frequency of spontaneous DNA mutations, which lead to genomic instability (Storz et al. 1987). The most susceptible DNA base is guanine, which can base-pair with adenine to produce transverse mutations after DNA replication is finished. Guanine is changed into 8-oxo-7,8-dihydroguanine (Freudenthal et al. 2015). The RNA is vulnerable to oxidative damage, where 8-oxoguanine in RNA reduces translational fidelity and causes several illnesses (Chen et al. 2022; Hahm et al. 2022; Poulsen et al. 2012).

The general structure and function of proteins can be determined by the reaction of oxidation products with side chains of amino acids (Davies 2016; Griffiths et al. 2014; Kim et al. 2015). The process of oxidative protein folding, which normally involves the formation of disulfide bonds between cysteine residues in proteins, takes place in the endoplasmic reticulum (ER) (Oka and Bulleid 2013). Numerous enzymes are involved in this process, including peroxiredoxin IV, glutathione-dependent peroxidases 7 and 8 (GPx7 and GPx8), the protein disulfide isomerase family of dithiol-disulfide oxidoreductases, and the pathways of ER oxidoreduction 1 (Ero1). After Ero1's formation of disulfide bonds, a molecule of H_2O_2 is created and is captured in the ER by GPx8 activity (Ramming et al. 2014). To preserve physiological redox equilibrium, GPx7 functions as a new oxidative stress sensor and controls thiol-containing proteins (Kanemura et al. 2020). Unfolded protein response (UPR) pathway activation occurs in response to an increase in ER stress and unfolded protein accumulation. Redox equilibrium in UPR is restored by the integration of additional redox signals with chaperone activity. Most importantly, cell death occurs eventually when UPR is unable to maintain redox equilibrium (Hartl et al. 2011).

Under oxidative stress conditions, lipids can also undergo oxidation, and the outcomes of oxidized lipids can function as ligands for the peroxisome proliferator-activated receptor (Davies et al. 2001). Lipid hydroperoxides, hydroxides, epoxides, malondialdehydes, and various other products are among them (Niki 2014; Spickett and Pitt 2015).

Like other molecules, carbohydrates can have their structures and functions disrupted by oxidation processes. As an example, breaks in DNA strands can occur when the sugar backbone of DNA is modified. Reactive carbonyls are a kind of oxidation product produced by free carbs (Robertson 2004). In the same way, oxidative stress can cause a rise in nonenzymatic glycosylation, glycation, and GlcNAcylation (Sies et al. 2017).

1.8 Mechanisms of Redox Reaction Regulation in Cellular Processes

1.8.1 Redox Signaling

Redox reactions are essential for controlling various elements of a wide range of biological processes, including those involving oxidant and redundant molecules. Under physiological conditions, oxidants are created in low, properly regulated amounts. Significantly, at these low levels, oxidants can play a key role as signaling molecules in the control of critical physiological functions, such as immunological response, autophagy, cell division, inflammation, and inflammation (Finkel 2011). For instance, ROS are primarily produced as byproducts of cellular metabolic reactions, and the main players responsible for ROS generation are xanthine oxidase, lipoxygenase, cyclooxygenases, peroxisomes, ER, mitochondrial respiratory chain, and mitochondrial metabolism (Bhardwaj and He 2020).

ROS are involved in many physiological processes, including the regulation of vasotone, immune responses, and several others (Liu et al. 2003). At physiologically low concentrations, ROS also function as second messengers in several major signaling pathways, including cell differentiation, growth, and death (Zhang et al. 2016). As mentioned, an excess of ROS and other reactive species is produced when oxidants and antioxidants are out of balance (Alfadda and Sallam 2012; Finkel 2011). This condition is known as oxidative stress, which subsequently damages cellular components (Davies et al. 2001; Finkel 2011). A member of the family of Cys-based protein tyrosine phosphatases (PTPs), protein tyrosine phosphatase 1B (PTP1B), possesses HCX5R(S/T) motifs in its active site. PTPs are easily oxidized by endogenous H_2O_2 produced by the activation of NADPH oxidases mediated by epidermal growth factor (EGF) because to the re-active thiol with low pKa in their active site Redox signaling pathway is shown in Fig. 1.2 (Zheng et al. 1998).

Reversible and irreversible redox processes could modify the structure and function of proteins and provide a means of regulating protein activation and

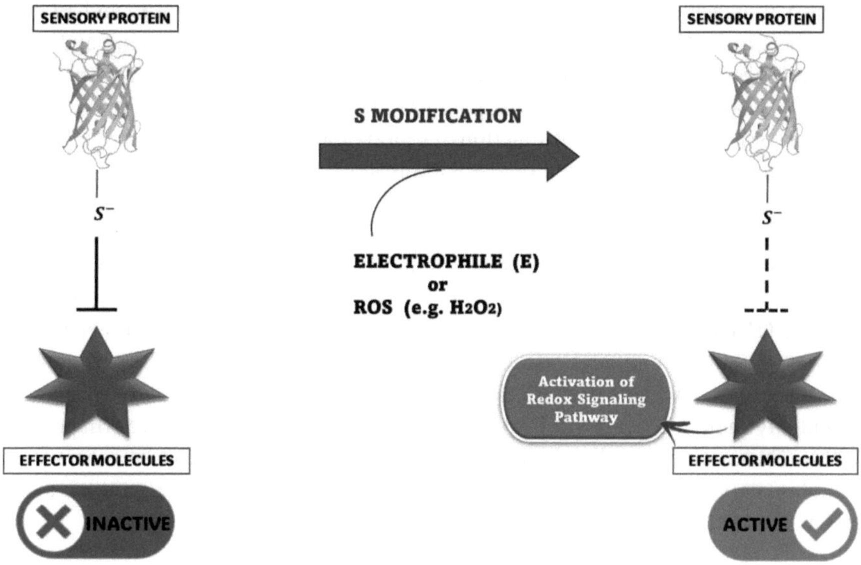

Fig. 1.2 The redox signaling pathway. It consists of targeted effector molecules and sensor proteins with deprotonated thiol groups. Sensor proteins suppress effector molecules in the basal state, which are activated when blocked by S-modification in electrophiles or reactive oxygen species. These pathways are triggered by the activation of effector molecules like kinases or transcription factors

signaling. This is a common method of controlling the actions of many transcription factors and enzymes. For instance, the reduction-oxidation state of cysteine residues in the DNA-binding domain of the heterodimeric transcription factor activator protein-1 (AP-1), which is composed of the Fos and Jun proteins, affects the protein's ability to perform its role (Abate et al. 1990; Liu et al. 2005). According to Paik et al. (2017), oxidative stress also regulates hypoxia-inducible factor 1α, another transcription factor. Comparably, the redox status affects enzyme activity; for instance, by thiol/disulfide redox modifications of amino acid residues (Bindoli and Rigobello 2013; D'Autreaux and Toledano 2007). Classical examples of enzymes regulated by redox status include Apoptosis Signal-regulating Kinase 1 (ASK1), protein tyrosine kinase, AMP-activated protein kinase, Src kinase, and EGF receptor (EGFR) (Heppner et al. 2018; Truong and Caroll 2013). H_2O_2 is an uncharged molecule and regulates enzyme activity and protein function by oxidatively altering protein residues. Its slow reaction rate constant (1–10 M^{-1} s^{-1}) makes it ideal for redox sensing and signaling (Sobotta et al. 2015). However, it possesses an exceptional ability to diffuse away from its site of generation, via transport via H_2O_2 aquaporins or peroxiporins, to locations where it modifies the activity of more reactive target proteins (Bienert et al. 2007; Prata et al. 2019). However, nothing is known about the process by which H_2O_2 oxidizes regulatory proteins (Winterbourn and Hampton 2015). Like other ROS, H_2O_2 can function as a signaling molecule when it is generated because of several receptors, including PDGFR and EGFR, being activated. Nevertheless, redox sensing by these receptors' cysteine and tyrosine residues as well as other enzymes in the signal transduction pathway is also a part of H_2O_2 signaling through this pathway (Di Marzo et al. 2018; Finkel 2011). Another oxidative switch that functions as an oxidative stress sensor is the human DJ-1 protein, which is encoded by the Parkinson's disease protein 7 gene and exists in both oxidized and reduced forms (Zhang et al. 2020). Thiol-based redox signaling can be modulated by other reactive species, including sulfide and nitrogen reactive species like nitric oxide (NO). Like ROS and H_2O_2, NO is a reactive species that serves as a key signaling molecule. NO is a free radical that can react with $O_2^{\bullet -}$ to produce ONOO−, or peroxynitrite. Through nitration processes, this molecule can convert tyrosine amino acid residues in proteins into 3-nitrotyrosine. Furthermore, ONOO- can behave as a biological oxidant, influencing mitochondrial activity and potentially inducing death via its oxidation and nitration processes (Beckman et al. 1990). Thus, redox signaling involves H_2O_2 NO and sulfide reactive species, which can ultimately alter the function of biological components, particularly proteins like enzymes and transcription factors (Beckman et al. 1990).

1.9 Molecular Redox Switches: Key Regulators of Transcription Factor Activity

Molecular redox switches coordinate their actions within the cell to regulate its oxidative state. The transcription factors: nuclear factor-E2/Kelch-like ECH-associated protein-1 (Nrf2/Keap1) and NF-κB, which are known to be important master regulators of a wide range of biological functions in mammalian cells, are regulated by eukaryotic redox switches.

Oxidants and electrophiles can activate the transcription factor Nrf2, which then causes the expression of a group of protective detoxifying enzymes and antioxidants. Enzymes involved in drug metabolism and deposition, including as glutathione S-transferase, NADPH:quinone oxidoreductase 1, and cytochrome P450 CYP2A5, are among the Nrf2 target genes. Nrf2 target genes also include those that encode proteins and enzymes including SOD3, GPx2, and peroxiredoxin-1 and 6 that are involved in oxidant signaling and antioxidant defense (Ma 2013). Another intriguing transcriptional target of Nrf2 is heme oxygenase 1 (HO-1), which degrades heme and possesses antioxidant and anti-inflammatory properties (Araujo et al. 2012). While Keap1 keeps Nrf2 in the cytoplasm, Nrf2 transcriptional activity is inhibited under unstressed situations. Additionally, Keap1 encourages Nrf2's ubiquitination, which leads to the proteasome's breakdown of it (Antelmann and Helmann 2011; Itoh et al. 1997; Ma 2013; Nguyen et al. 2004). Under stressful situations, Keap1's cysteinyl residues undergo modification to generate intra- and intermolecular disulfide bonds; as a result, Keap1 is no longer able to regulate Nrf2. As a result, Nrf2 is free to migrate to the nucleus and accumulate. To stimulate the production of its target genes, Nrf2 heterodimerizes with tiny Maf proteins in the nucleus. The heterodimer then attaches to the antioxidant response element, a DNA sequence (Antelmann and Helmann 2011; Ma 2013; Nguyen et al. 2004).

To increase oxidative stress, antioxidant-free radical-scavenging systems must be induced and regulated, and Nrf2 is thus active at extreme oxidation states. Likewise, an early aging phenotype can result from any abnormality in Nrf2 signaling. In fact, Nrf2's transcriptional activity can be inhibited by trapping it at the nuclear periphery by a mutation in lamin A, which codes for a key architectural protein of the nucleus. As an outcome, Nrf2 signaling will be disrupted, resulting in a persistent oxidative stress state that ultimately manifests as an early aging phenotype (Kubben et al. 2016). Mice lacking Nrf2 are more vulnerable to oxidative stress and several chemical toxins (Kensler et al. 2007; Klaassen and Reisman 2010; Ma 2013; Walters et al. 2008). Consequently, elevating Nrf2 activity can shield animal models from harm brought on by oxidative stress (Talalay et al. 2003). The processes underlying Nrf2's decrease of function in cancer remain unclear. Numerous diseases and associated complications have been linked to genetic differences in Nrf2 (Chen et al. 2019; Korytina et al. 2019; Xu et al. 2016; Yamaguchi et al. 2019; Zazueat et al. 2022).

Pharmaceutical Nrf2 activators such as dimethyl fumarate (marketed as BG-12 or Tecfidera by Biogen) and Oltipraz (4-methyl-5(pyrazinyl-2)-1–2-dithiole-3-thione) seem like promising therapeutic tools in the treatment of various diseases, especially those supported by inflammation and oxidative stress. Furthermore, it seems that BG-12 inhibits platelet activity and thrombus formation (Chu et al. 2023).

Overall, by detecting oxidative stress and controlling key transcription factors that can alter the transcriptional profiles of the cell to counteract oxidative stress, molecular redox switches regulate the oxidative state of the cell.

1.10 The Impact of Reactive Oxygen Species (ROS) on Human Health: Mechanisms and Consequences

The body develops OS when there is an imbalance between the generation and removal of free radicals. This means that the body's antioxidant system is overwhelmed by oxidants, which subsequently damage DNA, proteins, lipids, and membranes among other cellular structures, as shown in Fig. 1.3 (Dorge 2002; Hajam et al. 2020; Lobo et al. 2010; Uthra et al. 2017; Birben et al. 2012; Pham-Huy et al. 2008). Lipid peroxidation is a well-known biomarker for the assessment of (OS) (LPO). Because of the peroxidative damage that free radicals cause to the polyunsaturated lipid membrane during OS, the rate of LPO increases rapidly. In addition, OS changes the structure of proteins, reducing their functional efficiency and changing the activities of the enzymes (Birben et al. 2012).

Numerous human illnesses, including cancer, are linked to DNA damage and mutations caused by OS. OS causes lesions in DNA, indicating that double-strand

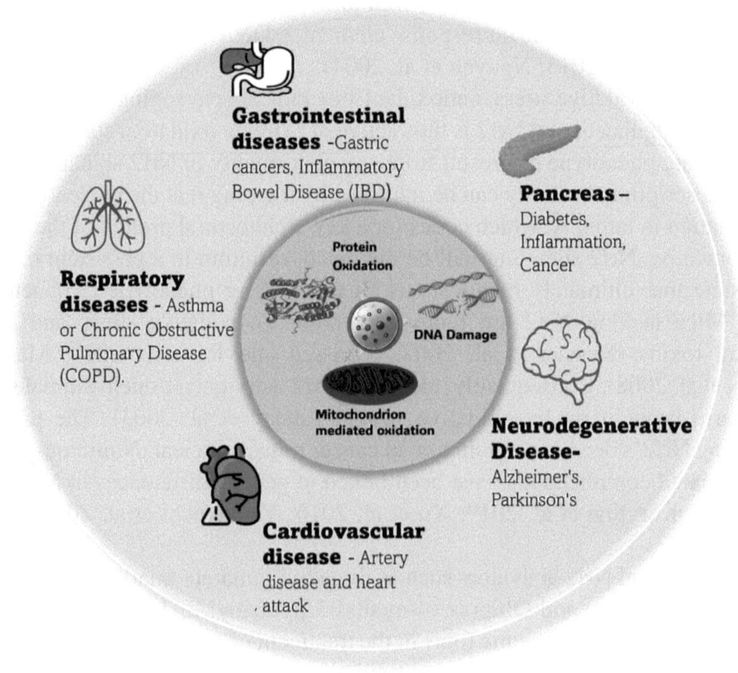

Fig. 1.3 Oxidative stress damage at cellular level and further to the different parts of the body

breaks (DSBs) in DNA may result in gene and chromosome mutations caused by ROS-induced OS (Valavanidis et al. 2013; Maechler et al. 2001).

Alterations in several organ systems are caused by ROS-induced OS, and these changes are explained as follows:

1. ROS-induced neurotoxicity alters neurons by causing the cellular membrane to permeabilize, the neuronal membrane's excitability to diminish, and the KATP pump to become activated.
2. Due to mitochondrial damage, perforations in the mitochondria, the release of cytochrome c, and the initiation of the apoptotic cascade, ROS-induced oxidative stress leads to cardiac myopathy.
3. Because they directly participate in filtration and metabolism, the liver and kidneys are the main organs attacked by ROS. ROS damage the hepatocyte membranes of the liver, causing deuteriation. This, in turn, causes the hepatocyte to accumulate collagen, ultimately leading to liver cirrhosis and fibrosis. Furthermore, insufficient oxidation of biomolecules results in immunological responses in the liver and lipoapoptosis in hepatic cells. ROS-induced oxidative stress in the kidneys primarily starts the synthesis of several pro-inflammatory cytokines, which starts nephrotic inflammation and finally affects renal functioning (Lobo et al. 2010).

1.11 The Role of Oxidative Stress in Disease Development: Mechanisms and Implications

Oxidative stress is known to contribute to disease through two main processes. First, when oxidative stress occurs, reactive species are produced, including •OH, ONOO−, and HOCl. These species directly oxidize macromolecules, such as membrane lipids, structural proteins, enzymes, and nucleic acids, resulting in abnormal cell activity and death. Redox signaling abnormalities are the second mechanism of oxidative stress. The second messengers can be oxidants, especially hydrogen peroxide produced by cells in response to physiological stimulation (Forman et al. 2010). In oxidative stress, redox signaling might go incorrect due to non-physiological H_2O_2 generation (Sies et al. 2017). In certain diseases, like diabetes, both forms of oxidative stress mechanisms can be present. In this case, advanced glycation products build up and abnormal activation of stress signaling pathways results in complications related to diabetes (Evans et al. 2002). Also, the rise in H_2O_2 generation and iron release from proteins under O_2• − induced oxidative stress (Liochev and Fridovich 1994) and ONOO− (Castro et al. 2019) leads to a significant increase in the synthesis of lipid peroxidation products, such as 4-hydroxy-2-nonenal (HNE), which can potentially result in abnormal signaling (Zhang and Forman 2017).

Many different diseases have been linked to oxidative stress. The pathologies have been classified into two groups based on the role that oxidative stress played in their etiology. The first group includes toxicities caused by radiation and paraquat as well as atherosclerosis. The second group includes oxidative stress as a secondary factor in the progression of diseases like COPD, hypertension, and Alzheimer's disease. However, it is merely preliminary because the exact impact of oxidative stress in numerous illnesses remains unknown (Forman and Zhang 2021).

1.11.1 Antioxidants as Therapeutic Agents

There are several antioxidant treatment approaches under investigation, some of which are amid clinical trials. Among these is the elimination of O_2• − prior to its reaction with •NO to produce ONOO− ($O_2^{\bullet-}$ + $^{\cdot}$No → $ONOO^-$) and H_2O_2 removal before the formation of •OH ($H_2O_2 + Fe^{2+} \rightarrow OH^- + Fe^{3+} + {}^{\cdot}OH$) or HOX ($H_2O_2 + X^- + H^+ \rightarrow HOX + H_2O$); using precursors to increase GSH; boosting antioxidant enzyme production, especially by activating NRF2; and inhibiting NOXs (NADPH $+2O_2$ + $NADP^+$+ H + $2o_2^{\bullet-}$) dietary antioxidant supplementation, mitochondrial antioxidant defense, and, lastly, suppression of abnormal redox signaling, as shown in Fig. 1.4 (Forman and Zhang 2021).

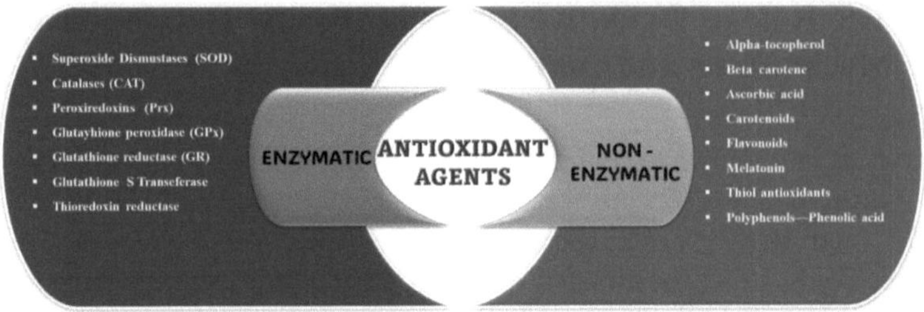

Fig. 1.4 Antioxidant agents which are divided into enzymatic and non-enzymatic

1.12 The Role of Oxidative Stress in Neurodegenerative Disease Progression

1.12.1 Alzheimer's Disease

Neurodegenerative illnesses affect older populations' life spans as well as their quality of life.

The effects of neurodegenerative disorders on the elderly are extensive and varied. Parkinson's disease (PD), Huntington's disease (HD), Alzheimer's disease (AD), and increasing memory loss are the main threats posed by these brain illnesses. Alzheimer's disease (AD) is among the most severe neurodegenerative conditions that impact the aged population. Certain regions of the brain experience brain atrophy and neuronal or synaptic loss because of the buildup of amyloid beta protein during AD. AD is primarily characterized by brain plaques and tangles, and it also affects the neocortex and the hippocampal regions. An imbalance in the oxidant-antioxidant systems may dictate the extent of cellular injury. Two of the main indicators of neurodegeneration are oxidative stress and mitochondrial dysfunction (Zhao and Zhao 2013) and contributes significantly to the development of neuronal degeneration in AD (Gandhi and Abramov 2012). The disruption of redox reactions or a decline in the activity of antioxidant enzymes like superoxide dismutase and catalase, and a drop in antioxidant levels (ascorbic acid and tocopherol) encourage the excess production of ROS and increased Aβ accumulation in AD patients, is what causes this ROS-mediated neuronal damage (Radi et al. 2014). Aβ plaques are accumulated in the Golgi apparatus and endoplasmic reticulum's extracellular area. The first stages of AD are characterized by a rise in Aβ plaque deposition, which also elevates oxidative stress and ultimately results in mitochondrial dysfunction (Matsuoka et al. 2001; Murakami et al. 2011). Increased expression of amyloid mutant amyloid precursor protein and presenilin-1 increases the levels of hydrogen peroxide (H_2O_2) and lipid peroxidation (Radi et al. 2014; Chen et al. 2008). In addition, it promotes AD patients' tau protein to be phosphorylated. Furthermore, in neural cells, mitochondrial dysfunction also suppresses the expression of electron transport chain (ETC) genes, which oversee pyruvate, cytochrome oxidase, and α-ketoglutarate dehydrogenase activation (Moon and Paek 2015; Blesa et al.

2015). Neurofibrillary tangles accumulate because of ROS-induced inactivation of antioxidant enzyme activities, such as superoxide dismutase (SOD), activation of p38, and phosphorylation of tau protein. GSK3β is activated as an outcome of the wnt pathway being inactivated, which in turn promotes the phosphorylation of tau proteins.

1.12.2 Parkinson's Disease

Parkinson's disease is a degenerative disease of the nervous system that affects movement. Tremors, slowness of movement, tense muscles, an unsteady gait, and issues with balance and coordination are typical symptoms. Slowly, the symptoms begin, sometimes with a tremor in one hand that is hardly perceptible. In addition to causing tremors, the disease frequently results in stiffness or slowed mobility. ROS are produced by a few processes and sources, including aging, calcium, iron, neuro-inflammatory cells, mitochondrial malfunction, and the metabolism of dopamine itself. DJ-1, PINK1, parkin, alpha-synuclein, and LRRK2 are among the gene products implicated in the development of Parkinson's disease (PD). These gene products also have complex effects on mitochondrial function, which exacerbate ROS generation and increase OS susceptibility. Additionally, OS affects the ubiquitin-proteasome system and mitophagy, two processes involved in cellular homeostasis. All these pathways work together to promote neurodegeneration in Parkinson's disease (PD) as a result of OS, which destroys key cellular pathogenetic proteins and produces an excess of reactive oxygen species (ROS). Reduced dopamine levels in the brain's nigrostriatal dopaminergic pathway and distinct neuronal death of dopaminergic neurons in the substantia nigra pars compacta (SNc) are characteristics of Parkinson's disease (PD) (Schapira 2008; Franco-Iborra et al. 2016). Parkinson's disease (PD) is believed to have its root cause in a pathophysiological situation created by OS (Franco-Iborra et al. 2016; Lin and Beal 2006).

A clinical parkinsonian phenotype can be caused by inhibitors like 1-methyl-4-phenyl 1,2,3,4-tetrahydrpyridine (MPTP) and its metabolites, 1-methyl4-phenylpyridinium (MPP+) of Complex I. These inhibitors can also cause nigral degeneration with cytoplasmic α-synuclein (Buneeva et al. 2020).

Putative kinase was linked to the ancient forms of Parkinson's disease, and it is induced by genetic mutations in proteins such α-synuclein, parkin, phosphatase, and tensin homolog (Franco-Iborra et al. 2016). Mutations in these genes have been shown to increase ROS-induced oxidative stress and impact mitochondrial functioning (Franco-Iborra et al. 2016).

Furthermore, dopaminergic neurons were destroyed by ROS. The amount of reduced glutathione (GSH) in the substantia nigra pars compacta of Parkinson's disease (PD) is lowered by ROS-induced oxidative stress (Oakley et al. 2007). As an outcome of the breakdown of protein and membrane polyunsaturated lipids, as well as the fragmentation of DNA, elevated oxidative stress quickens the pace of lipid peroxidation. In the end, all these processes cause dopaminergic neurons to be damaged, which impairs body part coordination.

1.13 Regulatory Roles of ROS in Cell Signaling Networks

ROS originate from two primary sources: the oxidative metabolism of mitochondria and the processes of cellular response to bacteria, xenobiotics, and cytokines. During these activities, ROS are generated as a component of either cellular defense mechanisms or signal transduction pathways (Finkel 2011). Superoxide radical anion ($O_2^{\bullet -}$) is the first ROS that complexes I and III of the respiratory chain of mitochondria generate. Catalase or glutathione peroxidase convert the superoxide radical to hydrogen peroxide, which is subsequently converted to water and oxygen by SOD. Another source of ROS in cells is NADPH oxidases (NOX) (Jomova et al. 2023).

1.14 Biomarkers of Oxidative Stress in Metabolic Syndrome: A Pathway to Improve Therapeutics

Antioxidant system molecules alter with redox stress and molecules changed by ROS in the microenvironment are typical oxidative stress biomarkers (Ho et al. 2013). Risk factors cause cell damage and the clinical manifestations of metabolic syndrome by interfering with cell signaling pathways, raising inflammatory markers, lipid peroxides, and free radicals. It is postulated that inflammatory indicators and oxidative stress play a role in the development of metabolic syndrome (Vona et al. 2019; Thomas et al. 2022). The best method for estimating the level of oxidative stress in vivo is to quantify biomarkers. The metabolic syndrome's total antioxidant capacity can also be used for assessing oxidative stress (Thomas et al. 2022; Marrocco et al. 2017). In disorders where both are suspected to be involved, the isoprostanes (IsoP) derived from arachidonic acid, specifically 8-iso prostaglandin-F2alpha (8-isoPGF2), may be a useful tool for concurrently examining oxidative stress and inflammation (Thomas et al. 2022; Mocan et al. 2013). Metabolic syndrome is associated with pro-inflammatory states and poor health, and it plays a part in the intricate process that underlies cardiometabolic illnesses. It is marked by elevated oxidative stress biomarkers and decreased antioxidant capacity (Thomas et al. 2022; Mocan et al. 2013; Espinosa-Moncada et al. 2018; Chang and Namkung 2021; Coji et al. 2021; Schonknecht et al. 2021).

Biomarkers of lipid peroxidation, protein and amino acid oxidation, and DNA oxidation are typical oxidative stress markers (Ho et al. 2013). Lipid peroxidation, an indicator of oxidative stress, can be detected by markers such as malondyadehide (MDA), 4-hydroxy-2-nonenal (4-HNE), thiobarbituric acid-reactive substances (TBARS), and F2-isoproteines. Protein oxidation is indicated by advanced glycation end products (AGEs), oxidized low-density lipoprotein (ox-LDL), and advanced oxidation proteins. DNA oxidation markers include 5-chlorouracil, 5-chlorocytosine, and 8-oxo-2 deoxyguanosine (8-0xo-Dg) (Ho et al. 2013; Cojic et al. 2021; Sangaleti et al. 2021).

Xanthine oxidase, gamma-glutamyl transferase (GGT), myeloperoxidase (MPO), NOX, and NOS are the markers linked to the production of ROS (Marrocco et al. 2017). An enzyme called gamma-glutamyl transferase (GGT) is present in the kidney, pancreas, liver, spleen, heart, and brain, among other organs. It turns precursors back into the antioxidant and metabolic substrate glutathione (GSH). An increased GGT is predictive with metabolic syndrome, diabetes, high blood pressure, and stroke risk (Jeong et al. 2019; Jeong et al. 2021). Glycoprotein A (GPA), C-reactive protein (CRP), ferritin, and uric acid are examples of non-enzymatic indicators. CRP is a non-specific biomarker that is used to diagnose and categorize inflammatory conditions, including rheumatic diseases, as well as to evaluate disease activity (Jeong et al. 2019). Elevated CRP is associated with metabolic syndrome, diabetes, and dyslipidemia (Jeong et al. 2019). Metabolic syndrome patients exhibit elevated serum ferritin levels without transferrin saturation. Insulin resistance and liver damage are two markers of oxidative stress that have a favorable correlation with serum ferritin (Lindberg et al. 2017).

Adipokines like adiponectin and lectins are helpful indicators that are also substantially positively related with metabolic syndrome. Adiponectin levels are expressed by adipose tissue, and the degree of adiposity is inversely correlated with the levels (Ryo et al. 2004; Ahonen et al. 2012; Lindberg et al. 2017). A commonly known and recognized marker for diabetes and the metabolic syndrome is adiponectin. Some have proposed using serum levels of adiponectin as a marker for metabolic syndrome, and others have linked its development to decreased levels (Ryo et al. 2004; Ahonen et al. 2012; Lindberg et al. 2017). Abdominal obesity, insulin resistance, and the hormone leptin have been related. The hormone leptin controls the way energy is metabolized. Elevated leptin levels have been suggested as a possible indicator of the onset of metabolic syndrome (Madeira et al. 2017; Lee and Shin 2020).

1.15 Oxidative Stress Interventions: Understanding the Challenges and Limitations of Current Therapeutic Strategies

Many diseases and toxicities include oxidative stress as a component of their underlying pathology. The antioxidant defenses and techniques indicated here significantly prevent or reduce pathology. Still, a few limitations make it challenging for people to use antioxidant techniques therapeutically. This limitation is likely the most important element that is frequently disregarded when evaluating antioxidant defenses in clinical trials. The difficulty is figuring out how many antioxidant techniques can be created to help with some of the disease's symptoms, if not the root cause. The commercialization of goods that include small molecules that are chemical antioxidants but do not act as such in vivo will not provide a meaningful benefit beyond what can be obtained from small molecules that induce antioxidant enzymes found in a sufficient diet. It will make it more difficult to create and gain over the public for truly effective treatments (Forman and Zhang 2021).

1.16 Conclusion

Essential components of the human body that affect both physiological and pathological processes are oxidants and oxidative stress. The body naturally produces reactive oxygen species (ROS) and other oxidants during regular cellular metabolism. These molecules are essential for immunological response, cell communication, and the control of numerous biological pathways. However, oxidative stress results when the body's antioxidant defenses are overcome by the synthesis of these oxidants, damaging lipids, proteins, and DNA in cells.

While the production of reactive oxygen species (ROS) and the body's capacity to eliminate these hazardous intermediates or reverse the damage they cause are out of balance, oxidative stress results. Free radicals like superoxide ($O_2^{\bullet-}$), hydroxyl radicals (•OH), and non-radical molecules like hydrogen peroxide (H_2O_2) are examples of chemically reactive compounds containing oxygen that are known as ROS. Even while ROS are necessary for physiological processes like immune response, cell communication, and apoptosis (planned cell death), an excessive build-up of ROS causes oxidative stress, which harms key biological components.

One important element linked to the pathogenesis of neurodegenerative illnesses is oxidative stress. It happens when the body's capacity to detoxify reactive oxygen species (ROS) or repair the harm they cause is out of balance with the rate at which they are produced. Key cellular constituents including lipids, proteins, and DNA can be harmed by ROS, which can result in cellular malfunction and death—two conditions that are indicative of neurodegeneration.

Under neurodegenerative disorders such as Parkinson's and Alzheimer's, oxidative stress is a key factor in the development of neuronal damage. Because the brain has a high metabolic rate and uses a lot of oxygen, it is especially susceptible to oxidative damage from free radicals produced during cellular respiration. Additionally, because neurons are post-mitotic cells, the effects of oxidative damage are amplified because they are not readily regenerable.

Numerous diseases, including cancer, diabetes, Alzheimer's and Parkinson's disease, and neurological disorders, are strongly associated with this oxidative damage and its progression. Sustaining cellular homeostasis and general health requires a careful balance between oxidants and antioxidants. A tipping point toward oxidative stress can be reached by a disturbance of this equilibrium, whether by increased generation of ROS or weaker antioxidant defenses. This can lead to persistent inflammation, cellular malfunction, and the advancement of illness. Furthermore, oxidative stress can be made worse by environmental variables like pollution, UV radiation, and bad lifestyle decisions like smoking or eating poorly. Comprehending these mechanisms facilitates the development of treatment therapies that attempt to reduce oxidative damage, including antioxidant supplements, lifestyle adjustments, and medications that target oxidative pathways. Investigations into the connection between oxidants and oxidative stress may result in novel approaches to the treatment or prevention of illnesses brought on by oxidative stress, hence extending human lifespans and health.

References

Abate C, Patel L, Rauscher FJ III, Curran T (1990) Redox regulation of fos and Jun DNA-binding activity in vitro. Science 249(4973):1157–1161

Ahonen TM, Saltevo JT, Kautiainen HJ, Kumpusalo EA, Vanhala MJ (2012) The association of adiponectin and low-grade inflammation with the course of metabolic syndrome. Nutr Metab Cardiovasc Dis 22(3):285–291

Alfadda AA, Sallam RM (2012) Reactive oxygen species in health and disease. Biomed Res Int 2012(1):1

Antelmann H, Helmann JD (2011) Thiol-based redox switches and gene regulation. Antioxid Redox Signal 14(6):1049–1063

Araujo JA, Zhang M, Yin F (2012) Heme oxygenase-1, oxidation, inflammation, and atherosclerosis. Front Pharmacol 3:119

Beckman JS, Beckman TW, Chen J, Marshall PA, Freeman BA (1990) Apparent hydroxyl radical production by peroxynitrite: implications for endothelial injury from nitric oxide and superoxide. Proc Natl Acad Sci 87(4):1620–1624

Bhardwaj V, He J (2020) Reactive oxygen species, metabolic plasticity, and drug resistance in cancer. Int J Mol Sci 21(10):3412

Bienert GP, Møller AL, Kristiansen KA, Schulz A, Møller IM, Schjoerring JK, Jahn TP (2007) Specific aquaporins facilitate the diffusion of hydrogen peroxide across membranes. J Biol Chem 282(2):1183–1192

Bindoli A, Rigobello MP (2013) Principles in redox signaling from chemistry to functional significance. Antioxid Redox Signal 18(13):1557–1593

Birben E, Sahiner UM, Sackesen C, Erzurum S, Kalayci O (2012) Oxidative stress and antioxidant defense. World Allergy Organ J 5:9–19

Blesa J, Trigo-Damas I, Quiroga-Varela A, Jackson-Lewis VR (2015) Oxidative stress and Parkinson's disease. Front Neuroanat 9:91

Brigelius-Flohé R, Maiorino M (2013) Glutathione peroxidases. Biochim Biophys Acta 1830(5): 3289–3303

Buneeva O, Fedchenko V, Kopylov A, Medvedev A (2020) Mitochondrial dysfunction in Parkinson's disease: focus on mitochondrial DNA. Biomedicine 8(12):591

Castro L, Tórtora V, Mansilla S, Radi R (2019) Aconitases: non-redox iron-sulfur proteins sensitive to reactive species. Acc Chem Res 52(9):2609–2619

Chang JS, Namkung J (2021) Effects of exercise intervention on mitochondrial stress biomarkers in metabolic syndrome patients: a randomized controlled trial. Int J Environ Res Public Health 18(5):2242

Chen L, Na R, Gu M, Richardson A, Ran Q (2008) Lipid peroxidation up-regulates BACE1 expression in vivo: a possible early event of amyloidogenesis in Alzheimer's disease. J Neurochem 107(1):197–207

Chen SK, Hsu CH, Tsai ML, Chen RH, Drummen GP (2013) Inhibition of oxidative stress by low-molecular-weight polysaccharides with various functional groups in skin fibroblasts. Int J Mol Sci 14(10):19399–19415

Chen S, Pan H, Chen Y, Lu L, He X, Chen H, Chen R, Zhan S, Tang S (2019) Association between genetic polymorphisms of NRF2, KEAP1, MAFF, MAFK and anti-tuberculosis drug-induced liver injury: a nested case-control study. Sci Rep 9(1)
Chen X, Yu H, Li Z, Ye W, Liu Z, Gao J, Wang Y, Li X, Zhang L, Alenina N, Bader M, Ding H, Li P, Aung LHH (2022) Oxidative RNA damage in the pathogenesis and treatment of type 2 diabetes. Front Physiol 13
Chu X, Zhang J, Li Y, Yuan K, Wang X, Gui X, Sun Y, Geng C, Ju W, Xu M, Li Z, Zeng L, Xu K, Qiao J (2023) Dimethyl fumarate possesses antiplatelet and antithrombotic properties. Int Immunopharmacol 120:110381
Cojic M, Kocic R, Klisic A, Kocic G (2021) The effects of vitamin D supplementation on metabolic and oxidative stress markers in patients with type 2 diabetes: a 6-month follow-up randomized controlled study. Front Endocrinol 12
D'Autréaux B, Toledano MB (2007) ROS as signalling molecules: mechanisms that generate specificity in ROS homeostasis. Nat Rev Mol Cell Biol 8(10):813–824
Davies MJ (2016) Protein oxidation and peroxidation. Biochem J 473(7):805–825
Davies SS, Pontsler AV, Marathe GK, Harrison KA, Murphy RC, Hinshaw JC, Prestwich GD, Hilaire AS, Prescott SM, Zimmerman GA, McIntyre TM (2001) Oxidized alkyl phospholipids are specific, high affinity peroxisome proliferator-activated receptor gamma ligands and agonists. J Biol Chem 276(19):16015–16023
Di Marzo N, Chisci E, Giovannoni R (2018) The role of hydrogen peroxide in redox dependent signaling: homeostatic and pathological responses in mammalian cells. Cells 7(10):156
Droge W (2002) Free radicals in the physiological control of cell function. Physiol Rev 82(1):47–95
Dziezak J (1986) Preservatives: antioxidants. The ultimate answer to oxidation. Food Technology (Chicago) 40(9):94–102
Espinosa-Moncada J, Marín-Echeverri C, Galvis-Pérez Y, Ciro-Gómez G, Aristizábal JC, Blesso CN, Fernandez ML, Barona-Acevedo J (2018) Evaluation of agraz consumption on adipocytokines, inflammation, and oxidative stress markers in women with metabolic syndrome. Nutrients 10(11):1639
Evans JL, Goldfine ID, Maddux BA, Grodsky GM (2002) Oxidative stress and stress-activated signaling pathways: a unifying hypothesis of type 2 diabetes. Endocr Rev 23(5):599–622
Finkel T (2011) Signal transduction by reactive oxygen species. J Cell Biol 194(1):7–15
Flohé L (2020) Looking back at the early stages of redox biology. Antioxidants 9(12):1254
Forman HJ, Zhang H (2021) Targeting oxidative stress in disease: promise and limitations of antioxidant therapy. Nat Rev Drug Discov 20(9):689–709
Forman HJ, Maiorino M, Ursini F (2010) Signaling functions of reactive oxygen species. Biochemistry 49(5):835–842
Franco-Iborra S, Vila M, Perier C (2016) The Parkinson disease mitochondrial hypothesis: where are we at? Neuroscientist 22(3):266–277
Freudenthal BD, Beard WA, Perera L, Shock DD, Kim T, Schlick T, Wilson SH (2015) Uncovering the polymerase-induced cytotoxicity of an oxidized nucleotide. Nature 517(7536):635–639
Gandhi S, Abramov AY (2012) Mechanism of oxidative stress in neurodegeneration. Oxidative Med Cell Longev 2012:1
Griffiths HR, Dias IH, Willetts RS, Devitt A (2014) Redox regulation of protein damage in plasma. Redox Biol 2:430–435
Hahm JY, Park J, Jang ES, Chi SW (2022) 8-Oxoguanine: from oxidative damage to epigenetic and epitranscriptional modification. Exp Mol Med 54(10):1626–1642
Hajam YA, Rai S, Ghosh H, Basheer M (2020) Combined administration of exogenous melatonin and insulin ameliorates streptozotocin induced toxic alteration on hematological parameters in diabetic male Wistar rats. Toxicol Rep 7:353–359
Hartl FU, Bracher A, Hayer-Hartl M (2011) Molecular chaperones in protein folding and proteostasis. Nature 475(7356):324–332

Heppner DE, Dustin CM, Liao C, Hristova M, Veith C, Little AC, Ahlers BA, White SL, Deng B, Lam YW, Li J, van der Vliet A (2018) Direct cysteine sulfenylation drives activation of the Src kinase. Nat Commun 9(1):4522

Ho E, Karimi Galougahi K, Liu CC, Bhindi R, Figtree GA (2013) Biological markers of oxidative stress: applications to cardiovascular research and practice. Redox Biol 1(1):483–491

Itoh K, Chiba T, Takahashi S, Ishii T, Igarashi K, Katoh Y, Oyake T, Hayashi N, Satoh K, Hatayama I, Yamamoto M, Nabeshima Y (1997) An Nrf2/small Maf heterodimer mediates the induction of phase II detoxifying enzyme genes through antioxidant response elements. Biochem Biophys Res Commun 236(2):313–322

Jeong H, Baek SY, Kim SW, Park EJ, Lee J, Kim H, Jeon CH (2019) C-reactive protein level as a marker for dyslipidaemia, diabetes and metabolic syndrome: results from the Korea National Health and Nutrition Examination Survey. BMJ Open 9(8):e029861

Jeong JH, Jung S, Kim KN (2021) Considering serum alanine aminotransferase and gamma-glutamyltransferase levels together strengthen the prediction of impaired fasting glucose risk: a cross-sectional and longitudinal study. Sci Rep 11(1):3333

Jomova K, Raptova R, Alomar SY, Alwasel SH, Nepovimova E, Kuca K, Valko M (2023) Reactive oxygen species, toxicity, oxidative stress, and antioxidants: chronic diseases and aging. Arch Toxicol 97(10):2499–2574

Jones DP (2006) Redefining oxidative stress. Antioxid Redox Signal 8(9–10):1865–1879

Kanemura S, Sofia EF, Hirai N, Okumura M, Kadokura H, Inaba K (2020) Characterization of the endoplasmic reticulum-resident peroxidases GPx7 and GPx8 shows the higher oxidative activity of GPx7 and its linkage to oxidative protein folding. J Biol Chem 295(36):12772–12785

Kensler TW, Wakabayashi N, Biswal S (2007) Cell survival responses to environmental stresses via the Keap1-Nrf2-ARE pathway. Annu Rev Pharmacol Toxicol 47:89–116

Kim HJ, Ha S, Lee HY, Lee KJ (2015) ROSics: chemistry and proteomics of cysteine modifications in redox biology. Mass Spectrom Rev 34(2):184–208

Kıran TR, Otlu O, Karabulut A (2023) Oxidative stress and antioxidants in health and disease. J Lab Med 47(1):1–11

Klaassen CD, Reisman SA (2010) Nrf2 the rescue: effects of the antioxidative/electrophilic response on the liver. Toxicol Appl Pharmacol 244(1):57–65

Korytina GF, Akhmadishina LZ, Aznabaeva YG, Kochetova OV, Zagidullin NS, Kzhyshkowska JG, Zagidullin SZ, Viktorova TV (2019) Associations of the NRF2/KEAP1 pathway and antioxidant defense gene polymorphisms with chronic obstructive pulmonary disease. Gene 692:102–112

Kubben N, Zhang W, Wang L, Voss TC, Yang J, Qu J, Liu GH, Misteli T (2016) Repression of the antioxidant NRF2 pathway in premature aging. Cell 165(6):1361–1374

Lee KW, Shin D (2020) Prospective associations of serum adiponectin, leptin, and leptin-adiponectin ratio with incidence of metabolic syndrome: the Korean Genome and Epidemiology Study. Int J Environ Res Public Health 17(9):3287

Liguori I, Russo G, Curcio F, Bulli G, Aran L, Della-Morte D, Gargiulo G, Testa G, Cacciatore F, Bonaduce D, Abete P (2018) Oxidative stress, aging, and diseases. Clin Interv Aging 13:757–772

Lin MT, Beal MF (2006) Mitochondrial dysfunction and oxidative stress in neurodegenerative diseases. Nature 443(7113):787–795

Lindberg S, Jensen JS, Bjerre M, Frystyk J, Flyvbjerg A, Jeppesen J, Mogelvang R (2017) Low adiponectin levels at baseline and decreasing adiponectin levels over 10 years of follow-up predict risk of the metabolic syndrome. Diab Metab 43(2):134–139

Liochev SI, Fridovich I (1994) The role of O2.- in the production of HO.: in vitro and in vivo. Free Radic Biol Med 16(1):29–33

Liu Y, Zhao H, Li H, Kalyanaraman B, Nicolosi AC, Gutterman DD (2003) Mitochondrial sources of H_2O_2 generation play a key role in flow-mediated dilation in human coronary resistance arteries. Circ Res 93(6):573–580

Liu H, Colavitti R, Rovira II, Finkel T (2005) Redox-dependent transcriptional regulation. Circ Res 97(10):967–974

Lobo B, Patil A, Phatak A, Chandra N (2010) Free radicals, antioxidants and functional foods: impact on human health. Pharmacogn Rev 4(8):118–126

Lubos E, Loscalzo J, Handy DE (2011) Glutathione peroxidase-1 in health and disease: from molecular mechanisms to therapeutic opportunities. Antioxid Redox Signal 15(7):1957–1997

Lushchak VI (2014) Classification of oxidative stress based on its intensity. EXCLI J 13:922–937

Ma Q (2013) Role of Nrf2 in oxidative stress and toxicity. Annu Rev Pharmacol Toxicol 53:401–426

Madeira I, Bordallo MA, Rodrigues NC, Carvalho C, Gazolla F, Collett-Solberg P, Medeiros C, Bordallo AP, Borges M, Monteiro C, Ribeiro R (2017) Leptin as a predictor of metabolic syndrome in prepubertal children. Archiv Endocrinol Metab 61(1):7–13

Maechler P, Wollheim CB (2001) Mitochondrial function in normal and diabetic beta-cells. Nature 414(6865):807–812

Marrocco I, Altieri F, Peluso I (2017) Measurement and clinical significance of biomarkers of oxidative stress in humans. Oxidative Med Cell Longev 2017

Matés JM, Pérez-Gómez C, Núñez de Castro I (1999) Antioxidant enzymes and human diseases. Clin Biochem 32(8):595–603

Matsuoka Y, Picciano M, La Francois J, Duff K (2001) Fibrillar beta-amyloid evokes oxidative damage in a transgenic mouse model of Alzheimer's disease. Neuroscience 104(3):609–613

McCord JM, Fridovich I (1969) Superoxide dismutase. An enzymic function for erythrocuprein (hemocuprein). J Biol Chem 244(22):6049–6055

McCord JM, Fridovich I (1988) Superoxide dismutase: the first twenty years (1968–1988). Free Radic Biol Med 5(5–6):363–369

Mocan M, Vesa Ş, Suciu Ş, Blaga SN (2013) Systemic markers of oxidative stress in relation to metabolic syndrome components. Clujul Medical (1957) 86(3):227–234

Moon HE, Paek SH (2015) Mitochondrial dysfunction in Parkinson's disease. Exp Neurobiol 24(2): 103–116

Murakami K, Murata N, Noda Y, Tahara S, Kaneko T, Kinoshita N, Hatsuta H, Murayama S, Barnham KJ, Irie K, Shirasawa T, Shimizu T (2011) SOD1 (copper/zinc superoxide dismutase) deficiency drives amyloid β protein oligomerization and memory loss in mouse model of Alzheimer disease. J Biol Chem 286(52):44557–44568

Nandi A, Yan LJ, Jana CK, Das N (2019) Role of catalase in oxidative stress- and age-associated degenerative diseases. Oxid Med Cell Longev 2019

Nguyen T, Yang CS, Pickett CB (2004) The pathways and molecular mechanisms regulating Nrf2 activation in response to chemical stress. Free Radic Biol Med 37(4):433–441

Niki E (2014) Biomarkers of lipid peroxidation in clinical material. Biochim Biophys Acta 1840(2): 809–817

Oakley AE, Collingwood JF, Dobson J, Love G, Perrott HR, Edwardson JA, Elstner M, Morris CM (2007) Individual dopaminergic neurons show raised iron levels in Parkinson disease. Neurology 68(21):1820–1825

Oka OB, Bulleid NJ (2013) Forming disulfides in the endoplasmic reticulum. Biochim Biophys Acta 1833(11):2425–2429

Paik JY, Jung KH, Lee JH, Park JW, Lee KH (2017) Reactive oxygen species-driven HIF1α triggers accelerated glycolysis in endothelial cells exposed to low oxygen tension. Nucl Med Biol 45:8–14

Pham-Huy LA, He H, Pham-Huy C (2008) Free radicals, antioxidants in disease and health. Int J Biomed Sci 4(2):89–96

Poulsen HE, Specht E, Broedbaek K, Henriksen T, Ellervik C, Mandrup-Poulsen T, Tonnesen M, Nielsen PE, Andersen HU, Weimann A (2012) RNA modifications by oxidation: a novel disease mechanism? Free Radic Biol Med 52(8):1353–1361

Prata C, Hrelia S, Fiorentini D (2019) Peroxiporins in cancer. Int J Mol Sci 20(6):1371

Radi E, Formichi P, Battisti C, Federico A (2014) Apoptosis and oxidative stress in neurodegenerative diseases. J Alzheimer's Dis 42(Suppl 3):S125–S152

Rahman I, Biswas SK, Kode A (2006) Oxidant and antioxidant balance in the airways and airway diseases. Eur J Pharmacol 533(1–3):222–239

Rajasekaran NS, Connell P, Christians ES, Yan LJ, Taylor RP, Orosz A, Zhang XQ, Stevenson TJ, Peshock RM, Leopold JA, Barry WH, Loscalzo J, Odelberg SJ, Benjamin IJ (2007) Human alpha B-crystallin mutation causes oxido-reductive stress and protein aggregation cardiomyopathy in mice. Cell 130(3):427–439

Ramming T, Hansen HG, Nagata K, Ellgaard L, Appenzeller-Herzog C (2014) GPx8 peroxidase prevents leakage of H2O2 from the endoplasmic reticulum. Free Radic Biol Med 70:106–116

Robertson RP (2004) Chronic oxidative stress as a central mechanism for glucose toxicity in pancreatic islet beta cells in diabetes. J Biol Chem 279(41):42351–42354

Ryo M, Nakamura T, Kihara S, Kumada M, Shibazaki S, Takahashi M, Nagai M, Matsuzawa Y, Funahashi T (2004) Adiponectin as a biomarker of the metabolic syndrome. Circ J 68(11): 975–981

Sandstrom PA, Buttke TM (1993) Autocrine production of extracellular catalase prevents apoptosis of the human CEM T-cell line in serum-free medium. Proc Natl Acad Sci USA 90(10): 4708–4712

Sangaleti CT, Katayama KY, De Angelis K, Lemos de Moraes T, Araújo AA, Lopes HF, Camacho C, Bortolotto LA, Michelini LC, Irigoyen MC, Olofsson PS, Barnaby DP, Tracey KJ, Pavlov VA, Consolim Colombo FM (2021) The cholinergic drug galantamine alleviates oxidative stress alongside anti-inflammatory and cardio-metabolic effects in subjects with the metabolic syndrome in a randomized trial. Front Immunol 12

Schapira AH (2008) Mitochondria in the aetiology and pathogenesis of Parkinson's disease. Lancet Neurol 7(1):97–109

Schonknecht YB, Crommen S, Stoffel-Wagner B, Coenen M, Fimmers R, Stehle P, Ramirez A, Egert S (2021) APOE ε4 is associated with postprandial inflammation in older adults with metabolic syndrome traits. Nutrients 13(11):3924

Sies H (1993) Strategies of antioxidant defense. Eur J Biochem 215(2):213–219

Sies H, Berndt C, Jones DP (2017) Oxidative stress. Annu Rev Biochem 86(1):715–748

Snezhkina AV, Kudryavtseva AV, Kardymon OL, Savvateeva MV, Melnikova NV, Krasnov GS, Dmitriev AA (2019) ROS generation and antioxidant defense systems in normal and malignant cells. Oxidative Med Cell Longev 2019(1):1

Sobotta MC, Liou W, Stöcker S, Talwar D, Oehler M, Ruppert T, Scharf AN, Dick TP (2015) Peroxiredoxin-2 and STAT3 form a redox relay for H2O2 signaling. Nat Chem Biol 11(1): 64–70

Spickett CM, Pitt AR (2015) Oxidative lipidomics coming of age: advances in analysis of oxidized phospholipids in physiology and pathology. Antioxid Redox Signal 22(18):1646–1666

Stephenson E, Nathoo N, Mahjoub Y, Dunn JF, Yong VW (2014) Iron in multiple sclerosis: roles in neurodegeneration and repair. Nat Rev Neurol 10(8):459–468

Storz G, Christman MF, Sies H, Ames BN (1987) Spontaneous mutagenesis and oxidative damage to DNA in *Salmonella typhimurium*. Proc Natl Acad Sci USA 84(24):8917–8921

Strzelak A, Ratajczak A, Adamiec A, Feleszko W (2018) Tobacco smoke induces and alters immune responses in the lung triggering inflammation, allergy, asthma and other lung diseases: a mechanistic review. Int J Environ Res Public Health 15(5):1033

Talalay P, Dinkova-Kostova AT, Holtzclaw WD (2003) Importance of phase 2 gene regulation in protection against electrophile and reactive oxygen toxicity and carcinogenesis. Adv Enzym Regul 43:121–134

Thomas MS, Huang L, Garcia C, Sakaki JR, Blesso CN, Chun OK, Fernandez ML (2022) The effects of eggs in a plant-based diet on oxidative stress and inflammation in metabolic syndrome. Nutrients 14(12):2548

Truong TH, Carroll KS (2013) Redox regulation of protein kinases. Crit Rev Biochem Mol Biol 48(4):332–356

Uthra C, Shrivastava S, Jaswal A, Sinha N, Reshi MS, Shukla S (2017) Therapeutic potential of quercetin against acrylamide induced toxicity in rats. Biomed Pharmacother 86:705–714
Valavanidis A, Vlachogianni T, Fiotakis K, Loridas S (2013) Pulmonary oxidative stress, inflammation and cancer: Respirable particulate matter, fibrous dusts and ozone as major causes of lung carcinogenesis through reactive oxygen species mechanisms. Int J Environ Res Public Health 10(9):3886–3907
Valko M, Leibfritz D, Moncol J, Cronin MT, Mazur M, Telser J (2007) Free radicals and antioxidants in normal physiological functions and human disease. Int J Biochem Cell Biol 39(1):44–84
Vona R, Gambardella L, Cittadini C, Straface E, Pietraforte D (2019) Biomarkers of oxidative stress in metabolic syndrome and associated diseases. Oxid Med Cell Longev 2019
Walters DM, Cho HY, Kleeberger SR (2008) Oxidative stress and antioxidants in the pathogenesis of pulmonary fibrosis: a potential role for Nrf2. Antioxid Redox Signal 10(2):321–332
Weaver K, Skouta R (2022) The selenoprotein glutathione peroxidase 4: from molecular mechanisms to novel therapeutic opportunities. Biomedicine 10(4):891
Winterbourn CC, Hampton MB (2015) Redox biology: signaling via a peroxiredoxin sensor. Nat Chem Biol 11(1):5–6
Xu X, Sun J, Chang X, Wang J, Luo M, Wintergerst KA, Miao L, Cai L (2016) Genetic variants of nuclear factor erythroid-derived 2-like 2 associated with the complications in Han descents with type 2 diabetes mellitus of Northeast China. J Cell Mol Med 20(11):2078–2088
Yamaguchi Y, Kamai T, Higashi S, Murakami S, Arai K, Shirataki H, Yoshida KI (2019) Nrf2 gene mutation and single nucleotide polymorphism rs6721961 of the Nrf2 promoter region in renal cell cancer. BMC Cancer 19(1):1137
Zazueta C, Jimenez-Uribe AP, Pedraza-Chaverri J, Buelna-Chontal M (2022) Genetic variations on redox control in cardiometabolic diseases: the role of Nrf2. Antioxidants 11(3):507
Zelko IN, Mariani TJ, Folz RJ (2002) Superoxide dismutase multigene family: a comparison of the CuZn-SOD (SOD1), Mn-SOD (SOD2), and EC-SOD (SOD3) gene structures, evolution, and expression. Free Radic Biol Med 33(3):337–349
Zhang ZY (1998) Protein-tyrosine phosphatases: biological function, structural characteristics, and mechanism of catalysis. Crit Rev Biochem Mol Biol 33(1):1–52
Zhang H, Forman HJ (2017) 4-hydroxynonenal-mediated signaling and aging. Free Radic Biol Med 111:219–225
Zhang J, Wang X, Vikash V, Ye Q, Wu D, Liu Y, Dong W (2016) ROS and ROS-mediated cellular signaling. Oxid Med Cell Longev 2016
Zhang L, Wang J, Wang J, Yang B, He Q, Weng Q (2020) Role of DJ-1 in immune and inflammatory diseases. Front Immunol 11:994
Zhao Y, Zhao B (2013) Oxidative stress and the pathogenesis of Alzheimer's disease. Oxidative Med Cell Longev 2013:1
Zhu Y, Wang G, Ni H, Xiao A, Cai H (2014) Cloning and characterization of a new manganese superoxide dismutase from deep-sea thermophile *Geobacillus* sp. EPT3. World J Microbiol Biotechnol 30(4):1347–1357

Chapter 2
Enumerating the Role of Oxidants and Antioxidants in Animal Health

Lakshita Pandey, Kunal, Ruchika Sharma, Rajni Mathur, S. Latha, and Anoop Kumar

Abbreviations

ABTS	2,2′-Azino-bis (3-ethylbenzothiazoline-6-sulfonic acid)
AGEs	Advanced glycation end products
ATP	Adenosine triphosphate
CAT	Catalase
CVD	Cardiovascular diseases
DCM	Dilated cardiomyopathy
DPPH	2,2-Diphenyl-1-picrylhydrazyl
EGCG	Epigallocatechin gallate
GPx	Glutathione peroxidase
GR	Glutathione reductase
GSH	Glutathione
GSH	Glutathione sulfhydryl
GSSG	Glutathione disulfide
LDL	Low-density lipoprotein
MDA	Malondialdehyde
NAC	N-acetylcysteine
NADPH	Nicotinamide adenine dinucleotide phosphate
NCDs	Non-communicable diseases

L. Pandey · Kunal · R. Mathur · S. Latha · A. Kumar (✉)
Department of Pharmacology, Delhi Institute of Pharmaceutical Sciences and Research (DIPSAR), Delhi Pharmaceutical Sciences and Research University (DPSRU), New Delhi, India
e-mail: anoopdpsru@dpsru.edu.in

R. Sharma
Centre for Precision Medicine and Pharmacy (CPMP), Delhi Pharmaceutical Sciences and Research University (DPSRU), New Delhi, India

N. Srivastava, S. A. Saraf (eds.), *Pharmacological Strategies for Prevention and Treatment of Oxidative Stress*, Oxidative Stress in Applied Basic Research and Clinical Practice, https://doi.org/10.1007/978-3-032-19837-2_2

NPSH	Nonprotein thiol molecule
RAGE	Receptor for advanced glycation end products
RNS	Reactive nitrogen species
ROS	Reactive oxygen species
SOD	Superoxide dismutase
TBA	Thio barbituric acid
UFA	Unsaturated fatty acid
VEGF	Vascular endothelial growth factor

2.1 Introduction

Numerous biological activities rely greatly on the oxidative quality of oxygen, which is necessary for life. Conversely, oxygen may exacerbate oxidative stress in cells. When the cell's mitochondria use oxygen to make adenosine triphosphate (ATP), free radicals are produced. Reactive oxygen species (ROS) and reactive nitrogen species (RNS) are produced by the cellular redox process. Both positive and negative effects on humans can result from these reactive species. The careful balancing act between these two opposing impacts is undoubtedly an important aspect of existence. At low-to-moderate concentrations, reactive species help the immune system and cellular redox signaling; but, at high concentrations, they lead to oxidative stress, a hazardous process that can alter the structure and function of cells. By providing free radicals with an electron, antioxidants can counteract their attack and reduce the amount of harm they can do. These antioxidants primarily scavenge free radicals to prevent or delay cellular damage (Halliwell 1995).

Antioxidants are substances that shield the body from oxidative damage by preventing, reducing, delaying, or eliminating the action of free radicals and oxidants (Lobo et al. 2010). The definition of an antioxidant was given by Halliwell and Gutteridge (1995) as "any substance that at low concentrations compared to the substrates to be oxidized significantly slows down or prevents the oxidation of that substrate." The term "any substance that delays, prevents, or reverses oxidative damage to a target molecule" was eventually added to the definition, though, to cover molecules that fix oxidative system damage (Halliwell and Gutteridge 1995). Effective antioxidants have the ability to hinder free radical formation and inhibit oxidation reactions, while also interfering with the self-oxidization chain reaction producing these free forms. In addition, they serve as metal chelators and reducing agents, allowing the hydroperoxides to be changed into stable molecules. Some antioxidants have the ability to chelate metals, stabilizing metal prooxidants in the process. Due to the depletion of antioxidants, with more reactive oxygen species present or RNS experience oxidative/nitrosative stress considering on early studies that states biological function of antioxidants focuses on their capacity to stop rancidity, which is brought on by the oxidation of unsaturated fats (German 1999). The body combats oxidative damage through two main processes. In the first, enzymes like superoxide dismutase (SOD), catalase (CAT), and others scavenge free radicals and reactive species; in the second, free radicals are removed from electron donors such glutathione (GSH), tocopherols, ascorbic acid, thioredoxin, etc. Pro-oxidant metals bind to particular metal-binding proteins, such as haptoglobin, metallothionein, transferrin, and iron (Kurutas 2015). These mechanisms can also occur through other means like electrolyte bonds. Foods that include antioxidants are better able to withstand oxidative reactions that might otherwise degrade their quality, while several dietary supplements have been promoted as antioxidants, only a few of these products have been shown to enhance health. An antioxidant's impact on health is contingent upon its systemic bioavailability, its capacity to function as intended, and the concentration at which it may be reached to different organ sites.

2.2 Free Radicals and Oxidative Stress

2.2.1 The Origins of Free Radicals

Free radicals are molecules with one or more unpaired electrons housed in their outer shell, which explains why they are more reactive (Sen and Chakraborty 2011). Free radicals are created during the mitochondrial process of cellular respiration, which produces ATP (Pham-Huy et al. 2008). More specifically, reactive oxygen species and reactive nitrogen species are the main byproducts of this process that are produced as free radicals. They can act as reducing agents or oxidants. Furthermore, the metabolism of smooth muscle cells, platelets, macrophages, arachidonic acid, and purine nucleotides can produce free radicals. A major source of free radicals may be lipid peroxidation. When a pathogen enters the body, it can produce ROS and other associated oxy-radicals. Stress and deteriorating mental health can occasionally be sources of free radicals. This occurs when catecholamines and cortisol, two hormones that mediate stress, break down into harmful free radicals. These are a few endogenous processes that produce free radicals.

In addition, free radicals may be produced by an external source. Pollutants include trihalomethanes and chloroform in water and ozone, carbon monoxide, and formaldehyde in the air can create free radicals. In addition, a diet high in processed foods, alcohol, hydrogenated vegetable oil, sugar, coffee, and other things can be a major source of free radical production. Free radical synthesis is accelerated by a number of drugs, such as bleomycin, chlorpromazine, and nitrofurantoin. Tobacco smoking and vehicle exhaust emissions are two factors that might cause the production of dangerous free radicals.

2.2.2 The Dual Function of Radicals

Reactive oxygen species (ROS) and reactive nitrogen species (RNS) have been demonstrated to be beneficial to human health when present in low-to-medium concentrations. All stages of cell maturation i.e. abnormal cell growth, proliferation, differentiation, and death-benefit from their presence. They also participate in phagocytosis process, which destroys microbes, to protect the human body (Pham-Huy et al. 2008). In addition to functioning as energy producers, ROS/RNS also contributes to the eradication of cancerous cells, the activation of transcription factors in genes, and the elucidation of xenobiotics via cytochrome P450 metabolism (Devasagayam et al. 2004). One type of RNS called NO is important for controlling blood pressure and avoiding thrombosis and platelet aggregation. Additionally, free radicals were used by a number of cellular signaling pathways, including growth factor signaling, calcium ion release from endoplasmic reticulum storage, and activation of non-receptor tyrosine kinases, which relied on free radicals to function.

However, the elimination of free radicals is hampered if the concentration rises above a particular point, which might result in oxidative or nitrosative stress. Cellular organelles are destroyed by this stress, which also harms the entire cell's structure, including all biomolecules like proteins, nucleotides, and carbohydrates. Additionally, oxidative stress results in genetic mutations and lipid peroxidation, primarily of unsaturated lipids, to which neurons are particularly vulnerable (Franklin 2011).

2.2.3 *Reactive Oxygen Species and Reactive Nitrogen Species*

Reactive oxygen species (ROS) and reactive nitrogen species (RNS) are examples of oxidants, with a brief half-life. In addition to free radicals, ROS are composed of neutral molecules and ions. Superoxide anion radical ($O_2{\bullet}_{-}$), peroxyl radical (ROO•), lipid peroxyl (LOO•), hydroxyl radical (•OH), and hydroperoxide (HO_2) are examples of free radicals. Singlet oxygen ($^1O_2$), hydrogen peroxide (H_2O_2), nitrous acid (HNO_2), dinitrogen trioxide (N_2O_3), hypochlorous acid (HOCl), lipid peroxide (LOOH), hypobromous acid (HOBr), nitrosyl cation (NO+), nitroxyl anion (NO), dinitrogen tetroxide (N_2O_4), and peroxynitrite (ONOO−) are a few examples of non-radical species. Nitric oxide (NO-), nitrosyl anion (NO), nitrogen dioxide (NO_2), nitrile cation (NO^{2+}), and peroxynitrite (ONOO) are examples of reactive nitrogen species (RNS) (Wiseman and Halliwell 1996).

2.2.3.1 Superoxide Anion Radical ($O_2{\bullet}_{-}$)

Superoxide anion radical is formed when molecular oxygen accepts electrons in an antibonding orbital. It is produced by either enzymatic or non-enzymatic processes during oxidative phosphorylation, which is what produces ATP, or when electrons connect directly to molecules of oxygen (Babcock 1999). The enzymes xanthine oxidase-dependent on lipoxygenase, oxygenases dependent on cytochrome P450, cyclooxygenase, and nicotinamide adenine dinucleotide phosphate (NADPH) oxidase are involved in the formation of superoxide (Inoue et al. 2003). A non-radical peroxide ion with a fragile oxygen-oxygen bond is created when an electron is given to the superoxide radical, and two more electrons provide the formation of two oxide ions. By reduction with two electrons, molecular oxygen yields hydrogen peroxide (H_2O_2), and with four electrons yields water (Babcock 1999).

2.2.3.2 Peroxyl Radical (ROO•)

Alkyl peroxide (ROOH) has the capacity to release hydrogen and produce peroxyl radicals when exposed to heat, radiation, transition metal ions, or other oxidants (Gutowski and Kowalczyk 2013). Alkoxy and peroxyl radicals are also produced when proteins and nucleic acids are oxidized (Augusto and Miyamoto 2011). This results in the production of an alkyl radical, which sets off a chain reaction by reacting with oxygen to form more peroxyl radicals. They have the ability to remove electrons from molecules with low reduction potential and form bonds with biomolecules such as the -SH groups of DNA and albumin (León-Carmona and Galano 2011).

2.2.3.3 Hydroxyl Radical (HO•)

The two main reactions that form hydroxyl radicals are the Fenton reaction (Fe^{2+} + $H_2O_2 \rightarrow Fe^{3+} + OH^- +$ HO•) and the second is the Haber–Weiss reaction in which the first step is ferric ion combine with superoxide to form ferrous (Fe^{3+} + $•O_2-$ $\rightarrow Fe^{2+} + O_2$) and the second step is $Fe^{2+} + H_2O_2 \rightarrow Fe^{3+} + OH^- +$ •OH, which is similar to Fenton reaction (Knight 1999). Because it has the highest reduction potential therefore can readily react with everything inside the live body. The hydroxyl radical that is produced is the most reactive of all the free radicals (Korycka-Dahl and Richardson 1978). It reacts with the C-H bond of saturated compounds and extracts the hydrogen atom to form a free radical, and then they react with oxygen to form other free radicals (Korycka-Dahl and Richardson 1978). Additionally, it can be readily bound with a double bond and is in charge of deactivating several mitochondrial enzymes and oxidizing proteins (Davies 2005). Hydroxyl radicals are produced and attack any species within a few nanometers of their source. Their rates of reaction range from 109 to 1010/M/s for both organic and inorganic compounds. All of these pathways demonstrated the detrimental effects of free radical production (Min and Ahn 2005).

2.2.3.4 Hydrogen Peroxide (H_2O_2)

When two identical superoxide anion radical molecules go through a disproportionation reaction or when oxygen molecules take up electrons and reduce, hydrogen peroxide is created. The other way that hydrogen peroxide is produced is when enzymes like glucose oxidase, D-amino acid oxidase, and urate oxidase, which are often present in peroxisomes and mitochondria microsomes, transfer two electrons to molecular oxygen (Winterbourn 2013). It can readily permeate the cell membrane since it is lipid soluble and non-free radical, but it cannot oxidize or damage any components of the cell, including lipids, nucleic acids, and proteins. Fenton reaction results in homolytic fission which leads to the formation of hydroxyl radical, which is then accountable for the harmful consequences of H_2O_2 (Choe and Min 2006). In biological systems, hydrogen peroxide can react with superoxide anion, HOCl, or chloramines to form singlet oxygen, which can subsequently have detrimental effects of its own (Stief 2003).

2.2.3.5 Hydroperoxyl Radical (HO^2)

Either hydroperoxide is broken up or a proton is added to the superoxide anion radical to generate a hydroperoxyl radical (De Grey 2002). The hydroperoxyl radical is essential to the chemistry of lipid peroxidation. Its ability to extract hydrogen atoms from arachidonic, linolenic, and linoleic fatty acids suggests that it plays a role in the initiation of lipid oxidation. This makes it a far stronger oxidant than superoxide anion. Oxidative stress is caused by Fenton or Haber–Weiss reactions, which are sparked when redox metals combine with hydroperoxyl radical to form hydrogen peroxide (Bielski et al. 1985).

2.2.3.6 Hypochlorous Acid (HOCl)

When internal inflammation arises, myeloperoxidase in neutrophils and macrophages catalyzes the reaction between hydrogen peroxide and chloride anion to produce hypochlorous acid. Along with other biological substances like tryptophan, urate, ascorbate, and pyridine nucleotides, it can oxidize thiols (Winterbourn and Kettle 2000). DNA may also be chlorinated by HOCl. It can convert amines to chloramines, cholesterol and unsaturated lipids to chlorohydrins, tyrosyl residues to ring chlorinated compounds, and more (Prütz 1996).

2.2.3.7 Nitric Oxide (NO)

Nitric oxide, another name for nitrogen monoxide (NO), is a free radical with one unpaired electron. Biological tissues produce nitric oxide through the breakdown of S-nitrosothiols in the presence of metal ions, the reaction of H_2O_2 with arginine, or specific nitric oxide synthases (Ghafourifar and Cadenas 2005; Singh et al. 1996). Nitrogen dioxide, also known as dinitrogen trioxide, is created when nitric oxide combines with molecular oxygen and nitrogen. Nitric oxide is a hazardous oxidizing and nitrosating agent (Radi 2004). Lipid peroxidation is initiated, and ascorbic acid and uric acid concentrations are reduced when NO is introduced into human blood plasma. The oxidized and nitrated products of free lipids and esterified cholesterol are produced during lipid oxidation by the fast reaction of species obtained from nitric oxide in cell membranes and lipoproteins with fatty acids and lipid peroxyl radicals (Repetto et al. 2012).

2.2.3.8 Nitrogen Dioxide (NO_2)

Nitrogen dioxide is created in contaminated air and smoke when peroxyl radicals and NO mix (Noguchi and Niki 2019). Its oxidizing power is medium, falling between that of nitric oxide and peroxynitrite. However, upon dimerization, it forms the extremely reactive dinitrogen tetroxide (N_2O_4). It causes ascorbic acid to oxidize, which is followed by lipid peroxidation and an increase in oxidative stress within the body (Papas 1999).

2.2.3.9 Peroxynitrite (ONOO)

Nitric oxide and superoxide anion combine to generate peroxynitrite (ONOO) (Beckman and Koppenol 1996). Because peroxynitrite readily diffuses across cell membranes, it can oxidize lipids, methionine residues, tyrosine in proteins, and DNA to nitroguanine (Douki and Cadet 1996). Low-density lipoprotein (LDL) is oxidized, and tissue damage is caused by peroxynitrite, which appears to be produced at inflammatory areas (Knight 1999). It is extremely poisonous and can react directly with CO_2 to generate peroxynitrous acid (ONOOH) or other highly reactive nitroso peroxo-carboxylates ($ONOOCO_2$). These compounds can then hemolyze further to form OH and NO_2 or reorganize to form NO_3 (Beckman and Koppenol 1996).

2.2.4 Oxidative Stress Lead Molecular Damage

Five percent of the oxygen we breathe in is reduced and transformed into dangerous free radicals like hydroxyl, peroxyl, and hydrogen peroxide (Gupta and Sharma 2006). When the body is exposed to these radicals repeatedly, biomolecules are destroyed, resulting in tissue damage. The process that occurs when important molecules are destroyed by life-sustaining oxygen and turn toxic for the human body is known as the "Oxygen Paradox." Oxidative stress causes molecular damage by membrane lipid peroxidation, which is followed by protein oxidation, DNA damage, and disruption of the cell's reducing equivalents. These events result in cell death and changed signaling pathways (Fang et al. 2002).

Lipid peroxidation is a mechanism that produces toxic by-products mainly by peroxidation of unsaturated fatty acid. The process starts when a free radical attaches with the methylene group (CH_2) of UFA, and there is hydrogen abstraction from the group and a lipid radical (•CH) is formed with unpaired electron on the carbon atom.

Fig. 2.1 The lipid peroxidation process

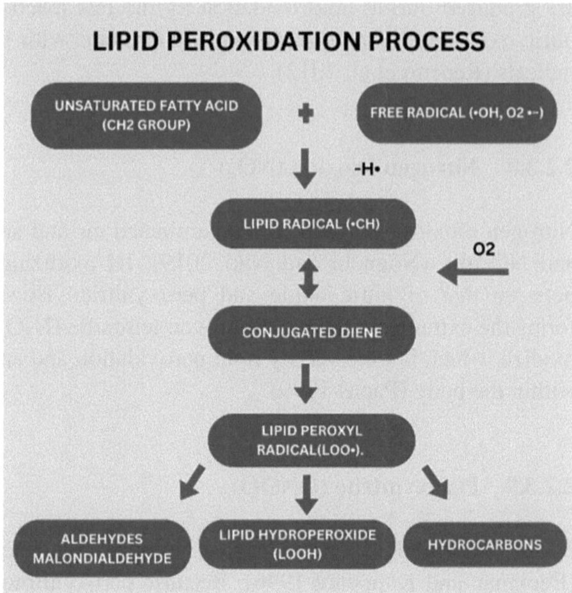

This species is stabilized by molecular rearrangement to form a conjugated diene, and this is attacked by oxygen molecule and leads to the production of lipid peroxyl radical (LOO•). Lipid peroxidation is sustained by the formation of lipid hydroperoxide (LOOH), which is created when peroxyl radicals break off hydrogen atoms from other lipid molecules. Lipid peroxyl can also be transformed into aldehydes, particularly malondialdehyde and hydrocarbons. This malondialdehyde (MDA) alters the purine and pyrimidine nitrogen bases and amino group of amino acids and modifies their structure, respectively. Additionally, in the presence of RNS, it can attack the lipid peroxyl radicals, resulting in the formation of secondary nitrogen-containing products (LOONO), and this is where the reaction terminates. As the propagation stage progresses, lipid hydroperoxide (LOOH) divides into alkoxyl and hydroxyl radicals before becoming peroxide bridged dimer (LOOL), which is when the chain breaks (Fig. 2.1).

In the presence of free radicals, proteins can also undergo modification. When RNS and ROS bind to amino acids containing sulfur, proteins become inactivated, denatured, and cross-linked, which eventually results in the formation of protein hydroperoxide. The nitration mechanism will work to change the structure of proteins that have aromatic rings in them. When free radicals are present, amino acids, including cysteine, methionine, tryptophan, tyrosine, and phenylalanine, oxidize. However, these oxidized amino acids are inert, do not damage the organism, and flush out quickly. Certain oxidized proteins play a role in the etiology of certain diseases as well as the ageing process.

DNA is also vulnerable to the attack of free radical as it can be either mitochondrial or nuclear DNA. Hydroxyl is the radical that can break phosphodiester bonds, alter nitrogen bases, and oxidize sugar moiety. Purines when attacked by hydroxyl lead to the formation of 8-hydroxydeoxyguanosine (8-OHdG), formamidopyrimidines, 8-hydroxydeoxyadenosine, etc. Pyrimidine also gets destroyed by hydroxyl radical and products such as thymine glycol, uracil glycol, urea residue, 5-hydroxydeoxyuridine,5-hydroxydeoxycytidine, hydantoin, and others. Further fragmentation of DNA, which leads to programmed cell death, occurs by ADP-ribose synthetase enzyme, which is again activated by the free radical. As the mitochondrial DNA also gets damaged, the products of the respiratory chain so formed are also defective.

2.3 The Role of Oxidants in Animal Health

2.3.1 Cancer

Oxidative stress has been connected to the development and metastasis of cancer in animals. Reactive nitrogen species and reactive oxygen species can directly damage DNA, causing mutations that initiate the carcinogenesis process. Apart from causing genetic harm, ROS can also trigger signaling pathways like AP-1 and NF-κB, which encourage angiogenesis, invasion, and tumor growth. For instance, research on mammary tumors in dogs has revealed diminished antioxidant defenses like glutathione peroxidase and superoxide dismutase, as well as increased levels of oxidative biomarkers such as MDA. Oxidative stress has also been linked to bovine ocular squamous cell carcinoma and equine melanoma, indicating that ROS plays a common function in a variety of animal malignancies (Cannavò et al. 2019; Tsujita and Plummer 2010).

Oxidative stress in cancer has been widely studied using animal models. The processes by which ROS contribute to the beginning and progression of cancer have been clarified by using rodent models, such as chemically generated mammary tumors in rats and hepatocellular carcinoma in mice. These models show that oxidative stress affects the tumor microenvironment by regulating angiogenesis and inflammatory responses in addition to causing mutations (Reuter et al. 2010). Because they resemble human cancers in both biology and course, spontaneous tumors such as osteosarcomas and lymphomas in dogs are useful models. Research employing these models has demonstrated how antioxidants, including dietary polyphenols and N-acetylcysteine (NAC), can be used therapeutically to reduce oxidative damage and decrease the growth of tumors (Ladas et al. 2004).

In animal models, antioxidant treatments that target ROS have demonstrated promise in slowing the growth of tumors and enhancing their results. For example, supplementing chickens with selenium has been linked to decreased oxidative damage and fewer cases of liver tumors caused by aflatoxin. Likewise, in canine models, vitamin E and coenzyme Q10 treatment have shown advantages in tumor management and oxidative stress markers.

2.3.2 *Cardiovascular Diseases*

Animal cardiovascular conditions such as heart failure, atherosclerosis, and pulmonary hypertension are significantly influenced by oxidative stress. The development of the disease is accelerated by the overproduction of reactive nitrogen species (RNS) and ROS, which causes endothelial dysfunction, inflammation, and cardiac remodeling.

2.3.2.1 Oxidative Stress Mechanisms in Cardiovascular Disorders Endothelial Dysfunction

ROS like hydrogen peroxide and superoxide anion react to generate peroxynitrite, a strong oxidant, which lowers the bioavailability of nitric oxide (NO). Endothelial function, a defining feature of cardiovascular disorders, is compromised by this process. Research in bovine and canine models has shown that afflicted animals had higher ROS levels and less NO-mediated vasodilation (Dhalla et al. 2000).

2.3.2.2 Inflammation and Cardiac Remodeling

ROS trigger pro-inflammatory signaling pathways, such as NF-κB, which causes adhesion molecules and cytokines to be upregulated. In animal models of heart failure, this inflammatory response leads to cardiac hypertrophy and fibrosis.

2.3.2.3 Mitochondrial Dysfunction

A key characteristic of heart failure and ischemic heart disease is the generation of ROS by the mitochondria. In animal studies, mitochondrial malfunction exacerbates cardiac damage by causing energy deficits and cardiomyocyte death (Madamanchi and Runge 2007).

2.3.3 *Diabetes*

Animal diabetes pathogenesis has been found to be significantly influenced by oxidative stress, especially when insulin resistance and pancreatic beta-cell dysfunction are present. The production of reactive oxygen species (ROS) during hyperglycemia damages cells and makes the consequences of diabetes worse.

2.3.3.1 Diabetes-Related Oxidative Stress Mechanisms

2.3.3.1.1 Beta-Cell Vulnerability

Because pancreatic beta cells have low quantities of antioxidant enzymes such glutathione peroxidase (GPx), catalase, and superoxide dismutase (SOD), they are naturally vulnerable to oxidative damage. Because of this susceptibility, they are more likely to sustain harm from ROS generated during the metabolism of glucose. Reduced antioxidant defenses and elevated lipid peroxidation in pancreatic tissue have been linked to reduced insulin production in diabetic animal models (Robertson et al. 2004).

2.3.3.1.2 Advanced Glycation End Products (AGEs)

AGEs are generated when hyperglycemia takes place. Reactive oxygen species (ROS) are subsequently produced when these AGEs interact with their receptor (RAGE). This raises oxidative stress and promotes inflammation, endothelial dysfunction, and tissue damage in diabetic mice. Elevated AGE levels in diabetic dogs and cats have been associated with microvascular issues (Singh et al. 2014).

2.3.3.1.3 Mitochondrial Dysfunction

The mitochondria's overproduction of ROS during hyperglycemia is one of the primary mechanisms of oxidative stress in diabetes. When too much ROS damages mitochondrial integrity and function, insulin-producing beta cells undergo death, and peripheral organs acquire insulin resistance (Lowell and Shulman 2005).

2.3.3.1.4 Polyol Pathway Activation

Excess glucose is transferred into the polyol pathway in diabetic animals, increasing the synthesis of fructose and sorbitol. This weakens the cellular antioxidant defense mechanism by depleting NADPH, which is necessary for the regeneration of reduced glutathione (GSH) (Obrosova 2005).

2.3.3.2 Complications of Diabetes and Oxidative Stress

2.3.3.2.1 Diabetic Neuropathy

Oxidative stress contributes to the degeneration of nerves in diabetic neuropathy. Animal models of diabetes, such as streptozotocin-induced diabetic rats, have shown reduced antioxidant enzyme activity and elevated levels of lipid peroxidation products in their brain tissues (Vincent et al. 2004).

2.3.3.2.2 Diabetic Nephropathy

By causing glomerular and tubular damage, ROS is a major contributor to the development of diabetic nephropathy. Research on diabetic mice has shown thickening of the glomerular basement membrane, albuminuria, and enhanced ROS generation in renal tissues (Forbes et al. 2008).

2.3.3.2.3 Diabetic Retinopathy

Through ROS-mediated retinal cell death and disruption of the blood-retinal barrier, oxidative stress has been linked to retinal injury in animal models. Malondialdehyde (MDA) and other oxidative stress indicators have been found to be elevated in diabetic rats, underscoring the role ROS plays in retinal degeneration (Kowluru and Chan 2007).

2.3.4 *Classification of Antioxidants*

Antioxidants are substances that, at low concentrations, strongly prevent or postpone the oxidative process, even though they are regularly oxidized themselves. By preserving redox balance, endogenous and exogenous antioxidants work together to neutralize free radicals and shield the body from them. These antioxidants fall into a number of different types. In Table 2.1, antioxidants are categorized in depth (Valko et al. 2007).

One significant endogenous antioxidant enzyme that comes in multiple common forms is superoxide dismutase (SOD). The body uses these proteins as its first defense against ROS, which scavenges superoxide radicals, and they contain Cu, Zn, Mn, Fe, or Ni. SOD aids in the transformation of $O_2^{\bullet-}$ to H_2O_2 and O_2 (Jacob 1995). Three different kinds of superoxide dismutase are found in humans: extracellular SOD3, mitochondrial SOD2, and cytoplasmic SOD1.

Catalase (CAT) is the most prevalent and effective enzyme present in cells. The enzyme can react with H_2O_2 because of its tetramer structure, which is made up of four polypeptide chains and four porphyrin heme groups. Millions of H_2O_2 molecules can be converted into O_2 and H_2O per second by a single catalase molecule. The cytoplasm of cells contains glutathione peroxidase (GPx), which shields the cell from oxidative damage brought on by H_2O_2 and stops H_2O_2 from producing hydroxyl radicals (Sies 1997). It consists of four protein subunits with one selenium atom in each of its active sites. GPx removes H_2O_2 by coupling its reduction to water with GSH oxidation. The flavoprotein enzyme glutathione reductase (GR) is a significant cellular antioxidant that is required for the GSH conversion. When the enzyme GR is present, glutathione disulfide (GSSG), the oxidized form of glutathione, is reduced to glutathione sulfhydryl (GSH), which needs NADPH as an electron

Table 2.1 Antioxidants classification

A. Classification based upon their nature
1. Enzymatic antioxidants
Catalase (CAT), superoxide dismutase (SOD), glutathione peroxidase (GPx), and glutathione reductase (GR)
2. Non-enzymatic antioxidants
(a) Metabolic antioxidants
Reduced glutathione (SOD), lipoic acid, L-arginine, transferrin, coenzyme Q10, bilirubin, uric acid, metal-chelating protein, etc.
(b) Nutrient antioxidants
Vitamin E, vitamin C, flavonoids, carotenoids, trace metals (Se, Mn, Zn), omega-3 and omega-6 fatty acid, etc.
B. Classification based upon source
1. Endogenous antioxidants Bilirubin, lipoic acid, glutathione, N-acetyl cysteine, uric acid, NADPH and NADH, ubiquinone (coenzyme Q10), enzyme (SOD, CAT, GPx, GR)
2. Dietary antioxidants Vitamin C, vitamin E, ß-carotene and other carotenoids and oxycarotenoids (lycopene and lutein), polyphenols (flavonoids, flavones, flavanols, proanthocyanidins)
3. Metal-binding protein
Albumin (Cu), ceruloplasmin (Cu), metallothionein (Cu), ferritin (Fe), myoglobin (Fe), transferrin (Fe)
C. Classification based upon mechanism of action
1. Catalytic systems to neutralize or divert CAT, ROS SOD, GPx
2. Self-suicidal and chain-breaking antioxidants scavenge and destroy ROS Vitamin C, vitamin E, glutathione, uric acid, flavonoids
3. Binding/inactivation of metal ions prevent formation of ROS by the Haber–Weiss reaction Catechins, ferritin, caeruloplasmin
4. Quenching ROS, chemical traps/sinks to "absorb" energy Carotenoids, anthocyanidins

donor. An important general indicator of an organism's oxidative stress is the ratio of GSH to GSSG; an excessively high concentration of GSSG can oxidatively damage numerous enzymes.

Large amounts of the powerful antioxidant tripeptide glutathione are present in the nuclei (3–15 mM), mitochondria (5–11 mM), and cytoplasm (1–11 mM), which is the primary soluble antioxidant that is not enzymatic and exists in these cell compartments. The redox state of essential protein sulfhydryls is maintained by GSH in the nucleus, which is required for DNA expression and repair (Fang et al. 2002). Glutamate, glycine, and cysteine combine to form the principal intracellular non-protein thiol molecule (NPSH). GSH is essential for the upkeeping of -SH groups in other molecules, including proteins, the regulation of the thiol-disulfide status of cells, and the body's elimination of impurities and free radicals. Methionine and cysteine, two amino acids high in sulfur, function as precursors to GSH but also produce -SH groups to interact with the $OH^{\bullet}$ radical and H_2O_2, potentially preventing tissue injury.

Through the catalytic activity of GPx, GSH has the potential to directly scavenge singlet oxygen and $OH^{\bullet}$ radicals, as well as detoxify lipid peroxides and hydrogen peroxides. Glutathione (GSH) is engaged in the movement of amino acids across the membrane of the plasma and in the regeneration of several important antioxidants. Additionally, it controls the amounts of vitamins C and E. For example, glutathione can convert semidehydroascorbate to ascorbate, which lowers the tocopherol radical of vitamin E either directly or indirectly. Another sulfur-containing natural antioxidant is lipoic acid, also referred to as a "thiol" or "biothiol." Because dihydrolipoic acid, a reduced version of lipoic acid, may inhibit free radicals in lipid and aqueous environments, it is known as a "universal antioxidant." Lipoic acid may potentially act as an antioxidant by chelating with pro-oxidant metals. Exogenous antioxidants are also necessary for the effective working of the body's endogenous antioxidants.

Numerous enzymes in the human body work against the reactive species that are produced. These consist of CAT, SOD, GPx, and glutathione reductase (GRx). SOD, CAT, and GPx activity is the first line of antioxidant defense and a crucial part of the host biological system's overall defensive mechanisms (Ighodaro and Akinloye 2018).

2.3.4.1 Superoxide Dismutase

Irvin Fridovich made the discovery of SOD in 1968. O_2 to H_2O_2 is catalyzed by this enzyme. SODs are members of the multimeric metalloenzyme family and are found in the cytosol and mitochondria. They are separated into families: Cu-SOD, Cu-Zn-SOD, Mn-SOD, Fe-SOD, and Ni-SOD. There are several places where these enzymes can be found. For example, MnSOD is primarily found in the bacterial mitochondria's matrix and cytoplasm, prokaryotes contain Fe-SOD and some plants, and Cu-Zn-SOD is largely found in the cytosol and chloroplast of eukaryotic cells (Duke and Salin 1985). One of these enzymes is Cu-Zn-SOD which is well known for offering a robust defense against intracellular oxygen toxicity. According to studies, Cu-Zn-SOD is a necessary and unique enzyme for aerobic life (Peskin et al. 1977). Studies show that in prokaryotes and eukaryotes, including higher plants, SODs help shield proteins and enzymes against oxygen toxicity (Bowler et al. 1992).

2.3.4.2 Catalase

CAT is regarded as a significant antioxidant enzyme and is found in the majority of cells. The primary site of CAT is the peroxisomes in mammalian cells. Although it is also present in mitochondria, animal CAT is mostly located in peroxisomes. Four identical 60 kDa subunits arranged tetrahedrally make up this tetrameric enzyme. Every subunit contains NADPH and an active heme group (Kirkman and Gaetani 1984; Martínez-Cayuela 1995). CAT converts H_2O_2 into H_2O and O_2. CAT has two different enzymatic activities based on the H_2O_2 content (Amir Aslani and Ghobadi 2016). When H_2O_2 concentrations are high, CAT functions catalytically, removing

H_2O_2 by combining with O_2 and H_2O. However, when a suitable H donor is present, such as methanol, ethanol, phenol, and others, it functions peroxidically at low H_2O_2 concentrations, oxidizing its substrate and removing H_2O_2. This process is known as a peroxidatic reaction.

2.3.4.3 Glutathione Peroxidase and Glutathione Reductase

The GPx enzyme family requires reduced GSH as its substrate. They offer an additional layer of protection against oxidative damage. It is composed of four identical subunits, each containing a selenocysteine (Sec) residue and weighing 21 kDa (Lubos et al. 2011). GPx is a catalyst that reduces both organic and inorganic H_2O_2 to H_2O and the related alcohols by using GSH as a cofactor (Birben et al. 2012). The preservation of polyunsaturated fatty acids in cell membranes depends on it and functions as a multicomponent antioxidant defense mechanism (Gathwala and Aggarwal 2016). Compared to CAT, GPx exhibits a greater affinity for the substrate and is known to lower fatty acid hydroperoxides (Sharma et al. 2012). The majority of animal cells carry both GPx and CAT. There are two forms of GPX: selenium dependent and selenium independent.

Both the cytosol and the mitochondria contain them. Mammalian tissues include four main types of selenium-dependent GPx isozymes: (a) classical GPx (GPx1), (b) gastrointestinal GPx (GPx2), (c) plasma GPx (GPx3), and (d) phospholipid GPx (PHGPx4 or GPx4). Most of GPx1 is found in the lung, liver, kidney, and red blood cells (RBC). The majority of GPx2 and GPx3 locations are in organs such as the kidney, placenta, lung, epididymis, vas deferens, seminal vesicle, heart, and muscle, despite their widespread distribution (Margis et al. 2008) GPx4 is widely distributed throughout many tissues. It has been discovered that GPxs are the first enzyme to activate in response to elevated ROS levels, and they have a major role in the detoxification of ROS (Halliwell and Gutteridge 2015).

A cytosolic protein called glutathione reductase exhibits a pattern of distribution in cells that is comparable to that of GPx. It uses NADPH to lower oxidized glutathione (GSSG). Keeping the GSH/GSSG ratio constant is one of GRx's key roles. Lipid peroxidation, protein denaturation, and DNA damage are typically caused by increases in the quantity of GSSG (oxidized glutathione) within the cell (Zitka et al. 2012).

2.3.4.4 Glutathione

γ-Glutamylcysteinylglycine, also known as glutathione, is the primary nonenzymatic antioxidant that is involved in numerous cellular processes. It's an widely used tripeptide that can be found in either its oxidized (GSSG) or reduced (GSH) form and controls intracellular redox balance. The tripeptide GSH, which is composed of glutamic acid, glycine, and cysteine, is present in numerous body tissues at comparatively high amounts. At millimolar concentrations, it is present in nearly every form of cell compartment, including the cytosol, ER, mitochondria, and vacuoles (Jiménez et al. 1998). GSH scavenges a variety of ROS, like H_2O_2, O_2^{-}·, and OH· (Misak et al. 2018). One of GSH's primary functions as an antioxidant is the ascorbate-GSH cycle (Noctor and Foyer 1998), which allows it to recover ascorbic acid. GSH functions as a barrier against oxidation caused by hydroperoxide and is a powerful xenobiotic detoxifier.

It is essential for lowering oxidative stress, preserving redox equilibrium, improving metabolic detoxification, and regulating immune system. Suboptimal or insufficient GSH levels have been linked to a number of chronic, age-related illnesses, including those associated with neurodegeneration, mitochondrial malfunction, and even cancer. More people are becoming aware of its ability to enhance the liver's capacity to convert and eliminate toxins such as mercury and persistent organic pollutants, thus lowering the body's toxin levels (Minich and Brown 2019).

2.3.4.5 Ascorbic Acid

Ascorbic acid, another name for vitamin C, is made up of two main parts: L-dehydroascorbic acid and L-ascorbic acid. Both have antioxidant properties, can be absorbed through the gastrointestinal tract, and can undergo enzymatic conversion in vivo. Ascorbic acid can be readily oxidized to yield dehydroascorbic acid. This water-soluble antioxidant is essential for biological fluids that the body needs for normal metabolic functions. It is a highly efficient free radical scavenger that protects from low-density lipoprotein (LDL) (Harats et al. 1998) integrity and keeps vitamin E levels in cell membranes constant. By scavenging reactive oxygen and nitrogen species, it shields macromolecules like proteins, lipids, and DNA from oxidative degradation. Ascorbic acid quenches O_2^{-}·, ·OH, and O_2, and reduces H_2O_2 to H_2O via the ascorbate peroxidase process, in addition to primarily scavenging ONOO−, NO, and HOCl (Noctor and Foyer 1998). Apart from its ability to scavenge free radicals, ascorbic acid also aids in the restoration of small molecules like β-carotene, GSH, urate, and α-tocopherol, enabling them to function as antioxidants as well (Uttara et al. 2009). It is a strong antioxidant that suppresses the production of 8-hydroxydeoxyguanosine (8-OHdG) in nucleic acids and can reduce or completely stop lipid peroxidation caused by H_2O_2. Because the body lacks the glucuronic pathway, it is unable to make ascorbic acid.

2.3.4.6 Tocopherols

The four naturally occurring forms of tocopherols are alpha (α), beta (β), gamma (γ), and delta (δ) tocopherols; the most common of these isoforms, α-tocopherol (αT), is also frequently called vitamin E. Each tocopherol isoform is composed of a chromanol ring and a 16-carbon phytyl-like side chain (Jiang 2014). Eight chemically different analogs of vitamin E have been identified, comprising tocopherols (T) (α, β, γ, and δ) and tocotrienols (T3) (α, β, γ, and δ). These analogs are all now referred to as vitamin E, even though α-tocopherol was the first to be found. While tocotrienols are unsaturated (contain three double bonds) and have an isoprenoid side chain, tocopherols are saturated forms of vitamin E (Ahsan et al. 2014). αT, an endogenous antioxidant obtained by the body from diet.

Vitamin E can be found in foods like nuts, seeds, vegetable oil, whole grains, and more. It functions as a break in the chain that results in lipid peroxidation of cell membranes and is an efficient lipid-soluble antioxidant present in the plasma membrane (Uttara et al. 2009). By giving a hydrogen atom to lipoperoxyl radicals, it can scavenge them directly and shield cell membranes from lipid peroxidation (Martínez-Cayuela 1995). The oxidized tocopheryl radical is subsequently transformed back to tocopherol via a mechanism assisted by ascorbic acid (Traber and Stevens 2011). Although αT is the most common and well-researched form, other isoforms with antioxidative and anti-inflammatory qualities, such as γ-tocopherol, γ-tocotrienol, and δ-tocopherol, are also involved in the pathophysiology of various illnesses (Mathur et al. 2015).

2.3.4.7 β-Carotene

Since humans cannot produce β-carotene, which is essential for many biological functions in the body, these molecules must be received through diet or medication. β-Carotene was originally extracted from plants using physicochemical methods, primarily from carrots. However, these days, chemical synthesis is the primary method used to create β-carotene (Bogacz-Radomska and Harasym 2018). Vitamin A that is fat-soluble is obtained from the β-carotene and was initially described by Monaghan and Schmitt (1932). This antioxidant is the most powerful scavenger of singlet oxygen radicals and protects lipids against rancidity.

Because fat-soluble carotenoids, such as β-carotene, can be transformed into retinol, the active form of vitamin A that is essential for vision, they are considered pro-vitamins. It is regarded as the greatest singlet oxygen quencher and a potent antioxidant. It shields lipid-rich tissues from free radical damage. Another carotenoid with antioxidant properties is lycopene. It was discovered to be helpful against prostate cancer.

2.3.4.8 Thioredoxin

Thioredoxin is a common cellular protein disulfide reductase that donates electrons to a number of other enzymes, such as ribonucleotide reductase, methionine sulfoxide reductase, and thioredoxin peroxidase (Arnér and Holmgren 2000). With only 12 kDa of molecular weight and 105 amino acid residues, thioredoxin is a potent antioxidant protein. Because it maintains proteins in their reduced form, signal transduction pathways' redox reactions are regulated (Kaimul et al. 2007).

It uses two cysteine residues in a dithiol–disulfide exchange process to perform the activities of an intracellular reductase (Holmgren 1985). Reduced thioredoxin creates dithiols in the active site, while oxidized thioredoxin adds disulfide linkages.

To lower the disulfide and oxidize itself, reduced thioredoxin delivers a hydrogen ion (H+) to the targeted oxidized protein (Engelman et al. 2016). Nitrosylation can also occur with thioredoxin. Certain malignancies, such as hepatocellular carcinoma and cervical tumors, have elevated thioredoxin levels (Nishiyama et al. 2001). Plasma levels of thioredoxin function as a biomarker for disorders linked to oxidative stress, including metabolic syndrome (Miyazaki et al. 1998). Increased vascular endothelial growth factor (VEGF), resistance to treatment, and tumor angiogenesis are caused by the transcription of hypoxia-inducible factor-1α (HIF-1α) being activated. Many novel roles of thioredoxin have been found, leading to the protein's reputation as a "moonlighting protein."

2.3.5 Sources of Antioxidants

Fruits and vegetables (FAV) are among the food plants that have been shown to have health-improving benefits (Nishiyama et al. 2001). Certain fruits and vegetables (FAV) have been linked to antioxidants, including citrus fruits (orange, grapefruit, lime, and lemon), pomegranates, grapes, apples, dates, green and yellow vegetables (peppers), cabbage, carrots, bananas, strawberries, and dark leafy greens. Antioxidants have the ability to reduce the risk of chronic diseases by working both additively and synergistically (Eastwood 1999). Because of this, fruits and vegetables provide protection against chronic conditions such as blood-related illnesses, diabetes, strokes, cancer, cerebrovascular, ophthalmic, and neurological diseases, and blood-related illnesses (Serna-Saldivar 2016). Numerous epidemiological studies have shown strong evidence that the natural chemicals found in FAV are linked to its capacity to treat most of these health issues. An estimated 11% of strokes and 31% of ischemic heart disease globally are attributed to low FAV intake (Jideani et al. 2021). For the prevention of stroke, high blood pressure, cardiovascular disorders, and other micronutrient-related deficiencies, the joint FAO/WHO consultation report on food, nutrition, and the prevention of chronic diseases recommended a minimum daily consumption of 400–500 g of FAV. So, a major risk factor for a number of non-communicable diseases (NCDs) linked to nutrition is an inadequate intake of FAV. Antioxidant components (natural radical terminators), such as vitamins A, C (ascorbic acid), E (α-tocopherol), β- and α-carotene, and glutathione, have been widely implicated in the protective action of FAV. Additional antioxidants found in FAV include phenolic and polyphenolic chemicals, terpenoids, sulfur-containing compounds, alkaloids, and polyphenols, which prevent oxidative damage by scavenging free radical activity. These bioactive, non-nutritive plant substances, often known as phytochemicals, help to terminate chain reactions by eliminating free radical intermediates (Pisoschi and Negulescu 2012). Carotenoids are an important bioactive found in FAV that are particularly good at preventing singlet oxygen-induced oxidation, according to Niki and Noguchi. Another class of bioactives are plant metabolites called polyphenolic flavonoids (Block et al. 1992), which have a variety of pharmacological and biological effects. It has been demonstrated that flavonoids, which include anthocyanins, anthocyanidins, flavanols, flavones, and flavanones, have anti-inflammatory, antioxidant, antimutagenic, and anticarcinogenic qualities (Steinmetz and Potter 1996).

Consuming fruits and vegetables separately has been shown to deliver recognized bioactives with positive health effects. On the other hand, processing, which is one way to increase FAV consumption in the synergistic enhancement of bioactive's impacts on molecular pathways is crucial for disease prevention, and this combination will further strengthen their benefits. Combining foods, such as salads, prepared meals, and mixtures of fresh fruit or vegetable items, will enhance the synergistic benefits of different phytochemicals.

2.3.6 High-Antioxidants-Rich Fruit and Vegetables

Roots, stalks, bulbs, tubers, leaves, stems, fruits, and flowers are among the edible plant parts. These include fruits and vegetables. Fruits have different antioxidant components depending on how many of them are consumed. FAV also contains unique antioxidants such as phenolic compounds, carotenoids, and vitamins C and E. Further studies have demonstrated that when ingested in appropriate quantities, FAV plays a significant role in providing vitamins, carotenoids, polyphenols, glucosinolates, saponins, and sterols. Many tropical and subtropical fruits, such as guava, pineapple, mango, passion fruit, litchi, papaya, Passiflora, kiwano, carambola, banana, feijoa, star fruit, lime, and longan, are utilized as components in meals eaten in North America and Europe. In addition to raising blood antioxidant levels, consuming FAV, according to Burton-Freeman (2010), balances the harmful effects of high-fat and high-carb meals. According to taxonomy and as demonstrated in the research of Carlsen et al. (2010), the antioxidant content of FAV in milligrams per 100 grams of fresh food is shown in Table 2.2. Citrus fruits are rich in nutrients and antioxidants.

Citrus fruits are high in ascorbic acid and include limonoids, flavones, hydrocinnamates, and flavanone glycosides (hesperidin, narirutin, and naringenin). The analysis of a variety of Fijian fruits, vegetables, and other easily accessible foods, green leafy vegetables had the most significant capacity for antioxidants, with fruits and root crops coming in second. Furthermore, the high content of total polyphenols in the banana ('Muomva-red cultivar') suggests that the fruit may contain bionutrients with important health and medicinal uses. The Opuntia cactus is known to contain a distinct class of flavonoids, the types and concentrations of which vary according to the variety, tissue type, and maturity. According to some studies, the Cactaceae family of plants contains dihydroflavonols, flavanones, flavanonols, and flavonol 3-O-glycosides (quercetin, kaempferol, and isorhamnetin). Most research on the flavonoids found in Opuntia cactus has concentrated on removing them from the tissue of the flowers.

Table 2.2 Comparative representation of distinct phytochemicals found in fruits and vegetables (FAVs) and their antioxidant concentration

Fruit and vegetables	Phytochemical class	Antioxidants	References
Apricot	Tetrapenoids Flavonoids	α-Carotene, β-carotene, lycopene Quercetin, kaempferol	
Banana	Phenolic acids Flavonoids	Gallic acids Catechin, epicatechin	
Grape	Tetrapenoids	α-Carotene, β-carotene, lycopene	
Orange	Monoterpenoids Tetrapenoids	Limonene α-Carotene, β-carotene, lycopene	
Vegetables Asparagus	Triterpenoids	Campesterol	
Broccoli	Tetrapenoids Triterpenoids Flavonoids	α-Carotene, β-carotene, lycopene α-Tocopherol, β-tocopherol, β-tocotrienols Cyanidin	Jideani et al. (2021)
Cabbage	Glucosinolates	Glucobrassicin Sulphoraphane	
Eggplant	Phenolic acids Flavonoids	Chlorogenic acids, caffeoylquinic acid Kaempferol, quercetin, apigenin	
Onion	Triterpenoids Flavonoids	Campesterol, β-sitosterol Cyanidin, delphinidin	

2.3.7 *Mechanism of Action of Antioxidants*

Antioxidants are crucial in the body's battle against oxidative stress. There are various types of mechanisms involved in the functioning of antioxidants. Unpaired electrons in atoms, molecules, or ions make them highly reactive in chemical reactions with other molecules, known as free radicals. Within the biological system, oxygen, nitrogen, and sulfur molecules frequently produce free radicals. Three categories of molecules are used to categorize these free radicals: reactive oxygen species (ROS), reactive nitrogen species (RNS), and reactive sulfur species (RSS). ROS include hydrogen peroxide (H_2O_2), hypochlorous acid (HOCl), singlet oxygen ($1O_2$), peroxynitrite (ONOO}), and free radicals such as superoxide anion ($O_2^{\bullet}$), perhydroxyl radical ($HO_2^{\bullet}$), hydroxyl radical (·OH), and nitric oxide. Nitric oxide is the source of RNS, which are created when it reacts with $O_2^{\bullet}$ to create ONOO}. When thiols react with ROS, RSS are readily produced. During cellular metabolism and functional processes, ROS are generated, but they also have significant effects on cell signaling, ion transport, gene expression, and death. ROS, due to their tiny size and high reactivity, can harm a variety of molecules, including DNA, RNA, proteins, and lipids. The most efficient oxidant is $OH^{\bullet}$. ROS can target double bonds in unsaturated fatty acids, side chains of amino acids in proteins, and bases in nucleic acids.

2.3.7.1 Free Radical Scavenging

The hunt for safe and natural antioxidants has significantly increased in recent years, particularly those produced from plants. Spectrophotometric techniques such as O_2 • and ·OH, DPPH, ABTS {·, and N, N-dimethyl-p-phenylenediamine dihydrochloride cation radical (DMPD {·) are commonly used to assess antioxidant capability in meals, beverages, and vegetable extracts. The most popular techniques for assessing a compound's antioxidant activity are DPPH and ABTS scavenging because of their straightforward, quick, accurate, and repeatable processes. Therefore, before conducting additional study in cellular lines and/or animal models for antioxidant studies, radical scavenging assays in cell-free systems are often considered by researchers. Free radicals are extremely reactive molecules or atoms that are unstable due to their unpaired electrons (Zhao et al. 2008). If their levels are not regulated, they can damage cells and play a part in certain illnesses. The main mechanism by which antioxidants combat free radicals is scavenging. Free radicals are neutralized by antioxidants by giving them electrons or hydrogen atoms. This stabilizes the radicals, preventing them from damaging biological constituents such proteins, lipids, and DNA by oxidation.

2.3.7.2 Metal Ion Chelation

It is well known that transition metals like iron and copper exacerbate oxidative stress. Metals (Fe^{2+} and Cu^{+}) mix with hydrogen peroxide, a result of O_2-• dismutation by SOD, to form the highly reactive ·OH (Eq. 2.1) (Dairam et al. 2008). When Cu reacts with H_2O_2, it produces more singlet oxygen than ·OH, in contrast to iron. Oxidation converts Fe^{2+} and Cu^{+} to Fe^{3+} and Cu^{2+} correspondingly. They are reduced in the presence of oxidized metal ions Fe3+} and Cu2+~, cellular reductants like NADH, and reducing agents like ascorbic acid (AscH}. This enables the recycling to combine with another H_2O_2 molecule to produce the ·OH radical (Eq. 2.2) (Delanty and Dichter 2000). Because of their extreme reactivity, hydroxyl radicals can react directly with proteins and other macromolecules, causing lipid peroxidation, cross-linking, and the formation of carbonyls (aldehydes and ketones). By reducing their activity, chelating metal ions can reduce the generation of ROS.

$$Fe^{2+}(\text{or } Cu^{+}) + H_2O_2 \rightarrow Fe^{3+}\ (\text{or } Cu^{2+}) + \cdot OH + OH^{-} \qquad (2.1)$$

$$Fe^{3+}(\text{or } Cu^{2+}) + AscH \rightarrow Fe^{2+} + (\text{or } Cu^{+}) + Asc^{-} + H^{+} \qquad (2.2)$$

Curcumin, capsaicin, and S-allylcysteine were investigated for their antioxidant and iron-binding qualities by Dairam et al. (2008). These substances have strong O_2• and DPPH scavenging capabilities. The absorbance of the ferrous iron–ferrozine combination at 562 nm shows that curcumin has chelated ferrous ions. In particular, curcumin is added to a $FeCl_2$ solution. The addition of 5 mM ferrozine starts the process. After shaking the mixture well, set it aside at room temperature for

10 minutes. A spectrophotometer is then used to measure the solution's absorbance at 562 nm. One can calculate the proportion of suppression of ferrozine-Fe^{2+} complex production (Ak and Gülçin 2008).

Lipid peroxidation is started for the tissue's ion chelating ability by soaking the tissue homogenate in $FeSO_4$ or $FeCl_2$ and H_2O_2 for one hour at 37 °C with or without antioxidants. When TBA is added, protein-bound malondialdehyde (MDA), a byproduct of lipid peroxidation, is removed from the solution as the MDA-TBA adduct. The amount of this adduct is measured by its absorbance at 532 nm (Zheng et al. 2006). Antioxidants such as curcumin, capsaicin, and S-allylcysteine can have their Fe^{2+} and Fe^{3+} binding activities determined by electrochemical examination. Curcumin and capsaicin molecules may lessen the amount of lipid peroxidation caused by Fe^{2+} because they can bind to Fe^{2+} and Fe^{3+} and stop the redox cycling of iron. Compared to Fe^{2+} (Battin et al. 2006), Cu^+ produces harmful ·OH radicals 60 times more quickly. Studies have shown the ability of resveratrol, an antioxidant rich in selenium, to efficiently chelate copper (made in situ with copper/ascorbic acid) and shield DNA from ·OH radical damage (formed by copper/H_2O_2) It is because the protection of oxidative DNA damage requires selenium-metal coordination. Cu (bipy)22+ 2{ (bipy {2,2}-bipyridine) is used as the copper source to validate this method. When H_2O_2 is present, ascorbic acid converts Cu $(bipy)_2$ 2{to Cu $(bipy)_2^{2+}$}, which destroys DNA (Kang et al. 2007). The damage is not inhibited by the addition of the selenium antioxidant, suggesting that the antioxidant does not coordinate with Cu $(bipy)_2^{2+}$. The fact that the damage was not stopped by the selenium antioxidant suggests that the antioxidant is unable to attach to Cu (bipy) 22+ because the bipy ligands fully coordinate the Cu^{2+} ion. It has been discovered that the complex's wide absorption bands in the UV-vis region directly demonstrate antioxidant-metal coordination (Kang et al. 2006). However, it was previously believed that ginsenosides' aglycone's Fe^{2+} chelating activity rather than a direct contact with •OH, was responsible for their ·OH radical scavenging activity, as measured by EPR spectroscopy (Fogliano et al. 1999).

2.3.7.3 Enzyme Inhibition

Enzyme inhibition is one of the crucial mechanisms of antioxidants. NADPH oxidases are a class of enzymes linked to the plasma membrane that produce $O_2^{\bullet}$ by transferring an electron from the cytosolic source NADPH to an extracellular O_2. An enzyme called XO helps the body produce uric acid by facilitating the conversion of xanthine and hypoxanthine to uric acid, which yields H_2O_2 and $O_2^{\bullet}$ and increases an organism's oxidative level. These enzymes have been shown in a variety of physiological and pathological settings and are the primary producers of free radicals. Along with a number of other enzymes such as monooxygenases, NADH oxidase, and cyclooxygenases, $O_2^{\bullet}$ is also a by-product of mitochondrial respiration. The immune system uses this physiologically very poisonous substance to eliminate invasive bacteria. The enzyme NADPH oxidase produces a considerable amount of O_2• in phagocytes, which is used to destroy infections through oxygen-dependent processes. For an organism to defend against infections while preserving tissue function, controlled production of reactive oxygen derivatives during respiratory burst is essential. Excessive ROS can cause oxidative stress, including LDL oxidation. Metabolic syndrome patients have been linked to higher phagocytic NADPH oxidase activity and oxidized LDL levels. To reduce negative consequences of oxidative stress in hemodialysis patients, techniques such as modulating NADPH oxidase and supplementing with antioxidants have been recommended.

2.3.8 Role of Antioxidants in Animal Health

Due to the complex interaction that may be played by antioxidants, veterinary science has a very high consideration on this topic. It refers to redox signaling, a communication process mediated by ROS or reactive nitrogen species (RNS) but is also known to possibly produce oxidative stress when unregulated. Numerous animal illnesses, including metabolic, reproductive, immunological, and age-related disorders, are influenced by oxidative stress (Sies et al. 2017). Understanding and leveraging the role of antioxidants in mitigating oxidative damage has led to advancements in animal health management.

Antioxidants scavenge ROS and RNS, protecting cellular constituents (lipids, proteins, and DNA) from oxidative damage. Supplementation of antioxidants in the diet has also been promising, showing potential for improving livestock productivity, disease incidences, and longevity in companion animals (Corino and Rossi 2021). Antioxidants derived from nature, such as from plants and herbs and antioxidant feed additives, are an environmentally friendly and effective strategy to enhance animal health. Common antioxidants like quercetin, lycopene, coenzyme Q10, and vitamins C and E have demonstrated pharmacological potential in preventing and treating oxidative stress-related disorders in animals (Corino and Rossi 2021).

2.3.8.1 Antioxidants Against Cardiovascular Disease in Animals

In animals, oxidative stress contributes to cardiovascular diseases (CVD), with dilated cardiomyopathy (DCM), congestive heart failure, and heartworm disease being examples (Młynarska et al. 2024). There is role of ROS and RNS in myocardial cellular injury, then cause cellular homeostasis disturbance, and ultimately cardiac dysfunction. Reactive species are responsible for oxidative alterations in proteins and lipids, affecting the integrity of cellular membranes and altering the enzymatic functions that are important for the activity of the myocardium (Corino and Rossi 2021).

Oxidative stress in animals exacerbates endothelial dysfunction, a primary cause of cardiovascular disease, by decreasing nitric oxide availability, increasing vascular inflammation, and accelerating lipid peroxidation. For example, it has been discovered that dogs with DCM have increased oxidative stress markers and weaker antioxidant defenses (Freeman et al. 2005). Antioxidant supplements such as vitamin E and coenzyme Q10 have been demonstrated to improve heart function by lowering oxidative damage and boosting myocardial energy production (Meagher 2001). However, when high doses of certain antioxidants increase oxidative stress rather than lessen it, this is known as pro-oxidant effects (Halliwell 2007). This highlights the need for species-specific research and tailored antioxidant therapies in veterinary medicine.

2.3.8.2 Antioxidants Against Diabetes in Animals

In companion animals, diabetes mellitus is common, especially in dogs and cats. Its pathogenesis is significantly influenced by oxidative stress, which also contributes to insulin resistance, beta-cell dysfunction, and consequences, such as retinopathy, nephropathy, and neuropathy (Chansaisakorn et al. 2009). Through processes such as mitochondrial oxidative phosphorylation, glucose autooxidation, and polyol pathway activation, hyperglycemia in diabetic animals increases the generation of ROS. These actions disturb the delicate balance between the animal's antioxidant defense mechanisms and ROS generation (Corino and Rossi 2021). By decreasing oxidative damage and enhancing cellular insulin signaling, antioxidants can aid in the management of diabetes and associated consequences in animals. For example, it has been demonstrated that giving diabetic dogs vitamin E supplements improves their glucose metabolism and reduces oxidative stress indicators (Chansaisakorn et al. 2009). In animal models of diabetes, polyphenols derived from plants, such as flavonoids and tannins, have also shown promise in reducing oxidative stress (Corino and Rossi 2021). In veterinary settings, synthetic antioxidants such as melatonin and alpha-lipoic acid have also been investigated for their potential to prevent diabetes complications. Moreover, for better therapeutic outcomes, antioxidant therapy can be combined with more conventional diabetes therapies like insulin or oral hypoglycemic medications. Antioxidants may enhance insulin production and slow the course of illness by lowering oxidative damage to pancreatic beta cells (Chansaisakorn et al. 2009).

2.3.8.3 Antioxidants Against Cancer in Animals

Oxidative stress has a major role in the pathophysiology of cancer, which is one of the main causes of death for companion animals. ROS and RNS create an environment that is conducive to neoplastic transformation and tumor development by causing DNA damage, activating oncogenes, and suppressing tumor suppressor genes (Halliwell and Gutteridge 2015). For example, decreased antioxidant levels and increased oxidative stress indicators are linked to mammary cancers in dogs and cats (Freeman et al. 2005). Antioxidants have a variety of uses in the prevention and treatment of cancer in animals. By lowering inflammation, improving DNA repair, and modifying signal transduction pathways linked to cell division and death, they can stop the growth of tumors (Corino and Rossi 2021). Certain antioxidants, such as vitamin E, selenium, and ß-carotene, have shown promising results in lowering oxidative DNA damage and inhibiting the formation of tumors (Halliwell 2007). In animal models, green tea polyphenols like epigallocatechin gallate (EGCG) have shown antitumor properties by increasing apoptosis of cancer cells and preventing angiogenesis (Meagher 2001).

In addition to their direct anticancer effects, antioxidants can help lessen the negative effects of conventional cancer treatments like radiation and chemotherapy. Excessive ROS production from these therapies damages the healthy tissues around the tumor. The general quality of life for animals receiving cancer treatment can be e

nhanced by antioxidants such as vitamin C and E, which can shield healthy cells from oxidative damage brought on by therapy (Salami et al. 2016).

2.3.9 Antioxidants in Livestock Health and Productivity

Oxidative stress has a major effect on livestock's immunity, reproductivity, and productivity. Oxidative stress is made worse by environmental conditions such as heat, stress, malnutrition, and illnesses. As a result, dairy cows produce less milk, poultry produce fewer eggs, and breeding animals do less well reproductively (Corino and Rossi 2021). Because antioxidants increase cellular resilience and lessen the impacts of environmental stresses, hence they are essential for decreasing these consequences. For example, it has been demonstrated that giving dairy cows vitamin E and selenium supplements enhances their immune system and reproductive capabilities (Corino and Rossi 2021). Poultry hatchability rates can be increased, and egg yolk quality improved by antioxidants such as lycopene and carotenoids (Salami et al. 2016). Antioxidants in the feed improve growth performance and lower morbidity in pigs by lowering oxidative stress during weaning stage (Corino and Rossi 2021). As sustainable substitutes for artificial feed additives, natural antioxidants derived from medicinal plants like turmeric (curcumin), oregano (carvacrol), and rosemary (rosmarinic acid) are being investigated more and more. By lowering oxidative damage, these natural substances not only increase production but also raise the quality of animal products, including milk, meat, and eggs (Halliwell and Gutteridge 2015).

References

Ahsan H, Ahad A, Iqbal J, Siddiqui WA (2014) Pharmacological potential of tocotrienols: a review. Nutr Metab (Lond). https://doi.org/10.1186/1743-7075-11-52

Ak T, Gülçin İ (2008) Antioxidant and radical scavenging properties of curcumin. Chem Biol Interact 174:27–37. https://doi.org/10.1016/j.cbi.2008.05.003

Amir Aslani B, Ghobadi S (2016) Studies on oxidants and antioxidants with a brief glance at their relevance to the immune system. Life Sci 146:163–173. https://doi.org/10.1016/j.lfs.2016.01.014

Arnér ESJ, Holmgren A (2000) Physiological functions of thioredoxin and thioredoxin reductase. Eur J Biochem 267:6102–6109. https://doi.org/10.1046/j.1432-1327.2000.01701.x
Augusto O, Miyamoto S (2011) Oxygen radicals and related species. In: Principles of free radical biomedicine, vol 1, pp 19–42
Babcock GT (1999) How oxygen is activated and reduced in respiration. Proc Natl Acad Sci USA 96:12971–12973. https://doi.org/10.1073/pnas.96.23.12971
Battin EE, Perron NR, Brumaghim JL (2006) The central role of metal coordination in selenium antioxidant activity. Inorg Chem 45:499–501. https://doi.org/10.1021/ic051594f
Beckman JS, Koppenol WH (1996) Nitric oxide, superoxide, and peroxynitrite: the good, the bad, and ugly. Am J Phys 271:C1424–C1437. https://doi.org/10.1152/ajpcell.1996.271.5.C1424
Bielski BHJ, Cabelli DE, Arudi RL, Ross AB (1985) Reactivity of HO2/O− 2 radicals in aqueous solution. J Phys Chem Ref Data 14:1041–1100
Birben E, Sahiner UM, Sackesen C, Erzurum S, Kalayci O (2012) Oxidative stress and antioxidant defense. World Allergy Organ J 5:9–19. https://doi.org/10.1097/WOX.0b013e3182439613
Block G, Patterson B, Subar A (1992) Fruit, vegetables, and cancer prevention: a review of the epidemiological evidence. Nutr Cancer 18:1–29. https://doi.org/10.1080/01635589209514201
Bogacz-Radomska L, Harasym J (2018) β-Carotene—properties and production methods. Food Qual Saf 2:69–74. https://doi.org/10.1093/fqsafe/fyy004
Bowler C, Montagu MV, Inze D (1992) Superoxide dismutase and stress tolerance. Annu Rev Plant Physiol Plant Mol Biol 43:83–116. https://doi.org/10.1146/annurev.pp.43.060192.000503
Burton-Freeman B (2010) Postprandial metabolic events and fruit-derived phenolics: a review of the science. Br J Nutr 104:S1–S14. https://doi.org/10.1017/S0007114510003909
Cannavò SP, Tonacci A, Bertino L, Casciaro M, Borgia F, Gangemi S (2019) The role of oxidative stress in the biology of melanoma: a systematic review. Pathol Res Pract 215:21–28. https://doi.org/10.1016/j.prp.2018.11.020
Carlsen MH, Halvorsen BL, Holte K, Bøhn SK, Dragland S, Sampson L, Willey C, Senoo H, Umezono Y, Sanada C, Barikmo I, Berhe N, Willett WC, Phillips KM, Jacobs DR, Blomhoff R (2010) The total antioxidant content of more than 3100 foods, beverages, spices, herbs and supplements used worldwide. Nutr J 9:3. https://doi.org/10.1186/1475-2891-9-3
Chansaisakorn W, Sriphavatsarakorn P, Sopakdittapong P, Trisiriroj M, Pondeenana S, Buranakarl C (2009) Oxidative stress and intraerythrocytic concentrations of sodium and potassium in diabetic dogs. Vet Res Commun 33:67–75. https://doi.org/10.1007/s11259-008-9073-7
Choe E, Min DB (2006) Mechanisms and factors for edible oil oxidation. Compr Rev Food Sci Food Saf 5:169–186
Corino C, Rossi R (2021) Antioxidants in Animal Nutrition. Antioxidants 10:1877. https://doi.org/10.3390/antiox10121877
Dairam A, Fogel R, Daya S, Limson JL (2008) Antioxidant and iron-binding properties of curcumin, capsaicin, and *S* -Allylcysteine reduce oxidative stress in rat brain homogenate. J Agric Food Chem 56:3350–3356. https://doi.org/10.1021/jf0734931
Davies MJ (2005) The oxidative environment and protein damage. Biochim Biophys Acta 1703: 93–109. https://doi.org/10.1016/j.bbapap.2004.08.007
De Grey ADNJ (2002) HO2•: the forgotten radical. DNA Cell Biol 21:251–257
Delanty N, Dichter MA (2000) Antioxidant therapy in neurologic disease. Arch Neurol 57. https://doi.org/10.1001/archneur.57.9.1265
Devasagayam TPA, Tilak JC, Boloor KK, Sane KS, Ghaskadbi SS, Lele RD (2004) Free radicals and antioxidants in human health: current status and future prospects. Japi 52:4
Dhalla NS, Temsah RM, Netticadan T (2000) Role of oxidative stress in cardiovascular diseases. J Hypertens 18:655–673. https://doi.org/10.1097/00004872-200018060-00002
Douki T, Cadet J (1996) Peroxynitrite mediated oxidation of purine bases of nucleosides and isolated DNA. Free Radic Res 24:369–380. https://doi.org/10.3109/10715769609088035
Duke MV, Salin ML (1985) Purification and characterization of an iron-containing superoxide dismutase from a eucaryote, Ginkgo biloba. Arch Biochem Biophys 243:305–314. https://doi.org/10.1016/0003-9861(85)90800-8

Eastwood MA (1999) Interaction of dietary antioxidants in vivo: how fruit and vegetables prevent disease? QJM 92:527–530. https://doi.org/10.1093/qjmed/92.9.527

Engelman R, Ziv T, Arnér ESJ, Benhar M (2016) Inhibitory nitrosylation of mammalian thioredoxin reductase 1: molecular characterization and evidence for its functional role in cellular nitroso-redox imbalance. Free Radic Biol Med 97:375–385. https://doi.org/10.1016/j.freeradbiomed.2016.06.032

Fang Y-Z, Yang S, Wu G (2002) Free radicals, antioxidants, and nutrition. Nutrition 18:872–879. https://doi.org/10.1016/S0899-9007(02)00916-4

Fogliano V, Verde V, Randazzo G, Ritieni A (1999) Method for measuring antioxidant activity and its application to monitoring the antioxidant capacity of wines. J Agric Food Chem 47:1035–1040. https://doi.org/10.1021/jf980496s

Forbes JM, Coughlan MT, Cooper ME (2008) Oxidative stress as a major culprit in kidney disease in diabetes. Diabetes 57:1446–1454. https://doi.org/10.2337/db08-0057

Franklin JL (2011) Redox regulation of the intrinsic pathway in neuronal apoptosis. Antioxid Redox Signal 14:1437–1448

Freeman LM, Rush JE, Milbury PE, Blumberg JB (2005) Antioxidant status and biomarkers of oxidative stress in dogs with congestive heart failure. J Vet Intern Med 19:537–541. https://doi.org/10.1111/j.1939-1676.2005.tb02724.x

Gathwala G, Aggarwal R (2016) Selenium supplementation for the preterm Indian neonate. Indian J Public Health 60:142. https://doi.org/10.4103/0019-557X.184571

German, J.B., 1999. Food processing and lipid oxidation. Pp. 23–50. https://doi.org/10.1007/978-1-4615-4853-9_3

Ghafourifar P, Cadenas E (2005) Mitochondrial nitric oxide synthase. Trends Pharmacol Sci 26: 190–195. https://doi.org/10.1016/j.tips.2005.02.005

Gupta VK, Sharma SK (2006) Plants as natural antioxidants

Gutowski M, Kowalczyk S (2013) A study of free radical chemistry: their role and pathophysiological significance. Acta Biochim Pol 60:1–16

Halliwell B (1995) How to characterize an antioxidant: an update. Biochem Soc Symp 61:73–101. https://doi.org/10.1042/bss0610073

Halliwell B (2007) Oxidative stress and cancer: have we moved forward? Biochem J 401:1–11. https://doi.org/10.1042/BJ20061131

Halliwell B, Gutteridge JMC (1995) The definition and measurement of antioxidants in biological systems. Free Radic Biol Med 18:125–126. https://doi.org/10.1016/0891-5849(95)91457-3

Halliwell B, Gutteridge JMC (2015) Free radicals in biology and medicine. Oxford University Press. https://doi.org/10.1093/acprof:oso/9780198717478.001.0001

Harats D, Chevion S, Nahir M, Norman Y, Sagee O, Berry E (1998) Citrus fruit supplementation reduces lipoprotein oxidation in young men ingesting a diet high in saturated fat: presumptive evidence for an interaction between vitamins C and E in vivo. Am J Clin Nutr 67:240–245. https://doi.org/10.1093/ajcn/67.2.240

Holmgren A (1985) THIOREDOXIN. Annu Rev Biochem 54:237–271. https://doi.org/10.1146/annurev.bi.54.070185.001321

Ighodaro OM, Akinloye OA (2018) First line defence antioxidants-superoxide dismutase (SOD), catalase (CAT) and glutathione peroxidase (GPX): their fundamental role in the entire antioxidant defence grid. Alexandria Journal of Medicine 54:287–293. https://doi.org/10.1016/j.ajme.2017.09.001

Inoue M, Sato EF, Nishikawa M, Park A-M, Kira Y, Imada I, Utsumi K (2003) Mitochondrial generation of reactive oxygen species and its role in aerobic life. Curr Med Chem 10:2495–2505. https://doi.org/10.2174/0929867033456477

Jacob RA (1995) The integrated antioxidant system. Nutr Res 15:755–766. https://doi.org/10.1016/0271-5317(95)00041-G

Jiang Q (2014) Natural forms of vitamin E: metabolism, antioxidant, and anti-inflammatory activities and their role in disease prevention and therapy. Free Radic Biol Med 72:76–90. https://doi.org/10.1016/j.freeradbiomed.2014.03.035

Jideani AIO, Silungwe H, Takalani T, Omolola AO, Udeh HO, Anyasi TA (2021) Antioxidant-rich natural fruit and vegetable products and human health. Int J Food Prop 24:41–67. https://doi.org/10.1080/10942912.2020.1866597

Jiménez A, Hernández JA, Pastori G, del Río LA, Sevilla F (1998) Role of the ascorbate-glutathione cycle of mitochondria and peroxisomes in the senescence of pea leaves. Plant Physiol 118: 1327–1335. https://doi.org/10.1104/pp.118.4.1327

Kaimul AM, Nakamura H, Masutani H, Yodoi J (2007) Thioredoxin and thioredoxin-binding protein-2 in cancer and metabolic syndrome. Free Radic Biol Med 43:861–868. https://doi.org/10.1016/j.freeradbiomed.2007.05.032

Kang KS, Kim HY, Pyo JS, Yokozawa T (2006) Increase in the free radical scavenging activity of ginseng by heat-processing. Biol Pharm Bull 29:750–754. https://doi.org/10.1248/bpb.29.750

Kang KS, Kim HY, Baek SH, Yoo HH, Park JH, Yokozawa T (2007) Study on the hydroxyl radical scavenging activity changes of ginseng and Ginsenoside-Rb2 by heat processing. Biol Pharm Bull 30:724–728. https://doi.org/10.1248/bpb.30.724

Kirkman HN, Gaetani GF (1984) Catalase: a tetrameric enzyme with four tightly bound molecules of NADPH. Proc Natl Acad Sci 81:4343–4347. https://doi.org/10.1073/pnas.81.14.4343

Knight, J.A., 1999. Biochemistry of free radicals and oxidative stress. Free radicals, antioxidants, ageing and disease; Knight, JA, Ed 21–43

Korycka-Dahl MB, Richardson T (1978) Activated oxygen species and oxidation of food constituents. CRC Crit Rev Food Sci Nutr 10:209–241. https://doi.org/10.1080/10408397809527250

Kowluru RA, Chan P-S (2007) Oxidative stress and diabetic retinopathy. Exp Diabetes Res 2007: 43603. https://doi.org/10.1155/2007/43603

Kurutas EB (2015) The importance of antioxidants which play the role in cellular response against oxidative/nitrosative stress: current state. Nutr J 15:71. https://doi.org/10.1186/s12937-016-0186-5

Ladas EJ, Jacobson JS, Kennedy DD, Teel K, Fleischauer A, Kelly KM (2004) Antioxidants and cancer therapy: a systematic review. J Clin Oncol 22:517–528. https://doi.org/10.1200/JCO.2004.03.086

León-Carmona JR, Galano A (2011) Is caffeine a good scavenger of oxygenated free radicals? J Phys Chem B 115:4538–4546. https://doi.org/10.1021/jp201383y

Lobo V, Patil A, Phatak A, Chandra N (2010) Free radicals, antioxidants and functional foods: impact on human health. Pharmacogn Rev 4:118. https://doi.org/10.4103/0973-7847.70902

Lowell BB, Shulman GI (2005) Mitochondrial dysfunction and type 2 diabetes. Science 307:384–387. https://doi.org/10.1126/science.1104343

Lubos E, Loscalzo J, Handy DE (2011) Glutathione Peroxidase-1 in health and disease: from molecular mechanisms to therapeutic opportunities. Antioxid Redox Signal 15:1957–1997. https://doi.org/10.1089/ars.2010.3586

Madamanchi NR, Runge MS (2007) Mitochondrial dysfunction in atherosclerosis. Circ Res 100: 460–473. https://doi.org/10.1161/01.RES.0000258450.44413.96

Margis R, Dunand C, Teixeira FK, Margis-Pinheiro M (2008) Glutathione peroxidase family—an evolutionary overview. FEBS J 275:3959–3970. https://doi.org/10.1111/j.1742-4658.2008.06542.x

Martínez-Cayuela M (1995) Oxygen free radicals and human disease. Biochimie 77:147–161. https://doi.org/10.1016/0300-9084(96)88119-3

Mathur P, Ding Z, Saldeen T, Mehta JL (2015) Tocopherols in the prevention and treatment of atherosclerosis and related cardiovascular disease. Clin Cardiol 38:570–576. https://doi.org/10.1002/clc.22422

Meagher E (2001) Antioxidant therapy and atherosclerosis animal and human studies. Trends Cardiovasc Med 11:162–165. https://doi.org/10.1016/S1050-1738(01)00105-0

Min B, Ahn DU (2005) Mechanism of lipid peroxidation in meat and meat products-a review. Food Sci Biotechnol 14:152–163

Minich DM, Brown BI (2019) A review of dietary (Phyto)nutrients for glutathione support. Nutrients 11:2073. https://doi.org/10.3390/nu11092073

Misak A, Grman M, Bacova Z, Rezuchova I, Hudecova S, Ondriasova E, Krizanova O, Brezova V, Chovanec M, Ondrias K (2018) Polysulfides and products of H 2 S/S-nitrosoglutathione in comparison to H 2 S, glutathione and antioxidant Trolox are potent scavengers of superoxide anion radical and produce hydroxyl radical by decomposition of H 2 O 2. Nitric Oxide 76:136–151. https://doi.org/10.1016/j.niox.2017.09.006

Miyazaki K, Noda N, Okada S, Hagiwara Y, Miyata M, Sakurabayashi I, Yamaguchi N, Sugimura T, Terada M, Wakasugi H (1998) Elevated serum level of thioredoxin in patients with hepatocellular carcinoma. Biotherapy 11:277–288. https://doi.org/10.1023/A:1008032703468

Młynarska E, Hajdys J, Czarnik W, Fularski P, Leszto K, Majchrowicz G, Lisińska W, Rysz J, Franczyk B (2024) The role of antioxidants in the therapy of cardiovascular diseases—a literature review. Nutrients 16:2587. https://doi.org/10.3390/nu16162587

Monaghan BR, Schmitt FO (1932) THE EFFECTS OF CAROTENE AND OF VITAMIN A ON THE OXIDATION OF LINOLEIC ACID. J Biol Chem 96:387–395. https://doi.org/10.1016/S0021-9258(18)76278-9

Nishiyama A, Masutani H, Nakamura H, Nishinaka Y, Yodoi J (2001) Redox regulation by Thioredoxin and Thioredoxin-binding proteins. IUBMB Life 52:29–33. https://doi.org/10.1080/15216540252774739

Noctor G, Foyer CH (1998) ASCORBATE AND GLUTATHIONE: keeping active oxygen under control. Annu Rev Plant Physiol Plant Mol Biol 49:249–279. https://doi.org/10.1146/annurev.arplant.49.1.249

Noguchi N, Niki E (2019) Chemistry of active oxygen species and antioxidants. In: Antioxidant status, diet, nutrition, and health. CRC Press, pp 1–20

Obrosova IG (2005) Increased sorbitol pathway activity generates oxidative stress in tissue sites for diabetic complications. Antioxid Redox Signal 7:1543–1552. https://doi.org/10.1089/ars.2005.7.1543

Papas AM (1999) Diet and antioxidant status. Food Chem Toxicol 37:999–1007. https://doi.org/10.1016/s0278-6915(99)00088-5

Peskin AV, Koen YM, Zbarsky IB (1977) Superoxide dismutase and glutathione peroxidase activities in tumors. FEBS Lett 78:41–45. https://doi.org/10.1016/0014-5793(77)80268-8

Pham-Huy LA, He H, Pham-Huy C (2008) Free radicals, antioxidants in disease and health. Int J Biomed Sci 4:89

Pisoschi AM, Negulescu GP (2012) Methods for Total antioxidant activity determination: a review. Biochemistry & Analytical Biochemistry 01. https://doi.org/10.4172/2161-1009.1000106

Prütz WA (1996) Hypochlorous acid interactions with thiols, nucleotides, DNA, and other biological substrates. Arch Biochem Biophys 332:110–120. https://doi.org/10.1006/abbi.1996.0322

Radi R (2004) Nitric oxide, oxidants, and protein tyrosine nitration. Proc Natl Acad Sci USA 101: 4003–4008. https://doi.org/10.1073/pnas.0307446101

Repetto M, Semprine J, Boveris A (2012) Lipid peroxidation: chemical mechanism, biological implications and analytical determination. Lipid Peroxid 1:3–30

Reuter S, Gupta SC, Chaturvedi MM, Aggarwal BB (2010) Oxidative stress, inflammation, and cancer: how are they linked? Free Radic Biol Med 49:1603–1616. https://doi.org/10.1016/j.freeradbiomed.2010.09.006

Robertson RP, Harmon J, Tran POT, Poitout V (2004) Beta-cell glucose toxicity, lipotoxicity, and chronic oxidative stress in type 2 diabetes. Diabetes 53(Suppl 1):S119–S124. https://doi.org/10.2337/diabetes.53.2007.s119

Salami SA, Guinguina A, Agboola JO, Omede AA, Agbonlahor EM, Tayyab U (2016) Review: in vivo and postmortem effects of feed antioxidants in livestock: a review of the implications on authorization of antioxidant feed additives. Animal 10:1375–1390. https://doi.org/10.1017/S1751731115002967

Sen S, Chakraborty R (2011) The role of antioxidants in human health

Serna-Saldivar SO (2016) Cereal grains. CRC Press. https://doi.org/10.1201/9781439882092

Sharma P, Jha AB, Dubey RS, Pessarakli M (2012) Reactive oxygen species, oxidative damage, and antioxidative defense mechanism in plants under stressful conditions. J Bot 2012:1–26. https://doi.org/10.1155/2012/217037

Sies H (1997) Oxidative stress: oxidants and antioxidants. Exp Physiol 82:291–295. https://doi.org/10.1113/expphysiol.1997.sp004024

Sies H, Berndt C, Jones DP (2017) Oxidative stress. Annu Rev Biochem 86:715–748. https://doi.org/10.1146/annurev-biochem-061516-045037

Singh RJ, Hogg N, Joseph J, Kalyanaraman B (1996) Mechanism of nitric oxide release from S-nitrosothiols. J Biol Chem 271:18596–18603. https://doi.org/10.1074/jbc.271.31.18596

Singh VP, Bali A, Singh N, Jaggi AS (2014) Advanced glycation end products and diabetic complications. Korean J Physiol Pharmacol 18:1–14. https://doi.org/10.4196/kjpp.2014.18.1.1

Steinmetz KA, Potter JD (1996) Vegetables, fruit, and cancer prevention. J Am Diet Assoc 96:1027–1039. https://doi.org/10.1016/S0002-8223(96)00273-8

Stief TW (2003) The physiology and pharmacology of singlet oxygen. Med Hypotheses 60:567–572. https://doi.org/10.1016/s0306-9877(03)00026-4

Traber MG, Stevens JF (2011) Vitamins C and E: beneficial effects from a mechanistic perspective. Free Radic Biol Med 51:1000–1013. https://doi.org/10.1016/j.freeradbiomed.2011.05.017

Tsujita H, Plummer CE (2010) Bovine ocular squamous cell carcinoma. Vet Clin North Am Food Anim Pract 26:511–529. https://doi.org/10.1016/j.cvfa.2010.08.003

Uttara B, Singh A, Zamboni P, Mahajan R (2009) Oxidative stress and neurodegenerative diseases: a review of upstream and downstream antioxidant therapeutic options. Curr Neuropharmacol 7:65–74. https://doi.org/10.2174/157015909787602823

Valko M, Leibfritz D, Moncol J, Cronin MTD, Mazur M, Telser J (2007) Free radicals and antioxidants in normal physiological functions and human disease. Int J Biochem Cell Biol 39:44–84. https://doi.org/10.1016/j.biocel.2006.07.001

Vincent AM, Russell JW, Low P, Feldman EL (2004) Oxidative stress in the pathogenesis of diabetic neuropathy. Endocr Rev 25:612–628. https://doi.org/10.1210/er.2003-0019

Winterbourn CC (2013) The biological chemistry of hydrogen peroxide. Methods Enzymol 528:3–25. https://doi.org/10.1016/B978-0-12-405881-1.00001-X

Winterbourn CC, Kettle AJ (2000) Biomarkers of myeloperoxidase-derived hypochlorous acid. Free Radic Biol Med 29:403–409. https://doi.org/10.1016/s0891-5849(00)00204-5

Wiseman H, Halliwell B (1996) Damage to DNA by reactive oxygen and nitrogen species: role in inflammatory disease and progression to cancer. Biochem J 313(Pt 1):17–29. https://doi.org/10.1042/bj3130017

Zhao F, Liu Z-Q, Wu D (2008) Antioxidative effect of melatonin on DNA and erythrocytes against free-radical-induced oxidation. Chem Phys Lipids 151:77–84. https://doi.org/10.1016/j.chemphyslip.2007.10.002

Zheng L, Wei Q, Cai Y, Fang J, Zhou B, Yang L, Liu Z (2006) DNA damage induced by resveratrol and its synthetic analogues in the presence of Cu (II) ions: mechanism and structure-activity relationship. Free Radic Biol Med 41:1807–1816. https://doi.org/10.1016/j.freeradbiomed.2006.09.007

Zitka O, Skalickova S, Gumulec J, Masarik M, Adam V, Hubalek J, Trnkova L, Kruseova J, Eckshulager T, Kizek R (2012) Redox status expressed as GSH:GSSG ratio as a marker for oxidative stress in paediatric tumour patients. Oncol Lett 4:1247–1253. https://doi.org/10.3892/ol.2012.931

Chapter 3
Understanding Different Signaling Involved During Oxidative Stress

Pallavi Prakash Shelke, Ramya Buradi, and Awanish Mishra

Abbreviations

ACSL4	Acyl-CoA synthetase long-chain family member 4
AP-1	Activator protein 1
APE1	Apurinic/apyrimidinic endonuclease 1
ATM	X-linked inhibitor of apoptosis protein
ATR	Ataxia Telangiectasia and Rad3-related
BER	Base excision repair
CEBPD	CCAAT/Enhancer-Binding Protein Delta
COX	Cyclooxygenase
CYP	Cytochromes p450
DR4	Death receptor 4 (TRAIL-R1)
eNOS	Endothelial nitric oxide synthase
ERK	Extracellular-regulated kinase
FLIP	Fas-associated death domain (FADD)-like interleukin-1-beta-converting enzyme-inhibitory protein
FOXO	Forkhead box O
GPx	Glutathione peroxidases
GSK-3β	Glycogen synthase kinase 3 beta
H_2O_2	Hydrogen peroxide
HIF-1	Hypoxia-inducible factor 1
HOCl	Hypochlorous acid

P. P. Shelke · R. Buradi · A. Mishra (✉)
Department of Pharmacology and Toxicology, National Institute of Pharmaceutical Education and Research (NIPER)—Guwahati, Changsari, Kamrup, Assam, India
e-mail: awanish@niperguwahati.in

N. Srivastava, S. A. Saraf (eds.), *Pharmacological Strategies for Prevention and Treatment of Oxidative Stress*, Oxidative Stress in Applied Basic Research and Clinical Practice, https://doi.org/10.1007/978-3-032-19837-2_3

IKKs	Inhibitor of nuclear factor-κB kinase
IL-1	Interleukin-1
IL-6	Interleukin - 6
iNOS	Inducible nitric oxide synthase
IκB	Inhibitor of nuclear factor-kappa B
JNK1/2/3	c-Jun N-terminal Kinase 1, 2, and 3
Keap1	Kelch-like ECH-associated protein 1
LOX	Lipoxygenases
LPS	Lipopolysaccharide
MAPK	Mitogen-activated protein kinase
MDM2	Mouse double minute 2
MEKK	Mitogen-activated protein kinase kinase kinase
mTOR	Mammalian target of rapamycin
mTORC1	Mammalian target of rapamycin complex 1
NADPH	Nicotinamide adenine dinucleotide phosphate hydrogen
NF-κB	Nuclear factor kappa-light-chain-enhancer of activated B cells
nNOS	Neuronal nitric oxide synthase
NOX	NADPH oxidase
NOXA	Phorbol-12-myristate-13-acetate-induced protein 1
NRF-2	Nuclear factor erythroid 2-related factor 2
OGG1	8-Oxoguanine DNA glycosylase 1
$ONOO^-$	Peroxynitrite
p38-MAPK	p38 Mitogen-activated protein kinase
p53	Tumor protein 53
PDGF	Platelet-derived growth factor
PI3K/AKT	Phosphatidylinositol 3-kinase/protein kinase B signaling pathway
PIP2	Phosphatidylinositol 4,5-bisphosphate
PIP3	Phosphatidylinositol (3,4,5)-trisphosphate
PKC	Protein kinase C
PTEN	Phosphatase and tensin homolog
PUMA	p53 upregulated modulator of apoptosis
RNS	Reactive nitrogen species
ROS	Reactive oxygen species
RyR	Ryanodine receptor
Sp-1	Specificity protein 1
TGF-β	Transforming growth factor–beta
TNF-α	Tumor necrosis factor alpha
TRAIL	Tumor necrosis factor-related apoptosis-inducing ligand
ULK1	Unc-51-like kinase 1
UPR	Unfolded protein response
XIAP	X-linked inhibitor of apoptosis protein
XO	Xanthine oxidase

3.1 Introduction

The phrase "oxidative stress" "has been formulated as a concept in redox biology and medicine," according to its intuitive definition from thirty years ago (Sies 2015). In essence, the open metabolic system creates a baseline redox tone by maintaining a steady-state redox balance at a specific fixed point. A stress reaction is brought on by any deviation from this equilibrium, which is viewed as a stressor. There are physiological deviations known as "oxidative eustress" and supraphysiological deviations known as "oxidative distress," while a movement to the opposite side of the equilibrium is called "reductive stress." The definition of oxidative stress implies these ideas (Sies et al. 2017). An integral component of redox regulation and physiological redox signaling is oxidative eustress (Sies 2017; Sies and Jones 2020). The hydroxyl radical ($OH^{\cdot}$), peroxides, singlet oxygen ($^1O_2$), and superoxide anion ($O_2^{\cdot-}$) are examples of reactive oxygen species (ROS) (Chen et al. 2012). These substances can damage DNA (DNA base damage, DNA single-strand and double-strand breaks, DNA and protein crosslinks, and DNA and chromosomal abnormalities), lipid membranes, collagen structures, mitochondrial function, and signals that promote cell survival and proliferation (Athar 2002; Pai et al. 2014). The perturbed redox homeostasis shifts the normal physiology towards pathological condition (illustrated in Fig. 3.1).

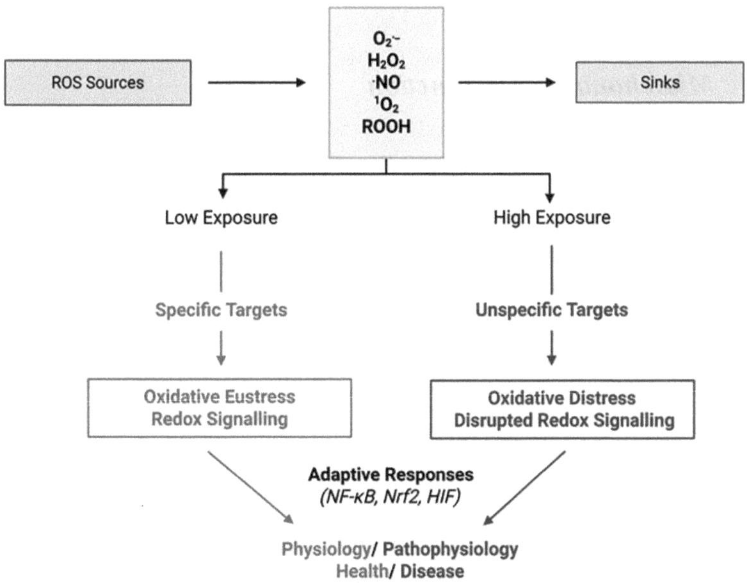

Fig. 3.1 Oxidative stress arises from an imbalance between oxidant production and removal. Endogenous and exogenous sources generate oxidants, crucial for redox signaling at low levels (oxidative eustress). Excessive oxidant exposure disrupts signaling and damages biomolecules (oxidative distress) (Sies et al. 2017). Adaptive responses influence the outcome, determining health or disease states. (Created in BioRender. Shelke, P. (2025) https://BioRender.com/w7fsrbm)

3.2 Reactive Oxygen Species (ROS) and Their Sources

Oxidative stress is a result of many molecules having oxidizing abilities (Sies et al. 2017). These ROS include hydroxyl radical (HO·), superoxide anions (O^{2-}), nitric oxide (NO), and lipid radicals, which present unpaired electrons being considered as free radicals (Radi 2018). ROS that having oxidizing properties but is not free radicals include hydrogen peroxide (H_2O_2), peroxynitrite ($ONOO^-$), and hypochlorous acid (HOCl) (Phaniendra et al. 2015). The cellular environment, exposure duration, concentration, and compartmentalization are some of the variables that affect ROS's two different functions. Three primary species of ROS are O_2, HO, and H_2O_2.

3.2.1 Intracellular Species of ROS

NADPH oxidases and mitochondria are the main producers of intracellular ROS (Holmström and Finkel 2014). Surprisingly, the two sources cooperate to promote the gradual onset of oxidative stress (Daiber et al. 2021). Additional sources of ROS include the endoplasmic reticulum, cytochrome p450, xanthine oxidase (XO), uncoupled NO synthase, peroxidases, and cyclooxygenases (Llanas-Cornejo and Husi 2017; Förstermann et al. 2017).

3.2.2 Mitochondrial Dysfunction

The oxidative phosphorylation processes that produce adenosine triphosphate (ATP), or energy, are carried out by the cytoplasmic organelles known as mitochondria. These structures are composed of two membranes: an outer membrane that separates the mitochondrion from the cytoplasm, and an inner membrane that produces the mitochondrial cristae. The latter displays the limits of the matrix that contains the mitochondrial DNA (mtDNA) (Taanman 1999). Four major multi-subunit enzyme complexes are found in the mitochondrial inner membrane, where oxidative phosphorylation occurs: NADH: ubiquinone oxidoreductase (complex I), succinate dehydrogenase (complex II), coenzyme Q cytochrome c reductase (complex III), and cytochrome C oxidase (complex IV) (Ashrafi and Schwarz 2013). Under stressful conditions mitochondria produce more ROS which is linked to the release of inflammatory cytokines (Missiroli et al. 2020). In instances where TNF-α is present, it can reduce complex I's regular activity, resulting in mitochondrial malfunction and eventual cell death (Prajapati et al. 2015). Furthermore it has been shown that mitochondrial malfunction heightens the cellular susceptibility to various pro-inflammatory cytokines and activate cellular inflammatory pathways, ultimately heightening cellular damage and ROS generation (van Horssen et al. 2019).

3.2.3 NADPH Oxidases

NADPH oxidases (NOX) are an important source of ROS, especially in the cardiovascular system (Montezano et al. 2014). NOX mediates the overproduction of ROS, which impacts cardiovascular pathophysiology (Zhang et al. 2020). Furthermore, either increased NOX activity or overexpression is associated with a number of age-related diseases (Touyz et al. 2019). NOX1, NOX2, NOX3, NOX4, NOX5, Duox1, and Duox2 are the seven isoforms of NOX that have been found. Their activation, location, and type of ROS generation differ (Fukai and Ushio-Fukai 2020). All of these NOX family isoforms are transmembrane proteins that reduce the oxygen content of superoxide anions (NOX1–3 and NOX5) or hydrogen peroxide (NOX4) by transporting electrons across cellular membranes.

The first stage in the production of ROS is NADPH binding to the dehydrogenase domain. After leaving the NADPH substrate, electrons are moved to the FAD cofactor and then to the two heme groups in the transmembrane domain. Oxygen is the last acceptor on the opposite side of the membrane, where it is reduced to produce superoxide or hydrogen peroxide (Magnani and Mattevi 2019). Numerous stimuli, including NADPH oxidase, Toll-like receptor agonists, leptin, cytokines (PDGF, IL-6, TGF-β, and TNF-α), and hormones including angiotensin 2, aldosterone, and endothelin-1, can cause the production of ROS (Krylatov et al. 2021).

3.2.4 Xanthine Oxidase

Xanthine oxidase (XO) catalyzes the oxidative hydroxylation of hypoxanthine to xanthine and xanthine to uric acid, the last two stages in the purine metabolism of humans (Schmidt et al. 2019). O_2 quickly absorbs electrons from hypoxanthine and xanthine oxidation by XO to generate O_2 (superoxide anion) and H_2O_2 (George and Struthers 2009). Aging is associated with oxidative damage, immunosenescence, and inflammation, all of which raise XO levels (Battelli et al. 2020). Moreover, XO seems to be a sign of endothelial dysfunction, a critical process that leads to several age-related ailments, such as cardiovascular diseases (CVDs) (Battelli et al. 2020). On the other hand, XO inhibitors, such as allopurinol, prevent oxidative stress and atrial remodeling (Yang et al. 2018).

3.2.5 Uncoupled NO Synthase

The NO synthase (NOS) enzyme has three different isoforms that mammals can use to create NO. According to Förstermann et al. (2017), the isozymes are called neuronal nNOS (also called NOS I), inducible form iNOS (also called NOS II), and endothelial eNOS (also called NOS III). Produced by uncoupled eNOS, superoxide scavenges NO to form $ONOO^-$, a strong and transient oxidant that stimulates oxidation and nitration processes that affect a variety of biomolecules, including proteins, DNA, and lipids (Bartesaghi and Radi 2018). Additionally, $ONOO^-$ oxidizes BH4 to BH3 and subsequently to BH2, which binds to eNOS with a high affinity without increasing its catalytic activity (Bendall et al. 2014). Furthermore, $ONOO^-$ reduces endothelium transport of L-arginine, the sole substrate for eNOS, and speeds up the rate of L-arginine outflow. Moreover, reduced glutathione (GSH), an endogenous scavenger crucial to the cell's defense against ROS, is directly oxidized by $ONOO^-$. Further, $ONOO^-$ also nitrates SOD, making it inactive, which lowers the quantity of antioxidant defense mechanisms within cells and raises superoxide levels.

3.2.6 Endoplasmic Reticulum

The endoplasmic reticulum (ER) is responsible for the formation of cellular ROS in eukaryotic cells, and ER oxirreductin-1 is the primary generator of these ROS (de Almeida et al. 2022). A healthy amount of ROS is necessary for the creation of disulfide bonds, an essential step in the machinery that folds proteins (Igbaria et al. 2019). Thus, for protein folding to be successful in the ER, the right organelle environment—including redox potential and metabolic requirements—is necessary. However, ER disruption can cause unfolded and misfolded proteins, which can trigger the unfolded protein response (UPR), which is crucial for fixing protein dysfunction.

Additionally, ROS production can be initiated by the release of Ca^{2+} from the ER (Görlach et al. 2015). The thiol groups that sense redox are in charge of the ryanodine receptor (RyR), a Ca^{2+} channel located in the ER. When RyR is triggered by the oxidation of its thiol sites, the ER can release calcium. The ER also expresses NOX4, one of the NADPH oxidase isoforms, which contributes to the oxidative stress cycle there (Lee et al. 2013).

3.2.7 Cytochrome p450

In phase I first-pass processes, the monooxygenase enzymes called cytochromes p450 (CYP) are metabolically active in the metabolism of endogenous substances, dietary additives, and medications (Guengerich 2019). By using an oxygen molecule to attach this enzyme to its substrate and create the oxi complex, CYP-catalyzed monooxygenation processes take place (He et al. 2017). ROS are created when the oxi complex undergoes electron transfer from NADPH to a peroxide complex, which is rapidly protonated by releasing a water molecule to create a highly reactive intermediate product. This intermediate product then absorbs neighboring hydrogen atoms from the substrate. The substrate exits the active site, and the enzyme returns to resting after these free radicals are produced (Manikandan and Nagini 2018). Even when, as in this redox system, the stereometric molar link between oxygen intake and hydroxylated products is not reached, the release of superoxide anions and hydrogen peroxide results in an excess of ROS production. Accordingly, the cytotoxicity caused by the generation of ROS is also greatly influenced by xenobiotic metabolism (Gómez-Tabales et al. 2020).

3.2.8 Peroxidases

Cellular peroxide radicals are eliminated by a class of enzymes called peroxidases. Glutathione peroxidases (GPx) and catalase are such peroxidase enzymes found in mammals (Ighodaro and Akinloye 2018). However, haloperoxidases, a different class of peroxidases, mediate interactions between hydrogen peroxide and halide ions (Menon et al. 2022). This class includes the three enzymes known to exist in animals: lactoperoxidase, myeloperoxidase (MPO), and eosinophil peroxidase. According to Moldogazieva et al. (2020), these enzymes can oxidize iodide, bromide, and chloride ions to create reactive halogen species like HOCl, which boost immunity and aid in the removal of a number of diseases (Ulfig and Leichert 2021). But if it continues, it increases oxidative stress and ultimately damages proteins, lipids, and DNA.

3.2.9 Cyclooxygenases

The cyclooxygenase (COX) enzyme catalyzes the metabolism of arachidonic acid by producing precursors of prostaglandins, which in turn promote inflammation, the generation of ROS, and lipid oxygenation (Tada and Suzuki 2016). There are two isoforms of COX: COX-1 and COX-2. COX-1 is a constitutively generated, ubiquitous isoform that is engaged in physiological activities in all tissues. COX-2 is expressed by the gastrointestinal tract, brain, lungs, kidney, and thymus, and it can be produced in a number of tissues as a result of inflammatory processes (Mitchell and Kirkby 2019). COX-2 is an enzyme whose expression is controlled by NF-κB, AP-1, and CREB, which can work together or independently to transcriptionally transcribe the enzyme. Additionally, the positive regulation of COX-2 also depends on increased ROS generation (Kang et al. 2006). Increased NADPH oxidase activity causes this event to occur, which in turn activates the Akt and MAPK pathways that support the production of COX-2 (Kim et al. 2015). In addition to being triggered by NF-κB, COX-2-mediated proinflammatory indicators in cellular senescence also play a role in regulating mitochondrial function and ROS generation via positive feedback loops involving p38MAPK, TGF-β, and mTOR, leading to telomeric dysfunction and the build-up of senescent cells(Jurk et al. 2014).

3.2.10 Lipid Oxidases

Lipoxygenases (LOX) are a type of iron-atom-based enzymes that oxidize polyunsaturated fatty acids, mainly arachidonic acid, to produce a variety of hydroperoxides as a byproduct of the ROS process (Zhong et al. 2019).The formation of ROS occurs when this enzyme interacts with its substrate and lowers the Fe^{3+} found in activated LOX. 5-Lipooxygenase (5-LOX), 12-lipoxygenase (12-LOX), and 15-lipoxygenase (15-LOX) are the most well-known members of this family, based on the carbon atom of arachidonic acid where oxygen is inserted. LOX is primarily involved in the synthesis of prostaglandins, thromboxane, and leukotrienes. They are also in responsible of lipid peroxidation in cell membranes (Khosravi et al. 2019). LOX is primarily involved in the synthesis of prostaglandins, thromboxane, and leukotrienes. They are also in responsible of lipid peroxidation in cell membranes (Khosravi et al. 2019).

3.2.11 Extracellular Sources of ROS

Diet, alcohol, drugs (halothane, doxorubicin, and metronidazole), industrial solvents, heavy metals (Fe, Cu, Co, and Cr), transition metals (Cd, Hg, Pb, and As), air pollution, physical stressors (UV, X-rays, and others), and lifestyle choices are some of the exogenous sources of ROS. These stressors work together to produce oxidative stress and inflammation, which are harmful to people's health, especially in the elderly. For example, air pollution contains a number of dangerous elements, such as metals and other molecules like ROS and RNS, which can lead to the development of disease. These harmful substances can cause inflammation and localized or systemic oxidative stress when they are consumed or inhaled by the body (Mudway et al. 2020). Furthermore, by promoting oxidative damage in proteins, lipids, and DNA, chemotherapy and radiation can raise ROS and harm the vascular, gastrointestinal, and hematological systems. Conversely, N-acetylcysteine (NAC), an antioxidant, can lessen or even completely avoid radiation-induced cell death in mice (Jia et al. 2010).

3.3 Cellular Response to Oxidative Stress

Oxidant damage at the cellular level causes a variety of reactions including proliferation, senescence, growth arrest, and cell death.

3.3.1 Apoptosis and Oxidative Stress

Due to its potential to produce early apoptosis and mitochondrial malfunction, oxidative stress induction by medication has become a popular approach in cancer therapy (Ding et al. 2009; Jeong et al. 2010; Widodo et al. 2010; Lee et al. 2013; Samhan-Arias et al. 2004). Surprisingly, oxidative stress may regulate many mediators and affect apoptosis. For example, oxidative stress may activate protein kinase C (PKC). PKC may be involved in the regulation of class III protein deacetylase sirtuin 1 (SIRT1). SIRT1 is a vital survival protein that regulates oxidative damage. PKC may modulate class III protein deacetylase SIRT1 (Thompson et al. 2014). SIRT1 is a vital survival molecule that regulates oxidative stress (Salminen et al. 2013). Moreover, mitogen-activated protein kinase (MAPK) activation results in the PKC-MAPK pathway, which is activated by PKC. Extracellular-regulated kinase (ERK) is one MAPK that may be activated by SIRT1 overexpression (Zhao et al. 2012), which may prevent apoptosis (Cho and Klemke 2000). As an alternative, via controlling the expression of PI3K/Akt, another MAPK p38 may promote apoptosis (Dong et al. 2016; Rane et al. 2010).

3.3.2 Autophagy and Oxidative Stress

The intracellular catabolic process known as autophagy generates energy by breaking down long-lived proteins and dysfunctional organelles for recycling when nutrients are limited or under stress (Mizushima and Klionsky 2007). Oxidative stress controls autophagy, which impacts cell survival and proliferation. Oxidative stress may affect autophagy and its capacity to regulate a variety of mediators. Oxidative stress may regulate MAPK to modify autophagy (Zhou et al. 2015). Autophagy (ULK1) is inhibited by p38 MAPK, which activates Unc-51-like kinase 1 (He et al. 2017). Furthermore, MAPK may interact with AMPK, which may then cooperate with ULK1 and mTORC1 to regulate autophagy. Additionally, MAPK could interact with AMPK, which might work with mTORC1 and ULK1 to control autophagy (Lee et al. 2010; Holczer et al. 2020). Thus, ROS play a crucial role in controlling autophagy.

3.3.3 Ferroptosis and Oxidative Stress

A unique kind of cell death known as ferroptosis is linked to lipid peroxidation, iron absorption triggered by oxidative stress, and downregulation of glutathione peroxidase 4 (GPX4) (Zhu et al. 2020). Ferroptosis is mediated by a number of different proteins, including GPX4, ACSL4, prostaglandin-endoperoxide synthase 2 (PTGS2; COX2), and CHAC1 (Chen et al. 2021b). In order to prevent stroke-induced ferroptosis and restore neurological function, acyl-CoA synthetase long-chain family member 4 (ACSL4) is downregulated (Chen et al. 2021a). In a similar vein, ACSL4 knockdown increases neuronal survival and reduces oxidative stress [58]. N-acetylcysteine reverses the expression of PTGS2 caused by oxidative stress via activating transforming growth factor-β-activated kinase 1 (TAK1), which phosphorylates MAPK and NF-κB (Onodera et al. 2015). Moreover, ferroptosis and the corresponding cell death are caused by MAPK activation (Xie et al. 2016). For instance, ferroptosis is brought on by extreme cold stress (Hattori et al. 2017). Eratin is a strong inducer of ferroptosis that also stimulates p38 after lipid peroxidation and is linked to MAPK activation (Hattori et al. 2017; Antonuccio et al. 2006).

3.3.4 Necroptosis and Oxidative Stress

Necroptosis can be carried on by oxidative stress, which can control many mediators (Choi et al. 2009) (He et al. 2017). Low plasma membrane integrity and cellular enlargement are hallmarks of necrotic cells (Troyano et al. 2001). A predetermined kind of necrosis is called necroptosis. Modulators for the necroptotic process include receptor-interacting serine/threonine-protein kinase 1 (RIPK1), RIPK3, and mixed lineage kinase domains like pseudokinase (MLKL) (Weinlich et al. 2017).

3.3.5 DNA Damage Response and Oxidative Stress

DNA damage and the DNA damage response (DDR) are brought on by oxidative stress in cancerous cells (Kello et al. 2020; González-Quiroz et al. 2020). Multiple mediators are impacted by oxidative stress and DDR responses, including p38 and MAPK ERK 1/2. Furthermore, during DDR, oxidative stress activates ERK, which prevents cell growth (Rezatabar et al. 2019).

Similarly, oxidative stress can alter several DNA repair mechanisms. For example, oxidative stress inhibits 8-oxoguanine DNA glycosylase 1 (OGG1), which initiates base excision repair (BER) (Bravard et al. 2015). The redox protein thioredoxin has the ability to down-regulate a BER protein known as apurinic/apyrimidinic endonuclease 1 (APE1) (Luo et al. 2005). Cross-complementing 1 (XRCC1) inhibits the downregulation of excision repair that depends on MAPK p38 DNA repair function (Tsai et al. 2012). Thus, several MAPK-related signaling pathways are impacted by oxidative stress and DNA repair.

3.4 Cellular Signaling Pathways Activated During Oxidative Stress

Several cellular signaling pathways are activated during oxidative stress (as illustrated in Fig. 3.2 and listed in Table 3.1) and are discussed under the following subheadings.

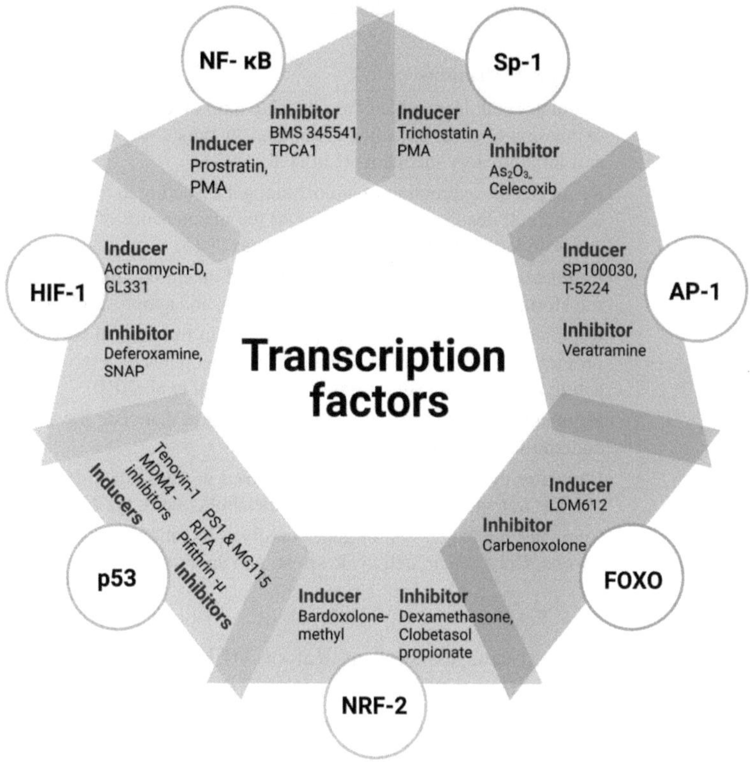

Fig. 3.2 Important transcription factors involved in oxidative stress along with their modulators. (Created in BioRender. Buradi, R. (2026) https://BioRender.com/i89p381)

3.4.1 MAPK Pathway

MAPK (Mitogen-Activated Protein Kinase) is a crucial signaling pathway that controls various cell functions, including proliferation, differentiation, migration, senescence, and apoptosis (Sun et al. 2015), as well as the response to oxidative stress. When a stimulus activates the topmost kinase in the MAPK cascade, MAPKKK, it triggers the phosphorylation and activation of the downstream MAPKK. The active MAPKK then activates and phosphorylates the bottom-tier kinase, which in turn activates cellular responses by phosphorylating various downstream substrates, such as structural proteins, protein kinases, transcription factors, and other enzymes (Zhang and Zhang 2022).

Activation Mechanisms The activation of the MAPK pathway by oxidative stress can occur through several mechanisms: *Direct activation by ROS*: can directly modify MAPK proteins, leading to their activation. *Activation of upstream kinases*:

Table 3.1 Role of transcription factors in oxidative stress

#	Transcription factors	Cellular mechanisms
1	NF- κB	Regulates cellular mechanisms related to oxidative stress by controlling the expression of genes encoding antioxidant enzymes, anti-apoptotic proteins, and inflammatory mediators (Liu et al. 2017)
2	Sp-1	Regulates the expression of antioxidant genes, acting as a protective factor against cell death, and contributing to the maintenance of cellular homeostasis under oxidative conditions (Kurutas 2015)
3	AP-1	Regulates the expression of genes involved in antioxidant defense, cellular transformation, proliferation, differentiation, and apoptosis (Ye et al. 2014)
4	FOXO	Regulates the expression of genes involved in free radical scavenging and promoting apoptosis when oxidative stress becomes excessive, essentially acting as a cellular "stress sensor" (Ambrogini et al. 2010)
5	NRF-2	Regulates and activates the expression of genes that code for antioxidant and anti-inflammatory proteins (Ma 2013)
6	HIF-1	Regulates the expression of genes involved in antioxidant defense mechanisms, mitochondrial function, and metabolic pathways (Bae et al. 2024)
7	p53	Regulates the expression of genes involved in antioxidant defense mechanisms, DNA repair, cell cycle arrest, and apoptosis (Liu and Xu 2011)

ROS can activate upstream kinases, such as Raf or MEKK, initiating the MAPK cascade. *Indirect activation through other signaling pathways*: Oxidative stress can trigger other signaling pathways, such as the PI3K/Akt pathway, which can subsequently activate MAPK (Pinal et al. 2018).

Cellular Responses Mediated by MAPK The activation of MAPK pathways in response to oxidative stress leads to a complex array of cellular responses, including *Cell proliferation*: MAPK pathways, particularly the ERK pathway, are crucial for the regulation of cell proliferation. They transduce signals from growth factors and other stimuli, leading to cellular growth and division. *Inflammation:* MAPKs are also involved in mediating inflammatory responses. They can be activated by pro-inflammatory cytokines and contribute to the signaling cascades that lead to inflammation. *Apoptosis*: Certain MAPK pathways, such as p38 and JNK, are involved in the regulation of apoptosis (programmed cell death). They are activated in response to stress signals and contribute to the decision of a cell to undergo apoptosis (Zhang and Liu 2002).

3.4.2 NF-κB Pathway

Nuclear factor kappa-light-chain-enhancer of activated B cells, abbreviated as NF-κB, is a crucial signaling cascade that controls inflammation, immunological responses, cell survival, and proliferation (Yu et al. 2020). NF-κB represents a family of inducible transcription factors. RelA (p65), RelB, c-Rel, NF-κB1 (p50), and NF-κB2 (p52) are the five transcription factors that make up the NF-κB family. Together, these proteins create dimers, which are normally kept in the cytoplasm by IκB inhibitory proteins (Liu et al. 2017). The canonical and non-canonical NF-κB pathways are two distinct pathways that are activated by different processes and stimuli and are involved in the activation of NF-κB (Yu et al. 2020).

Activation Mechanisms The activation of the NF-κB pathway, by oxidative stress can occur through several mechanisms: *Canonical Pathway: Cytokines* (e.g., TNF-α, IL-1), lipopolysaccharide (LPS), and other inflammatory signals activate the canonical pathway. *IκB Phosphorylation and Degradation:* Stimuli activate IκB kinases (IKKs), which phosphorylate IκB proteins. Phosphorylated IκB is then ubiquitinated and degraded by the proteasome. *NF-κB Nuclear Translocation:* Released NF-κB dimers (typically p65/p50) translocate to the nucleus and bind to specific DNA sequences called NF-κB response elements (Li and Verma 2002). *Non-canonical Pathway:* CD40, BAFF, and other signals activate the non-canonical pathway. *NF-κB-inducing kinase Activation:* leading to the phosphorylation and processing of NF-κB2 (p100). *RelB Activation:* Processed p100 (p52) forms a heterodimer with RelB, which translocates to the nucleus and activates target genes (Liu et al. 2017; Sun 2012). *Alternative Activation*: Various stimuli, including oxidative stress and DNA damage, can activate NF-κB through alternative mechanisms (Lingappan 2018).

Cellular responses mediated by NF-κB pathway include majorly inflammation and apoptosis. *Inflammation*: The NF-κB pathway is activated by pro-inflammatory cytokines such as TNF-α and IL-1, leading to the expression of various inflammatory mediators, including IL-6, which contributes to the regulation of inflammation (Noort et al. 2014). *Anti-apoptotic gene activation*: The NF-κB pathway promotes cell survival by inducing the expression of various anti-apoptotic proteins, such as FLIP, Bcl-X_L, and XIAP. These proteins inhibit components of the apoptotic machinery, thereby preventing programmed cell death in response to various stimuli, including cytokines and stress signals (Luo et al. 2005). *Activation of pro-apoptotic genes*: NF-κB can activate genes encoding pro-apoptotic proteins, such as FasL and TRAIL. TRAIL is induced by interferons and other stimuli, leading to apoptosis in tumor cells through the engagement of death receptors DR4 and DR5 (Allen and El-Deiry 2012).

3.4.3 PI3K/Akt Pathway

The PI3K/Akt signaling system, also known as the Akt pathway, is a crucial intracellular signaling cascade that controls metabolism, growth, survival, and cell division. It begins when phosphoinositide 3-kinase (PI3K) is triggered by a number of stimuli, such as growth factors, hormones, and oxidative stress, which is one kind of cellular stressor. Phosphatidylinositol (4,5)-bisphosphate (PIP2) is converted to phosphatidylinositol (3,4,5)-trisphosphate (PIP3) by PI3K, which then recruits and activates Akt at the plasma membrane. The PI3K/Akt pathway modulates, inhibits, and activates oxidative stress levels, among other interactions with oxidative stress. Akt expression may be inhibited or activated by oxidative stress, which can result in a variety of intricate cellular reactions like autophagy and death (Shiau et al. 2022).

Activation Mechanisms Oxidative stress can activate the PI3K/Akt pathway through several mechanisms. *Direct activation by ROS*: can directly activate PI3K, initiating the pathway. *Indirect activation through receptor tyrosine kinase*: Growth factors attach to receptor tyrosine kinases, which cause them to become autophosphorylated. This activation of PI3K then changes PIP2 into PIP3. *Inhibition of negative regulators*: Oxidative stress can inhibit phosphatases like PTEN, a negative regulator of PI3K, leading to increased PI3K activity (Kitagishi and Matsuda 2013).

Cellular Responses Mediated by PI3K/Akt pathway includes apoptosis, cell proliferation, and autophagy. *Apoptosis:* The PI3K/Akt pathway can inhibit apoptosis by triggering XIAP to suppress caspase-3 activity. It can also reduce the expression of caspase-9 and caspase-3. *Cell proliferation:* The PI3K/Akt pathway can regulate tumor proliferation by affecting mTOR, GSK-3β, and NF-κB (Liu et al. 2020). *Autophagy:* The PI3K/Akt/mTOR pathway suppresses autophagy when activated, as mTORC1 acts as a major negative regulator of the autophagic process. Conversely, inhibition of mTORC1 promotes autophagy, facilitating the clearance of damaged proteins and cellular debris (Rajendran et al. 2024).

3.4.4 NRF2-Keap1 Pathway

The NRF2-Keap1 pathway is a critical cellular defense mechanism that regulates the expression of antioxidant and cytoprotective genes in response to oxidative stress and environmental insults. NRF2 (Nuclear factor erythroid 2-related factor 2) is a transcription factor that, under normal conditions, is kept in the cytoplasm by its inhibitor Keap1 (Kelch-like ECH-associated protein 1) (Sykiotis 2021). To improve cellular defense against oxidative damage, NRF2 separates from this complex in response to oxidative stress or electrophilic modification of particular cysteine residues in Keap1, moves to the nucleus, and triggers the transcription of genes with antioxidant response elements in their promoter regions (Kim and Vaziri 2010).

Activation Mechanisms Oxidative stress can activate the NRF2-Keap1 pathway through several mechanisms. *Keap1 Regulation:* NRF2 is primarily regulated by Keap1, which binds to Nrf2's ETGE and DLG motifs, leading to its ubiquitination and degradation. *Oxidative Stress Response:* Oxidative stress or electrophiles modify cysteine residues in Keap1, causing a conformational change that disrupts NRF2-Keap1 binding and prevents NRF2 ubiquitination. *p62 Interaction:* The protein p62 sequesters Keap1 in autophagosomes, reducing NRF2 ubiquitination and increasing NRF2 stability, thus activating NRF2 target genes (Jaramillo and Zhang 2013).

Cellular responses mediated by NRF2-Keap1 pathway include antioxidant defense, detoxification, and modulation of inflammatory response. *Antioxidant Defense:* The NRF2-Keap1 pathway enhances cellular antioxidant defense by activating genes that encode antioxidant enzymes like glutathione peroxidase and superoxide dismutase. This process helps neutralize ROS and protects cells from oxidative damage. *Detoxification:* NRF2 regulates the expression of phase II detoxifying enzymes, such as glutathione S-transferases and NAD(P)H: quinone oxidoreductase 1, facilitating the detoxification of harmful compounds. This mechanism reduces cellular toxicity and promotes the elimination of toxic metabolites. *Inflammatory Response Modulation:* The pathway modulates inflammatory responses by regulating anti-inflammatory cytokines and inhibiting pro-inflammatory mediators. This balance helps maintain tissue homeostasis and protects against chronic inflammatory diseases (Song et al. 2021).

3.4.5 p53 Pathway

The p53 pathway, often referred to as the "Guardian of the Genome" (Borrero and El-Deiry 2021), is a critical signaling network that plays a central role in maintaining cellular integrity by regulating various processes such as DNA repair, cell cycle progression, and apoptosis (Lahalle et al. 2021). The TP53 gene, which codes for the p53 protein—a crucial transcription factor that coordinates a number of cellular functions in response to stress signals, is the primary regulator of this pathway. Additionally, this pathway interacts with angiogenesis, autophagy, and senescence, underscoring the complexity and importance of p53 in preserving cellular integrity and averting cancer. The intricate relationships between p53 and other proteins and pathways that impact cellular metabolism and the production of ROS have been revealed by recent studies. Developing therapeutic techniques that improve cellular resilience and encourage healthy responses to stress requires an understanding of the complexities of the p53 pathway (Borrero and El-Deiry 2021).

Activation Mechanisms The activation mechanisms of the p53 signaling pathway involve several post-translational modifications and interactions with various proteins in response to cellular stress, such as DNA damage. Here are the key activation mechanisms: *Phosphorylation:* p53 is phosphorylated at multiple serine residues, which enhances its stability and transcriptional activity. For instance, phosphorylation at Ser-15 by ATM and ATR kinases is crucial for activating p53 in response to DNA damage. *Acetylation:* Acetylation of p53 at specific lysine residues within its C-terminal regulatory domain (e.g., K370, K372, K373, K381, K382, and K386) plays a significant role in modulating its transcriptional activity and stability. This modification is a response to genotoxic stress and helps fine-tune p53's tumor suppressor functions. *MDM2 Interaction:* The interaction between p53 and MDM2, an E3 ubiquitin ligase, is critical for regulating p53 levels. Under normal conditions, MDM2 promotes the ubiquitination and degradation of p53. However, in response to stress signals, p53 is stabilized, allowing it to activate target genes involved in cell cycle arrest, DNA repair, and apoptosis (Marei et al. 2021).

Cellular responses mediated by p53 pathway include cell cycle arrest, apoptosis, and senescence. *Cell Cycle Arrest:* p53 can induce cell cycle arrest at the G1/S or G2/M checkpoints, allowing time for DNA repair before the cell proceeds to division. This is achieved through the transcriptional activation of target genes such as p21, which inhibits cyclin-dependent kinases (Wang 2021). *Apoptosis:* In cases where DNA damage is irreparable, p53 can trigger programmed cell death (apoptosis). This is mediated through the activation of pro-apoptotic genes such as PUMA and NOXA, and the phosphorylation of p53 at Ser-46, which enhances its apoptotic function (Lees et al. 2021). *Senescence:* p53 can also induce cellular senescence, a state of permanent cell cycle arrest that prevents the proliferation of damaged cells. This response is crucial for tumor suppression, as it limits the growth of potentially cancerous cells (Mijit et al. 2020).

3.5 Conclusion

A key idea in redox biology and medicine, oxidative stress is defined as the disturbance of the steady-state redox balance in cells. The dual character of ROS in cellular activities is demonstrated by the fact that this imbalance can appear as either oxidative eustress or oxidative distress. Sources such as mitochondria, NADPH oxidases, xanthine oxidase, and the endoplasmic reticulum contribute substantially to ROS production, which, when in excess, can lead to significant cellular damage, including lipid peroxidation, DNA damage, and, ultimately cellular dysfunction or death. Numerous processes, including apoptosis, autophagy, ferroptosis, and necroptosis, are part of the cellular responses to oxidative stress. These pathways are all controlled by important mediators and signaling systems, such as the PI3K/Akt pathway, NF-κB route, MAPK pathway, etc. Developing treatment approaches for various oxidative damage-related diseases, especially age-related disorders and inflammation, requires an understanding of the complex balance of ROS and the cellular mechanisms activated during oxidative stress. Future research in this area might lead to novel strategies for reducing oxidative damage and improve cellular resilience against stress-induced pathologies.

References

Allen JE, El-Deiry WS (2012) Regulation of the human trail gene. Cancer Biol Ther 13:1143–1151

Almeida-Nunes DL, Mendes-Frias A, Silvestre R, Dinis-Oliveira RJ, Ricardo S (2022) Immune tumor microenvironment in ovarian cancer ascites. Int J Mol Sci 23:10692

Ambrogini E, Almeida M, Martin-Millan M, Paik J-H, Depinho RA, Han L, Goellner J, Weinstein RS, Jilka RL, O'brien CA (2010) Foxo-mediated defense against oxidative stress in osteoblasts is indispensable for skeletal homeostasis in mice. Cell Metab 11:136–146

Antonuccio P, Minutoli L, Romeo C, Nicòtina PA, Bitto A, Arena S, Altavilla D, Zuccarello B, Polito F, Squadrito F (2006) Lipid peroxidation activates mitogen-activated protein kinases in testicular ischemia-reperfusion injury. J Urol 176:1666–1672

Ashrafi G, Schwarz T (2013) The pathways of mitophagy for quality control and clearance of mitochondria. Cell Death Different 20:31–42

Athar M (2002) Oxidative stress and experimental carcinogenesis. Indian J Exp Biol 40:656–667

Bae T, Hallis SP, Kwak M-K (2024) Hypoxia, oxidative stress, and the interplay of HIFs and NRF2 signaling in cancer. Exp Mol Med 56:501–514

Bartesaghi S, Radi R (2018) Fundamentals on the biochemistry of peroxynitrite and protein tyrosine nitration. Redox Biol 14:618–625

Battelli MG, Bortolotti M, Bolognesi A, Polito L (2020) Pro-aging effects of xanthine oxidoreductase products. Antioxidants 9:839

Bendall JK, Douglas G, McNeill E, Channon KM, Crabtree MJ (2014) Tetrahydrobiopterin in cardiovascular health and disease. Antioxid Redox Signal 20:3040–3077

Borrero LJH, El-Deiry WS (2021) Tumor Suppressor P53: biology, signaling pathways, and therapeutic targeting. Biochim Biophys Acta 1876:188556

Bravard A, Auvré F, Fantini D, Bernardino-Sgherri J, Sissoëff L, Daynac M, Xu Z, Etienne O, Dehen C, Comoy E (2015) The prion protein is critical for DNA repair and cell survival after genotoxic stress. Nucleic Acids Res 43:904–916

Chen L, Hu JY, Wang SQ (2012) The role of antioxidants in photoprotection: a critical review. J Am Acad Dermatol 67:1013–1024

Chen J, Yang L, Geng L, He J, Chen L, Sun Q, Zhao J, Wang X (2021a) Inhibition of acyl-CoA Synthetase long-chain family member 4 facilitates neurological recovery after stroke by regulation ferroptosis. Front Cell Neurosci 15:632354

Chen X, Comish PB, Tang D, Kang R (2021b) Characteristics and biomarkers of ferroptosis. Front Cell Dev Biol 9:637162

Cho SY, Klemke RL (2000) Extracellular-regulated kinase activation and CAS/CRK coupling regulate cell migration and suppress apoptosis during invasion of the extracellular matrix. J Cell Biol 149:223–236

Choi K, Kim J, Kim GW, Choi C (2009) Oxidative stress-induced necrotic cell death via mitochondira-dependent burst of reactive oxygen species. Curr Neurovasc Res 6:213–222

Daiber A, Steven S, Euler G, Schulz R (2021) Vascular and cardiac oxidative stress and inflammation as targets for cardioprotection. Curr Pharm Des 27:2112–2130

Ding H, Han C, Guo D, Chin Y-W, Ding Y, Kinghorn AD, D'ambrosio SM (2009) Selective induction of apoptosis of human Oral cancer cell lines by avocado extracts via a ROS-mediated mechanism. Nutr Cancer 61:348–356

Dong Y, Yin S, Song X, Huo Y, Fan L, Ye M, Hu H (2016) Involvement of ROS-P38-H2AX Axis in novel curcumin analogues-induced apoptosis in breast cancer cells. Mol Carcinog 55:323–334

Förstermann U, Xia N, Li H (2017) Roles of vascular oxidative stress and nitric oxide in the pathogenesis of atherosclerosis. Circ Res 120:713–735

Fukai T, Ushio-Fukai M (2020) Cross-talk between NADPH oxidase and mitochondria: role in ROS signaling and angiogenesis. Cells 9:1849

George J, Struthers A (2009) The role of urate and xanthine oxidase in vascular oxidative stress: future directions. Ther Clin Risk Manag 5:799–803

Gómez-Tabales J, García-Martín E, Agúndez JA, Gutierrez-Merino C (2020) Modulation of CYP2C9 activity and hydrogen peroxide production by cytochrome b 5. Sci Rep 10:15571

González-Quiroz M, Blondel A, Sagredo A, Hetz C, Chevet E, Pedeux R (2020) When endoplasmic reticulum proteostasis meets the DNA damage response. Trends Cell Biol 30:881–891

Görlach A, Dimova EY, Petry A, Martínez-Ruiz A, Hernansanz-Agustín P, Rolo AP, Palmeira CM, Kietzmann T (2015) Reactive oxygen species, nutrition, hypoxia and diseases: Problems solved? Redox Biol 6:372–385

Guengerich FP (2019) Inhibition of drug metabolizing enzymes. Handbook of Drug Metabolism, Third Edition. CRC Press

Hattori K, Ishikawa H, Sakauchi C, Takayanagi S, Naguro I, Ichijo H (2017) Cold stress-induced Ferroptosis involves the ASK 1-P38 pathway. EMBO Rep 18:2067–2078

He L, He T, Farrar S, Ji L, Liu T, Ma X (2017) Antioxidants maintain cellular redox homeostasis by elimination of reactive oxygen species. Cell Physiol Biochem 44:532–553

Holczer M, Hajdú B, Lőrincz T, Szarka A, Bánhegyi G, Kapuy O (2020) Fine-tuning of AMPK–Ulk1–Mtorc1 regulatory triangle is crucial for autophagy oscillation. Sci Rep 10:17803

Holmström KM, Finkel T (2014) Cellular mechanisms and physiological consequences of redox-dependent signalling. Nat Rev Mol Cell Biol 15:411–421

Igbaria A, Merksamer PI, Trusina A, Tilahun F, Johnson JR, Brandman O, Krogan NJ, Weissman JS, Papa FR (2019) Chaperone-mediated reflux of secretory proteins to the cytosol during endoplasmic reticulum stress. Proc Natl Acad Sci U S A 116:11291–11298

Ighodaro O, Akinloye O (2018) First line defence antioxidants-superoxide dismutase (SOD), catalase (CAT) and glutathione peroxidase (GPX): Their fundamental role in the entire antioxidant defence grid. Alex J Med 54:287–293

Jaramillo MC, Zhang DD (2013) The emerging role of the Nrf2–Keap1 signaling pathway in cancer. Genes Dev 27:2179–2191

Jeong JC, Jang SW, Kim TH, Kwon CH, Kim YK (2010) Mulberry fruit (Moris Fructus) extracts induce human glioma cell death in vitro through ROS-dependent mitochondrial pathway and inhibits glioma tumor growth in vivo. Nutr Cancer 62:402–412

Jia D, Koonce NA, Griffin RJ, Jackson C, Corry PM (2010) Prevention and mitigation of acute death of mice after abdominal irradiation by the antioxidant N-acetyl-cysteine (NAC). Radiat Res 173:579–589

Jurk D, Wilson C, Passos JF, Oakley F, Correia-Melo C, Greaves L, Saretzki G, Fox C, Lawless C, Anderson R (2014) Chronic inflammation induces telomere dysfunction and accelerates ageing in mice. Nat Commun 5:4172

Kang SM, Lim S, Song H, Chang W, Lee S, Bae SM, Chung JH, Lee H, Kim HG, Yoon DH (2006) Allopurinol modulates reactive oxygen species generation and Ca2+ overload in ischemia-reperfused heart and hypoxia-reoxygenated cardiomyocytes. Eur J Pharmacol 535:212–219

Kauppinen A (2013) Crosstalk between oxidative stress and SIRT1: impact on the aging process. Int J Mol Sci 14:3834–3859

Kello M, Takac P, Kubatka P, Kuruc T, Petrova K, Mojzis J (2020) Oxidative stress-induced DNA damage and apoptosis in clove buds-treated MCF-7 cells. Biomolecules 10:139

Khosravi M, Poursaleh A, Ghasempour G, Farhad S, Najafi M (2019) The effects of oxidative stress on the development of atherosclerosis. Biol Chem 400:711–732

Kim HJ, Vaziri ND (2010) Contribution of impaired CFf2-Keap1 pathway to oxidative stress and inflammation in chronic renal failure. Am J Physiol 298:F662–F671

Kim HG, Kim YR, Park JH, Khanal T, Choi JH, Do MT, Jin SW, Han EH, Chung YH, Jeong HG (2015) Endosulfan induces COX-2 expression via NADPH oxidase and the ROS, MAPK, and AKT pathways. Arch Toxicol 89:2039–2050

Kitagishi Y, Matsuda S (2013) Redox regulation of tumor suppressor PTEN in cancer and aging. Int J Mol Med 31:511–515

Krylatov AV, Tsibulnikov SY, Mukhomedzyanov AV, Boshchenko AA, Goldberg VE, Jaggi AS, Erben RG, Maslov LN (2021) The role of natriuretic peptides in the regulation of cardiac tolerance to ischemia/reperfusion and postinfarction heart remodeling. J Cardiovasc Pharmacol Ther 26:131–148

Kurutas EB (2015) The importance of antioxidants which play the role in cellular response against oxidative/Nitrosative stress: current state. Nutr J 15:1–22

Lahalle A, Lacroix M, De Blasio C, Cissé MY, Linares LK, Le Cam L (2021) The P53 pathway and metabolism: the tree that hides the Forest. Cancer 13:133

Lee JW, Park S, Takahashi Y, Wang H-G (2010) The association of AMPK with ULK1 regulates autophagy. PLoS One 5:E15394

Lee J-C, Hou M-F, Huang H-W, Chang F-R, Yeh C-C, Tang J-Y, Chang H-W (2013) Marine algal natural products with anti-oxidative, anti-inflammatory, and anti-cancer properties. Cancer Cell Int 13:1–7

Lees A, Sessler T, Mcdade S (2021) Dying to survive—the P53 paradox. Cancer 13:3257

Li Q, Verma IM (2002) NF-Kb regulation in the immune system. Nat Rev Immunol 2:725–734

Lingappan K (2018) NF-Kb in oxidative stress. Curr Opin Toxicol 7:81–86

Liu D, Xu Y (2011) P53, oxidative stress, and aging. Antioxid Redox Signal 15:1669–1678

Liu T, Zhang L, Joo D, Sun S-C (2017) NF-Kb Signaling In Inflammation. Signal Transduct Target Ther 2:1–9

Liu R, Chen Y, Liu G, Li C, Song Y, Cao Z, Li W, Hu J, Lu C, Liu Y (2020) PI3k/AKT pathway as A key link modulates the multidrug resistance of cancers. Cell Death Dis 11:797

Llanas-Cornejo D, Husi H (2017) CVD and oxidative stress. J Clin Med 6:E22–E22

Luo J-L, Kamata H, Karin M (2005) IKK/NF-Kb signaling: balancing life and death–A new approach to cancer therapy. J Clin Invest 115:2625–2632

Ma Q (2013) Role of Nrf2 in oxidative stress and toxicity. Annu Rev Pharmacol Toxicol 53:401–426

Magnani F, Mattevi A (2019) Structure and mechanisms of ROS generation by NADPH oxidases. Curr Opin Struct Biol 59:91–97

Manikandan P, Nagini S (2018) Cytochrome P450 structure, function and clinical significance: a review. Curr Drug Targets 19:38–54

Marei HE, Althani A, Afifi N, Hasan A, Caceci T, Pozzoli G, Morrione A, Giordano A, Cenciarelli C (2021) P53 signaling in cancer progression and therapy. Cancer Cell Int 21:703

Menon BR, Richmond D, Menon N (2022) Halogenases for biosynthetic pathway engineering: Toward new routes to naturals and non-naturals. Catal Rev 64:533–591

Mijit M, Caracciolo V, Melillo A, Amicarelli F, Giordano A (2020) Role of P53 in the regulation of cellular senescence. Biomolecules 10:420

Missiroli S, Genovese I, Perrone M, Vezzani B, Vitto VA, Giorgi C (2020) The role of mitochondria in inflammation: from cancer to neurodegenerative disorders. J Clin Med 9:740

Mitchell JA, Kirkby NS (2019) Eicosanoids, prostacyclin and cyclooxygenase in the cardiovascular system. Br J Pharmacol 176:1038–1050

Mizushima N, Klionsky DJ (2007) Protein turnover via autophagy: implications for metabolism. Annu Rev Nutr 27:19–40

Moldogazieva NT, Mokhosoev IM, Terentiev AA (2020) Metabolic heterogeneity of cancer cells: an interplay between HIF-1, GLUTs, and AMPK. Cancers 12:862

Montezano AC, Touyz RM (2014) Reactive oxygen species, vascular Noxs, and hypertension: focus on translational and clinical research. Antioxid Redox Signal 20:164–182

Mudway I, Kelly F, Holgate S (2020) Oxidative stress in air pollution research. Free Radic Biol Med 151:2

Noort AR, Van Zoest KP, Weijers EM, Koolwijk P, Maracle CX, Novack DV, Siemerink MJ, Schlingemann RO, Tak PP, Tas SW (2014) NF-Kb-inducing kinase is a key regulator of inflammation-induced and tumour-associated angiogenesis. J Pathol 234:375–385

Onodera Y, Teramura T, Takehara T, Shigi K, Fukuda K (2015) Reactive oxygen species induce Cox-2 expression via TAK1 activation in synovial fibroblast cells. FEBS Open Bio 5:492–501

Pai VV, Shukla P, Kikkeri NN (2014) Antioxidants in dermatology. Indian Dermatol Online J 5: 210–214

Phaniendra A, Jestadi DB, Periyasamy L (2015) Free radicals: properties, sources, targets, and their implication in various diseases. Indian J Clin Biochem 30:11–26

Pinal N, Martín M, Medina I, Morata G (2018) Short-term activation of the Jun N-terminal kinase pathway in apoptosis-deficient cells of drosophila induces tumorigenesis. Nat Commun 9:1541

Prajapati P, Sripada L, Singh K, Bhatelia K, Singh R, Singh R (2015) TNF-A regulates MiRNA targeting mitochondrial complex-I and induces cell death in dopaminergic cells. Biochim Biophys Acta 1852:451–461

Radi R (2018) Oxygen radicals, nitric oxide, and peroxynitrite: redox pathways in molecular medicine. Proc Natl Acad Sci 115:5839–5848

Rajendran P, Sekar R, Dhayasankar PS, Ali EM, Abdelsalam SA, Balaraman S, Chellappan BV, Metwally AM, Abdallah BM (2024) PI3k/AKT signaling pathway mediated autophagy in oral carcinoma-A comprehensive review. Int J Med Sci 21:1165

Rane MJ, Song Y, Jin S, Barati MT, Wu R, Kausar H, Tan Y, Wang Y, Zhou G, Klein JB (2010) Interplay between AKT and P38 MAPK pathways in the regulation of renal tubular cell apoptosis associated with diabetic nephropathy. Am J Physiol 298:F49–F61

Rezatabar S, Karimian A, Rameshknia V, Parsian H, Majidinia M, Kopi TA, Bishayee A, Sadeghinia A, Yousefi M, Monirialamdari M (2019) RAS/MAPK signaling functions in oxidative stress, DNA damage response and cancer progression. J Cell Physiol 234:14951–14965

Salminen A, Kaarniranta K, kauppinen A (2013) Crosstalk between oxidative stress and SIRT1: impact on the aging process. Int J Mol Sci 14:3834–3859

Samhan-Arias AK, Martın-Romero FJ, Gutiérrez-Merino C (2004) Kaempferol blocks oxidative stress in cerebellar granule cells and reveals a key role for reactive oxygen species production at the plasma membrane in the commitment to apoptosis. Free Radic Biol Med 37:48–61

Schmidt HM, Kelley EE, Straub AC (2019) The impact of xanthine oxidase (XO) on hemolytic diseases. Redox Biol 21:101072

Shiau J-P, Chuang Y-T, Tang J-Y, Yang K-H, Chang F-R, Hou M-F, Yen C-Y, Chang H-W (2022) The impact of oxidative stress and AKT pathway on cancer cell functions and its application to natural products. Antioxidants 11:1845

Sies H (2015) Oxidative stress: A concept in redox biology and medicine. Redox Biol 4:180–183

Sies H (2017) Hydrogen peroxide as A central redox signaling molecule in physiological oxidative stress: oxidative eustress. Redox Biol 11:613–619

Sies H, Jones DP (2020) Reactive oxygen species (ROS) as pleiotropic physiological signalling agents. Nat Rev Mol Cell Biol 21:363–383

Sies H, Berndt C, Jones DP (2017) Oxidative stress. Annu Rev Biochem 86:715–748

Song M-Y, Lee D-Y, Chun K-S, Kim E-H (2021) The role of NRF2/KEAP1 signaling pathway in cancer metabolism. Int J Mol Sci 22:4376

Sun Y, Liu WZ, Liu T, Feng X, Yang N, Zhou HF (2015) Signaling pathway of MAPK/ERK in cell proliferation, differentiation, migration, senescence and apoptosis. J Recept Signal Transduct Res 35:600–604

Sun SC (2012) The noncanonical NF-Kb pathway. Immunol Rev 246:125–140

Sykiotis GP (2021) Keap1/Nrf2 signaling pathway. MDPI

Taanman J-W (1999) The mitochondrial genome: structure, transcription, translation and replication. Biochim Biophys Acta 1410:103–123

Tada Y, Suzuki J-I (2016) Oxidative stress and myocarditis. Curr Pharm Des 22:450–471

Thompson AM, Wagner R, Rzucidlo EM (2014) Age-related loss of SirT1 expression results in dysregulated human vascular smooth muscle cell function. Am J Physiol Heart Circ Physiol 307:H533–H541

Touyz RM, Anagnostopoulou A, Camargo LL, Rios FJ, Montezano AC (2019) Vascular biology of superoxide-generating NADPH oxidase 5—implications in hypertension and cardiovascular disease. Antioxid Redox Signal 30:1027–1040

Troyano A, Fernández C, Sancho P, De Blas E, Aller P (2001) Effect of glutathione depletion on antitumor drug toxicity (apoptosis and necrosis) in U-937 human promonocytic cells: the role of intracellular oxidation. J Biol Chem 276:47107–47115

Tsai MS, Weng SH, Chen HJ, Chiu YF, Huang YC, Tseng SC, Kuo YH, Lin YW (2012) Inhibition of p38 MAPK-dependent excision repair cross-complementing 1 expression decreases the DNA repair capacity to sensitize lung cancer cells to etoposide. Mol Cancer Ther 11:561–571

Ulfig A, Leichert LI (2021) The effects of neutrophil-generated hypochlorous acid and other hypohalous acids on host and pathogens. Cell Mol Life Sci 78:385–414

Van Horssen J, Van Schaik P, Witte M (2019) Inflammation and mitochondrial dysfunction: A vicious circle in neurodegenerative disorders? Neurosci Lett 710:132931
Wang Z (2021) Regulation of cell cycle progression by growth factor-induced cell signaling. Cells 10:3327
Weinlich R, Oberst A, Beere HM, Green DR (2017) Necroptosis in development, inflammation and disease. Nat Rev Mol Cell Biol 18:127–136
Widodo N, Priyandoko D, Shah N, Wadhwa R, Kaul SC (2010) Selective killing of cancer cells by Ashwagandha leaf extract and its component Withanone involves ROS signaling. PLoS One 5: E13536
Xie Y, Hou W, Song X, Yu Y, Huang J, Sun X, Kang R, Tang D (2016) Ferroptosis: process and function. Cell Death Different 23:369–379
Ye N, Ding Y, Wild C, Shen Q, Zhou J (2014) Small molecule inhibitors targeting activator protein 1 (AP-1). J Med Chem 57:6930–6948
Yang Y, Zhao J, Qiu J, Li J, Liang X, Zhang Z, Zhang X, Fu H, Korantzopoulos P, Letsas K (2018) Xanthine oxidase inhibitor allopurinol prevents oxidative stress-mediated atrial remodeling in alloxan-induced diabetes mellitus rabbits. J Am Heart Assoc 7:e008807
Yu H, Lin L, Zhang Z, Zhang H, Hu H (2020) Targeting NF-Kb pathway for the therapy of diseases: mechanism and clinical study. Signal Transduct Target Ther 5:209
Zhang Z, Nian Q, Chen G, Cui S, Han Y, Zhang J (2020) Klotho alleviates lung injury caused by paraquat via suppressing ROS/P38 MAPK-Regulated inflammatory responses and apoptosis. Oxid Med Cell Longev 2020:1854206
Zhang M, Zhang S (2022) Mitogen-activated protein kinase cascades in plant signaling. J Integr Plant Biol 64:301–341
Zhang W, Liu HT (2002) MAPK signal pathways in the regulation of cell proliferation in mammalian cells. Cell Res 12:9–18
Zhao Y, Luo P, Guo Q, Li S, Zhang L, Zhao M, Xu H, Yang Y, Poon W, Fei Z (2012) Interactions between SIRT1 and MAPK/ERK regulate neuronal apoptosis induced by traumatic brain injury in vitro and in vivo. Exp Neurol 237:489–498
Zhong Q, Wei B, Wang S, Ke S, Chen J, Zhang H, Wang H (2019) The antioxidant activity of polysaccharides derived from marine organisms: An overview. Mar Drugs 17:674
Zhou YY, Li Y, Jiang WQ, Zhou LF (2015) MAPK/JNK signalling: a potential autophagy regulation pathway. Biosci Rep 35:e00199
Zhu J, Xiong Y, Zhang Y, Wen J, Cai N, Cheng K, Liang H, Zhang W (2020) The molecular mechanisms of regulating oxidative stress-induced ferroptosis and therapeutic strategy in tumors. Oxidative Med Cell Longev 2020:8810785

Chapter 4
Mechanistic Insights of Oxidative Stress in Inflammation

Jay Prakash Dewangan, Mansi Singh, and Rahul Shukla

Abbreviations

ALE	*Apios americana* leaf extract
AP-1	Activator protein 1
ApoD	Apolipoprotein D
BMI	Body mass index
CAT	catalase
COPD	Chronic obstructive pulmonary disease
COX	Cyclooxygenase
DAMPs	Danger-associated molecular patterns
DCs	Dendritic cells
ETC	Electron transport chain
GPx	Glutathione peroxidase
HF	Heart failure
HMGB1	Intracellular high mobility group box protein 1
HOCl	Hypochlorous acid
IFN-γ	Interferon gamma
iNos	Inducible nitric oxide synthase
JAK-STAT	Janus kinase and signal transducer of activation pathway
LDL	Low-density lipoprotein
LPC	Lysophosphatidylcholine
LPT	Lipoteichoic acid
MEK/ERK	Mitogen-activated protein kinase/extracellular signal regulated kinase
MMP	Matrix metalloproteinase

J. P. Dewangan · M. Singh · R. Shukla (✉)
Department of Pharmaceutics, National Institute of Pharmaceutical Education and Research-Raebareli (NIPER-R), Lucknow, UP, India
e-mail: rahul.shukla@niperraebareli.edu.in

N. Srivastava, S. A. Saraf (eds.), *Pharmacological Strategies for Prevention and Treatment of Oxidative Stress*, Oxidative Stress in Applied Basic Research and Clinical Practice, https://doi.org/10.1007/978-3-032-19837-2_4

NF-κβ	Nuclear factor kappa β
NHEJ	Non-equivalent ends joining
NODs	Nucleotide binding oligomerization domain protein like receptors
NOX	NADPH oxidase
PAMPs	Pathogen associated molecular patterns
PCR	Polymer chain reaction
PD	Parkinson's disease
PRR	Pattern recognition receptors
RERG	RAS like estrogen-regulated growth inhibitor
ROS	Reactive oxygen species
SASP	Senescence-associated secretory phenotype
SOD	Superoxide dismutase
TGF-β	Transforming growth factor beta
TLRs	Toll-like receptors
TNF- α	Tumor necrosis factor alpha
TRAF	Tumor necrosis factor receptor-associated factor
UCB	Unconjugated bilirubin
XO	Xanthine oxidase

4.1 Introduction

Practically all illnesses and aging in particular are closely correlated with two phenomena: oxidative stress and inflammation. An essential mechanism in many physiological processes, oxidative stress is connected to the redox state, including cell communication, hypothalamic appetite regulation, and adaptations to physical activity. The well-established significance of inflammatory mediators lies in physiological processes such as production of gastric mucus, which shields the stomach, and stem cell-mediated tissue repair. On the other hand, these two events cause harm when they are not controlled (Gambini and Stromsnes 2022). An imbalance between the generation of reactive metabolites and free radicals, sometimes referred to as oxidants or ROS, and their removal by antioxidants, which serve as protective mechanisms. Important biomolecules and cells are harmed by this imbalance, which may have an effect on the entire body. When intracellular and extracellular environmental conditions change, the normal metabolism of cells produces ROS, which are essential for initiating signaling pathways in both plant and animal cells. The great majority of ROS produced by cells originates from the respiratory chain in the mitochondria. ROS, which include hydroxyl radical (OH•), hydrogen peroxide (H_2O_2), superoxide ($O_2^{\bullet -}$), and organic peroxides, are spontaneously formed by aerobic organisms when molecular oxygen is biologically reduced during endogenous metabolic activities (Reuter et al. 2010). Many acute and chronic clinical disorders are centered around inflammation, which is a fundamental biological process. Numerous cellular and humoral components are involved in intricate inflammatory networks that coordinate vital elements. Cancer, Diabetes, Osteoarthritis, PD, and HF are a few examples. Consequently, there has been a consistent increase in clinical research in the subject of inflammation, as seen by the 4,15,000 references that were found while searching Medline for the phrase "Inflammation," more than 2,15,000 of which are less than 10 years old. Any change in tissue integrity triggers inflammation, which is the body's response to restore tissue homeostasis through the activation of numerous healing processes. In order to avoid uncontrollably amplifying the initial incendiary and shifting the focus from tissue healing to non-target damage and the evolution of illness, these processes must be properly regulated (Lugrin et al. 2014).

4.2 Oxidative Stress: ROS as Key Participant

Usually, an unbalanced interaction between protective qualities of the antioxidant and the harmful effects of pro-oxidant stress is often employed to describe as oxidative stress. Pro-oxidants are substances that either produce ROS or suppress antioxidant systems in order to cause oxidative stress. Current research indicates that the lack of balance between antioxidants and pro-oxidants and resulting tissue damage are less significant factor in oxidative stress than the redox signaling disruption (Jones 2006; Puglia and Powell 1984). But according to the revised fundamental definition, a mismatch relation between oxidants and antioxidants known as oxidative stress can lead to molecular destruction as well as an interruption of oxidation-reduction reaction (Sies 2018). ROS is the collective term used to refer to oxidants, including free radicals and ions (Lushchak 2014).

ROS are frequently produced by the organism in limited amounts endogenously and are essential components involved in the control of cellular homeostasis mechanisms. They are essential for gene expression regulation, signal transduction, and the start of receptor activation (Kumar and Pandey 2015). NADPH oxidase (NOX), cyclooxygenase (COX), xanthine oxidase (XO), cytochrome P450 monooxygenase, uncoupled nitric oxide synthetase (e-NOS), peroxidases, myo-peroxidases (MPO), and lipo-oxygenase are examples of enzymatic sources of ROS. Whereas nonenzymatic sources entail free iron, electron transport chain (ETC) leakage, xenobiotics, etc. (Dan Dunn et al. 2015). Hypochlorous acid (HOCl), OH•, $O_2•^-$, and H_2O_2 are the primary sources of ROS. Molecular oxygen is reduced to produce $O_2•^-$ by NOX or XO, e-NOS, or the mitochondrial ETC (Mittal et al. 2014). The principal origin of ROS within the plasma membrane is attributed to the generation of $O_2•^-$ by membrane-bound NOX; this approach aids in the destruction of bacterial invaders (Rada and Leto 2008). Since they use more than 90% of the oxygen resources in a cell, mitochondria are thought to be the main redox-active domain (Lambert and Brand 2009). The prevalent ROS is $O_2•^-$, which undergoes conversion by mitochondrial superoxide dismutase (SOD) to H_2O_2. Subsequently, H_2O_2 further transforms into potent oxidants such as OH• (Younus 2018). Additionally, ROS can be produced by exposure to environmental elements like ultraviolet

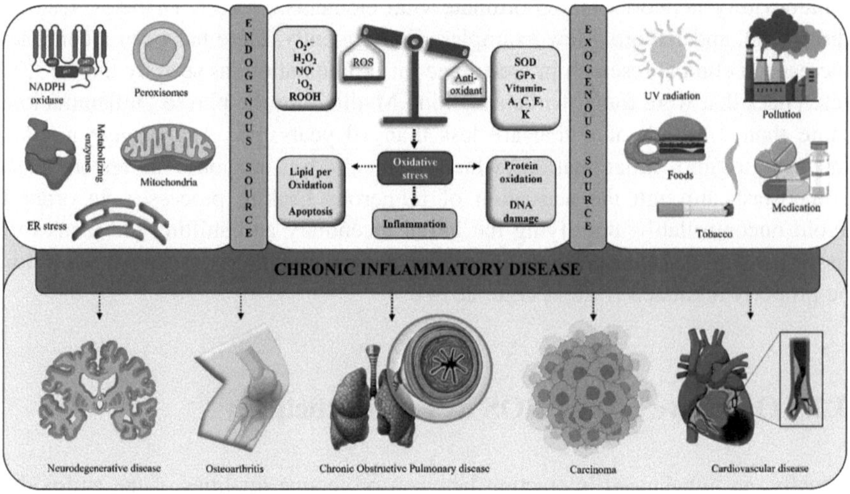

Fig. 4.1 Imbalance between ROS and antioxidants. Sources of ROS and contribution of oxidative stress in chronic inflammatory disease

radiation (UV), ionizing radiation (IR), biological creatures, hazardous substances, foods, medications, and narcotics like cigarette smoke and alcohol (Aranda-Rivera et al. 2022) (Fig. 4.1).

4.3 Inflammation: As a Defense Mechanism

The body utilizes inflammation as a defensive mechanism in response to undesirable stimuli. Inflammation functions as eradicating pathogens, promoting tissue healing, and restoring physiological equilibrium (Chen et al. 2018). It is described as a main immune system response to tissue damage brought on by a variety of harmful substances, including infections, irritants, or cellular damage, with the goal of restoring homeostasis (Ranneh et al. 2017). Inflammation, which is positioned as an adaptive response to preserving tissue integrity, is highlighted by the close relationship that exists between the onset of inflammation and the physiological homeostasis (Medzhitov 2010). Any host with a functioning innate immune system is triggered within minutes by an appropriate stimulant, such as a microbial infection, external invaders, or any irritant (internal or external) (Akira et al. 2006). When a pathogen is encountered by the innate immune system, it triggers an instantaneous response of inflammation, which leads to leukocyte migration, vascular leakage, and systemic vasodilation that coordinated set of actions of this response (Clark and Kupper 2005). The identification of microbial pathogens is a crucial first step for this response, and this identification is aided by pattern recognition receptors (PRRs) like toll-like receptors (TLRs) and nucleotide-binding oligomerization domain protein-like receptors (NODs). These PRRs are specifically bound to pathogen-associated molecular patterns (PAMPs) indicated by microorganisms. Furthermore, PRRs are able to identify endogenous danger-associated molecular patterns (DAMPs) generated from cells of the host that have been injured. Because of their dual recognition ability, PRRs may effectively respond to internal tissue damage as well as invasive pathogens (Beutler 2009; Janeway 1989). The expression of PRRs by a variety of immunological cells, such as neutrophils, macrophages, dendritic cells (DCs), and monocytes, allows for early identification of infections (Kumar et al. 2011). Additionally, an inflammatory multiprotein complex found in the cytosol, the "Inflammasome," cannot be assembled without nucleotide-binding and leucine-rich repeat containing protein (NLR). Typically, this complex consists of an effector protein called pro-caspase-1 and NLR acting as a sensor protein, including NLRP1, NLRP3, or NLRC4 (Davis et al. 2011). A caspase-1-dependent processing is required for the release of active pro-inflammatory cytokines include IL-18 and IL-1β. Caspase-1 also cleaves IL-33, another cytokine that stimulates T helper 2 (TH2) cell responses (Schmitz et al. 2005). TLRs and NLRs activation sets off signaling cascades that are essential for inflammation, including nuclear factor kappa β (NF-κβ). Apart from the NF-κβ pathway, there exist two more mechanisms that have the ability to both induce and intensify inflammation. These comprise the Janus kinase and signal transducer of activation pathway (JAK-STAT) and p38 mitogen-activated protein kinase pathway (p38 MAPK). When these pathways are activated collectively, a range of inflammatory mediators are produced, which include chemokines (CCL2 and CXCL8), cytokines like IL-1, tumor necrosis factor alpha (TNF-α), and IL-6, along with prostaglandins (Gaestel et al. 2009; Hu et al. 2021). Prostaglandins are an essential pro-inflammatory mediator that assists in blood flow regulation and vasodilation. They are produced by the COX enzyme from arachidonic acid (Ranneh et al. 2017). It is believed that two of the most significant inflammatory indicators in cells are COX-2 and NO. In which NO can produce important pro-inflammatory cytokines such as IL-1, TNF- α, and IL-8 and serves a crucial function in the pathophysiological processes underlying inflammation from a variety of lymphocytes and leukocytes (Corriveau et al. 1998). It is widely acknowledged that NF-κβ plays a crucial function in controlling genes that encode growth factors, adhesion molecules, chemokines, pro-inflammatory cytokines, and inducible enzymes. As a result, it has been suggested that NF-κβ has become a potentially effective promising target for the management of inflammation-associated disorders (Ghosh et al. 1998). Like NF-κβ, STATs (such as STAT-1, STAT-2, STAT-3, STAT-4, and STAT-6) are also involved in immunological and inflammatory processes and have been shown to have multiple roles in immunological and inflammation responses by means of coordinated action of anti- and pro-inflammatory cytokines (Kaplan 2013). Tumor necrosis factor receptor-associated factor (TRAF) can be stimulated by pro-inflammatory cytokines (TNF-α, interferon gamma (IFN-γ), interleukins, and ROS) by binding to corresponding receptors, through the p38-JNK MAPK cascade (Yang et al. 2014). The removal of pathogens and cell debris is usually a sign that the acute inflammatory response is coming to an end; this process is called inflammation resolution

(Buckley et al. 2013). A multitude of mechanisms work in concert to bring about resolution; these include the change in macrophage activation states, the process through which pro-inflammatory prostaglandins are changed into anti-inflammatory lipoxins that aid in resolution, the availability of glucocorticoids, and the actions of transforming growth factor beta (TGF-β), as well as IL-10 that reduces inflammation. These actions are also aided by purines, oxygenated lipids, neurotransmitters, nitrated lipids, and gaseous signals. T-lymphocytes, macrophages, and myeloid cells all produce TGF-β and IL-10, and these molecules efficiently inhibit pro-inflammatory signaling that is mediated by TLRs (Sugimoto et al. 2016).

4.4 Correlation of Oxidative Stress and Inflammation

Numerous advancements of inflammatory conditions are significantly influenced by generation of ROS, functioning as an agent or signaling molecule that promotes inflammation (Beckman 1996). The invading agents are eliminated by cells of phagocytic origin (such as macrophages and neutrophils), which generate significant quantities of ROS (OH•, H_2O_2, HOCl, NO, and $O_2{\bullet}^-$) and reactive nitrogen species during the inflammatory process. Localized oxidative stress and tissue damage can be triggered by reactive species that are produced excessively under pathological inflammatory conditions and some of which leave phagocytic cells by leaking (Halliwell et al. 1988). The generation of ROS that occurs during the oxidative burst is not caused by mitochondria, but rather by the controlled activation of NOX proteins, which are NADPH oxidases. The three major NOX isoforms in the bloodstream, NOX1, NOX2, and NOX4, have a strong association with vascular injury caused by inflammation (Burtenshaw et al. 2017). PAMPs containing microorganisms are identified by particular PRRs, growth regulators, or complement elements, which then activate the NOX2 complex (Jiang et al. 2011). Similarly, when a phagocyte is exposed to cytokines that promote inflammation, such as IL-1β, TNF-α, or IFN-γ, it triggers NOX-2 complex activation, which greatly enhances the levels of ROS within the cell (Singel and Segal 2016). The vital role that mitochondrial ROS plays in important molecular processes and signaling pathways occurs in inflammatory cells. NF- κβ and Activator protein 1 (AP-1), in the context of lysophosphatidylcholine (LPC), are two examples of these, which then drives to activation of endothelium (Forrester et al. 2018). Typically, NF-κβ is located in the cytoplasm of the cell in a phase of inactivity, the location where it is linked to cytosolic proteins like Iκβ kinase (IKK). ROS alters IKK by phosphorylating them with a serine kinase that is well known as NF-κβ kinase inhibitor. This modification enables deterioration of IKK, causing NF-κβ to be released. Subsequently, NF-κβ is able to move into the nucleus on its own or in coordination with different transcription factors. NF-κβ triggers expression of multiple genes that produce inflammatory proteins within the nucleus (Flohé et al. 1997). Additionally, mitochondrial ROS is essential for regulating the function of inflammasomes (Frank and Vince 2019).

The body's innate immune system is activated by an oligomeric molecular complex known as NLRP3 inflammasome, resulting in production of IL-18 and IL-1 β (Schroder and Tschopp 2010).

Thioredoxin-interacting protein breaks down in response to ROS during oxidative stress, which normally inhibits the endogenous antioxidant thioredoxin. Instead, it binds with NLRP3, leading to the NLRP3 activation (Zhang et al. 2022). Furthermore, it has been shown that DNA base modifications brought on by ROS cause inflammation (Aguilera-Aguirre et al. 2014). Overexposure to ROS causes necrosis in the cells, which in turn triggers intracellular high mobility group box protein 1 (HMGB1) to be passively released into the extracellular environment (Scaffidi et al. 2002). HMGB1 is a nuclear non-histone protein that is widely distributed and engages in DNA replication, transcription, and repair by binding to DNA and possesses the capacity to engage with diverse components such as DNA, lipoteichoic acid (LPT), lipopolysaccharides, and IL-1. These interactions lead to activation of responses that promote inflammation via different PRRs and TLRs, which sets off various processes, including immune tolerance, chemotaxis, and autophagy, along with inflammation (Andersson and Tracey 2011; Janko et al. 2014). HMGB1 is composed of two DNA-binding patterns known as box A and B, along with an acidic C-terminal tail. Cys-23 and Cys-45, two adjacent cysteines in box A and Cys-106, a single cysteine at position 106 in box B, have been identified. Depending upon particular redox scenarios, three distinct redox states are possible for HMGB1, including thiol HMGB1, disulfide HMGB1, and sulfonic HMGB1. The only form of HMGB1 that possesses the capability to attach to TLR4 and stimulate innate immune responses is disulfide HMGB1 (Tang et al. 2012). Consequently, inflammation is stimulated by the generation of cytokines, chemokines, growth regulators, and adhesion molecules via ROS (Fig. 4.2).

4.5 Contribution of Oxidative Stress in Chronic Inflammatory Diseases

4.5.1 Osteoarthritis

One important component in the onset of osteoarthritis is oxidative stress, which is caused by chondrocytes producing excessive amounts of ROS (Lepetsos and Papavassiliou 2016). In numerous cell types, NOXs produce O_2^- and H_2O_2, which are necessary for the proper signal transduction of growth factors and cytokines (Holmström and Finkel 2014). Further synthesis of ROS as a result of NOX4 initiation exacerbates the degenerative process of cartilage (Rousset et al. 2015). The production of COX-2, inducible nitric oxide synthase (iNos), COX-2, IL-6, and their by-products PGE2 and NO, as well as other pro-inflammatory phenotypic changes in chondrocytes, are all influenced by elevated ROS levels, which also activate other redox-sensitive transcription factors. Furthermore, an

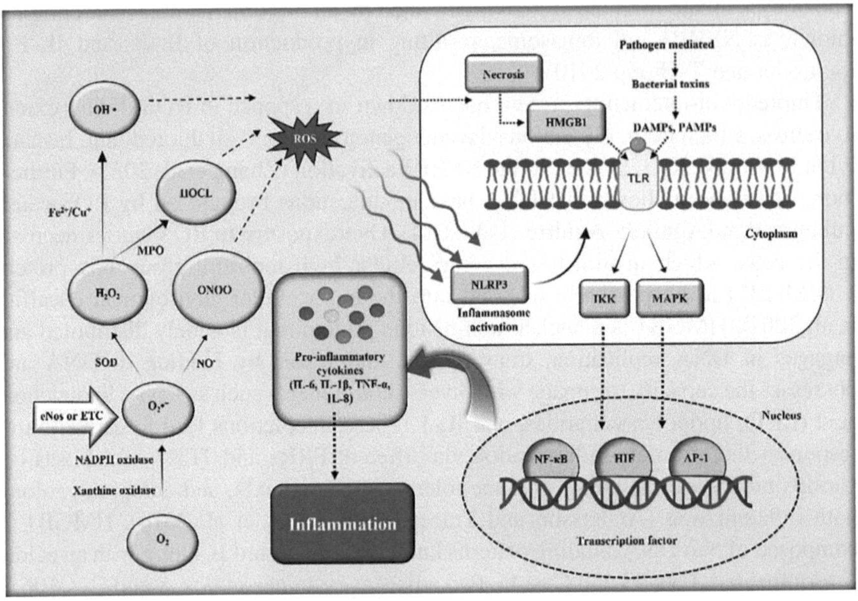

Fig. 4.2 Reactive oxygen species triggers the oxidative stress that plays an important role in molecular processes and signaling pathways in presence of inflammatory mediators like IL-6, IL-1β, TNF-α, and IL-8 and causes inflammation

elevated pro-oxidant load can inhibit the phosphatidyl inositol 3 kinase and also hinder the production of proteoglycan by triggering mitogen-activated protein kinase/extracellular signal-regulated kinase (MEK/ERK) (Yin et al. 2009). NO is generated in synovium as well as cartilage and inflammatory factors like TNF- α, IFN-γ, IL-17, and IL-1β are among those mediated by this NO. Additionally, it invokes chondrocyte apoptosis, constrains the biosynthesis of proteoglycan and collagen, and also causes pain (Suantawee et al. 2015). Moreover, the overproduction of NO by iNOS causes damage to the cartilage by raising the expression of matrix metalloproteinase (MMP), along with reducing the synthesis of proteoglycan and collagen (Lepetsos and Papavassiliou 2016).

4.5.2 Cancer

According to new findings, inflammation and oxidative stress are two interrelated cellular effects of a biological defensive system that may promote several carcinogenic processes, including angiogenesis, migration, metabolic reconfiguration, and escape from programmed cell death. A consistent pattern of oxidative stress and chronic inflammation leads to a significant cellular signaling response and subsequent effects in the progression of cancer (Vaidya et al. 2020). Aberrant redox homeostasis is seen in cancer cells, yet although ROS are pro-tumorigenic, excessive

ROS concentrations are lethal. In particular, high ROS production is associated with tumor cells' hyperproliferation; however, these cells are suited to flourish in environments where this oxidative load causes redox equilibrium to change from a reduced state. In order to do this, tumor cells boost their antioxidant defenses to maximize ROS-driven proliferation. At the same time, they avoid reaching ROS thresholds that would result in apoptosis, premature death, or ferroptosis. Since redox modification underlies tumor cells' ability to withstand elevated ROS levels, the study examined the biochemistry of thioredoxins, reduced glutathione, NADPH as well as the pathway that controls its production under typical physiological states throughout their initial development, metastatic phases of cancer, and the event of a recurrence following therapy. They also talk about how the late phases of cancer's evolution are heterogeneous, which may help to explain the seemingly odd impact of antioxidants on emergence and its metastatic spread for various cancer types (Hayes et al. 2020). Inflammation and infection make up around 25% of the variables that cause cancer. Tumors associated with inflammation are often characterized by mutagenic DNA lesions. Precancerous conditions and tumors are primarily caused by infection (e.g., Epstein-Barr virus-infected nasal congestion carcinoma, *Helicobacter pylori*-infected gastric cancer, human papilloma virus-related cervical cancer, and *Helicobacter pylori*-infected gastric malignancies) and also by pro-inflammatory factors. It's interesting to note that a number of their investigations revealed that aggressive clinical aspects of cancer arise as a result of DNA damage linked to inflammation in cancer stem-like cells. Inflammatory reactive oxygen/nitrogen radicals attack not just DNA but also macromolecules like proteins and lipids, causing malfunction. They used 2D Oxyblot and MALDI-TOF/TOF to find cancer tissues including proteins that have been oxidatively damaged. For instance, iron produced by oxidatively damaged transferrin has the potential to mediate Fenton reactions that create more energetic oxygen species. Oxidative stress may be increased as a result of an antioxidative protein malfunction. Such biomacromolecule damage may set off an endless cascade of oxidative stress, which in turn may cause cancer. Particularly in malignancies linked to inflammation, epigenetic changes, including methylation of DNA and microRNA dysregulation, are essential components of the carcinogenesis process. They discovered many tumor suppressor gene candidates that were differentially methylated. They found that nasopharyngeal cancer specimens obtained from tissue biopsy had increased RAS-like estrogen-regulated growth inhibitor (RERG) through the use of polymer chain reaction (PCR). The methods for cancer diagnosis could facilitate the development from this finding (Murata 2018).

4.5.3 Parkinson's Disease

Numerous studies have been carried out with encouraging outcomes to determine the role of ROS in the development of neurodegenerative diseases. Results have shown that ROS do not appear to play a causative role in many disorders because of oxidative damage and their interaction with mitochondria, which may exacerbate disease development. In contrast to other neurodegenerative diseases, PD ranks among the top two conditions that affect people over 60. According to statistics, PD is uncommon in people under 50; in those over 65, the prevalence is about 2%, and in those over 80, the prevalence is more than twice as high, or 5% (Teleanu et al. 2022). Based on data from ClinicalTrials.gov, a total of 16 studies have looked at the link between oxidative stress and PD. Out of the entire number of these trials, eight studies have been completed, seven are registered with unclear status, and one was dropped in 2020 as a result of the COVID-19 pandemic. Three studies have published their findings. To evaluate the tolerability of the iron chelator deferiprone, the first study (NCT01539837) looked at the ability to remove iron from different parts of the brain, as well as the potential to treat PD symptoms. The results showed that deferiprone medicine has short-term safety and reduced iron in specific brain areas, without having a negative impact on mood or cognitive function. However, in order to thoroughly evaluate the neuroprotective benefits of deferiprone, these studies should be followed by longer-term clinical trials in the future. On the other hand, the second research (NCT01470027) examined whether dietary N-acetylcysteine supplementation may lessen a glutathione deficiency in the live brain of PD patients. In contrast, the most recent trial (NCT01467960) assessed the efficacy of using human serum Apolipoprotein D (ApoD) as a measure of the oxidative stress and inflammatory vicious loop for the early detection of PD (Martin-Bastida et al. 2017).

4.5.4 Hypercholesterolemia-Induced Inflammation

In both Europe and the United States, hypercholesterolemia is the predominant risk factor linked to atherosclerosis. As a result, identifying the alterations in the structure and functionality of the major arteries brought about by long-term increases in blood cholesterol levels has received a lot of attention. Increased platelet-endothelial cell adhesion, increased rolling, adhering, and emigration leukocytes, and enhanced oxygen radical production by endothelial cells are among the inflammatory features observed in hypercholesterolemic microvessels. Animals with high cholesterol also have an overreaction in their microcirculation to inflammatory stimuli that are applied experimentally (Stokes et al. 2002). Although it despite receiving a lot of attention, the notion that inflammation plays a critical role in the pathophysiology of atherosclerosis and its effects is still not commonly used in therapeutic settings. The inflammatory cellular and molecular mechanisms that underpin atherosclerosis have been clarified by experimental research. Some important inflammatory processes have been brought to light by the realization that atherogenesis is a dynamic process, not a sickness, in which cholesterol or calcium is stored. For instance, mononuclear phagocytes are involved in every stage of this illness, demonstrating the connection between atherosclerosis and inflammation. From a clinical standpoint, using inflammation to target therapies, modify recommendations, and integrate into routine practice is now possible. There is unquestionable evidence that low-density lipoprotein (LDL) cholesterol contributes to the development of atherosclerosis. We possess very powerful techniques to reduce LDL and thereby events. Events still happen, though, even with a significant LDL decrease. Part of this remaining risk may be explained by inflammation. In patients receiving effective treatment with LDL-lowering medications, an anti-inflammatory strategy has recently been demonstrated to improve outcomes. Anti-inflammatory treatments are currently being pursued in this setting through a series of trials. The identification and management of residual inflammatory risk holds great potential for advancing preventive cardiology. Examining the potential reasons and therapeutic implications of the association between inflammation and atherogenesis is the aim of this concise study (Geovanini and Libby 2018).

4.5.5 Heart Failure

The pathophysiology of HF has been demonstrated to be significantly influenced by oxidative stress, which is defined as an excess of ROS generation in comparison to antioxidant defense. At lower concentrations, it modifies intracellular processes, such as redox signaling, slightly at greater concentrations; however, it results in cellular malfunction and damage. Numerous ROS comes from intracellular sources. While normal antioxidant enzyme activities are maintained, failing hearts produce more ROS within their mitochondria. Prolonged elevations in ROS synthesis within the mitochondria result in a disastrous cascade of damage to the mitochondrial DNA (mtDNA), functional deterioration, more ROS production, and cellular harm. By altering proteins essential to the excitation-contraction connection, ROS directly reduces contractile performance. Furthermore, ROS mediates apoptosis by activating a wide range of transcription factors and hypertrophic signaling kinases. Additionally, they promote the growth of cardiac fibroblasts and turn on matrix metalloproteinases, which modify the extracellular matrix. These molecular routes impact the onset, progression, and outcome of maladaptive remodeling and HF. Insulin resistance and exercise intolerance are associated with skeletal muscle dysfunction in HF, which is also a result of oxidative stress. As a result, oxidative stress affects both the skeletal muscle and the heart pathophysiology of HF. Enhanced comprehension of these systems could facilitate the creation of innovative and efficacious therapeutic approaches to address HF (Tsutsui et al. 2011). Numerous experimental investigations have revealed that HF is connected to an increased production of ROS. Most experimental research utilizing several types of HF animal models, including our own, was done on young animals without concomitant risk factors like hypertension. Nonetheless, they have continuously offered strong proof that oxidative stress is higher in HF and plays a role in the onset and advancement of the condition. As a result, despite the fact that animal models of HF only partially replicate the clinical characteristics observed in patients, they believe the oxidative stress level is higher in both HF patients and animal models (Belch et al. 1991). The two primary animal models of HF employed in investigations were postinfarct HF in mice and rapid pacing-induced HF in dogs. The anatomical, functional, and hemodynamic features of both animals are comparable to those found in HF patients. According to Belch et al., malondialdehyde and left ventricular (LV) ejection fraction had a statistically significant negative connection ($r = -0.35$). The main biochemical indicators of ROS formation, lipoperoxides, and 8-iso-prostaglandin F2α, were demonstrated by Mallat et al. to be elevated in HF patients' plasma and pericardial fluid and to positively correlate with the severity of the illness (Mallat et al. 1998).

4.5.6 Chronic Obstructive Pulmonary Disease

COPD is a highly prevalent respiratory disorder characterised by persistent structural and functional alterations of the airways, primarily caused by long-term inhalation of harmful particles and gases. Tobacco smoke induces bronchial irritation in COPD that has been shown to remain long after quitting (Hikichi et al. 2019). It is quite likely that the generation of ROS both intracellularly and extracellularly accounts for the age-related alterations in COPD. Normal aging requires appropriate ROS levels, while excessive ROS levels activate P53, a tumor suppressor protein that causes an end to cell cycle and senescence. COPD's accelerated aging is mostly caused by the PI3K/AKT/mTOR pathway. ROS triggers PI3K activation, which is then followed by AKT and mTOR activation and sirtuin-1 (SIRT1) suppression. Histone deacetylase SIRT1 is linked to oxidative stress, inflammation, and cellular senescence. It regulates many transcription factors. Senescent cells, in contrast to apoptotic cells, secrete a large number of inflammatory substances that might impact other tissues and cells. The term "senescence-associated secretory phenotype" (SASP) refers to these senescent cells. p21 triggers SASP, which in turn triggers Janus-activated kinase p38, which is a protein activated by mitogen kinase. NF-κB activation triggers the synthesis of chemokines, MMPs, growth factors (e.g., TGF-β1), and cytokines that cause inflammation. Moreover, SASP plays a part in NRLP3 inflammasome's activation, which results in low-grade, persistent airway inflammation. Additionally, research has shown that cellular senescence markers and SIRT1 expression are both elevated in COPD patient lung tissue samples and airway epithelial cells exposed to CS (Acosta et al. 2013). Numerous characteristics of COPD are induced in vivo by oxidant exposure in animals, and multiple mouse models of COPD exist. All of the distinguishing characteristics of the human disease were created in recent mice models of COPD caused by second-hand smoke exposure, and they developed in a short amount of time (8–12 weeks). They exhibit impaired lung function, abnormal gas exchange, pulmonary emphysema, mucus hypersecretion, airway remodeling, and persistent airway and lung inflammation. Additionally, they show systemic and lung oxidative stress as well as mitochondrial abnormalities. Similar to people, mice with experimental COPD exhibit higher prone to colonization and bacterial and viral infections that cause exacerbations. As a result, these animal models are useful for testing novel treatments and clarifying the pathogenic pathways, such as oxidative stress. Carbonyl levels and amounts of antibodies against carbonyl-modified self-protein (MDA and 4-HNE) are considerably raised in COPD and relate to disease severity, as determined by studies of clinical specimens and animal models. There have also been reports of elevated protein carbonyl levels in the leg muscles in individuals with experimental COPD. Enhanced carbonylation has been correlated to decreased tolerance to exercise and lower quality of life in these animals. Wistar rats were exposed to cigarette smoke (20% and 60%, 30 minutes, twice daily, 45 days) in order to create an experimental model of lung injury, and DNA damage was measured. Exposure to cigarette smoke raised 8-OHdG and lowered DNA repair enzyme expression. 8- Mut T homolog 1 and oxyguanine DNA glycosylase. Cellular senescence, programmed cell death, and mutagenesis are caused by persistent DNA damage resulting from exposure to cigarette smoke. To learn more about the mechanism underlying DNA damage, repair, and senescence brought on by smoke exposure, genetically engineered mice are employed. Using these mice has shown the involvement of the poly (ADP-ribose) polymerase-1 pathway and p21CDKN1A (p21), a senescence marker, in DNA repair pathways in mice contaminated with cigarette smoke. Wild-type mice were exposed twice a day for 3 days, which resulted in double-strand breaks, substantial increases in gH2AX protein levels, and a reduction in non-equivalent ends joining (NHEJ) proteins (Ku70, Ku80). But p21 deletion (p21−/-mice) led to a reduction in smoke-induced DNA damage and impairment in NHEJ repair. In the LPS and elastase emphysema models in Wistar rats, DNA repair genes are increased in lung tissue in response to DNA damage. When combined, diverse COPD experimental models triggered by distinct stressors can cause DNA damage that results in lung problems (Barnes 2022).

4.6 Biomarkers

Biological markers, also known as biomarkers, are indicators of health that can be reliably and precisely evaluated and are observed from outside the patient. A biomarker is any distinguishing trait that may be systematically evaluated and examined as evidence of healthy physiological approaches, pathological sequence, or drug-related reactions to a therapy interference (Aronson and Ferner 2017). The measured response may include functional, physiological, and biotechnological features on a microscopic level, as well as subatomic interactions. Indicator molecules, sometimes referred to as molecular markers or signature molecules, are identifiable substances that can be used to indicate the presence or progression of a biological process or condition. Medical indications that fall within a wide category can be accurately and consistently measured from the patient. Biomarkers that propose early warning indications allow for an additional timely diagnosis or determination of the intended outcome. Biomarkers are used as an observable measure of a biological entity. This refers to a state that can signal a symptom or stage of a disease, either moderately or undetectably, depending on the conditions

(Mayeux 2004). Biomarkers are categorized into two overarching classifications: (A) biological indicators of exposure are utilized in the assessment of potential risks, while (B) biomarkers of illness are employed in the identification, detection, and tracking of diseases. Therefore, biomarkers have the potential to be used as targets in clinical studies, allowing for early identification of diseases and helping in the alleviation and protection of maladies. Oxidative stress marker represents a complex field of research where interconnections are starting to emerge. According to Hans Selye Fig. 4.2, stress is defined as a nonspecific reaction response of the body to any demand imposed upon it. More precisely, oxidative stress is an imbalance in the generation of ROS, different physiological processes, and a biological system's ability to neutralize these reactive chemicals. This imbalance can lead to a dangerous state that contributes to cellular damage (Thakkar et al. 2021).

Bonetta et al. explained that the objective of this study was to examine DNA damage, the frequency of micronuclei, and alterations in the nuclear structure of buccal cells among workers who manufactured pigment-grade TiO2, with 15 exposed and 20 not exposed. Additionally, they evaluated potential correlations between genotoxicity biomarkers and reported respiratory symptoms, as well as connections between biomarkers and quantification of biomarkers for oxidative stress and inflammation in urine and exhaled breath condensate. Even when workers were exposed to TiO_2 at levels within the established Occupational Exposure Limits, the frequency of micronuclei formation and elevated oxidative DNA damage were observed. Exploring the link between oxidative stress and inflammation indicators in urine and EBC and genotoxicity measures has been established, indicating that exposure to TiO_2 can impact the oxidative balance. Early respiratory symptoms were reported by workers with higher levels of genotoxic/oxidative stress indicators, suggesting that cellular modifications may be predictive of prospective health anomalies. These findings highlight the importance of assessing preliminary markers of health decline monitoring systems and addressing workplace potential safety considerations for titanium dioxide handling situations (Bonetta et al. 2024).

According to recent research, lower levels of unconjugated bilirubin (UCB) are linked to the opposite impact of elevated amounts, suggesting that UCB may operate as an inhibitory host factor toward the onset of noncommunicable illnesses (NCDs). The findings of this European study, which employed high-performance liquid chromatography to measure the UCB concentration in 2489 samples along with additional data obtained according to the MARK-AGE research records for analysis, provide further evidence that higher UCB concentrations are associated with a lesser likelihood of contracting noncommunicable illnesses and may potentially serve as a prognostic indicator of cellular aging because those with higher The UCB concentrations had better markers of oxidative stress and metabolic health. These results emphasize the necessity of monitoring patients with mildly elevated bilirubin levels and frequently looking into UCB, especially in the context of aging, as this illness impacts millions globally but has not received enough attention in clinical research or treatment up to this time (Schoissengeier et al. 2024).

There is little information on how small Particulate debris (PM2.5) and related respiratory system deposits affect cardiovascular health in the short term. Thirty college students participated in a panel study conducted in Hefei, China, to investigate the possible mediating roles of inflammation and the relationships between PM2.8 and its depositions at three respiratory tract positions, which correlate with different lag lengths and a number of cardiovascular indicators. Researchers found that even at PM2.8 levels below the Chinese Atmospheric Indoor Air Quality Standard, exposure to PM2.5 had a substantial concentration-response effect on the changes in biomarkers at certain lag days. For instance, with lag 0–1d, an increase in ADAMTS13 of −9.33% was linked. When the amount of PM2.5 rose on lag 5–6d, the related increases were 30.51%, 11.43%, and 20.05%, respectively. Also, compared to PM2.8, the effects of tracheobronchial region (TB) deposition on these biomarkers were more pronounced. They also found that there were stronger correlations between PM2.8 signals and physical activity, aberrant body mass index (BMI), and females. The results demonstrated that, even at concentrations below CAAQS Grade I, short-term exposure to PM2.8 and associated deposit levels in the respiratory tract were concentration-dependent and linked to compromised cardiovascular health by causing oxidative stress, inflammation, and blood coagulation in healthy young adults. PM2.5's effects on coagulation function and its respiratory tract depositions may be partially attributed to elevated serum levels of systemic inflammation (Yuan et al. 2024).

Another study explained that inflammation and oxidative stress are key components of the cataract causation network. The present investigation employed the dietary inflammatory index to conduct a thorough analysis of the nutritional status in terms of inflammation and oxidative damage, and to explore any potential correlation with the risk of cataract development. They propose that diets that are pro-inflammatory and pro-oxidant are linked to relevant blood biomarkers and raise the risk of cataract development.

There were 263 cataract patients and 326 healthy controls in the study. A genuine 168-item food frequency questionnaire was used to measure the individuals' dietary intake during the past 12 months. Based on FFQ, DII and DAI were computed (Vahid and Rahmani 2024).

Ma et al. studied that Astaxanthin enhanced markers of oxidative stress, according to earlier in vitro and animal research. To test this prediction, they conducted a comprehensive systemic review and meta-analysis of prior randomized controlled trials. They anticipated that astaxanthin would have the same effects in humans. From January 1970 to April 2021, Cochrane Library, Scopus databases, and PubMed were searched for relevant literature. The primary requirements for eligibility consist of a minimum of one week of astaxanthin-based intervention, a placebo control group, and the measurement of at least one popular oxidative stress and inflammation biomarkers as well as both before and after the intervention. There were 12 randomized controlled studies with a total of 380 participants. Astaxanthin considerably decreased blood malondialdehyde concentration in comparison to the placebo. There were just a few studies on astaxanthin's effects on various oxidative stress indicators. In overweight participants, astaxanthin administration appeared to lower serum isoprostane levels and to increase the enzyme superoxide dismutase activity. In people with type 2 diabetes, astaxanthin dramatically lowered blood interleukin-6 levels. Astaxanthin had no discernible effects on the Levels of TNF-α or blood C-reactive protein. According to the results of the current study, astaxanthin supplementation may help reduce oxidative stress and many inflammatory markers, particularly in persons with type 2 diabetes. Future studies ought to look into how astaxanthin affects type 2 diabetes (Ma et al. 2022).

Sierra et al. studied that the tropical fruit known as the golden berry (*Physalis peruviana*), which is high in antioxidants, has been suggested as having the ability to regulate lipid profiles in individuals with high cholesterol. One independent risk factor for cardiometabolic disorders is dyslipidemia. In addition to being engaged in intestinal permeability, inflammation, and redox balance, the gut microbiome is intimately associated with cardiometabolic risk. The effects of golden berries on a few of these variables, such as the human gut microbiome, have never been studied, and there are no instruments available to track dietary consumption or check compliance with regard to nutritional interventions, including this fruit. Eighteen adult men took part in a pre-post quasi-experimental dietary intervention wherein they were fed golden berries (Dorada variety) for a duration of 3 weeks. They measured the plasma marker of oxidative stress, inflammatory process, and gut permeability, evaluated potential exposure to biomarkers using a non-targeted metabolomics technique, and investigated the impact of fruit consumption on the gut microbiota using feces. Initially, kaemferol and syringic acid were identified as potential biomarkers of golden berries. After 3 weeks, the fruit intervention caused physiological changes in the participants, which included a decrease, and Interestingly, eating golden berries modulated each participant's gut microbiota in an individualised manner rather than producing a uniform effect. At the conclusion of the intervention and 1 month later, the gut microbiota compositions of a particular person were statistically closer to their own baseline than to a matched sample from another person. In addition to probable benefits of consuming golden berries associated with lowering the risk of cardiometabolic disease, this intervention found potential indicators of golden berry intake. Golden berries have the potential to positively modulate redox balance; however, further controlled clinical trials are needed to confirm this impact (Sierra et al. 2022).

Another study states that saliva and urine are examples of noninvasive biosamples that show promise in measuring oxidative stress and inflammation biomarkers. Few studies have looked into how such biomarkers would react to brief exposure to PM2.5. In Shanghai, China, we performed a 4-times examination longitudinal research with 40 healthy, nonsmoking people. Samples taken by hand were used to measure PM2.5 exposure. Next, measurements were made for five biomarkers, alpha-1 antitrypsin (A1AT) in saliva, total antioxidant capacity, C-reactive protein, total antioxidant capacity in urine, and 8-iso-Prostaglandin F2α. They used linear mixed-effect models to determine the short-term impact of individual PM2.5 exposure on salivary and urine indicators that controlled for potential meteorological, sociodemographic, and biomarker detection confounders. Additionally, we looked into sensitivity exposure windows for various biomarkers. Strong correlations were seen between PM2.5 exposure and biomarker levels. Furthermore, salivary inflammatory markers elicited faster responses than urine oxidative stress markers. This study demonstrated the impact of PM2.8 exposure on responses to oxidative stress and systemic inflammation using noninvasive sampling methods. Periods of 6–12 hours and 24 hours or more following PM2 exposure were particularly significant. A sensitive time window of five exposures may be used to measure the reactions of urine biomarkers, such as 8-iso-PGF2α, and salivary biomarkers, such as CRP and TNF (Zhu et al. 2021).

4.7 Antioxidants: As a Counteract Mechanism

A molecule that can stop or reduce macromolecule oxidation is known as an antioxidant. Antioxidants work by eliminating free radicals or preventing other oxidation reactions in order to interrupt these chain reactions (Duarte and Lunec 2005). Antioxidants are defined as substances that, when present in small amounts compared to the substances that can be oxidized, effectively slow down or prevent the oxidation of those substances. However, this definition was subsequently revised to encompass any substance that retards, inhibits, or eliminates oxidative harm to a specific molecule, in order to encompass the molecules that restore the oxidative damage to the system (Halliwell and Gutteridge 1995). Antioxidants have the ability to either reduce the amount of molecular oxygen or lower its concentration in a specific area. They can also eliminate pro-oxidative metal ions and capture highly ROS such as O_2^- or H_2O_2. Additionally, antioxidants can neutralize chain-initiating radicals like peroxyl (ROO•), alkoxyl (RO•), or •OH, interrupting quenching singlet oxygen ($^1O_2$) or the sequence of radical reactions (Pisoschi and Pop 2015). Antioxidants can be categorized as endogenous or exogenous, as obtained from outside sources such as food or dietary supplements. Their primary role is to get rid of ROS or free radicals by preventing their formation, eliminating them or linking to metal ions that trigger ROS production (Halliwell 2013). The intricate internal antioxidant defense system of the cell maintains redox homeostasis. It includes endogenous enzymatic antioxidants like SOD, catalase (CAT), glutathione peroxidase (GPx) as well as endogenous nonenzymatic antioxidants like proteins (e.g., ferritin, albumin, transferrin, and ceruloplasmin), glutathione and low molecular weight scavengers (e.g., coenzyme Q and uric acid), among others (Poljsak et al. 2013). Exogenous antioxidants, also known as dietary antioxidants, are primarily acquired through the consumption of fruits, vegetables, food supplements and derived products. The human body is unable to synthesize these antioxidants (Pisoschi et al. 2021). Ascorbic acid (vitamin C), tocopherols (vitamin E), vitamin A (carotenoids and retinoids), and polyphenols such as flavonoids, resveratrol and N-acetylcysteine. (Poljsak et al. 2013).

The antioxidants and their functions as inhibiting oxidants are discussed in Table 4.1.

Q. Chu et al. studied about *Apios americana* leaf extract (ALE), which is rich in flavonoids, and used a lipopolysaccharide-induced RAW264.7 cell model to identify the anti-inflammatory properties of ALE. ALE alleviated oxidative stress and impairment of mitochondria also reduce inflammatory cytokines and NO release by blocking NF-κB and MAPK pathway (Chu et al. 2019).

Table 4.1 The antioxidants and their function as inhibiting oxidants

Sr. no.	Antioxidants	Example	Function as inhibiting oxidants	References
Endogenous antioxidants				
1	Enzymatic antioxidants	Superoxide dismutase	Converted $O_2 \bullet^-$ into H_2O_2	McCord and Fridovich (1969)
2		Catalase	Facilitated breakdown of H_2O_2 H_2O and O_2	Kobayashi et al. (2005)
3		Glutathione peroxidase	Facilitated conversion of both organic and inorganic H_2O_2 into H_2O and associated alcohols	Birben et al. (2012)
4	Nonenzymatic antioxidants	Glutathione	Removed harmful foreign substances by participating in enzymatic activities	Farhat et al. (2018)
5		Proteins	Inhibits the process of iron-dependent lipid peroxidation and converts O_2 into H_2O	(Sequeira et al. 2012)
6		Uric acid	Lipoic acid scavenges H_2O_2, •OH, $^1O_2$, and lipid peroxides	Stinefelt et al. (2005)
7		Coenzyme Q	It scavenges both $O_2 \bullet^-$ and perferryl radical (Fe^{3+}- $O_2 \bullet^-$)	Beyer (1992)
Exogenous antioxidants				
8	Vitamin A	Carotenoids	Scavenges singlet oxygen and can also trap peroxyl radicals	Burton (1989)
9	Polyphenols	Flavonoids	Antioxidant activity manifested against $O_2 \bullet^-$	Hernández-Rodríguez et al. (2019)
10		Resveratrol	Removed some RONS such as •OH, $O_2 \bullet^-$, H_2O_2, and ONOO–	Holthoff et al. (2010), Jia et al. (2008), Ungvari et al. (2007)
11		N-acetylcysteine	Considered to originate from its ability to stimulate GSH synthesis, and scavenges ROS/RNS	Bajt (2004)
12	Vitamins C		Scavenges HOCl, $O_2 \bullet^-$, •OH, O_2, and reduces H_2O_2 to H_2O via ascorbate peroxidase reaction	Kojo (2004)
13	Vitamins E		Scavenges lipoperoxyl radicals and prevent cell membranes from lipid peroxidation	Liebler (1993)

The phenolic stilbene molecule called resveratrol, which is present in red grapes, effectively blocked initiation of NLRP3 which triggered autophagy that is investigated by in vivo and in vitro demonstration of this compound (resveratrol). Additionally, it was found that it also reduced inflammation of the liver in obesity model of mice that was brought on by high fatty diet (Yang and Lim 2014).

4.8 Conclusion

The key findings and ongoing studies on importance of ROS, inflammation, and the recurring process of oxidative stress and inflammation reinforcing each other were covered in this chapter. A number of diseases are thought to have developed as a result of this relationship including chronic disease linked to inflammation. Inflammatory mediators are released as a direct result of pro-inflammatory signaling pathways being by ROS produced throughout oxidative stress. On the other hand, inflammatory reactions have ability to greatly increase ROS production, which contributes to oxidative damage. Finding treatment strategies that work requires thorough understanding of the cellular processes controlling that connection. This is the potential for showing the progression of disease by focusing on specific signalling pathways associated with oxidative stress and inflammation. Likewise the discovery of novel antioxidants along with anti-inflammatory substances with synergistic effects may lead to inventive therapeutic approaches. Future studies should focus on elucidating the specific molecular processes underlying this specific correlation and converting these insights into clinically significant uses. By figuring out the complex processes that underlie inflammation and oxidative stress, scientists may be able to help develop targeted therapeutic approaches for dealing with a range of human diseases.

References

Acosta JC, Banito A, Wuestefeld T, Georgilis A, Janich P, Morton JP, Athineos D, Kang T-W, Lasitschka F, Andrulis M, Pascual G, Morris KJ, Khan S, Jin H, Dharmalingam G, Snijders AP, Carroll T, Capper D, Pritchard C, Inman GJ, Longerich T, Sansom OJ, Benitah SA, Zender L, Gil J (2013) A complex secretory program orchestrated by the inflammasome controls paracrine senescence. Nat Cell Biol 15:978–990. https://doi.org/10.1038/ncb2784

Aguilera-Aguirre L, Bacsi A, Radak Z, Hazra TK, Mitra S, Sur S, Brasier AR, Ba X, Boldogh I (2014) Innate inflammation induced by the 8-Oxoguanine DNA Glycosylase-1–KRAS–NF-κB pathway. J Immunol 193:4643–4653. https://doi.org/10.4049/jimmunol.1401625

Akira S, Uematsu S, Takeuchi O (2006) Pathogen recognition and innate immunity. Cell 124:783–801. https://doi.org/10.1016/j.cell.2006.02.015

Andersson U, Tracey KJ (2011) HMGB1 is a therapeutic target for sterile inflammation and infection. Annu Rev Immunol 29:139–162. https://doi.org/10.1146/annurev-immunol-030409-101323

Aranda-Rivera AK, Cruz-Gregorio A, Arancibia-Hernández YL, Hernández-Cruz EY, Pedraza-Chaverri J (2022) RONS and oxidative stress: an overview of basic concepts. Oxygen 2:437–478. https://doi.org/10.3390/oxygen2040030

Aronson JK, Ferner RE (2017) Biomarkers—a general review. Curr Protoc Pharmacol 76. https://doi.org/10.1002/cpph.19

Bajt ML (2004) Acetaminophen-induced oxidant stress and cell injury in cultured mouse hepatocytes: protection by N-acetyl cysteine. Toxicol Sci 80:343–349. https://doi.org/10.1093/toxsci/kfh151

Barnes PJ (2022) Oxidative stress in chronic obstructive pulmonary disease. Antioxidants 11:965. https://doi.org/10.3390/antiox11050965

Beckman JS (1996) Oxidative damage and tyrosine nitration from peroxynitrite. Chem Res Toxicol 9:836–844. https://doi.org/10.1021/tx9501445

Belch JJ, Bridges AB, Scott N, Chopra M (1991) Oxygen free radicals and congestive heart failure. Heart 65:245–248. https://doi.org/10.1136/hrt.65.5.245

Beutler BA (2009) TLRs and innate immunity. Blood 113:1399–1407. https://doi.org/10.1182/blood-2008-07-019307

Beyer RE (1992) An analysis of the role of coenzyme Q in free radical generation and as an antioxidant. Biochem Cell Biol 70:390–403. https://doi.org/10.1139/o92-061

Birben E, Sahiner UM, Sackesen C, Erzurum S, Kalayci O (2012) Oxidative stress and antioxidant defense. World Allergy Organ J 5:9–19. https://doi.org/10.1097/WOX.0b013e3182439613

Bonetta S, Macrì M, Acito M, Villarini M, Moretti M, Bonetta S, Bosio D, Mariella G, Bellisario V, Bergamaschi E, Carraro E (2024) DNA damage in workers exposed to pigment grade titanium dioxide (TiO2) and association with biomarkers of oxidative stress and inflammation. Environ Toxicol Pharmacol 105:104328. https://doi.org/10.1016/j.etap.2023.104328

Buckley CD, Gilroy DW, Serhan CN, Stockinger B, Tak PP (2013) The resolution of inflammation. Nat Rev Immunol 13:59–66. https://doi.org/10.1038/nri3362

Burtenshaw D, Hakimjavadi R, Redmond E, Cahill P (2017) Nox, reactive oxygen species and regulation of vascular cell fate. Antioxidants 6:90. https://doi.org/10.3390/antiox6040090

Burton GW (1989) Antioxidant action of carotenoids. J Nutr 119:109–111. https://doi.org/10.1093/jn/119.1.109

Chen L, Deng H, Cui H, Fang J, Zuo Z, Deng J, Li Y, Wang X, Zhao L (2018) Inflammatory responses and inflammation-associated diseases in organs. Oncotarget 9:7204–7218. https://doi.org/10.18632/oncotarget.23208

Chu Q, Yu X, Jia R, Wang Y, Zhang Y, Zhang S, Liu Y, Li Y, Chen W, Ye X, Zheng X (2019) Flavonoids from *Apios americana* Medikus leaves protect RAW264.7 cells against inflammation via inhibition of MAPKs, Akt-mTOR pathways, and Nfr2 activation. Oxidative Med Cell Longev 2019:1–14. https://doi.org/10.1155/2019/1563024

Clark R, Kupper T (2005) Old meets new: the interaction between innate and adaptive immunity. J Invest Dermatol 125:629–637. https://doi.org/10.1111/j.0022-202X.2005.23856.x

Corriveau CC, Madara PJ, Van Dervort AL, Tropea MM, Wesley RA, Danner RL (1998) Effects of nitric oxide on chemotaxis and endotoxin-induced Interleukin-8 production in human neutrophils. J Infect Dis 177:116–126. https://doi.org/10.1086/513829

Dan Dunn J, Alvarez LA, Zhang X, Soldati T (2015) Reactive oxygen species and mitochondria: a nexus of cellular homeostasis. Redox Biol 6:472–485. https://doi.org/10.1016/j.redox.2015.09.005

Davis BK, Wen H, Ting JP-Y (2011) The inflammasome NLRs in immunity, inflammation, and associated diseases. Annu Rev Immunol 29:707–735. https://doi.org/10.1146/annurev-immunol-031210-101405

Duarte TL, Lunec J (2005) ReviewPart of the series: from dietary antioxidants to regulators in cellular signalling and gene ExpressionReview: when is an antioxidant not an antioxidant? A review of novel actions and reactions of vitamin C. Free Radic Res 39:671–686. https://doi.org/10.1080/10715760500104025

Farhat Z, Browne RW, Bonner MR, Tian L, Deng F, Swanson M, Mu L (2018) How do glutathione antioxidant enzymes and total antioxidant status respond to air pollution exposure? Environ Int 112:287–293. https://doi.org/10.1016/j.envint.2017.12.033

Flohé L, Brigelius-Flohé R, Saliou C, Traber MG, Packer L (1997) Redox regulation of NF-kappa B activation. Free Radic Biol Med 22:1115–1126. https://doi.org/10.1016/S0891-5849(96)00501-1

Forrester SJ, Kikuchi DS, Hernandes MS, Xu Q, Griendling KK (2018) Reactive oxygen species in metabolic and inflammatory signaling. Circ Res 122:877–902. https://doi.org/10.1161/CIRCRESAHA.117.311401

Frank D, Vince JE (2019) Pyroptosis versus necroptosis: similarities, differences, and crosstalk. Cell Death Differ 26:99–114. https://doi.org/10.1038/s41418-018-0212-6

Gaestel M, Kotlyarov A, Kracht M (2009) Targeting innate immunity protein kinase signalling in inflammation. Nat Rev Drug Discov 8:480–499. https://doi.org/10.1038/nrd2829
Gambini J, Stromsnes K (2022) Oxidative stress and inflammation: from mechanisms to therapeutic approaches. Biomedicine 10:753. https://doi.org/10.3390/biomedicines10040753
Geovanini GR, Libby P (2018) Atherosclerosis and inflammation: overview and updates. Clin Sci 132:1243–1252. https://doi.org/10.1042/CS20180306
Ghosh S, May MJ, Kopp EB (1998) NF-κB AND REL PROTEINS: evolutionarily conserved mediators of immune responses. Annu Rev Immunol 16:225–260. https://doi.org/10.1146/annurev.immunol.16.1.225
Halliwell B (2013) The antioxidant paradox: less paradoxical now? Br J Clin Pharmacol 75:637–644. https://doi.org/10.1111/j.1365-2125.2012.04272.x
Halliwell B, Gutteridge JMC (1995) The definition and measurement of antioxidants in biological systems. Free Radic Biol Med 18:125–126. https://doi.org/10.1016/0891-5849(95)91457-3
Halliwell B, Hoult JR, Blake DR (1988) Oxidants, inflammation, and anti-inflammatory drugs. FASEB J 2:2867–2873. https://doi.org/10.1096/fasebj.2.13.2844616
Hayes JD, Dinkova-Kostova AT, Tew KD (2020) Oxidative stress in cancer. Cancer Cell 38:167–197. https://doi.org/10.1016/j.ccell.2020.06.001
Hernández-Rodríguez P, Baquero LP, Larrota HR (2019) Flavonoids. In: Bioactive compounds. Elsevier, pp 265–288. https://doi.org/10.1016/B978-0-12-814774-0.00014-1
Hikichi M, Mizumura K, Maruoka S, Gon Y (2019) Pathogenesis of chronic obstructive pulmonary disease (COPD) induced by cigarette smoke. J Thorac Dis 11:S2129–S2140. https://doi.org/10.21037/jtd.2019.10.43
Holmström KM, Finkel T (2014) Cellular mechanisms and physiological consequences of redox-dependent signalling. Nat Rev Mol Cell Biol 15:411–421. https://doi.org/10.1038/nrm3801
Holthoff JH, Woodling KA, Doerge DR, Burns ST, Hinson JA, Mayeux PR (2010) Resveratrol, a dietary polyphenolic phytoalexin, is a functional scavenger of peroxynitrite. Biochem Pharmacol 80:1260–1265. https://doi.org/10.1016/j.bcp.2010.06.027
Hu X, li J, Fu M, Zhao X, Wang W (2021) The JAK/STAT signaling pathway: from bench to clinic. Signal Transduct Target Ther 6:402. https://doi.org/10.1038/s41392-021-00791-1
Janeway CA (1989) Approaching the asymptote? Evolution and revolution in immunology. Cold Spring Harb Symp Quant Biol 54:1–13. https://doi.org/10.1101/SQB.1989.054.01.003
Janko C, Filipović M, Munoz LE, Schorn C, Schett G, Ivanović-Burmazović I, Herrmann M (2014) Redox modulation of HMGB1-related signaling. Antioxid Redox Signal 20:1075–1085. https://doi.org/10.1089/ars.2013.5179
Jia Z, Zhu H, Misra BR, Mahaney JE, Li Y, Misra HP (2008) EPR studies on the superoxide-scavenging capacity of the nutraceutical resveratrol. Mol Cell Biochem 313:187–194. https://doi.org/10.1007/s11010-008-9756-y
Jiang F, Zhang Y, Dusting GJ (2011) NADPH oxidase-mediated redox signaling: roles in cellular stress response, stress tolerance, and tissue repair. Pharmacol Rev 63:218–242. https://doi.org/10.1124/pr.110.002980
Jones DP (2006) Redefining oxidative stress. Antioxid Redox Signal 8:1865–1879. https://doi.org/10.1089/ars.2006.8.1865
Kaplan MH (2013) STAT signaling in inflammation. JAK-STAT 2:e24198. https://doi.org/10.4161/jkst.24198
Kobayashi M, Sugiyama H, Wang D-H, Toda N, Maeshima Y, Yamasaki Y, Masuoka N, Yamada M, Kira S, Makino H (2005) Catalase deficiency renders remnant kidneys more susceptible to oxidant tissue injury and renal fibrosis in mice. Kidney Int 68:1018–1031. https://doi.org/10.1111/j.1523-1755.2005.00494.x
Kojo S (2004) Vitamin C: basic metabolism and its function as an index of oxidative stress. Curr Med Chem 11:1041–1064. https://doi.org/10.2174/0929867043455567
Kumar S, Pandey A (2015) Free radicals: health implications and their mitigation by herbals. Br J Med Med Res 7:438–457. https://doi.org/10.9734/BJMMR/2015/16284

Kumar H, Kawai T, Akira S (2011) Pathogen recognition by the innate immune system. Int Rev Immunol 30:16–34. https://doi.org/10.3109/08830185.2010.529976

Lambert AJ, Brand MD (2009) Reactive oxygen species production by mitochondria, pp 165–181. https://doi.org/10.1007/978-1-59745-521-3_11

Lepetsos P, Papavassiliou AG (2016) ROS/oxidative stress signaling in osteoarthritis. Biochim Biophys Acta (BBA) - Mol Basis Dis 1862:576–591. https://doi.org/10.1016/j.bbadis.2016.01.003

Liebler DC (1993) The role of metabolism in the antioxidant function of vitamin E. Crit Rev Toxicol 23:147–169. https://doi.org/10.3109/10408449309117115

Lugrin J, Rosenblatt-Velin N, Parapanov R, Liaudet L (2014) The role of oxidative stress during inflammatory processes. Biol Chem 395:203–230. https://doi.org/10.1515/hsz-2013-0241

Lushchak VI (2014) Free radicals, reactive oxygen species, oxidative stress and its classification. Chem Biol Interact 224:164–175. https://doi.org/10.1016/j.cbi.2014.10.016

Ma B, Lu J, Kang T, Zhu M, Xiong K, Wang J (2022) Astaxanthin supplementation mildly reduced oxidative stress and inflammation biomarkers: a systematic review and meta-analysis of randomized controlled trials. Nutr Res 99:40–50. https://doi.org/10.1016/j.nutres.2021.09.005

Mallat Z, Philip I, Lebret M, Chatel D, Maclouf J, Tedgui A (1998) Elevated levels of 8-iso-prostaglandin F $_{2\alpha}$ in pericardial fluid of patients with heart failure. Circulation 97: 1536–1539. https://doi.org/10.1161/01.CIR.97.16.1536

Martin-Bastida A, Ward RJ, Newbould R, Piccini P, Sharp D, Kabba C, Patel MC, Spino M, Connelly J, Tricta F, Crichton RR, Dexter DT (2017) Brain iron chelation by deferiprone in a phase 2 randomised double-blinded placebo controlled clinical trial in Parkinson's disease. Sci Rep 7:1398. https://doi.org/10.1038/s41598-017-01402-2

Mayeux R (2004) Biomarkers: potential uses and limitations. NeuroRx 1:182–188. https://doi.org/10.1602/neurorx.1.2.182

McCord JM, Fridovich I (1969) Superoxide Dismutase. J Biol Chem 244:6049–6055. https://doi.org/10.1016/S0021-9258(18)63504-5

Medzhitov R (2010) Inflammation 2010: new adventures of an old flame. Cell 140:771–776. https://doi.org/10.1016/j.cell.2010.03.006

Mittal M, Siddiqui MR, Tran K, Reddy SP, Malik AB (2014) Reactive oxygen species in inflammation and tissue injury. Antioxid Redox Signal 20:1126–1167. https://doi.org/10.1089/ars.2012.5149

Murata M (2018) Inflammation and cancer. Environ Health Prev Med 23:50. https://doi.org/10.1186/s12199-018-0740-1

Pisoschi AM, Pop A (2015) The role of antioxidants in the chemistry of oxidative stress: a review. Eur J Med Chem 97:55–74. https://doi.org/10.1016/j.ejmech.2015.04.040

Pisoschi AM, Pop A, Iordache F, Stanca L, Predoi G, Serban AI (2021) Oxidative stress mitigation by antioxidants - an overview on their chemistry and influences on health status. Eur J Med Chem 209:112891. https://doi.org/10.1016/j.ejmech.2020.112891

Poljsak B, Šuput D, Milisav I (2013) Achieving the balance between ROS and antioxidants: when to use the synthetic antioxidants. Oxidative Med Cell Longev 2013:1–11. https://doi.org/10.1155/2013/956792

Puglia CD, Powell SR (1984) Inhibition of cellular antioxidants: a possible mechanism of toxic cell injury. Environ Health Perspect 57:307–311. https://doi.org/10.1289/ehp.8457307

Rada B, Leto TL (2008) Oxidative innate immune defenses by Nox/Duox family NADPH oxidases. In: Trends in innate immunity. KARGER, Basel, pp 164–187. https://doi.org/10.1159/000136357

Ranneh Y, Ali F, Akim AM, Hamid HA, Khazaai H, Fadel A (2017) Crosstalk between reactive oxygen species and pro-inflammatory markers in developing various chronic diseases: a review. Appl Biol Chem 60:327–338. https://doi.org/10.1007/s13765-017-0285-9

Reuter S, Gupta SC, Chaturvedi MM, Aggarwal BB (2010) Oxidative stress, inflammation, and cancer: how are they linked? Free Radic Biol Med 49:1603–1616. https://doi.org/10.1016/j.freeradbiomed.2010.09.006

Rousset F, Hazane-Puch F, Pinosa C, Nguyen MVC, Grange L, Soldini A, Rubens-Duval B, Dupuy C, Morel F, Lardy B (2015) IL-1beta mediates MMP secretion and IL-1beta neosynthesis via upregulation of p22phox and NOX4 activity in human articular chondrocytes. Osteoarthr Cartil 23:1972–1980. https://doi.org/10.1016/j.joca.2015.02.167

Scaffidi P, Misteli T, Bianchi ME (2002) Release of chromatin protein HMGB1 by necrotic cells triggers inflammation. Nature 418:191–195. https://doi.org/10.1038/nature00858

Schmitz J, Owyang A, Oldham E, Song Y, Murphy E, McClanahan TK, Zurawski G, Moshrefi M, Qin J, Li X, Gorman DM, Bazan JF, Kastelein RA (2005) IL-33, an Interleukin-1-like cytokine that signals via the IL-1 receptor-related protein ST2 and induces T helper type 2-associated cytokines. Immunity 23:479–490. https://doi.org/10.1016/j.immuni.2005.09.015

Schoissengeier V, Maqboul L, Weber D, Grune T, Bürkle A, Moreno-Villaneuva M, Franceschi C, Capri M, Bernhard J, Toussaint O, Debacq-Chainiaux F, Weinberger B, Gonos ES, Sikora E, Dollé M, Jansen E, Slagboom PE, Hervonnen A, Hurme M, Breusing N, Frank J, Bulmer AC, Wagner K-H (2024) Association between bilirubin and biomarkers of metabolic health and oxidative stress in the MARK-AGE cohort. iScience 27:110234. https://doi.org/10.1016/j.isci.2024.110234

Schroder K, Tschopp J (2010) The Inflammasomes. Cell 140:821–832. https://doi.org/10.1016/j.cell.2010.01.040

Sequeira S, Rao AV, Rao A (2012) Increased oxidative stress and altered antioxidants status in patients with chronic allergic rhinitis. Adv Biosci Biotechnol 03:951–956. https://doi.org/10.4236/abb.2012.327117

Sierra JA, Escobar JS, Corrales-Agudelo V, Lara-Guzmán OJ, Velásquez-Mejía EP, Henao-Rojas JC, Caro-Quintero A, Vaillant F, Muñoz-Durango K (2022) Consumption of golden berries (Physalis peruviana L.) might reduce biomarkers of oxidative stress and alter gut permeability in men without changing inflammation status or the gut microbiota. Food Res Int 162:111949. https://doi.org/10.1016/j.foodres.2022.111949

Sies H (2018) On the history of oxidative stress: concept and some aspects of current development. Curr Opin Toxicol 7:122–126. https://doi.org/10.1016/j.cotox.2018.01.002

Singel KL, Segal BH (2016) NOX2-dependent regulation of inflammation. Clin Sci 130:479–490. https://doi.org/10.1042/CS20150660

Stinefelt B, Leonard SS, Blemings KP, Shi X, Klandorf H (2005) Free radical scavenging, DNA protection, and inhibition of lipid peroxidation mediated by uric acid. Ann Clin Lab Sci 35:37–45

Stokes KY, Cooper D, Tailor A, Granger DN (2002) Hypercholesterolemia promotes inflammation and microvascular dysfunction: role of nitric oxide and superoxide1. Free Radic Biol Med 33: 1026–1036. https://doi.org/10.1016/S0891-5849(02)01015-8

Suantawee T, Tantavisut S, Adisakwattana S, Tanpowpong T, Tanavalee A, Yuktanandana P, Anomasiri W, Deepaisarnsakul B, Honsawek S (2015) Upregulation of inducible nitric oxide synthase and nitrotyrosine expression in primary knee osteoarthritis. J Med Assoc Thail 98 (Suppl 1):S91–S97

Sugimoto MA, Sousa LP, Pinho V, Perretti M, Teixeira MM (2016) Resolution of inflammation: what controls its onset? Front Immunol 7. https://doi.org/10.3389/fimmu.2016.00160

Tang D, Billiar TR, Lotze MT (2012) A Janus tale of two active high mobility group box 1 (HMGB1) redox states. Mol Med 18:1360–1362. https://doi.org/10.2119/molmed.2012.00314

Teleanu DM, Niculescu A-G, Lungu II, Radu CI, Vladâcenco O, Roza E, Costăchescu B, Grumezescu AM, Teleanu RI (2022) An overview of oxidative stress, Neuroinflammation, and neurodegenerative diseases. Int J Mol Sci 23:5938. https://doi.org/10.3390/ijms23115938

Thakkar H, Eerla R, Gangakhedkar S, Shah RP (2021) Exploring unexplored biomarkers of oxidative distress and their use. Adv Redox Res 3:100020. https://doi.org/10.1016/j.arres.2021.100020

Tsutsui H, Kinugawa S, Matsushima S (2011) Oxidative stress and heart failure. Am J Phys Heart Circ Phys 301:H2181–H2190. https://doi.org/10.1152/ajpheart.00554.2011

Ungvari Z, Orosz Z, Rivera A, Labinskyy N, Xiangmin Z, Olson S, Podlutsky A, Csiszar A (2007) Resveratrol increases vascular oxidative stress resistance. Am J Phys Heart Circ Phys 292: H2417–H2424. https://doi.org/10.1152/ajpheart.01258.2006

Vahid F, Rahmani D (2024) Pro-inflammatory and pro-oxidant diets are associated with increased odds of cataracts and serum biomarkers of inflammation and oxidative stress: hospital-based case-control study. Human Nutr Metab 35:200234. https://doi.org/10.1016/j.hnm.2023.200234

Vaidya FU, Chhipa AS, Sagar N, Pathak C (2020) Oxidative stress and inflammation can fuel cancer. In: Role of oxidative stress in pathophysiology of diseases. Springer Singapore, Singapore, pp 229–258. https://doi.org/10.1007/978-981-15-1568-2_14

Yang SJ, Lim Y (2014) Resveratrol ameliorates hepatic metaflammation and inhibits NLRP3 inflammasome activation. Metabolism 63:693–701. https://doi.org/10.1016/j.metabol.2014.02.003

Yang Y, Kim SC, Yu T, Yi Y-S, Rhee MH, Sung G-H, Yoo BC, Cho JY (2014) Functional roles of p38 mitogen-activated protein kinase in macrophage-mediated inflammatory responses. Mediat Inflamm 2014:1–13. https://doi.org/10.1155/2014/352371

Yin W, Park J-I, Loeser RF (2009) Oxidative stress inhibits insulin-like growth factor-I induction of chondrocyte proteoglycan synthesis through differential regulation of phosphatidylinositol 3-Kinase-Akt and MEK-ERK MAPK signaling pathways. J Biol Chem 284:31972–31981. https://doi.org/10.1074/jbc.M109.056838

Younus H (2018) Therapeutic potentials of superoxide dismutase. Int J Health Sci (Qassim) 12:88–93

Yuan Z, Miao L, Fang M, Chen P, Yang L, Jiang C, Wang H, Xu D, Lin Z (2024) A panel study of exposure to fine particulate matter and modeled respiratory deposition on biomarkers of inflammation, blood coagulation, and oxidative stress in healthy young adults in Hefei, China. Atmos Environ 329:120535. https://doi.org/10.1016/j.atmosenv.2024.120535

Zhang Y, Xing C-J, Liu X, Li Y-H, Jia J, Feng J-G, Yang C-J, Chen Y, Zhou J (2022) Thioredoxin-interacting protein (TXNIP) knockdown protects against sepsis-induced brain injury and cognitive decline in mice by suppressing oxidative stress and Neuroinflammation. Oxidative Med Cell Longev 2022:1–23. https://doi.org/10.1155/2022/8645714

Zhu X, Chen C, Zhang B, Ge Y, Wang W, Cai J, Kan H (2021) Acute effects of personal exposure to fine particulate matter on salivary and urinary biomarkers of inflammation and oxidative stress in healthy adults. Chemosphere 272:129906. https://doi.org/10.1016/j.chemosphere.2021.129906

Chapter 5
Pathological Remuneration of Oxidative Stress in Limiting Gut-Brain Axis

Rohit Kumar, Akash Kashyap, and Saba Naqvi

Abbreviations

5-HT	5-hydroxytryptamine (serotonin)
ANS	Autonomic nervous system
ASD	Autism spectrum disorder
Aβ	Amyloid-beta
BACE-1	β-amyloid precursor protein–cleaving enzyme 1
CFU	Colony-forming units
CHD	Coronary heart disease
CMDs	Cardiometabolic disorders
CNS	Central nervous system
CORT	Cortisol
CRF	Corticotropin-releasing factor
CVD	Cardiovascular disease
DG	Dentate gyrus
DUOX2	Dual oxidase 2
FMOs	Flavin-containing monooxygenases
FMT	Fecal microbiota transplantation
GBA	Gut brain axis
HPA Axis	Hypothalamic–pituitary–adrenal axis
HS	Hemorrhagic stroke
IBD	Inflammatory bowel disease
KynA	Kynurenic acid
LPS	Lipopolysaccharides

R. Kumar · A. Kashyap · S. Naqvi (✉)
Department of Pharmacology and Toxicology, National Institute of Pharmaceutical Education and Research (NIPER), Raebareli, Lucknow, UP, India
e-mail: saba.naqvi@niperraebareli.edu.in

N. Srivastava, S. A. Saraf (eds.), *Pharmacological Strategies for Prevention and Treatment of Oxidative Stress*, Oxidative Stress in Applied Basic Research and Clinical Practice, https://doi.org/10.1007/978-3-032-19837-2_5

MAO	Monoamine oxidase
MGBA	Microbiota–gut–brain axis
MS	Multiple sclerosis
NFTs	Neurofibrillary tangles
NF-κB	Nuclear factor kappa-light-chain enhancer of activated B cells
NMDA	N-methyl-D-aspartate
NOXs	NADPH oxidases
PARP	Poly (ADP-ribose) polymerase
PINK1	PTEN-induced kinase 1
RNS	Reactive nitrogen species
ROS	Reactive oxygen species
SCFAs	Short-chain fatty acids
SN	Substantia nigra
TBIs	Traumatic brain injuries
TGF-β	Transforming growth factor beta
TMAO	Trimethylamine N-oxide
VN	Vagus nerve

5.1 Introduction

Oxidative stress is a biochemical state defined by an imbalance between the generation of reactive oxygen species (ROS) and the body's antioxidant defenses. This imbalance can cause considerable cellular damage, impacting proteins, lipids, and DNA, potentially resulting in various clinical diseases. The gut, as a vital element of the human body, is essential for sustaining metabolic homeostasis and overall health. It is perpetually subjected to many oxidants and possible carcinogens originating from food sources and microbial activities, rendering it especially vulnerable to oxidative stress. Recent studies have progressively examined the effects of oxidative stress on the gastrointestinal tract, highlighting its significance for gut barrier integrity and the onset of gastrointestinal disorders. The gut microbiota, a diverse assemblage of bacteria inhabiting the intestines, has become a pivotal factor in this process. It facilitates the fermentation of indigestible carbohydrates into advantageous metabolites, including short-chain fatty acids (SCFAs), and also affects the host's immunological response and oxidative stress levels. The detrimental effects of oxidative stress in the gastrointestinal tract can be significant. Chronic oxidative stress correlates with heightened intestinal permeability, commonly known as "leaky gut," facilitating the translocation of detrimental chemicals, such as endotoxins, into systemic circulation. This may initiate systemic inflammation and facilitate the onset of certain illnesses, such as inflammatory bowel disease (IBD), metabolic syndrome, and neurodegenerative disorders. Moreover, the imbalance of gut microbiota, known as dysbiosis, can intensify oxidative stress, resulting in a detrimental cycle of inflammation and tissue injury. The communication routes between the gut and other bodily systems are essential for comprehending the effects of oxidative stress on general health. The gut-brain axis is a bidirectional communication network that encompasses biochemical, neurological, and immunological processes, enabling signal transfer between the gut and the brain. This axis is impacted by the gut microbiome, which generates chemicals that influence brain function and behavior. The enteric nervous system (ENS) and the vagus nerve are crucial in transmitting information on gut health to the central nervous system (CNS). Disruptions in these communication networks may result in modified stress responses and facilitate the emergence of mental health issues, underscoring the interrelation between gut health and overall well-being. Comprehending the interaction between oxidative stress, gut health, and communication channels is crucial for clarifying the mechanisms that lead to these adverse results. This chapter seeks to examine the origins and consequences of oxidative stress in the gastrointestinal tract, the influence of gut microbiota on these effects, the communication routes implicated, and the possible approaches for alleviating oxidative stress to enhance gut health.

5.2 Oxidative Stress and the Gut

Oxidative stress refers to a biochemical condition characterized by an imbalance between oxidants and antioxidants, where there is an excess of oxidants. It can be neurotoxic, causing damage to biomolecules such as proteins, lipids, and DNA (Deoxyribonucleic acid). This damage can lead to a wide range of cellular malfunctions and may ultimately result in cell death. This imbalance leads to disturbances in redox signaling and control, and can cause molecular damage (Choi et al. 2016). Oxidative stress is inherently linked to aerobic metabolism, as nearly all chemical reactions involving molecular oxygen produce short-lived, highly unstable, and reactive intermediate products called ROS (Arnér and Holmgren 2000). ROS are produced internally, such as during mitochondrial oxidative phosphorylation, or they can arise from interactions with external sources like xenobiotic compounds (Jones et al. 2012). The mitochondrial complexes I (NADH-coenzyme Q oxidoreductase) and III (cytochrome c oxidoreductase) are the primary sites of ROS production in cells, but complex II (succinate-Q oxidoreductase) is also involved (Babior et al. 1973). Recent data indicate that stress signaling is essential for maintaining the gut's mucosal homeostasis. It has been demonstrated that the gut microbiota, a crucial element of the gastrointestinal environment, supports intestinal health when it encounters physiological stress. Meanwhile, these bacteria may contribute to the pathogenesis of IBD (inflammatory bowel disease) by promoting cell death and differentiation, collapse of the epithelial barrier, and inflammatory response through intricate interactions with the stress pathways in the host cells. Targeting the gut microbiota and cellular stress signaling has recently been suggested as novel treatments (Cao 2018).

5.2.1 *Sources of Oxidative Stress in the Gut*

Cellular sources of ROS include several key components: (I) Mitochondrial Electron Transport Chain: The primary source of ROS, generated during electron transfer reactions. (II) NADPH Oxidases (NOXs): Found on either side of the plasma membrane, these enzymes also contribute to ROS production. (III) Smooth Endoplasmic Reticulum: Contains cytochrome P-450 and P-450b5 families, which are involved in detoxifying fat-soluble drugs and harmful metabolites, and can produce ROS in the process. (IV) Peroxisomes: These organelles are significant producers of hydrogen peroxide (H_2O_2) and are involved in its dismutation to water (H_2O) and oxygen (O_2), as well as in the β-oxidation of fatty acids. (V) Cytoplasmic Enzymes: Free enzymes in the cytoplasm, such as xanthine oxidase, aldehyde oxidase, flavoprotein dehydrogenase, and tryptophan dioxygenase, can generate ROS during their catalytic activities (Vona et al. 2021) (Fig. 5.1). The gastrointestinal tract serves as a major source of ROS. Various cell types within the gut mucosa generate ROS as part of normal physiological functions. Nonetheless, the gut mucosa is vulnerable to oxidants that may induce pathological conditions. Redox signaling, predominantly governed by NADPH oxidases (NOXs), is essential for the physiological roles of the gastrointestinal epithelium. Furthermore, commensal bacteria aid in epithelial homeostasis through ROS produced by NOX1 and dual oxidase 2 (DOUX2). Molecules from gut microbiota, such as N-butyrate, are crucial for regulating mitochondrial oxidative stress, modulating inflammatory responses, controlling pathogen proliferation and adherence, and enhancing metabolism and energy expenditure during physical activity (Clark and Mach 2017). NADPH oxidase is a key contributor to the generation of ROS and reactive nitrogen species (RNS) in the gastrointestinal tract. This enzyme group consists of multiple membrane-bound multimeric NOX isoforms and DUOX complexes, each located in distinct parts of the gut. DUOX complexes are distributed throughout the whole intestine, while NOX1 is confined to the ileum, cecum, and colonic epithelium. NOX2 is predominantly expressed by professional phagocytes, whereas NOX4 is present in epithelial

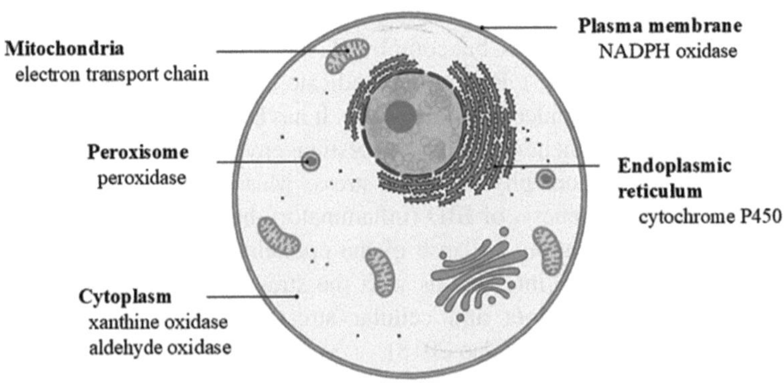

Fig. 5.1 Cellular source of ROS

cells, fibroblasts, and smooth muscle cells, with its expression enhanced by stimuli like transforming growth factor beta (TGF-β) and hypoxia (Mandal et al. 2010; Diebold et al. 2010). In the context of gastrointestinal tract pathologies, significant research has focused on mucosal changes related to oxidative stress. However, understanding of oxidative stress-induced muscle alterations is a more recent development. Most available literature comes from studies using animal models, and the role of oxidative stress in the development of muscular gastrointestinal diseases in humans remains underexplored (van der Vliet et al. 1989).

5.2.2 *Impact on Gut Barrier Integrity*

Bile, mucus, desquamated epithelial cells, various microbes and their fermentation products, undigested or unabsorbed food, and its metabolic products are all found in the human colon. It is thought that oxidants and carcinogens derived from bacteria and food present a continuous threat to the gut mucosa. In any case, repeated exposure Such difficult circumstances may result in unchecked production of free radicals, redox imbalance, and DNA damage, all of which might impact gut metabolic homeostasis and potentially lead to cancer (Bokoch and Gilman 1984). The gut mucosa is subjected to a variety of positive and moderating elements (such as exercise and a healthy diet) that might mitigate the negative consequences of harmful challenges, rather than being continuously exposed to detrimental challenges. Numerous endogenous and external factors have a range of detrimental, advantageous, and moderating effects on the antioxidant status, including food, aging, physical exercise, obesity, gut microbiota, inflammation, and many more. In this section, we particularly focus on the "gut microbiota" and how it relates to oxidative stress (Perše 2013). Variations in the variety and balance of the microbiota can result in physiological changes that are not exclusive to the gastrointestinal tract. Intestinal mucosa permeability regulates the mechanism by which the gut microbiota affects other body parts: a breakdown in this barrier function permits the gut microbiota and/or its endotoxins to pass through the intestinal epithelium and into systemic circulation, where they can reach various target organs. Additionally, this phenomena is referred to as "leaky gut" condition, and it allows the immune system and the entire body to be impacted by gut bacteria (Vajro et al. 2013).

5.3 Oxidative Stress Transmission from Gut to Brain

The gut-brain axis is a complex interplay that encompasses biochemical, neurological, and immunological processes that facilitate the transmission of stress from the gut to the brain. This relationship emphasizes the ways in which mental stress can impact digestive health and the ways in which gut health can impact mental health. An impaired intestinal barrier may allow a microbiota-induced pro-inflammatory

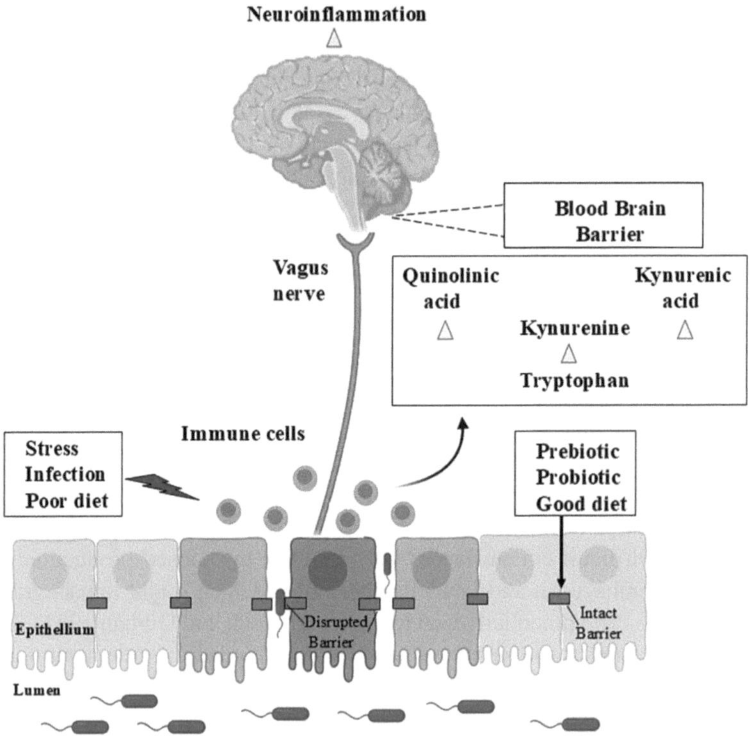

Fig. 5.2 Signaling pathways between intestinal microbes, the intestinal barrier, and the brain

condition with consequences for the brain (Fig. 5.2). The sequence of this process remains ambiguous. An elevation in intestinal permeability may precede mucosal inflammation, triggering the inflammatory response and resulting in a self-perpetuating cycle between inflammatory responses and barrier failure. This may subsequently sustain and intensify the low-grade inflammatory response. Conversely, systemic inflammation may enhance intestinal barrier permeability, facilitating the transfer of commensal microorganisms and exacerbating systemic inflammation. The origin of the low-grade inflammation associated with depression has not been attributed to a specific cause. Regardless of the order, these procedures could include the gut microbiota. Elevated IgA- and IgM-mediated immune responses targeting lipopolysaccharides (LPS) of specific commensal gram-negative gut bacteria have been seen in individuals with depression (Kelly et al. 2015). This system, which consists of immunological, hormonal, and neurological pathways, is essential for preserving homeostasis. One of the main factors affecting this axis is oxidative stress (OS), which is defined as an imbalance between the generation of ROS and antioxidant defenses. Because of its high oxygen consumption, the brain is especially prone to oxidative damage, which can result in neurodegenerative illnesses like Parkinson's and Alzheimer's (Dumitrescu et al. 2018).

5.3.1 Pathways of Communication

Clinically, the physiology and function of the brain and stomach often pertain to two separate systems, with several mechanisms available to enhance communication between them, including the immune system, the neurological system, and the neuroendocrine system.

5.3.1.1 Neural Pathway

The communication between the gut bacteria and the brain is primarily initiated by the ENS (Geng et al. 2022). Independent of the central nervous system, the ENS controls the physiology and function of the integument. It can also interact with the central nervous system (CNS) via the vagus nerve (VN), neuroendocrine pathways, and bacterial metabolites (Drokhlyansky et al. 2020). The ENS has a long-term bilateral connection with the microbiota since the neurological system and the microbiota have the most direct link. The microbiota depends on the ENS for growth, development, and the preservation of the ENS‘s structural and functional integrity. The ENS offers the microbiota a healthy living environment (Obata and Pachnis 2016). By activating 5-HT4 receptors and producing 5-hydroxytryptamine (5-HT), the gut microbiota can also aid in the maturation of the adult ENS. This serotonin, generated by gut bacteria, improves gastrointestinal transit speed and modifies the neuroanatomy of the ENS (Qiu et al. 2023). The link that joins the ENS and the CNS is called VN. The VN transmits the chemicals that the gut microbiota releases to the ENS to the CNS, where they are involved (Heiss and Olofsson 2019). The intestinal metabolism of glutamate usually sends gut luminal signals from VN to glutamatergic neurons in the dorsal hippocampus in milliseconds, enabling the recognition of gut stimuli with the temporal precision and topographical resolution of a synapse (Suarez et al. 2018). The neurons in the vagal ganglion can detect the mechanical distension of the gastrointestinal tract, innervate the intestinal villi, sense nutrients, and regulate intestinal motility in addition to acting as a bridge for signals to be transmitted from the ENS to the CNS. The regulation of eating habits and glucose homeostasis is significantly influenced by these vagal-mediated gastrointestinal hormone responses (Williams et al. 2016).

5.3.1.2 Immunological Pathway

Through interactions with gut immune mechanisms, the gut microbiota and brain neuronal function safeguard the ENS (Sharkey and Mawe 2023). Intestinal motility abnormalities and neuronal death from the ENS can be avoided by immune-related macrophages in the gut changing into protective phenotypes after gut infection. Through the modulation of associated immunological responses, ENS also influences the gut immunity's functionality. According to research, interleukin-18 (IL-18) signals from gut neurons can regulate the immunological barrier of the gut, which has a major impact on the mucosal barrier and the removal of invading microorganisms (Ahrends et al. 2021). Pathogenic microorganisms incorrectly colonize and translocate when the barrier is damaged, which results in neuroinflammation and the ensuing decline in cognitive performance (Travier et al. 2021). Researchers found that dysbiosis of the gut microbiota causes inflammation and increases the permeability of the irritable bowel (IB) and blood-brain barrier (BBB) in constipated mice. This leads to negative consequences in autoimmune encephalomyelitis by interfering with the production of helper and regulatory T lymphocytes in the neurological system (Zhao et al. 2023). Leaky gut syndrome (LGS) is caused by a variety of mechanisms, including disruptions to the gut microbiota in the GI tract, harm to enterocytes, and a weakening of the tight connections between enterocytes. Stress is also a significant factor in the pathophysiology of depression (Camilleri 2019). The development of this illness leads to the overactivation of the immune system due to the translocation of lipopolysaccharide-containing Gram-negative bacteria (Horn et al. 2022). Increases in pro-inflammatory cytokine concentration result from this stimulation, and an overabundance of these cytokines can be harmful to host cells, including CNS cells. One theory about the emergence of affective disorders is that the body becomes generally inflammatory due to the production of various cytokines, chemokines, and inflammatory mediators by immune cells (Jones et al. 2020).

5.3.1.3 Neuroendocrine Pathway

In neurogastroenterology, the connection between the brain and the gut is a key topic of study. The GI tract provides the brain with a steady stream of sensations and information, and the brain responds to target cells in the GI tract by initiating the proper integrated response (Mayer 2011). Information travels via gut hormones, immunological mediators, vagal and spinal afferent neurons, and microbiota-related signaling molecules from the gut to the brain, and by autonomic neurons and neuroendocrine components from the brain to the gut (Furness et al. 2013). Many concerns remain unanswered, such as how pathological situations disrupt the interaction between the central nervous system (CNS) and the peripheral nervous system. It is now well recognized that there is bidirectional cross-talk between the brain and the stomach. A common indication of malfunction in the brain-gut axis is seen in functional gastrointestinal diseases, such as both functional dyspepsia and irritable bowel syndrome (Mayer and Tillisch 2011). Because it influences immune system cells by regulating the secretion of cytokines that act on the HPA axis, cortisol is important for the endocrine mechanisms that regulate the gut-brain axis. It also has a major impact on the differentiation and function of the intestinal microbiota (Foster et al. 2021). One of the primary neuronal stimulants, glutamic acid (glutamate), is influenced by the intestinal microbiota in terms of both its amount and metabolism.

Glutaminase is involved in the synthesis of glutamic acid in glial cells generated from glutamine. The N-methyl-D-aspartate receptor (NMDA) is the most significant receptor in the glutamatergic system (Jewett and Thapa 2024).

5.3.1.4 The Hypothalamic-Pituitary-Adrenal (HPA) Axis

The neurological and endocrine systems of which the HPA axis plays a major role as well as the immune system are the main regulators of the cerebral-intestinal axis. The HPA axis primarily regulates the immune system by altering cytokine release (Gershon 1999). Between the brain and gut, the HPA axis serves as an endocrine communication channel that controls responses to stressors and stimuli (Keller et al. 2017; Bi et al. 2022). The neuropeptide corticotropin-releasing factor (CRF) is secreted by the paraventricular nucleus of the hypothalamus, serving as a traditional factor impacting the sympathetic and parasympathetic nervous systems, CNS, immunological, and metabolic reactions. It is discovered that the HPA axis's bidirectional communication between the gut and brain plays a vital part in preserving general health. Prior research revealed that the microbial-gut-brain route may involve the endocrine system, relationship during infancy, and the gut microbiota's makeup may impact the HPA axis's early development (Rosin et al. 2021; Ge et al. 2021). The neurological and endocrine systems of which the HPA axis plays a major role as well as the immune system are the main regulators of the cerebral-intestinal axis. The primary method by which the HPA axis is immune regulated is by altering release of cytokines. The identification of dysbiosis within the internal environment triggers the pituitary gland, which then triggers a cascade amplification effect. Adrenocorticotropic hormone is released in response to this, and this increases the synthesis of the adrenal glands' store of cortisol (CORT). This generates unfavorable feedback loop that includes the hippocampus, hypothalamus, and pituitary (Lightman et al. 2020). Regarding the gut-brain nexus, inflammation, or alterations in the stomach Stress can be released when the HPA axis is activated by bacteria hormones, like CORT. These hormones have an impact on several physiological bodily functions, particularly those connected to mood, thinking, and actions. The aberrant release of CRF is thought to provide the foundation of the atypical HPA axis. Its excessive activity may cause anxiety and depression (Burokas et al. 2017).

5.3.1.5 Kynurenine

Important physiologic roles for the kynurenine arm of the tryptophan metabolic pathway are played both in the peripheral and central nervous systems. In fact, it has been proposed that disruption of kynurenine metabolism underlies the emergence of gut-brain axis disorders ((Dehhaghi et al. 2019; Agus et al. 2018). Quinolinic acid and kynurenic acid (KynA) are two of the downstream by-products of kynurenine that are particularly significant from a pathological standpoint. Under typical circumstances, significant levels of quinolinic acid and KynA do not cross the BBB

(Schwarcz et al. 2012). Important physiologic roles for the kynurenine arm of the tryptophan metabolic pathway are played both in the peripheral and central nervous systems. In fact, it has been proposed that despite this, the BBB is permeable to kynurenine, which is mostly converted into KynA in astrocytes and then metabolized into quinolinic acid in microglia. Quinolinic acid is a neurotoxic pro-inflammatory chemical that functions as an agonist at the N-methyl-D-aspartate (NMDA) glutamate receptor (Lugo-Huitrón et al. 2013). Naturally occurring as a competitive antagonist, KynA exhibits a strong affinity for the glycine site connected to the NMDA receptor (Filpa et al. 2016). Neuroprotective in both the central nervous system and the extrasensory perception system, kynA shields neurons from glutamate-mediated excitotoxic damage (Filpa et al. 2015; Carpanese et al. 2014). The BBB's integrity and the maturation and activation states of astrocytes and microglia may be influenced by the microbiota.

5.3.2 *Brain Regions Impacted by Oxidative Stress*

The brain is the portion of the central nervous system that controls nearly every bodily function. The brain uses 20% of the total amount of glucose, oxygen, and fats that the body uses (Angelova et al. 2021), which is around ten times more than any other tissue. The likelihood of this organ creating free radicals, particularly ROS, is greatly increased by all of these conditions. In reaction to a physiological or pathological stimulus, various enzymes in brain cells, as well as the electron transport chain of mitochondria, continuously produce ROS. As a result, as part of the signaling mechanism, monoamine oxidase (MAO A and MAO B) uses catecholamines to create hydrogen peroxide (Vaarmann et al. 2010; Yamato et al. 2010). Biochemical studies reveal that certain neurons have varying degrees of susceptibility to oxidative stress. Research by Wang and Michaelis (2010) suggests that regions such as the hippocampus, amygdala, and cerebellar granule cells are particularly vulnerable to oxidative damage. Consequently, these areas are often among the first to experience a decline in their functionality. The hippocampal region of the brain is under significant metabolic alterations that ultimately impact neuronal connections and function. It is widely known that within the hippocampus, dentate gyrus cornu ammonis (CA3) system has structural flexibility and the ability to regenerate and remodel. Additionally, a number of studies have revealed that pyramidal cells of the dentate gyrus (DG) granule cells and CA3 are oxidative. Some have proposed that CA1 pyramidal cells are particularly vulnerable to oxidative damage in stress-prone locations (Salim 2017). There are differences in the distribution of ROS generation throughout the brain. The brain stem and cerebellar regions had the highest rate during resting conditions. Fascinatingly, the generation of ROS in glial cells was the primary factor influencing this outcome, whereas the generation of ROS by neurons in the same brain regions was not greater than in other brain regions. Production of ROS was not linked to the glia/neuron ratio, which is higher in certain areas due to the greater density of glial cells, and is smaller in the cerebellum than the cortex (Herculano-Houzel et al. 2006).

5.4 Pathological Outcome of Gut-Brain Axis Disruption

5.4.1 Neurodegenerative Diseases

The gut microbiota alters the gut-brain axis, providing new insights into the pathogenesis of numerous neurological disorders, which has been characterized as a paradigm shift in neuroscience. Psychological stress and inflammation are fundamental factors in the pathophysiology of disorders involving microbiota. Stress contributes to depression, schizophrenia, autism spectrum disorder (ASD), epilepsy, and migraine, while inflammation is implicated in depression, schizophrenia, ASD, Parkinson's disease, epilepsy, and migraine. Moreover, the aforementioned disorders frequently co-occur. Depression and ASD frequently co-occur with epilepsy. Depression frequently coexists with migraine. Furthermore, those afflicted with migraines exhibit an increased likelihood of gastrointestinal disorders (Socała et al. 2021).

5.4.1.1 Alzheimer Disease

Neurofibrillary tangles (NFTs) and amyloid-beta (Aβ) plaques are the two main indicators that contribute to the course of AD (Blennow and Zetterberg 2018). It has been suggested that the orbitofrontal, basal, and temporal neocortex regions of the human brain are where Aβ plaques first originate (Reiss et al. 2018; Mroczko et al. 2018). According to a recent study, aberrant cleavage of APP β- and γ-secretase results in the production of insoluble Aβ fibrils, which is the first step toward amyloid pathology (Correia et al. 2016; Tiwari et al. 2019). Subsequently, Aβ oligomerizes, diffuses into clefts in synapses, and tampers with synaptic transmission (Ihara and Washida 2018). It then polymerizes to form insoluble amyloid fibrils, which congregate to form plaques (Mroczko et al. 2018). Kinases are activated as a result of this polymerization (Reiss et al. 2018), which might hasten the hyperphosphorylation of tau protein linked with microtubules and its polymerization into insoluble NFTs (Crews and Masliah 2010). The synthesis of several neurotransmitters and neuromodulators, which impact gut-brain communication and brain function, can be influenced by gut bacteria (Burokas et al. 2017; Fjell et al. 2014). These produced neurotransmitters have the ability to pass through the intestinal mucosal layer and into the bloodstream (Mayer et al. 2014; Franceschi et al. 2019). The microbiota of elderly AD patients was shown to include fewer bacteria, which led to lower levels of butyrate (Rieder et al. 2017). This finding may have an impact on brain inflammation and the advancement of cognitive decline (Naveed et al. 2019). According to these results, the microbiota is thought to carry out a variety of essential tasks in human bodies, including the release of gases and metabolic by-products such as short-chain fatty acids (Rowland et al. 2018). The development of insulin resistance, which is linked to AD, can be brought on by changes in the gut microbiota that can activate pro-inflammatory cytokines and increase intestinal permeability (Ma et al. 2019). It's interesting to note that new research indicates AD development may begin in the stomach and subsequently migrate to the brain (Sun et al. 2020). Aβ1–42 oligomers were injected into the stomach wall of mice used in this investigation. It was noted that the amyloid moved from the colon to the brain during the course of a year. As a result, AD and neuroinflammation may be significantly influenced by the translocation of Aβ oligomers from the gut to the brain (Sun et al. 2020). A number of bacterial species, including *Salmonella enterica*, *Bacillus subtilis*, *Mycobacterium tuberculosis*, *Escherichia coli*, and *Staphylococcus aureus*, are capable of producing useful extracellular amyloid fibers (Cattaneo et al. 2017). By forming biofilms and forming strong bonds with one another, these amyloid proteins aid bacterial strains in fending off attacks from the immune system and the environment (Friedland and Chapman 2017). Although they differ from CNS amyloids in their main structure, bacterially produced amyloids exhibit similarities in their tertiary structure (Friedland 2015). When bacterial amyloid accumulates in the stomach, the immune system may become activated, potentially resulting in increased immunological responses and endogenous production of neuronal amyloid in the brain (Friedland 2015). Large amounts of lipopolysaccharides (LPSs) and amyloids can be excreted by bacteria that inhabit the gut microbiome. These substances may have a role in the pathophysiology of AD, particularly as people age and the GI tract epithelium and BBB become more permeable. According to certain theories, LPS and amyloids can either directly cross a damaged GI tract or blood-brain barrier, or they can do so indirectly by triggering cytokines or other minute pro-inflammatory molecules that are typically transited (Zhao and Lukiw 2015). A significant portion of the outer membrane of gram-negative bacteria contains LPS, often known as endotoxin. Furthermore, LPS is frequently utilized to cause inflammation in animals used in experiments. It has been noted that several inflammatory and degenerative characteristics of AD can be replicated by chronically infusing LPS into the fourth ventricle of rats (Hauss-Wegrzyniak et al. 2000).

5.4.1.2 Parkinson Disease

The gradual degradation of dopaminergic neurons in the substantia nigra (SN) pars compacta characterizes Parkinson's disease (PD), the second most common neurodegenerative condition after AD (Bekris et al. 2010). Its primary neuropathological feature is the presence of Lewy bodies, which are intracellular inclusions (LBs) and neurites (LNs) (Forno 1996), made up of aggregated and misfolded versions of the presynaptic protein α-synuclein (Spillantini et al. 1998). In fact, reports have indicated that mitochondrial DNA may reduce an organelle's qualitative capacity control systems and therefore make the original offense more severe, with a steady rise in metabolic dysfunction as well as oxidative stress (Murphy 2009). It has been discovered that aberrant ROS generation in the mitochondria of PTEN-induced kinase 1 (PINK1) mutant neurons inhibits the mitochondrial Na^{2+}/Ca^{2+} exchanger or glucose transporter, and antioxidants should stop it (Gandhi et al. 2009; Kostic et al. 2015). ROS from mitochondria is crucial not only in PINK1 pathology (mutation or deficiency) but also through the stimulation of PARKIN's mitochondrial recruitment in the physiology of PINK1/Parkin-related mitophagy (Xiao et al. 2017). Furthermore, it was recently stated that generation of ROS in mitochondria in both familial and sporadic Certain PD strains damaged DNA and triggered the poly ADP ribose polymerase (PARP) DNA repair process related with enzymes (Delgado-Camprubi et al. 2017). Furthermore, the formation of ROS was mediated by oligomeric α-synuclein (Fusco et al. 2017) without any involvement from the established enzymatic pathways that impacted mitochondrial function and triggered lipid peroxidation (Angelova et al. 2015). PD is characterized by an overall rise in end-product indicators of oxidative and nitrosative stress, which indicates severe damage to biomolecules, in addition to mitochondrial dysfunction and ROS generation. This clearly shows that PD and other neurodegenerative conditions are promoted by a cascade of events involving these three interrelated pathways (Floor and Wetzel 1998). Signs such as constipation, early vagus nerve involvement, and a close relationship between the stomach and the brain's dopamine-releasing system are everything in accordance with the theory that PD could come from the digestive system (Fitzgerald et al. 2019; Perez-Pardo et al. 2017). In a recent study, the brain tissue lysate of a PD was used to separate three different forms of α-syn (monomeric, aggregated, and oligomeric) and further introduced them into the mouse enteric nervous system (ENS) submucosal layer. This supported the hypothesis that α-synuclein travels from the ENS via the vagus nerve to the central nervous system. Furthermore, they observed that ⍰-synuclein was retrogradely transported from ENS to vagus nerve (VN) and brain with varied time, depending on type of alpha synuclein injected. This confirms α-synuclein spreads retrogradely from periphery to brain (Rietdijk et al. 2017). The bacteria also govern the proportional levels of neurotransmitters, such as glutamate, serotonin, γ-aminobutyric acid, norepinephrine, and dopamine, which are necessary for different manifestations of social behaviors (Elfil et al. 2020; Strandwitz et al. 2019; Levy et al. 2017). Recent in vivo studies demonstrate that gut microbiota (GM) alterations lead to α-synuclein aggregation in the ENS and brain inflammation-induced neuron loss (Chen et al. 2016) brain's intestinal inflammation-induced loss of neurons. GI infection increases intestinal epithelial permeability, allowing microorganisms and metabolites to enter circulation. This promotes α-synuclein accumulation and pro-inflammatory cytokine release. Pathological α-synuclein then spreads to the midbrain via ENS, while cytokines travel through vagus nerve (VN) or humoral routes to activate brain inflammation. These cytokines, combined with brain synucleinopathies, cause neuronal damage, death, and acute inflammatory reactions (Dogra et al. 2022).

5.4.1.3 Ischemic Stroke

Stroke, also known as hemorrhagic stroke (HS) or ischemic stroke (IS), is an acute cerebrovascular condition that can be brought on by either a sudden rupture of cerebral arteries or vascular blockage (Tuttolomondo 2020). The term "intestinal fora" refers to all microorganisms found in the human gastrointestinal system, mostly representing the phyla Bacteroidetes and Firmicutes, with between 15,000 and 36,000 species (Qin et al. 2010; Dinan and Cryan 2017; Mohajeri et al. 2018). Intestinal fora also include viruses and bacteriophages in addition to bacteria, Archaea, and eukaryotes (Breitbart et al. 2003). Numerous illnesses arise when the intestinal micro-ecosystems lose their equilibrium, and the central nervous system may also be affected (Sampson et al. 2016; Yang et al. 2020; Sharon et al. 2019). Furthermore, interactions between the autonomic and central nervous systems are possible through the enteric nervous system, sometimes referred to as the human "second brain" (Brunetti et al. 2019). A significant factor in the development of atherosclerotic plaques is intestinal fora. Atherosclerosis can be influenced by intestinal fora in three different ways:

I. The immune system is triggered by bacterial infections (Chistiakov et al. 2019).
II. The development of atherosclerotic plaque is influenced by intestinal fat and cholesterol metabolism (Vlacil et al. 2020).
III. By stimulating platelet activity, certain metabolites, such as trimethylamine N-oxide (TMAO), which is generated by intestinal fora, aid in the development of atherosclerotic plaque.

Intestinal microbiota influence atherosclerosis through the trimethylamine N-oxide (TMAO) pathway, the most direct mechanism (Wang et al. 2011). Trimethylamine (TMA), a gut microbial metabolite, is produced from dietary nutrients (choline, L-carnitine, phosphatidylcholine) by intestinal microbial TMA lyases (Hou et al. 2020). TMA is converted to TMAO by the host's hepatic flavin-containing monooxygenases (FMOs) (Fang et al. 2017). Risk factors for recurrent IS and cardiovascular events with elevated TMAO levels were strongly associated with the levels of pro-inflammatory monocytes (CD14++/CD16+) (Komaroff 2018). When TMAO stimulates platelets, there is a potential for thrombosis because of an increase in the release of Ca^{2+} from intracellular reserves and platelet-dependent hyperreactivity (Zhu et al. 2016). An increased risk of a first stroke was linked to raised trimethylamine-n-oxide levels, according to a clinical investigation conducted in the Chinese hypertensive population. Additionally, the greatest frequency of strokes was seen in individuals with low folic acid and high trimethylamine-*n*-oxide (Nie et al. 2018). Elevated levels of trimethylamine-*n*-oxide were linked to an increased risk of recurrent stroke in individuals following their initial stroke (Komaroff 2018).

5.4.1.4 Traumatic Brain Injury

Over 50 million traumatic brain injuries (TBIs) occur annually around the world, making TBIs a leading cause of mortality and disability among individuals of all ages (Maas et al. 2017). Stress-induced TBI affects the well-established autonomic nervous system (ANS) regulation of gastrointestinal function (Mayer 2000). After a TBI, systemic immune system activation may be a crucial factor in the development of GI dysfunction, and psychological factors may exacerbate the consequences of

TBI. Distress exacerbates GI symptoms. Brain-gut communication is crucial as sensory data from the gut lumen, which synchronizes activity across digestive tract regions (Carabotti et al. 2015). Since TBI has been demonstrated to cause harmful alterations in the GI tract, the gut has come to light as a key factor in the pathophysiology of the condition.

Through the gut-brain axis, these alterations eventually exacerbate brain inflammation and cognitive consequences (Hanscom et al. 2021). In addition to primary mechanical brain damage, TBI also causes secondary brain damage that occurs at the cellular and molecular level. These damages can result in metabolic abnormalities such as mitochondrial dysfunction, oxidative stress, inflammation, activation of microglia, and excitotoxicity, which can cause short-term or permanent cognitive impairments (Greve and Zink 2009). According to observations, gut dysbiosis also affects the pathophysiology of traumatic CNS damage, changing the permeability of the blood-brain barrier and activating microglia, which can result in severe repercussions (Hoban et al. 2018; Chu et al. 2019). A recent study revealed that mice given minor, repeated brain damage for 20 days developed mild, temporary intestinal dysbiosis, white matter damage, and cognitive decline (Angoa-Pérez et al. 2020). According to a new study, gut dysbiosis brought on by a wide range antibiotic before, during, and after TBI healing highlights the critical role that gut microbiota plays in this process and following a heightened loss of neurons due to TBI, inhibited neurogenesis, as well as changes to peripheral microglia and immunological reaction in addition to the alteration in fear memory reply (Celorrio et al. 2021).

5.4.1.5 Multiple Sclerosis

The CNS is affected by multiple sclerosis (MS), an autoimmune inflammatory disease that severely damages both gray and white matter. Pathogenesis is the process by which autoreactive immune cells enter the CNS across the BBB, which in turn sets up regional inflammatory reactions. This incendiary results in significant axonal degradation, demyelination, and the formation of characteristic lesions, all of which are common indicators of MS diagnosis (Dendrou et al. 2015). All stages of MS experience localized inflammation; however, it is most noticeable in the acute phase, during which peripheral-to-central nervous system autoreactive immune cells move. Th17 cells are a subset of these peripheral immune cells that are especially important in illness growth (Steinman 2001). Recent research suggests that dysbiosis of the gut, or an imbalance in the composition of the gut microbiota, may impact redox signaling and accelerate the development of neurological disorders through the gut-brain axis (Buga et al. 2023; Suganya and Koo 2020). This prompts questions about the potential role of the gut-brain axis in demyelination and neurodegeneration in MS. In the instance of MS, gut dysbiosis plays a key role in the illness's progression. One of the key factors affecting the connection between gut dysbiosis and MS is the production of ROS by the gut microbiota. Bacteria, both beneficial and harmful, alter their mitochondrial activity, which can either directly or indirectly create ROS by generating active metabolites like formyl peptides and short-chain fatty acids (SCFA) (Zorov et al. 2014). Pathogenic bacteria usually cause excessive ROS production, which can damage intracellular signaling pathways and exacerbate MS sufferers' inflammation. This inflammatory cascade, triggered by disease-causing pathogens, that may exacerbate MS progression (de Vos et al. 2022).

5.4.2 Mental Health Disorders

The pathogenesis of mental disorders involves a clear role for the gut microbiota in the biochemical signaling pathways between the gastrointestinal tract and the central nervous system. Disruptions in these communication pathways can contribute to the development of mental health issues. The gut microbiota may impact the host's mental health by modulating neurotransmitter levels (Liu et al. 2020).

5.4.2.1 Anxiety and Depression

The gut microbiome is now recognized as a crucial element of this cross-talk between the gut and the brain, and the name has been expanded to microbiome gut-brain axis (MGBA) in light of findings that the gut microbiome can impact symptoms of anxiety and depression. Additionally, stress modifies gut microbiota (Bailey et al. 2011; Bangsgaard Bendtsen et al. 2012; Galley et al. 2014), and the consequences of early Stress from life may persist throughout maturity on the microbiota (O'Mahony et al. 2009). The understanding of the gut microbiome's involvement in the synthesis of metabolic and endocrine products, as well as its interactions with the host immunological and neurological systems, has led to a surge in interest in recent decades. The assertion that gut problems are associated with and may be influenced by microbiota dates back to individuals with gut illnesses, such as irritable bowel syndrome and inflammatory bowel disease, which have a significant co-morbidity of anxiety and depressive disorders (Kurina et al. 2001; Addolorato et al. 1997; Masand et al. 1995). Stress-induced gastrointestinal damage and emotional behavior changes are primarily driven by inflammation, with gut microbiota which further modulating both systemic and local inflammatory responses. Increased levels of plasma inflammatory markers have been observed in anxiety and depression, alongwith interleukin-6 (IL-6) and tumor necrosis factor-α (TNF-α) (Vogelzangs et al. 2013) (Köhler et al. 2017; O'Brien et al. 2007). The dysregulation of the HPA axis in individuals suffering from depression is not a simple affair, as there is evidence of both elevated and decreased cortisol levels, along with other dysfunctions such as glucocorticoid resistance and a delayed return to baseline after acute stress. Anxiety and depression may arise as a result of a malfunctioning hypothalamic-pituitary-adrenal (HPA) axis and stress response (Arborelius et al. 1999). The HPA axis is activated by immune activity (Silverman et al. 2005; Turnbull and Rivier 1999; Silverman and Sternberg 2012), providing gut microbiota an indirect mechanism to modulate stress responses. Certain bacteria are known to create catecholamines, adrenaline, dopamine, and norepinephrine, among others. This is not likely to result in a significant rise in systemic catecholamine concentrations, since GF rats (Silverman and Sternberg 2012; Tsavkelova et al. 2006) generally been discovered to have higher HPA-axis indicators than rats with normal gut bacteria (Neufeld et al. 2011; Sudo et al. 2004). Understand the pathways that link psychological stress to brain and systemic inflammation to mood disorders, including anxiety and depression. Stress on the mind can make the intestines more permeable, which lets bacteria's lipopolysaccharides (sometimes referred to as endotoxins) enter the bloodstream and initiate an immune reaction. The central nervous system (CNS) may then become inflamed in many ways, which therefore promotes neurotoxicity and impacts mental health as well as impeding neurotransmitter (de Punder and Pruimboom 2015).

5.4.2.2 Autism Spectrum Disorder

Autism spectrum disorder (ASD), a group of heterogeneous neurodevelopmental disorders, is characterized by deficits in social communication and repetitive and stereotyped behaviors. Clinical evidence confirms gut microbiota dysbiosis in ASD patients, as their faecal samples show altered composition with elevated Firmicutes/ Bacteroidetes ratio (Vuong and Hsiao 2017). Compared to the general population, children with ASD have GI symptoms four times more frequently (Marler et al. 2017). These kids have a wide range of gastrointestinal symptoms, including gaseousness, constipation, diarrhea, bloating, reflux, vomiting, and food allergies (Fulceri et al. 2016; Iovene et al. 2017; Wakefield 2002). The increased prevalence of *Clostridia* species among autistic people indicates a potential role in the pathophysiology of ASD (Finegold et al. 2017). The brain's structural alterations and the less gregarious behaviors shown in germ-free mice suggest that the microbiota and the brain are functionally connected. The intensity of the ASD symptoms, and the mother Age is one of the six parameters influencing the gut microbiota's diversity. Reduced Bacteroides and Faecali bacterium and increased Clostridia levels has been observed in autistic children, whwereas, elevated Faecalibacterium levels were reported also reported in children suffering from severe autism (Ding et al. 2020). In ASD, increased intestinal permeability elevates circulating lipopolysaccharides (LPS) from gut bacteria, triggering systemic pro-inflammatory cytokine surges. Patients with ASD have been found to have elevated cytokines, especially in those mental regression subgroups. Normal neurodevelopment requires cytokines, and abnormalities in these molecules can have an influence on this procedure (Fobofou and Savidge 2022).

5.4.3 Potential for Systemic Effects

There exists a sophisticated mechanism for nutrition sensing and a combination of neuronal and hormonal signaling through the gut-brain axis and the heart. The gastrointestinal system and the brain interact closely to regulate energy balance and metabolism in response to food consumption. The central and enteric nervous systems evidently affect various facets of lipid and lipoprotein metabolism through bidirectional communication between the gut and the brain. Dysregulation of these pathways may predispose individuals to postprandial dyslipidemia and insulin resistance, hence increasing the risk of cardiometabolic disorders (CMDs). In diabetes dyslipidemia characterized by insulin resistance, the excessive generation of atherogenic chylomicron particles post-prandially in the intestine is significant, indicating that gut dysfunction, alongside liver impairment, also affects cardiovascular health (Singh et al. 2018).

5.4.3.1 Cardiovascular Health

The high death rate, rapid disease progression, and high incidence among the youth reflect the Indian picture for cardiovascular disease (CVD). The main causes of this endemic are urbanization, bad lifestyle choices, and diet. It has been demonstrated that the interaction between gut microbiota and stress causes an inflammatory response in the vascular cells, which results in a number of cardiovascular disorders and concomitant physiological reactions. Individuals suffering from cardiovascular diseases are recognized to be prone to depression and worry. Evaluations have demonstrated that the relationship between stress and the gut microbiota could emerge as a key element in identifying the burden on CVD (Foxx and Lowry 2020). Numerous studies have demonstrated the close relationship between depression and CVD. Furthermore, they share risk factors, including diabetes and obesity. Dysbiosis of the gut microbiota might be a possible biomarker for the onset of cardiovascular disease in T2D individuals (Tsai et al. 2021). CVD and depression share similar contributing factors variables such as the gut-brain axis, diabetes, inflammatory response, and damaging oxidation (Dhar and Barton 2016). Stress caused by oxidation has been shown to raise the possibility of depression in those with cardiovascular disease and for cardiac issues in individuals suffering from depression (Lin et al. 2019). The brain may encourage oxidative stress activity, which causes molecule damage and DNA damage peroxidation. Coronary heart disease (CHD), depression, and stroke share common risk factors and pathophysiological mechanisms. Abnormal interactions with the cardiovascular system and gastrointestinal disorders are largely caused by inflammation-mediated dysregulation. Immunomodulation and inflammatory pathway activation can be achieved in two main methods. Innate or adaptive immune system activation by direct stimulation of immune cells is one technique. The synthesis of bacterial metabolites such as Short Chain Fatty Acids (SCFAs) is the second way. The pathogenesis of CVDs, such as hypertension, atherosclerosis, heart failure, and myocardial infarction, is largely dependent on inflammation. Tissue damage caused by inflammation is the result of immune cell activation, accumulation, and inflammatory cytokine release during the inflammatory response. An increasing body of research indicates that gut microbes are crucial for the gut–heart axis's interaction with the cardiovascular system (Wang et al. 2023). Atherosclerosis is strongly correlated with gut dysbiosis (Rahman et al. 2022). Patients with atherosclerosis were also found to have high concentrations of the pro-inflammatory bacterium *Acidaminococcus* (Shen et al. 2021). Additionally, by altering other metabolite-driven pathways such as the TMAO and BA pathways, gut dysbiosis contributes to atherosclerosis. In the world, hypertension (HTN) is the most prevalent sign of CVDs. There is now clear evidence of a link between gut dysbiosis and HTN. Increased intestinal permeability results from this. Consequently, the more readily the contents of the stomach can permeate the permeable walls of the stomach. Increased bacterial translocation and dysbiosis occur due to this. Gut microbiota metabolites mediate blood pressure variations via host-microbiota interactions, exhibiting both immune-dependent and immune-independent mechanisms. Metabolites produced by gut microbes may be advantageous for SCFAs, for instance, arachidonic acid, indole-3-lactic acid, or harmful to control blood pressure (O'Donnell et al. 2023).

5.4.3.2 Metabolic Syndrome

The gut microbiota constitutes a complex physiological environment that affects host health. Changes in microbiome metabolism influence bodily homeostasis and have been linked to the onset of several neurodegenerative and neuropsychiatric illnesses in humans. There are several ways in which the microbiota can communicate with the host. The microbiota may create a variety of compounds by breaking down fibers and minerals, including (I) SCFA, of which succinate, propionate, and lactate are the most well-known; (II) indole and its derivatives; and (III) some neurohormones or mimetics, such as Gamma aminobutyric acid (GABA), caseinolytic peptidase B (ClpB) (Breton et al. 2016). SCFAs are frequently thought to act as a mediator between the favorable effects of the gut microbiota on human metabolism (Canfora et al. 2015). The fermentation of indigestible carbohydrates into the three most prevalent SCFAs, acetate, propionate, and butyrate, is largely dependent on the gut microbiota (Cummings 1981). The gut-brain axis is impacted by a number of variables, such as nutrition, lifestyle, anatomy (such as the impact of bariatric surgery), and genes, and in the last 10 years, the gut microbiota are acknowledged as significant participants in the control of these intestinal signals (Mayer et al. 2015). To understand the diverse mechanisms by which gut microbiota influences gut-brain communication and why this bidirectional axis requires fundamental understanding. Inflammation play a key role here and represents the body's natural response to stimuli such as infections, wounds, or toxins. On the other hand, elevated pro-inflammatory cytokine expression is also a characteristic of typical aging (Canfora et al. 2015). Such persistent activation is conceivable of pro-inflammatory cytokines to raise nitrosative and oxidative stress (Tansey et al. 2007) and raise the risk of neuropsychiatric conditions as depression and neurodegeneration (Tansey et al. 2007). Microbial metabolites are widely recognized as efficient immune cell activators. Increased intestinal permeability may contributes to hippocampal and cerebellar deficits, may result in Alzheimer's disease progression (Pistollato et al. 2016) (Akiyama et al. 2000).

5.5 Strategies for Mitigating Oxidative Stress in the Gut-Brain Axis

The connection between the gut and the brain is becoming more important as we learn about how they affect our overall health. When there is too much oxidative stress (damage caused by free radicals) in this system, it can lead to problems like depression, anxiety, brain diseases, and gut issues like IBD. Here are some simple ways to reduce oxidative stress and support both the gut and brain.

5.5.1 Nutritional Interventions

A balanced gut microbiome is crucial for good health, and it relies on a stable environment. This balance refers to the healthy interplay between oxidants and antioxidants in the body. When this balance is disrupted, often due to an imbalance in gut bacteria known as dysbiosis, it can lead to increased oxidative stress. There is growing interest in targeting the gut microbiota through diet and nutritional approaches (Tain and Hsu 2024).

5.5.1.1 Antioxidant-Rich Diets

Research on antioxidants and ROS has established a novel paradigm in human health, prioritizing disease prevention. The public is inundated every day with prevalent assertions regarding specific miraculous antioxidant vitamins or dietary supplements purported to combat free radicals, safeguard against diseases, and maintain perpetual youth. Medical professionals and nutritionists are frequently consulted for specific supplements, foods, or herbal remedies that should be incorporated into our diet to prevent diseases. The public appears to think that scientists hold the ultimate secret to eternal youth. Antioxidants may indirectly impede platelet activity by scavenging ROS, several of which modify platelet function. Although water-soluble and lipid-soluble antioxidants are situated in distinct subcellular sites, the antioxidant pathways in platelets are intricately interconnected. Antioxidants may indirectly impede platelet activity by modulating the metabolism of ROS, which significantly affect platelet function (Freedman 2008). Antioxidants are entities that counteract the effects of oxidants, or compounds that combat ROS to prevent the depletion of the body's natural resources. There are two classifications of antioxidants. (I) Natural Antioxidants: Living organisms have sophisticated internal defense mechanisms to protect cells from damage inflicted by free radicals. Catalase, glutathione peroxidase, and superoxide dismutase (SOD) are enzymatic antioxidants and reduced oxidative stress. They require micronutrients as cofactors for optimal outcomes, elements such as manganese, iron, copper, zinc, and selenium exhibit robust antioxidant defense mechanisms and catalytic activity (Halliwell and Gutteridge 1990; Halliwell 2001). (II) Exogenous Antioxidants (Nutritional Contribution): The dietary antioxidants that have undergone the most extensive research are vitamins C, E, and beta-carotene. Vitamin C is considered the most essential water-soluble antioxidant in extracellular fluids, as it may prevent ROS from oxidizing lipids in the aqueous phase before initiation. Vitamin E is a crucial lipid-soluble antioxidant, recognized as the most effective antioxidant within cell membranes, where it disrupts chain reactions to prevent lipid peroxidation from harming membrane fatty acids. Furthermore, beta carotene and other carotenoids provide antioxidant protection for lipid-rich tissues. Vegetables and fruits are superior sources of vitamin C and carotenoids, while whole grain cereals and high-quality vegetable oils are significant sources of vitamin E (Halliwell 1994).

5.5.1.2 Prebiotics and Probiotics

Modifying your diet to include more prebiotics and probiotics, which promote the growth of beneficial gut bacteria, may help alleviate symptoms of anxiety and depression while improving overall mood. This approach could be preferable to conventional medical treatments due to their potential negative side effects. The gut-brain axis involves complex pathways that utilize neuro-immune-endocrine signaling molecules to facilitate communication between the gastrointestinal system and the central nervous system (Taylor and Holscher 2020). Probiotics are described as "living bacteria that, when ingested in suitable concentrations, confer health advantages on the host." A minimum of 106–107 CFU/g of live probiotics is required for health benefits to be noticed, despite the absence of reliable information regarding minimum effective concentrations (Stavropoulou and Bezirtzoglou 2020). Probiotics are typically present in fermented dairy products, including yogurt, cultured buttermilk, and cheese. Other sources of bacterial fermentation comprise Japanese miso, tempeh, sauerkraut, beer, sourdough, bread, chocolate, kimchi, kefir, olives, and pickles. Probiotic strains appear to enhance the production of IgA by B cells, which aids in safeguarding the intestines by binding to harmful substances, thereby preventing their penetration of the gut lining and subsequent damage (Syngai et al. 2016).

5.5.2 Lifestyle Modifications

Lifestyle changes can help reduce oxidative stress in the gut-brain axis, improve overall health, and protect both gut and brain from damage. These small changes, when done consistently, can have a big impact on reducing inflammation, supporting cognitive function, and promoting better physical and mental well-being.

5.5.2.1 Stress Management Technique

Ayurveda emphasizes the significance of plants, herbs, spices, and meals in the treatment and prevention of diseases to sustain a healthy lifestyle (AlAli et al. 2021). Nutraceuticals comprise a diverse array of bioactive chemicals, dietary fibers, probiotics, and prebiotics present in food. Nutraceuticals exhibit antioxidant, antiaging, and anticancer capabilities, while also enhancing biochemical processes and structures (Zhou et al. 2020). Fruits, herbs, vegetables, and similar items include supplementary health advantages in addition to their inherent nutritional properties. Functional food encompasses cereal grains, legumes, lentils, fruits, dairy products, meat, nuts, and beans (Sikand et al. 2015). The inclusion of various active substances in functional foods enhances health and aids in disease prevention by supplying essential anti-inflammatory and intrinsic antioxidant qualities. Soybean includes saponins, which possess antioxidant qualities and improve enzymatic activity and immunological response (Singh et al. 2017). Herbs are the shortest type of plants that do not mature into a woody structure. Their leaves, roots, and stems are processed into a solution or desiccated for medicinal use. Numerous herbs utilized for their aromatic qualities in culinary applications possess antioxidant capabilities. Recombinant nutraceuticals in golden kiwi involve genes that promote the upregulation of ascorbic acid, lutein, and carotenoid synthesis, hence enhancing immunity. It also elevates the levels of vitamins, fiber, and potassium (Skinner et al. 2011). The human body possesses many antioxidant chemicals and enzymes that regulate the levels of reactive species. It comprises vitamin C, zinc, selenium, vitamin E, and enzymes such as glutathione peroxidases and catalases, whose principal function is to eliminate reactive species (AlAli et al. 2021).

5.5.2.2 Physical Activity

A single session of moderate-intensity exercise (<70% VO2max) offers numerous health benefits. One of the key advantages may be a reduction in intestinal permeability, which helps prevent pathogens from crossing the intestinal barrier and, in turn, reduces systemic inflammation. At the same time, moderate exercise also impacts the microbiota. These effects on the microbiota can be evaluated by assessing either its diversity or its functions. Intense exercise sessions (>70% VO2max) also offer health benefits. However, based on current studies, these sessions appear to cause more pronounced disruptions to the body's homeostasis compared to moderate-intensity exercise. In addition to exercise load, an individual's fitness level also plays a crucial role in shaping the microbiome. Current research suggests that fitness status is the more significant factor. Studies have shown that the microbiome of physically fit individuals tends to produce higher levels of butyrate, which lead to increased abundances of key butyrate-producing bacterial taxa within the Firmicutes phylum (Clauss et al. 2021).

5.5.3 *Pharmacological Approaches*

Pharmacological approaches for managing stress in the gut-brain axis typically focus on addressing the biochemical and physiological factors that contribute to stress and its impact on both the gut and brain. These approaches often involve medications or supplements that help modulate neurotransmitter systems and support gut health.

5.5.3.1 Antioxidant Supplementation

Antioxidant vitamins or supplements are said to combat free radicals, guard against illness, and prolong our lives. Physicians and nutritionists are frequently asked about specific foods and supplements. To prevent disease, a healthy diet combined with herbal remedies which provides essential vitamins (A, E, C) and minerals (selenium) from fresh fruits and vegetables, alongside phytonutrients naturally occurring plant compounds with potent antioxidant properties (Gupta and Prakash 2014). The gut-brain axis may provide new therapeutic opportunities as dietary polyphenolic chemicals alter the gut microbiota to produce therapeutically relevant polyphenolic compounds with increased blood-brain barrier permeabilities. As an illustration, the naturally existing polyphenolic curcumin effectively binds to amyloid-β, and it provides protection against AD. β-amyloid precursor protein–cleaving enzyme 1 (BACE-1) is inhibited by quercetin and similar flavonoids (Scalbert et al. 2005). Because they alter the nuclear factor-kappa β (NF-κB) signaling pathways, flavonoids and nonsteroidal anti-inflammatory drugs (NSAIDs) may be used as therapeutic targets for neurodegenerative illnesses, including AD and PD (Jha et al. 2019).

5.5.3.2 Targeted Therapies for Gut Microbiota Modulation

The gut microbiota is a modifiable target that can be changed and has the capacity to influence epigenetics; it may be possible to cure and reduce the symptoms of mental illnesses. Various prebiotics (diets rich in nondigestible fiber), probiotics (live bacteria), antibiotics, symbiotics (combinations of pre- and probiotics), postbiotics (products of bacterial fermentation such as short-chain fatty acids), and fecal microbiota transplantation (FMT) can be employed to alter the microbiota gut-brain-axis (MGBA). Since all of these methods are thought to enhance mental health by altering the microbiome, they might all be considered prospective psychobiotics. Probiotics are live microorganisms that have been shown to improve health; they are dosed in "colony-forming units" (CFU). Probiotics such as Lactobacillus and Bifidobacterium species have been shown in many studies to improve gut microbiota balance and support gastrointestinal health when administered in adequate amounts. In addition to bacterial probiotics, yeast strains such as Saccharomyces boulardii are also commonly used due to their beneficial effects in preventing and treating various gastrointestinal disorders. Psychobiotics were originally defined as probiotics that induce changes in mood, anxiety, and cognitive performance. The term "psychobiotics" encompasses any microbiota-targeted therapies, including probiotics and prebiotics, that affect the interaction between bacteria and the brain (Mörkl et al. 2020).

5.6 Conclusion

In summary, oxidative stress plays a pivotal role in the health of the gastrointestinal tract and has far-reaching implications for overall well-being. The imbalance between ROS and antioxidant defenses can lead to significant cellular damage, contributing to various pathological conditions, including inflammatory bowel disease, metabolic disorders, and neurodegenerative diseases. The gut microbiota emerges as a crucial mediator in this process, influencing both oxidative stress levels and the integrity of the gut barrier. Furthermore, the communication pathways between the gut and other systems, particularly the gut-brain axis, underscore the interconnectedness of gut health and mental well-being. Disruptions in these pathways can exacerbate the effects of oxidative stress, leading to a cycle of inflammation and disease. Addressing oxidative stress through dietary interventions, lifestyle modifications, and targeted therapies may offer promising strategies for enhancing gut health and preventing associated disorders. Continued research into the complex interactions between oxidative stress, gut microbiota, and communication pathways will be essential for developing effective interventions and improving health outcomes. Ultimately, fostering a balanced gut environment is key to promoting overall health and resilience against disease.

Acknowledgments The authors acknowledge the Department of Pharmaceuticals (DoP), Ministry of Chemicals and Fertilizers, Government of India, for their support. NIPER, Raebareli communication number for this manuscript is NIPER-R/Communication/862

References

Addolorato G, Capristo E, Stefanini GF, Gasbarrini G (1997) Inflammatory bowel disease: a study of the association between anxiety and depression, physical morbidity, and nutritional status. Scand J Gastroenterol 32(10):1013–1021

Agus A, Planchais J, Sokol H (2018) Gut microbiota regulation of tryptophan metabolism in health and disease. Cell Host Microbe 23(6):716–724

Ahrends T, Aydin B, Matheis F, Classon CH, Marchildon F, Furtado GC et al (2021) Enteric pathogens induce tissue tolerance and prevent neuronal loss from subsequent infections. Cell 184(23):5715–27.e12

Akiyama H, Barger S, Barnum S, Bradt B, Bauer J, Cole GM et al (2000) Inflammation and Alzheimer's disease. Neurobiol Aging 21(3):383–421

AlAli M, Alqubaisy M, Aljaafari MN, AlAli AO, Baqais L, Molouki A et al (2021) Nutraceuticals: transformation of conventional foods into health promoters/disease preventers and safety considerations. Molecules 26(9)

Angelova PR, Horrocks MH, Klenerman D, Gandhi S, Abramov AY, Shchepinov MS (2015) Lipid peroxidation is essential for α-synuclein-induced cell death. J Neurochem 133(4):582–589

Angelova PR, Esteras N, Abramov AY (2021) Mitochondria and lipid peroxidation in the mechanism of neurodegeneration: finding ways for prevention. Med Res Rev 41(2):770–784

Angoa-Pérez M, Zagorac B, Anneken JH, Briggs DI, Winters AD, Greenberg JM et al (2020) Repetitive, mild traumatic brain injury results in a progressive white matter pathology, cognitive deterioration, and a transient gut microbiota dysbiosis. Sci Rep 10(1):8949

Arborelius L, Owens MJ, Plotsky PM, Nemeroff CB (1999) The role of corticotropin-releasing factor in depression and anxiety disorders. J Endocrinol 160(1):1–12

Arnér ES, Holmgren A (2000) Physiological functions of thioredoxin and thioredoxin reductase. Eur J Biochem 267(20):6102–6109

Babior BM, Kipnes RS, Curnutte JT (1973) Biological defense mechanisms. The production by leukocytes of superoxide, a potential bactericidal agent. J Clin Invest 52(3):741–744

Bailey MT, Dowd SE, Galley JD, Hufnagle AR, Allen RG, Lyte M (2011) Exposure to a social stressor alters the structure of the intestinal microbiota: implications for stressor-induced immunomodulation. Brain Behav Immun 25(3):397–407

Bangsgaard Bendtsen KM, Krych L, Sørensen DB, Pang W, Nielsen DS, Josefsen K et al (2012) Gut microbiota composition is correlated to grid floor induced stress and behavior in the BALB/c mouse. PLoS One 7(10):e46231

Bekris LM, Mata IF, Zabetian CP (2010) The genetics of Parkinson disease. J Geriatr Psychiatry Neurol 23(4):228–242

Bi C, Guo S, Hu S, Chen J, Ye M, Liu Z (2022) The microbiota-gut-brain axis and its modulation in the therapy of depression: comparison of efficacy of conventional drugs and traditional Chinese medicine approaches. Pharmacol Res 183:106372

Blennow K, Zetterberg H (2018) Biomarkers for Alzheimer's disease: current status and prospects for the future. J Intern Med 284(6):643–663

Bokoch GM, Gilman AG (1984) Inhibition of receptor-mediated release of arachidonic acid by pertussis toxin. Cell 39(2 Pt 1):301–308

Breitbart M, Hewson I, Felts B, Mahaffy JM, Nulton J, Salamon P, Rohwer F (2003) Metagenomic analyses of an uncultured viral community from human feces. J Bacteriol 185(20):6220–6223

Breton J, Legrand R, Akkermann K, Järv A, Harro J, Déchelotte P, Fetissov SO (2016) Elevated plasma concentrations of bacterial ClpB protein in patients with eating disorders. Int J Eat Disord 49(8):805–808

Brunetti V, Vollono C, Testani E, Pilato F, Della MG (2019) Autonomic nervous system modifications during wakefulness and sleep in a cohort of patients with acute ischemic stroke. J Stroke Cerebrovasc Dis 28(6):1455–1462

Buga AM, Padureanu V, Riza AL, Oancea CN, Albu CV, Nica AD (2023) The gut-brain axis as a therapeutic target in multiple sclerosis. Cells 12(14)

Burokas A, Arboleya S, Moloney RD, Peterson VL, Murphy K, Clarke G et al (2017) Targeting the microbiota-gut-brain axis: prebiotics have anxiolytic and antidepressant-like effects and reverse the impact of chronic stress in mice. Biol Psychiatry 82(7):472–487

Camilleri M (2019) Leaky gut: mechanisms, measurement and clinical implications in humans. Gut 68(8):1516–1526

Canfora EE, Jocken JW, Blaak EE (2015) Short-chain fatty acids in control of body weight and insulin sensitivity. Nat Rev Endocrinol 11(10):577–591

Cao SS (2018) Cellular stress responses and gut microbiota in inflammatory bowel disease. Gastroenterol Res Pract 2018:7192646

Carabotti M, Scirocco A, Maselli MA, Severi C (2015) The gut-brain axis: interactions between enteric microbiota, central and enteric nervous systems. Ann Gastroenterol 28(2):203–209

Carpanese E, Moretto P, Filpa V, Marchet S, Moro E, Crema F et al (2014) Antagonism of ionotropic glutamate receptors attenuates chemical ischemia-induced injury in rat primary cultured myenteric ganglia. PLoS One 9(11):e113613
Cattaneo A, Cattane N, Galluzzi S, Provasi S, Lopizzo N, Festari C et al (2017) Association of brain amyloidosis with pro-inflammatory gut bacterial taxa and peripheral inflammation markers in cognitively impaired elderly. Neurobiol Aging 49:60–68
Celorrio M, Abellanas MA, Rhodes J, Goodwin V, Moritz J, Vadivelu S et al (2021) Gut microbial dysbiosis after traumatic brain injury modulates the immune response and impairs neurogenesis. Acta Neuropathol Commun 9(1):40
Chen SG, Stribinskis V, Rane MJ, Demuth DR, Gozal E, Roberts AM et al (2016) Exposure to the functional bacterial amyloid protein Curli enhances alpha-Synuclein aggregation in aged Fischer 344 rats and Caenorhabditis elegans. Sci Rep 6:34477
Chistiakov DA, Kashirskikh DA, Khotina VA, Grechko AV, Orekhov AN (2019) Immune-inflammatory responses in atherosclerosis: the role of myeloid cells. J Clin Med 8(11)
Choi S, Joo HK, Jeon BH (2016) Dynamic regulation of APE1/Ref-1 as a therapeutic target protein. Chonnam Med J 52(2):75–80
Chu C, Murdock MH, Jing D, Won TH, Chung H, Kressel AM et al (2019) The microbiota regulate neuronal function and fear extinction learning. Nature 574(7779):543–548
Clark A, Mach N (2017) The crosstalk between the gut microbiota and mitochondria during exercise. Front Physiol 8:319
Clauss M, Gérard P, Mosca A, Leclerc M (2021) Interplay between exercise and gut microbiome in the context of human health and performance. Front Nutr 8:637010
Correia SC, Perry G, Moreira PI (2016) Mitochondrial traffic jams in Alzheimer's disease - pinpointing the roadblocks. Biochim Biophys Acta 1862(10):1909–1917
Crews L, Masliah E (2010) Molecular mechanisms of neurodegeneration in Alzheimer's disease. Hum Mol Genet 19(R1):R12–R20
Cummings JH (1981) Short chain fatty acids in the human colon. Gut 22(9):763–779
Dehhaghi M, Kazemi Shariat Panahi H, Guillemin GJ (2019) Microorganisms, tryptophan metabolism, and kynurenine pathway: a complex interconnected loop influencing human health status. Int J Tryptophan Res 12:1178646919852996
Delgado-Camprubi M, Esteras N, Soutar MP, Plun-Favreau H, Abramov AY (2017) Deficiency of Parkinson's disease-related gene Fbxo7 is associated with impaired mitochondrial metabolism by PARP activation. Cell Death Differ 24(1):120–131
Dendrou CA, Fugger L, Friese MA (2015) Immunopathology of multiple sclerosis. Nat Rev Immunol 15(9):545–558
Dhar AK, Barton DA (2016) Depression and the link with cardiovascular disease. Front Psych 7:33
Diebold I, Flügel D, Becht S, Belaiba RS, Bonello S, Hess J et al (2010) The hypoxia-inducible factor-2alpha is stabilized by oxidative stress involving NOX4. Antioxid Redox Signal 13(4): 425–436
Dinan TG, Cryan JF (2017) The microbiome-gut-brain axis in health and disease. Gastroenterol Clin N Am 46(1):77–89
Ding X, Xu Y, Zhang X, Zhang L, Duan G, Song C et al (2020) Gut microbiota changes in patients with autism spectrum disorders. J Psychiatr Res 129:149–159
Dogra N, Mani RJ, Katare DP (2022) The gut-brain axis: two ways signaling in Parkinson's disease. Cell Mol Neurobiol 42(2):315–332
Drokhlyansky E, Smillie CS, Van Wittenberghe N, Ericsson M, Griffin GK, Eraslan G et al (2020) The human and mouse enteric nervous system at single-cell resolution. Cell 182(6):1606–22.e23
Dumitrescu L, Popescu-Olaru I, Cozma L, Tulbă D, Hinescu ME, Ceafalan LC et al (2018) Oxidative stress and the microbiota-gut-brain Axis. Oxid Med Cell Longev 2018:2406594
Elfil M, Kamel S, Kandil M, Koo BB, Schaefer SM (2020) Implications of the gut microbiome in Parkinson's disease. Mov Disord 35(6):921–933

Fang S, Wang M, Zheng Y, Zhou S, Ji G (2017) Acupuncture and lifestyle modification treatment for obesity: a meta-analysis. Am J Chin Med 45(2):239–254

Filpa V, Carpanese E, Marchet S, Prandoni V, Moro E, Lecchini S et al (2015) Interaction between NMDA glutamatergic and nitrergic enteric pathways during in vitro ischemia and reperfusion. Eur J Pharmacol 750:123–131

Filpa V, Moro E, Protasoni M, Crema F, Frigo G, Giaroni C (2016) Role of glutamatergic neurotransmission in the enteric nervous system and brain-gut axis in health and disease. Neuropharmacology 111:14–33

Finegold SM, Summanen PH, Downes J, Corbett K, Komoriya T (2017) Detection of Clostridium perfringens toxin genes in the gut microbiota of autistic children. Anaerobe 45:133–137

Fitzgerald E, Murphy S, Martinson HA (2019) Alpha-Synuclein pathology and the role of the microbiota in Parkinson's disease. Front Neurosci 13:369

Fjell AM, McEvoy L, Holland D, Dale AM, Walhovd KB (2014) What is normal in normal aging? Effects of aging, amyloid and Alzheimer's disease on the cerebral cortex and the hippocampus. Prog Neurobiol 117:20–40

Floor E, Wetzel MG (1998) Increased protein oxidation in human substantia nigra pars compacta in comparison with basal ganglia and prefrontal cortex measured with an improved dinitrophenylhydrazine assay. J Neurochem 70(1):268–275

Fobofou SA, Savidge T (2022) Microbial metabolites: cause or consequence in gastrointestinal disease? Am J Physiol Gastrointest Liver Physiol 322(6):G535–Gg52

Forno LS (1996) Neuropathology of Parkinson's disease. J Neuropathol Exp Neurol 55(3):259–272

Foster JA, Baker GB, Dursun SM (2021) The relationship between the gut microbiome-immune system-brain axis and major depressive disorder. Front Neurol 12:721126

Foxx CL, Lowry CA (2020) Chapter 11 - The microbiome-gut-brain axis: the missing link in depression. In: Chantler PD, Larkin KT (eds) Cardiovascular implications of stress and depression. Academic Press, pp 255–274

Franceschi F, Ojetti V, Candelli M, Covino M, Cardone S, Potenza A et al (2019) Microbes and Alzheimer' disease: lessons from H. pylori and GUT microbiota. Eur Rev Med Pharmacol Sci 23(1):426–430

Freedman JE (2008) Oxidative stress and platelets. Arterioscler Thromb Vasc Biol 28(3):s11–s16

Friedland RP (2015) Mechanisms of molecular mimicry involving the microbiota in neurodegeneration. J Alzheimer's Dis 45(2):349–362

Friedland RP, Chapman MR (2017) The role of microbial amyloid in neurodegeneration. PLoS Pathog 13(12):e1006654

Fulceri F, Morelli M, Santocchi E, Cena H, Del Bianco T, Narzisi A et al (2016) Gastrointestinal symptoms and behavioral problems in preschoolers with autism spectrum disorder. Dig Liver Dis 48(3):248–254

Furness JB, Rivera LR, Cho HJ, Bravo DM, Callaghan B (2013) The gut as a sensory organ. Nat Rev Gastroenterol Hepatol 10(12):729–740

Fusco G, Chen SW, Williamson PTF, Cascella R, Perni M, Jarvis JA et al (2017) Structural basis of membrane disruption and cellular toxicity by α-synuclein oligomers. Science 358(6369): 1440–1443

Galley JD, Yu Z, Kumar P, Dowd SE, Lyte M, Bailey MT (2014) The structures of the colonic mucosa-associated and luminal microbial communities are distinct and differentially affected by a prolonged murine stressor. Gut Microbes 5(6):748–760

Gandhi S, Wood-Kaczmar A, Yao Z, Plun-Favreau H, Deas E, Klupsch K et al (2009) PINK1-associated Parkinson's disease is caused by neuronal vulnerability to calcium-induced cell death. Mol Cell 33(5):627–638

Ge T, Yao X, Zhao H, Yang W, Zou X, Peng F et al (2021) Gut microbiota and neuropsychiatric disorders: implications for neuroendocrine-immune regulation. Pharmacol Res 173:105909

Geng ZH, Zhu Y, Li QL, Zhao C, Zhou PH (2022) Enteric nervous system: the bridge between the gut microbiota and neurological disorders. Front Aging Neurosci 14:810483

Gershon MD (1999) The second brain: a groundbreaking new understanding of nervous disorders of the stomach and intestine. (No Title)
Greve MW, Zink BJ (2009) Pathophysiology of traumatic brain injury. Mt Sinai J Med 76(2): 97–104
Gupta C, Prakash D (2014) Phytonutrients as therapeutic agents. J Complement Integr Med 11(3): 151–169
Halliwell B (1994) Free radicals, antioxidants, and human disease: curiosity, cause, or consequence? Lancet 344(8924):721–724
Halliwell B (2001) Role of free radicals in the neurodegenerative diseases: therapeutic implications for antioxidant treatment. Drugs Aging 18(9):685–716
Halliwell B, Gutteridge JM (1990) Role of free radicals and catalytic metal ions in human disease: an overview. Methods Enzymol 186:1–85
Hanscom M, Loane DJ, Shea-Donohue T (2021) Brain-gut axis dysfunction in the pathogenesis of traumatic brain injury. J Clin Invest 131(12)
Hauss-Wegrzyniak B, Vraniak PD, Wenk GL (2000) LPS-induced neuroinflammatory effects do not recover with time. Neuroreport 11(8):1759–1763
Heiss CN, Olofsson LE (2019) The role of the gut microbiota in development, function and disorders of the central nervous system and the enteric nervous system. J Neuroendocrinol 31(5):e12684
Herculano-Houzel S, Mota B, Lent R (2006) Cellular scaling rules for rodent brains. Proc Natl Acad Sci USA 103(32):12138–12143
Hoban AE, Stilling RM, Moloney G, Shanahan F, Dinan TG, Clarke G, Cryan JF (2018) The microbiome regulates amygdala-dependent fear recall. Mol Psychiatry 23(5):1134–1144
Horn J, Mayer DE, Chen S, Mayer EA (2022) Role of diet and its effects on the gut microbiome in the pathophysiology of mental disorders. Transl Psychiatry 12(1):164
Hou L, Zhang Y, Zheng D, Shi H, Zou C, Zhang H et al (2020) Increasing trimethylamine N-oxide levels as a predictor of early neurological deterioration in patients with acute ischemic stroke. Neurol Res 42(2):153–158
Ihara M, Washida K (2018) Linking atrial fibrillation with Alzheimer's disease: epidemiological, pathological, and mechanistic evidence. J Alzheimer's Dis 62(1):61–72
Iovene MR, Bombace F, Maresca R, Sapone A, Iardino P, Picardi A et al (2017) Intestinal dysbiosis and yeast isolation in stool of subjects with autism Spectrum disorders. Mycopathologia 182(3–4):349–363
Jewett BE, Thapa B (2024) Physiology, NMDA receptor. In: StatPearls. StatPearls Publishing Copyright © 2024, StatPearls Publishing LLC., Treasure Island
Jha NK, Jha SK, Kar R, Nand P, Swati K, Goswami VK (2019) Nuclear factor-kappa β as a therapeutic target for Alzheimer's disease. J Neurochem 150(2):113–137
Jones RM, Mercante JW, Neish AS (2012) Reactive oxygen production induced by the gut microbiota: pharmacotherapeutic implications. Curr Med Chem 19(10):1519–1529
Jones BDM, Daskalakis ZJ, Carvalho AF, Strawbridge R, Young AH, Mulsant BH, Husain MI (2020) Inflammation as a treatment target in mood disorders: review. BJPsych Open 6(4):e60
Keller J, Gomez R, Williams G, Lembke A, Lazzeroni L, Murphy GM Jr, Schatzberg AF (2017) HPA axis in major depression: cortisol, clinical symptomatology and genetic variation predict cognition. Mol Psychiatry 22(4):527–536
Kelly JR, Kennedy PJ, Cryan JF, Dinan TG, Clarke G, Hyland NP (2015) Breaking down the barriers: the gut microbiome, intestinal permeability and stress-related psychiatric disorders. Front Cell Neurosci 9:166028
Köhler CA, Freitas TH, Maes M, de Andrade NQ, Liu CS, Fernandes BS et al (2017) Peripheral cytokine and chemokine alterations in depression: a meta-analysis of 82 studies. Acta Psychiatr Scand 135(5):373–387
Komaroff AL (2018) The microbiome and risk for atherosclerosis. JAMA 319(23):2381–2382

Kostic M, Ludtmann MH, Bading H, Hershfinkel M, Steer E, Chu CT et al (2015) PKA phosphorylation of NCLX reverses mitochondrial calcium overload and depolarization, promoting survival of PINK1-deficient dopaminergic neurons. Cell Rep 13(2):376–386
Kurina LM, Goldacre MJ, Yeates D, Gill LE (2001) Depression and anxiety in people with inflammatory bowel disease. J Epidemiol Community Health 55(10):716–720
Levy M, Kolodziejczyk AA, Thaiss CA, Elinav E (2017) Dysbiosis and the immune system. Nat Rev Immunol 17(4):219–232
Lightman SL, Birnie MT, Conway-Campbell BL (2020) Dynamics of ACTH and cortisol secretion and implications for disease. Endocr Rev 41(3)
Lin D, Wang L, Yan S, Zhang Q, Zhang JH, Shao A (2019) The role of oxidative stress in common risk factors and mechanisms of cardio-cerebrovascular ischemia and depression. Oxidative Med Cell Longev 2019:2491927
Liu T, Pan X, Wang X, Feenstra KA, Huang Z (2020) Exploring the microbiota-gut-brain axis for mental disorders with knowledge graphs. J Artif Intell Med Sci 1
Lugo-Huitrón R, Ugalde Muñiz P, Pineda B, Pedraza-Chaverrí J, Ríos C, Pérez-de la Cruz V (2013) Quinolinic acid: an endogenous neurotoxin with multiple targets. Oxidative Med Cell Longev 2013:104024
Ma Q, Xing C, Long W, Wang HY, Liu Q, Wang RF (2019) Impact of microbiota on central nervous system and neurological diseases: the gut-brain axis. J Neuroinflammation 16(1):53
Maas AIR, Menon DK, Adelson PD, Andelic N, Bell MJ, Belli A et al (2017) Traumatic brain injury: integrated approaches to improve prevention, clinical care, and research. Lancet Neurol 16(12):987–1048
Mandal D, Fu P, Levine AD (2010) RETRACTED: REDOX regulation of IL-13 signaling in intestinal epithelial cells: usage of alternate pathways mediates distinct gene expression patterns. Cell Signal 22(10):1485–1494
Marler S, Ferguson BJ, Lee EB, Peters B, Williams KC, McDonnell E et al (2017) Association of rigid-compulsive behavior with functional constipation in autism spectrum disorder. J Autism Dev Disord 47(6):1673–1681
Masand PS, Kaplan DS, Gupta S, Bhandary AN, Nasra GS, Kline MD, Margo KL (1995) Major depression and irritable bowel syndrome: is there a relationship? J Clin Psychiatry 56(8): 363–367
Mayer EA (2000) The neurobiology of stress and gastrointestinal disease. Gut 47(6):861–869
Mayer EA (2011) Gut feelings: the emerging biology of gut-brain communication. Nat Rev Neurosci 12(8):453–466
Mayer EA, Tillisch K (2011) The brain-gut axis in abdominal pain syndromes. Annu Rev Med 62: 381–396
Mayer EA, Knight R, Mazmanian SK, Cryan JF, Tillisch K (2014) Gut microbes and the brain: paradigm shift in neuroscience. J Neurosci 34(46):15490–15496
Mayer EA, Tillisch K, Gupta A (2015) Gut/brain axis and the microbiota. J Clin Invest 125(3): 926–938
Mohajeri MH, La Fata G, Steinert RE, Weber P (2018) Relationship between the gut microbiome and brain function. Nutr Rev 76(7):481–496
Mörkl S, Butler MI, Holl A, Cryan JF, Dinan TG (2020) Probiotics and the microbiota-gut-brain axis: focus on psychiatry. Curr Nutr Rep 9(3):171–182
Mroczko B, Groblewska M, Litman-Zawadzka A, Kornhuber J, Lewczuk P (2018) Amyloid β oligomers (AβOs) in Alzheimer's disease. J Neural Transm (Vienna) 125(2):177–191
Murphy MP (2009) How mitochondria produce reactive oxygen species. Biochem J 417(1):1–13
Naveed M, Mubeen S, Khan A, Ibrahim S, Meer B (2019) Plasma biomarkers: potent screeners of Alzheimer's disease. Am J Alzheimers Dis Other Dement 34(5):290–301
Neufeld KM, Kang N, Bienenstock J, Foster JA (2011) Reduced anxiety-like behavior and central neurochemical change in germ-free mice. Neurogastroenterol Motil 23(3):255–264, e119

Nie J, Xie L, Zhao BX, Li Y, Qiu B, Zhu F et al (2018) Serum trimethylamine N-oxide concentration is positively associated with first stroke in hypertensive patients. Stroke 49(9): 2021–2028

O'Brien SM, Scully P, Fitzgerald P, Scott LV, Dinan TG (2007) Plasma cytokine profiles in depressed patients who fail to respond to selective serotonin reuptake inhibitor therapy. J Psychiatr Res 41(3–4):326–331

O'Donnell JA, Zheng T, Meric G, Marques FZ (2023) The gut microbiome and hypertension. Nat Rev Nephrol 19(3):153–167

O'Mahony SM, Marchesi JR, Scully P, Codling C, Ceolho AM, Quigley EM et al (2009) Early life stress alters behavior, immunity, and microbiota in rats: implications for irritable bowel syndrome and psychiatric illnesses. Biol Psychiatry 65(3):263–267

Obata Y, Pachnis V (2016) The effect of microbiota and the immune system on the development and organization of the enteric nervous system. Gastroenterology 151(5):836–844

Perez-Pardo P, Kliest T, Dodiya HB, Broersen LM, Garssen J, Keshavarzian A, Kraneveld AD (2017) The gut-brain axis in Parkinson's disease: possibilities for food-based therapies. Eur J Pharmacol 817:86–95

Perše M (2013) Oxidative stress in the pathogenesis of colorectal cancer: cause or consequence? Biomed Res Int 2013:725710

Pistollato F, Sumalla Cano S, Elio I, Masias Vergara M, Giampieri F, Battino M (2016) Role of gut microbiota and nutrients in amyloid formation and pathogenesis of Alzheimer disease. Nutr Rev 74(10):624–634

de Punder K, Pruimboom L (2015) Stress induces endotoxemia and low-grade inflammation by increasing barrier permeability. Front Immunol 6:223

Qin J, Li R, Raes J, Arumugam M, Burgdorf KS, Manichanh C et al (2010) A human gut microbial gene catalogue established by metagenomic sequencing. Nature 464(7285):59–65

Qiu S, Cai Y, Yao H, Lin C, Xie Y, Tang S, Zhang A (2023) Small molecule metabolites: discovery of biomarkers and therapeutic targets. Signal Transduct Target Ther 8(1):132

Rahman MM, Islam F, Or-Rashid MH, Mamun AA, Rahaman MS, Islam MM et al (2022) The gut microbiota (microbiome) in cardiovascular disease and its therapeutic regulation. Front Cell Infect Microbiol 12:903570

Reiss AB, Arain HA, Stecker MM, Siegart NM, Kasselman LJ (2018) Amyloid toxicity in Alzheimer's disease. Rev Neurosci 29(6):613–627

Rieder R, Wisniewski PJ, Alderman BL, Campbell SC (2017) Microbes and mental health: a review. Brain Behav Immun 66:9–17

Rietdijk CD, Perez-Pardo P, Garssen J, van Wezel RJ, Kraneveld AD (2017) Exploring Braak's hypothesis of Parkinson's disease. Front Neurol 8:37

Rosin S, Xia K, Azcarate-Peril MA, Carlson AL, Propper CB, Thompson AL et al (2021) A preliminary study of gut microbiome variation and HPA axis reactivity in healthy infants. Psychoneuroendocrinology 124:105046

Rowland I, Gibson G, Heinken A, Scott K, Swann J, Thiele I, Tuohy K (2018) Gut microbiota functions: metabolism of nutrients and other food components. Eur J Nutr 57(1):1–24

Salim S (2017) Oxidative stress and the central nervous system. J Pharmacol Exp Ther 360(1): 201–205

Sampson TR, Debelius JW, Thron T, Janssen S, Shastri GG, Ilhan ZE et al (2016) Gut microbiota regulate motor deficits and neuroinflammation in a model of Parkinson's disease. Cell 167(6): 1469–80.e12

Scalbert A, Johnson IT, Saltmarsh M (2005) Polyphenols: antioxidants and beyond. Am J Clin Nutr 81(1 Suppl):215s–217s

Schwarcz R, Bruno JP, Muchowski PJ, Wu HQ (2012) Kynurenines in the mammalian brain: when physiology meets pathology. Nat Rev Neurosci 13(7):465–477

Sharkey KA, Mawe GM (2023) The enteric nervous system. Physiol Rev 103(2):1487–1564

Sharon G, Cruz NJ, Kang DW, Gandal MJ, Wang B, Kim YM et al (2019) Human gut microbiota from autism spectrum disorder promote behavioral symptoms in mice. Cell 177(6):1600–18.e17

Shen X, Li L, Sun Z, Zang G, Zhang L, Shao C, Wang Z (2021) Gut microbiota and atherosclerosis-focusing on the plaque stability. Front Cardiovasc Med 8:668532

Sikand G, Kris-Etherton P, Boulos NM (2015) Impact of functional foods on prevention of cardiovascular disease and diabetes. Curr Cardiol Rep 17(6):39

Silverman MN, Sternberg EM (2012) Glucocorticoid regulation of inflammation and its functional correlates: from HPA axis to glucocorticoid receptor dysfunction. Ann N Y Acad Sci 1261:55–63

Silverman MN, Pearce BD, Biron CA, Miller AH (2005) Immune modulation of the hypothalamic-pituitary-adrenal (HPA) axis during viral infection. Viral Immunol 18(1):41–78

Singh B, Singh JP, Singh N, Kaur A (2017) Saponins in pulses and their health promoting activities: a review. Food Chem 233:540–549

Singh R, Fedacko J, Pella D, Jarcuska P, Mojto V, Al-bawareed O et al (2018) The gut-brain-axis and the heart. MOJ Public Health 7(3):129–138

Skinner MA, Loh JM, Hunter DC, Zhang J (2011) Gold kiwifruit (Actinidia chinensis 'Hort16A') for immune support. Proc Nutr Soc 70(2):276–280

Socała K, Doboszewska U, Szopa A, Serefko A, Włodarczyk M, Zielińska A et al (2021) The role of microbiota-gut-brain axis in neuropsychiatric and neurological disorders. Pharmacol Res 172:105840

Spillantini MG, Crowther RA, Jakes R, Hasegawa M, Goedert M (1998) Alpha-Synuclein in filamentous inclusions of Lewy bodies from Parkinson's disease and dementia with lewy bodies. Proc Natl Acad Sci USA 95(11):6469–6473

Stavropoulou E, Bezirtzoglou E (2020) Probiotics in medicine: a long debate. Front Immunol 11:2192

Steinman L (2001) Multiple sclerosis: a two-stage disease. Nat Immunol 2(9):762–764

Strandwitz P, Kim KH, Terekhova D, Liu JK, Sharma A, Levering J et al (2019) GABA-modulating bacteria of the human gut microbiota. Nat Microbiol 4(3):396–403

Suarez AN, Hsu TM, Liu CM, Noble EE, Cortella AM, Nakamoto EM et al (2018) Gut vagal sensory signaling regulates hippocampus function through multi-order pathways. Nat Commun 9(1):2181

Sudo N, Chida Y, Aiba Y, Sonoda J, Oyama N, Yu XN et al (2004) Postnatal microbial colonization programs the hypothalamic-pituitary-adrenal system for stress response in mice. J Physiol 558 (Pt 1):263–275

Suganya K, Koo BS (2020) Gut-brain axis: role of gut microbiota on neurological disorders and how probiotics/prebiotics beneficially modulate microbial and immune pathways to improve brain functions. Int J Mol Sci 21(20)

Sun Y, Sommerville NR, Liu JYH, Ngan MP, Poon D, Ponomarev ED et al (2020) Intra-gastrointestinal amyloid-β1-42 oligomers perturb enteric function and induce Alzheimer's disease pathology. J Physiol 598(19):4209–4223

Syngai GG, Gopi R, Bharali R, Dey S, Lakshmanan GA, Ahmed G (2016) Probiotics-the versatile functional food ingredients. J Food Sci Technol 53:921–933

Tain Y-L, Hsu C-N (2024) Nutritional approaches targeting gut microbiota in oxidative-stress-associated metabolic syndrome: focus on early life programming. Nutrients 16(5):683

Tansey MG, McCoy MK, Frank-Cannon TC (2007) Neuroinflammatory mechanisms in Parkinson's disease: potential environmental triggers, pathways, and targets for early therapeutic intervention. Exp Neurol 208(1):1–25

Taylor AM, Holscher HD (2020) A review of dietary and microbial connections to depression, anxiety, and stress. Nutr Neurosci 23(3):237–250

Tiwari S, Atluri V, Kaushik A, Yndart A, Nair M (2019) Alzheimer's disease: pathogenesis, diagnostics, and therapeutics. Int J Nanomedicine 14:5541–5554

Travier L, Alonso M, Andronico A, Hafner L, Disson O, Lledo PM et al (2021) Neonatal susceptibility to meningitis results from the immaturity of epithelial barriers and gut microbiota. Cell Rep 35(13):109319

Tsai HJ, Tsai WC, Hung WC, Hung WW, Chang CC, Dai CY, Tsai YC (2021) Gut microbiota and subclinical cardiovascular disease in patients with type 2 diabetes mellitus. Nutrients 13(8)

Tsavkelova EA, Klimova S, Cherdyntseva TA, Netrusov AI (2006) Hormones and hormone-like substances of microorganisms: a review. Prikl Biokhim Mikrobiol 42(3):261–268
Turnbull AV, Rivier CL (1999) Regulation of the hypothalamic-pituitary-adrenal axis by cytokines: actions and mechanisms of action. Physiol Rev 79(1):1–71
Tuttolomondo A (2020) Ischemic stroke pathogenesis: genetics, epigenetics and inflammation. Curr Pharm Des 26(34):4207–4208
Vaarmann A, Gandhi S, Abramov AY (2010) Dopamine induces Ca2+ signaling in astrocytes through reactive oxygen species generated by monoamine oxidase. J Biol Chem 285(32): 25018–25023
Vajro P, Paolella G, Fasano A (2013) Microbiota and gut-liver axis: their influences on obesity and obesity-related liver disease. J Pediatr Gastroenterol Nutr 56(5):461–468
Vlacil AK, Schuett J, Ruppert V, Soufi M, Oberoi R, Shahin K et al (2020) Deficiency of nucleotide-binding oligomerization domain-containing proteins (NOD) 1 and 2 reduces atherosclerosis. Basic Res Cardiol 115(4):47
van der Vliet A, Tuinstra TJ, Bast A (1989) Modulation of oxidative stress in the gastrointestinal tract and effect on rat intestinal motility. Biochem Pharmacol 38(17):2807–2818
Vogelzangs N, Beekman AT, de Jonge P, Penninx BW (2013) Anxiety disorders and inflammation in a large adult cohort. Transl Psychiatry 3(4):e249
Vona R, Pallotta L, Cappelletti M, Severi C, Matarrese P (2021) The impact of oxidative stress in human pathology: focus on gastrointestinal disorders. Antioxidants (Basel) 10(2)
de Vos WM, Tilg H, Van Hul M, Cani PD (2022) Gut microbiome and health: mechanistic insights. Gut 71(5):1020–1032
Vuong HE, Hsiao EY (2017) Emerging roles for the gut microbiome in autism spectrum disorder. Biol Psychiatry 81(5):411–423
Wakefield AJ (2002) The gut-brain axis in childhood developmental disorders. J Pediatr Gastroenterol Nutr 34(Suppl 1):S14–S17
Wang Z, Klipfell E, Bennett BJ, Koeth R, Levison BS, Dugar B et al (2011) Gut flora metabolism of phosphatidylcholine promotes cardiovascular disease. Nature 472(7341):57–63
Wang W, Zhu LJ, Leng YQ, Wang YW, Shi T, Wang WZ, Sun JC (2023) Inflammatory response: a crucial way for gut microbes to regulate cardiovascular diseases. Nutrients 15(3)
Williams EK, Chang RB, Strochlic DE, Umans BD, Lowell BB, Liberles SD (2016) Sensory neurons that detect stretch and nutrients in the digestive system. Cell 166(1):209–221
Xiao B, Goh JY, Xiao L, Xian H, Lim KL, Liou YC (2017) Reactive oxygen species trigger Parkin/PINK1 pathway-dependent mitophagy by inducing mitochondrial recruitment of Parkin. J Biol Chem 292(40):16697–16708
Yamato M, Kudo W, Shiba T, Yamada KI, Watanabe T, Utsumi H (2010) Determination of reactive oxygen species associated with the degeneration of dopaminergic neurons during dopamine metabolism. Free Radic Res 44(3):249–257
Yang X, Yu D, Xue L, Li H, Du J (2020) Probiotics modulate the microbiota-gut-brain axis and improve memory deficits in aged SAMP8 mice. Acta Pharm Sin B 10(3):475–487
Zhao Y, Lukiw WJ (2015) Microbiome-generated amyloid and potential impact on amyloidogenesis in Alzheimer's disease (AD). J Nat Sci 1(7)
Zhao N, Chen QG, Chen X, Liu XT, Geng F, Zhu MM et al (2023) Intestinal dysbiosis mediates cognitive impairment via the intestine and brain NLRP3 inflammasome activation in chronic sleep deprivation. Brain Behav Immun 108:98–117
Zhou W, Cheng Y, Zhu P, Nasser MI, Zhang X, Zhao M (2020) Implication of gut microbiota in cardiovascular diseases. Oxidative Med Cell Longev 2020:5394096
Zhu W, Gregory JC, Org E, Buffa JA, Gupta N, Wang Z et al (2016) Gut microbial metabolite TMAO enhances platelet Hyperreactivity and thrombosis risk. Cell 165(1):111–124
Zorov DB, Plotnikov EY, Silachev DN, Zorova LD, Pevzner IB, Zorov SD et al (2014) Microbiota and mitobiota. Putting an equal sign between mitochondria and bacteria. Biochemistry (Mosc) 79(10):1017–1031

Chapter 6
Enumerating the Role of Oxidative Stress in Cancer Progression

Reetika Tandon, Vijayshree S. Karankar, and Nidhi Srivastava

Abbreviations

ATP	Adenosine triphosphate
CLL	Chronic lymphocytic leukemia
CNPs	Cerium oxide nanoparticles
DIPG	Diffuse intrinsic pontine glioma
EGFR	Epidermal growth factor receptor
EMT	Epithelial-to-mesenchymal transition
GPXs	Glutathione peroxidase
GSH	Glutathione
HIF-1	Hypoxia-inducible factor 1
MAPKs	Mitogen-activated protein kinases
MDA	Malondialdehyde
NF-κB	Nuclear factor kappa-light-chain enhancer of activated B cells
NRF2	Nuclear erythroid 2-related factor 2
NSCLC	Non-small cell lung cancer
PDGFR	Platelet-derived growth factor receptors
PRXs	Peroxiredoxins
RNS	Reactive nitrogen species
ROS	Reactive oxygen species
SOD	Superoxide dismutase
TME	Tumor microenvironment

R. Tandon · V. S. Karankar · N. Srivastava (✉)
Department of Biotechnology, National Institute of Pharmaceutical Education and Research-Raebareli, Lucknow, India
e-mail: nidhi1.srivastava@niperraebareli.edu.in

N. Srivastava, S. A. Saraf (eds.), *Pharmacological Strategies for Prevention and Treatment of Oxidative Stress*, Oxidative Stress in Applied Basic Research and Clinical Practice, https://doi.org/10.1007/978-3-032-19837-2_6

6.1 Introduction

Despite the available treatment modalities and latest innovations in cancer research, it remains the most appalling disease worldwide (Siegel et al. 2024). Cancer is a genetic disorder resulting from both intrinsic variables (including inherited mutations, hormonal influences, and immune system abnormalities) and extrinsic/acquired causes (such as food choices, radiation exposure, pathogenic agents, and cigarette smoking). Both intrinsic and extrinsic factors regulate significant cellular components, including proto-oncogenes, tumor suppressor genes, and DNA repair genes, through cell intermediates (Sadikovic et al. 2008). Reactive oxygen species (ROS) serve as significant biological components generated in all aerobic organisms. At low concentrations, reactive oxygen species (ROS) demonstrate advantageous benefits; however, their excessive buildup leads to many diseases, including cancer (Acharya et al. 2010). Excessive ROS accumulation leads to oxidative stress, causing harm to cells and tissues. The harm causes somatic mutations and tumor transformation, disrupting the normal functioning of immune cells, leading to inflammation, cell death, and ultimately, tumor metastasis (Nathan and Cunningham-Bussel 2013) Oxidative stress, characterized by an imbalance promoting ROS over antioxidants, has been associated with neurodegenerative diseases, cardiovascular diseases, and various other maladies (Sies 2015). These relationships underscore the importance of maintaining homeostasis between the levels of oxidants and antioxidants. Cells exhibit intricate biochemical and genetic processes to sustain this equilibrium, and it is evident that their disruption can lead to significant pathogenic effects (Hayes et al. 2020). This chapter examines the role of oxidative stress in cancer development and its associated regulatory signaling mechanisms.

6.2 Role of Oxidants in Generation of Oxidative Stress

Cells produce ROS and reactive nitrogen species (RNS) as an inevitable byproduct of metabolism; although potentially deleterious, these species serve as intracellular signaling molecules (Halliwell and Gutteridge 2015). Cells possess a variety of antioxidant systems to sustain ROS/RNS signaling processes and prevent oxidative damage. Cells have indirect-acting antioxidant mechanisms that either reduce the production of oxidants or neutralize the reactive metabolites generated by them, in addition to direct-acting antioxidants. Oxidative stress, characterized by an excessive rise in ROS/RNS compared to antioxidant capability, is mitigated by the cell through numerous mechanisms (Hayes et al. 2020). ROS are relatively small molecules originating from oxygen, comprising oxygen radicals such as alkoxyl (RO•), peroxyl (RO_2•), hydroxyl (•OH), and superoxide (O_2•−). Additionally, they encompass multiple nonradicals, including hydrogen peroxide (H_2O_2), ozone (O_3), hypochlorous acid (HOCl), and singlet oxygen ($1O_2$) that function as oxidizing agents or readily transform into radicals (Prasad et al. 2017). Several endogenous and exogenous factors are responsible for ROS generation (Galadari et al. 2017).

Mitochondria have been shown to generate massive amounts of endogenous ROS that aid in intracellular oxidative stress. Predominantly, the electron transport chain produces ROS, which are located on the inner mitochondrial membrane during oxidative phosphorylation, a process that synthesizes adenosine triphosphate (ATP) from oxygen and simple sugars (Starkov 2008). Five protein complexes are involved in the process. F1F0-ATP synthase (complex V), cytochrome c oxidase (complex IV), ubiquinol: cytochrome c reductase (complex III), succinate: ubiquinone reductase (complex II), and NADH: ubiquinone reductase (complex I) are key components of the electron transport chain and ATP synthesis in cellular respiration (Schägger and Pfeiffer 2001). Reactive oxygen species (ROS) are primarily generated at complexes I and III, where significant alterations were observed in the potential power of electrons due to oxygen reduction (Balaban et al. 2005). Superoxide is swiftly converted to hydrogen peroxide in both the intermembrane space and matrix by superoxide dismutase 1 (SOD1) and SOD2, respectively. Superoxide and hydrogen peroxide are recognized as oxygen radicals generated by cellular mitochondria during this process (Li et al. 2013). Catalase (CAT), glutathione peroxidase (GPXs), and peroxiredoxins (PRXs) mitigate cellular damage by converting hydrogen peroxide (H_2O_2) into water and oxygen (Rhee et al. 2005). Six distinct types of Peroxiredoxins are localized in the endoplasmic reticulum, mitochondria, peroxisome, and cytosol, rendering them effective scavengers for ROS/H_2O_2 (Schieber and Chandel 2014).

Cancer cells indicate abnormal redox homeostasis; although reactive oxygen species (ROS) promote tumorigenesis, elevated ROS levels are cytotoxic (Reczek et al. 2017). Hyperproliferation of tumor cells is associated with elevated production of ROS. These cells adapt to conditions where oxidative stress disrupts redox balance by enhancing their antioxidant capacity. This adaptation allows tumor cells to optimize ROS-driven proliferation while preventing ROS levels from reaching thresholds that could induce different cell death processes such as apoptosis, senescence, or ferroptosis (Dodson et al. 2019; Redza-Dutordoir and Averill-Bates 2016). Cancer cells mitigate excessive intracellular ROS in response to oxidative damage by escalating the levels of enzymatic oxidants (Perillo et al. 2020). The gene transcription factor nuclear erythroid 2-related factor 2 (NRF2) is an essential controller of antioxidant defense in malignant cells. NRF2 is triggered and elevated in cancer to promote the survival of cancer cells; the cellular antioxidant network is predominantly monitored by NRF2 and the related genes (Sporn and Liby 2012). Cancer cells thus protect themselves from excessive ROS.

6.3 Role of Antioxidant Defense Mechanism in Cancer Progression

The reactive oxygen species (ROS) generation is lower in normal physiological processes. The ROS formation contributes to cellular homeostasis by maintaining oxidant and antioxidant production (Snezhkina et al. 2019). Both endogenous and

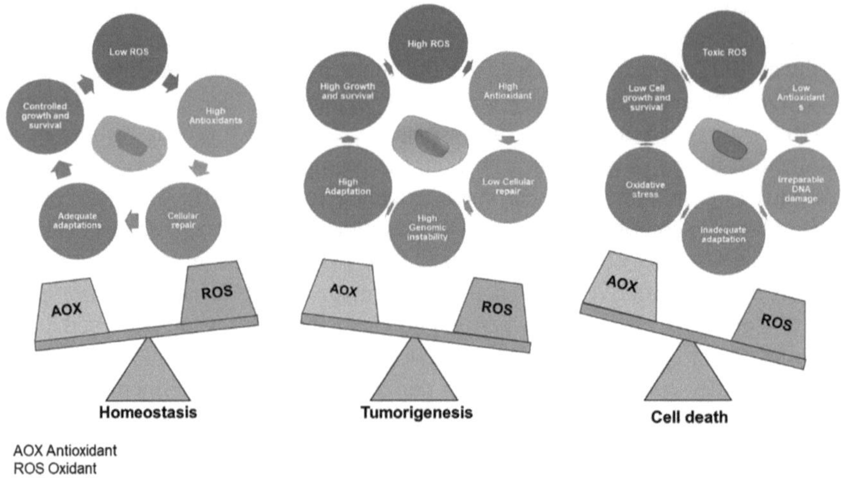

Fig. 6.1 Role of oxidative stress in cancer progression

exogenous sources are responsible for generating reactive species. Endogenous ROS are produced during cellular physiological processes by subcellular organelles and enzyme interactions. Intracellular ROS production is crucial for cellular signaling and activation of inflammatory cells (Jena et al. 2023). The mitochondria convert 2% of their consumed oxygen into superoxide. Peroxisomes are responsible for both generating and scavenging ROS. The exogenous species include extracellular pollutants and harmful heavy metals and pollutants. However, the excessive ROS production results in damaging macromolecules that lead to genomic instability and severe mutations (Afzal et al. 2023) (Fig. 6.1).

To combat oxidative stress, cells have many antioxidant mechanisms. Antioxidants are substance which handle oxidative stress and stop redox changes in the cellular environment. Antioxidants can be classified as either enzymatic or nonenzymatic (Jomova et al. 2024). Catalase (CAT), peroxidases, and the superoxide dismutase family are the main enzymatic antioxidants. The tripeptide glutathione (GSH), thioredoxin (Trx), and ascorbate are nonenzymatic antioxidants in cells that directly react with oxidants. Catalase enzyme is present in peroxisomes, which transforms the produced hydrogen peroxide into H2O and oxygen (Rajput et al. 2021). Glutathione peroxidase is selenium-dependent enzyme that reduces intracellular hydrogen peroxide and lipid hydroperoxides to water and corresponding alcohol. The superoxide dismutase family is a significant superoxide scavenger present in the extracellular matrix (SOD3), mitochondria (SOD2), and cytoplasm (SOD1) (Jomova et al. 2023).

GSH is produced in the cytosol at millimolar amounts; it has the ability to scavenge superoxide and hydroxyl radicals. GSH includes coordination sites for metal ions. Metal complexes, including RuIII, CoIII, and PtIV, undergo GSH-mediated metal ion reduction, which is necessary to release their anticancer potential. Elevated GSH levels have been linked to metallodrug resistance in cells, as higher GSH levels can neutralize the reactive metal species, reducing their effectiveness as anticancer agents. GSH pool depletion causes an increase in ROS level, severe oxidative stress, and apoptosis, and is associated with cell death (Forman et al. 2009). Trx is a small polypeptide with an active disulfide site, present in both mitochondria and cytosol. It reacts directly to oxidants and indirectly donates electrons to peroxidases, which convert hydrogen peroxide into water. Other than vitamin C, vitamin E, coenzyme Q, flavonoids, and carotene are vital antioxidants to neutralize the oxidative stress (Pannala and Dash 2015).

ROS's impact on cancer is a complicated process that depends on several factors, such as quantity, duration of exposure, and cellular location. In early stages of cancer, ROS can enhance abnormal genetic expression, activate antioxidant pathways (KEAP1), and transcription factors like nuclear factor erythroid 2-related factor 2 (Nrf2), which further boost cell survival and proliferation (Aggarwal et al. 2019). In late stages of cancer, ROS promotes metastasis through the signaling pathways. TME often upregulates antioxidant enzymes to buffer the damaging effects of ROS and maintain a redox balance favorable to survival (Chavda et al. 2022).

Moreover, oxidative stress is involved in various tumorigenic processes, including angiogenesis, invasiveness, and stemness. It activates signaling pathways like PI3K/Akt and NF-κB, supports cell proliferation, and contributes to therapeutic resistance. Conversely, excessive oxidative stress can lead to cell death through mechanisms such as apoptosis, necroptosis, and ferroptosis, particularly when antioxidant defenses are inadequate (Li et al. 2025).

6.4 Contribution of ROS Signaling in Cancer Progression

ROS play a critical part in cellular physiology by balancing cellular processes such as proliferation, differentiation, migration, and programmed cell death (Moloney and Cotter 2018). ROS play a crucial role as second messengers or in the modification (activation or inactivation) of proteins linked to oncogenesis and their progression. ROS largely regulates protein activity by oxidizing thiol groups in cysteine residues, which form disulfide bonds and cause particular protein changes that can either activate or deactivate the protein's function. ROS-regulated signaling pathways that are essential to tumor progression include p53, MAPKs, HIF-1α, PI3K, NRF2, and NF-κB pathways (Sarmiento-Salinas et al. 2021) (Fig. 6.2).

RelB, p65/RelA, c-Rel, p50/p105 (NF-κB1), and p52/p100 (NF-κB2) proteins are members of the NF-κB transcription factor family. Upon activation, these proteins assemble into dimers or heterodimers that control target gene transcription. The inhibitors of IκB and regulatory genes of the NF-κB family (p100 and p105) negatively regulate these proteins, while the NF-κB activators of I kappa-B kinase (IKK) family positively regulate them (Sarmiento-Salinas et al. 2021). The expression of the NF-κB protein has been shown to be induced by H_2O_2. Moreover, its activation adversely impacts pro-inflammatory cytokines associated with tumorigenesis (IL-1β and IL-6), genes involved in drug resistance and detoxification, as well as free radical scavengers (Li et al. 2015).

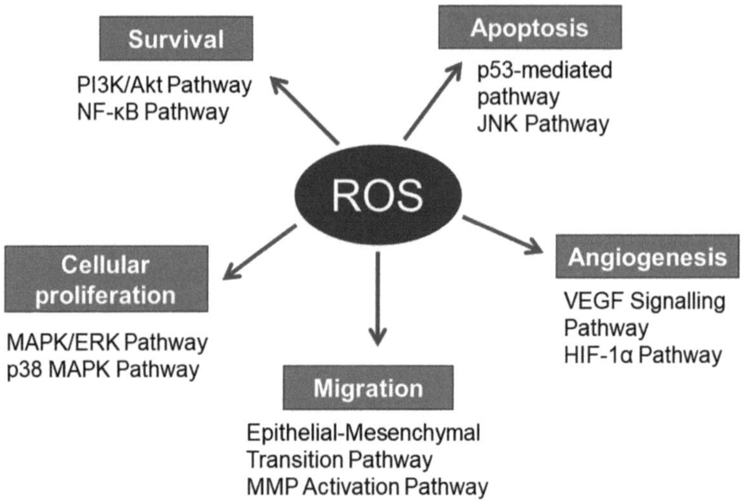

Fig. 6.2 ROS-driven signaling pathways in cancer

The administration of IL-1β, TNFα, or sodium arsenite to breast carcinoma cells results in the production of H_2O_2 and O_2^-, which subsequently activate NF-κB and enhance cellular proliferation (Ruiz-Ramos et al. 2009). The deletion of superoxide dismutase (SOD) leads to increased basal ROS quantities and elevated NF-κB activity in oral carcinoma. NF-κB signaling based on IKK is turned on by elevated cellular oxidative damage, which can occur through H_2O_2, rotenone-mediated O_2-, or inhibition of the glutathione system. Conversely, IKK-independent activation of NF-κB is initiated by the phosphorylation of IκBα at the tyrosine residue in response to ROS, resulting in the release of NF-κB. H_2O_2 increases the joining of IKKγ/NEMO, which fosters the activation and phosphorylation of the subunit of IKKβ. This indicates either the blockade of an IKK phosphatase or a direct oxidation of IKKβ, resulting in its deactivation and reactivation of the NF-κB pathway (Morgan and Liu 2011).

Hypoxia-inducible factor 1 (HIF-1) monitors cellular adaptation to hypoxic conditions, which serves as a critical regulator. It functions as a heterodimer composed of an alpha subunit and a beta subunit (Sarmiento-Salinas et al. 2021). The HIF-1 signaling pathway has been thoroughly investigated across multiple cancer types such as gastric (Wu et al. 2017), pancreatic (Movafagh et al. 2015), and prostate cancer (Zapatero et al. 2015), and it is recognized for its role in regulating the expression of genes associated with angiogenesis, metabolism, glucose transport, and cellular migration (Jiang et al. 2017). Reactive oxygen species (ROS) regulate through the induction of HIF-1α accumulation. HIF-1 signaling is regulated by mitochondrial O_2•- under oxygen-deficient conditions (Qutub and Popel 2008). Moreover, accumulation of HIF-1α is observed due to elevated levels of NO in both normal and oxygen-deficient conditions. The reason behind this is likely due to stabilization of HIF-1α by S-nitrosylation or PHD inhibition (Movafagh et al. 2015).

The MAPK family comprises ERK1/2, JNK, and p38. MAPK pathways serve as crucial intracellular signaling mechanisms that facilitate cell proliferation and differentiation. ROS have been shown to oxidize and deactivate MAPK phosphatases, while concurrently triggering the epidermal growth factor receptor (EGFR) and platelet-derived growth factor receptors (PDGFR) signaling in a ligand-independent manner through the RAS and ERK pathways (Lee et al. 2018). Exposure to oxidative conditions or stimuli that induce intracellular ROS formation triggers MAPK or EGF signaling, irrespective of EGF presence (León-Buitimea et al. 2012). Antioxidant therapy is recognized for its ability to inhibit MAPK activation following growth factor stimulation across several experimental contexts (Son et al. 2013). Transformation of RAS has been associated with the activation of NOX1 via a RAS-RAC1-NOX1 pathway (Bauer 2014).

6.5 Oxidative Stress in Specific Cancer Types

The expression levels of antioxidant enzymes, such as catalase, superoxide dismutase (SOD), and glutathione peroxidase, are often altered in various cancer types, contributing to oxidative stress and tumor progression (Ezema et al. 2024). Notably, many cancers exhibit reduced levels of these enzymes compared to normal cells, leading to a diminished capacity to neutralize ROS (Table 6.1). Most cancers exhibit reduced enzyme levels, impairing ROS neutralization and contributing to tumor progression. However, exceptions include elevated manganese superoxide dismutase (MnSOD) in glioblastoma, lung, and advanced prostate cancer, which supports metabolic reprogramming and tumor survival (Iqbal et al. 2024).

6.6 Revamping the ROS Generation in Cancer Therapeutic

ROS are crucial modulators in cancer therapy; elevated ROS levels in cancer cells promote tumorigenesis by fostering proliferation, migration, and angiogenesis (Nakamura and Takada 2021). However, excessive ROS can induce oxidative stress, leading to cellular damage and apoptosis. The delicate balance between ROS production and antioxidant defenses is a key area of exploration in cancer treatment. Blocking antioxidant systems weakens cancer cells' defenses, making them more vulnerable to ROS (Iqbal et al. 2024). This paradoxical role of ROS opens opportunities to either increase or decrease ROS levels for effective cancer treatment. Therapeutic strategies aimed at enhancing ROS generation are being explored to selectively induce cancer cell death (Galadari et al. 2017). Many cancer therapies, including chemotherapy, radiotherapy, and targeted therapies, utilize the tumor's dependence on ROS. However, these strategies are still in their early stages of development (Zhou et al., 2023).

Table 6.1 The expression of antioxidant defense enzymes in various cancers

Antioxidant enzyme	Cancer type	Key observations	Reference
CAT, SOD	Cervical cancer (Stages II-IV)	Significant elevation in enzymatic activity	Jelić et al. (2018)
GPx, GR		The activity of GPx and GR in cervical cancer was not prominently altered	
GST		Significantly higher activity compared to healthy controls	
SOD, CAT	Bladder cancer	Upregulation of SOD, increased activity in higher grade tumors	Lepara et al. (2023)
SOD, CAT, GST, GPx	Breast cancer	Increased SOD concentration; reduced CAT, GPx, and GST levels; increased susceptibility to oxidative damage	Hamed A. Hamed A. Hasan (2023)
SOD, Gpx, CAT	Thyroid cancer	Higher levels of SOD, GPx, and CAT; increased hydrogen peroxide; and possibly hydroxyl radical involvement	Sadani and Nadkarni (1996)
Mn-SOD	Brain cancer	Expression increases with tumor grade. Higher in glioblastoma than in astrocytomas	Shwetha et al. (2016)
Mn-SOD	Gastric cancer	Expression upregulated in tumors with lymph node metastases	Malafa et al. (2000)
Mn-SOD	Ovarian cancer	Significantly higher expression. Suppression of Mn-SOD leads to increased superoxide, cell proliferation, and aggressive tumor growth	Hu et al. (2005)
CuZn-SOD	Gastric cancer	Lower expression compared to normal mucosa	Monari et al. (2006)
CuZn-SOD	Colorectal cancer	More poorly differentiated tumors retain higher CuZn-SOD expression, whereas highly and moderately differentiated tumors show greater CuZn-SOD suppression	Van Driel et al. (1997)
Total SOD	Colorectal cancer	SOD activity is increased in stages I, II, and III of colorectal cancer. In stage VI, activity is markedly decreased compared to earlier stages	Maffei et al. (2011)
Sod	Colorectal cancer	SOD activity is increased in response to elevated oxidative stress in CRC	Zińczuk et al. (2019)
CAT, GPx		CAT and GPx activities are significantly reduced, indicating weakened antioxidant defenses	
SOD, CAT	Lung cancer	Reduced SOD activity, indicating inefficient superoxide radical detoxification	Zalewska-Ziob et al. (2019)
GPx		Significantly high activity in lung cancer tissues, regardless of histological type	
SOD, CAT	Chronic lymphocytic leukemia (CLL)	SOD activity decreased leads to reduced superoxide radical detoxification and increased oxidative stress	Oltra et al. (2001)
GPx		GPx activity increased leads to counteract oxidative stress by detoxifying peroxide	

6.7 Therapeutic Strategies for Reducing ROS in Early Cancer Stage

Several therapeutic approaches are being investigated to prevent the progression of preneoplastic lesions to cancer by reducing ROS levels. The intention is to neutralize excess ROS by delivering exogenous antioxidants or stimulating endogenous antioxidant capacities, thereby avoiding oxidative damage that can trigger the onset of cancer (Osama et al. 2023). These include using peptide-based molecules that act like antioxidant enzymes, delivering antioxidants directly to preneoplastic cells via lipid nanoparticles, and activating the Nrf2 pathway to enhance the body's natural antioxidant capacity (Karankar et al. 2025; Xu et al. 2024). Additional strategies involve employing exosomes to deliver antioxidant enzymes to cells, targeting mitochondria with specific antioxidants to prevent ROS overproduction, and using drugs to block lipid peroxidation (Shao et al. 2023). Together, these interventions aim to re-establish a healthy redox balance in preneoplastic cells, preventing tumor initiation.

6.8 Direct Use of ROS to Induce Cancer Cell Death

Recent studies highlight diverse methods to harness ROS for therapeutic purposes, including direct use of ROS for regulating inflammation and inhibiting cancer metastasis (Hu et al. 2024). Increasing ROS levels may trigger cancer cells to exceed their oxidative stress threshold, thus may result in selective cytotoxicity, as these cells often possess an altered redox microenvironment (Perillo et al. 2020). In photodynamic therapy (PDT), light-activated photosensitizing drugs produce ROS, which kill cancer cells. Porfimer Sodium (Photofrin) is a widely studied photosensitizer activated by red light from a laser, generating ROS to treat lung and esophageal cancer (Al-Jamal et al. 2025). PDT's therapeutic efficacy is further enhanced by its disruption of tumor blood vessels and immune system stimulation (Dudzik et al. 2024). Many chemotherapeutic agents, such as cisplatin, doxorubicin, paclitaxel, and bleomycin, target cancer cells by increasing ROS levels, thereby triggering apoptosis, causing DNA damage, disrupting cellular functions, and overcoming drug resistance associated with enhanced antioxidant defenses (Li et al. 2023). Conversely, while antioxidants are known to reduce oxidative stress, their use must be carefully considered as they can sometimes promote tumor progression and metastasis (Chandimali et al. 2025).

6.9 Targeting Antioxidant System

Targeting antioxidant pathways is an effective strategy to enhance ROS-induced cancer cell death. Inhibitors of SODs, catalase, and GPx may block ROS detoxification, thereby increasing the intracellular ROS burden and sensitizing cancer cells to apoptosis. Several small molecules that inhibit these antioxidant enzymes are under clinical investigation (Perillo et al. 2020). However, this approach presents complex challenges. Antioxidants protect normal cells by neutralizing elevated ROS and may inadvertently support tumor survival and resistance. This has led to the exploration of various antioxidant therapeutic strategies, including nonenzymatic antioxidants like NRF2 activators and vitamins, as well as enzymatic antioxidants such as NOX inhibitors and superoxide dismutase (SOD) mimetics (Herdiana et al. 2023; Luo et al. 2022; Singh and Manna 2022). However, the clinical efficacy of antioxidant therapies remains contentious. While antioxidants can protect normal cells from oxidative damage, their use in cancer treatment may inadvertently support tumor growth by enhancing the survival of cancer cells; their clinical efficacy remains debated (Woldeselassie and Tamene 2024).

Elevated ROS levels can drive cancer progression, promoting epithelial-to-mesenchymal transition (EMT) and immune evasion. Furthermore, some studies suggest that certain antioxidants may confer resistance to chemotherapy and immunotherapy by enabling cancer cells to withstand oxidative stress (Glorieux and Buc Calderon 2024). Thus, a balanced approach that prevents tumor adaptability, while selectively increasing ROS in cancer cells, is crucial. Ongoing research aims to refine context-dependent antioxidant modulation for effective cancer therapy (Shah and Rogoff 2021).

6.10 Redox-Responsive Drug Delivery System

Redox-active nanoparticles are revolutionizing cancer treatment by manipulating oxidative stress within tumor cells (Sanati et al. 2022). These nanoparticles, such as cerium oxide nanoparticles (CNPs), possess both pro-oxidative and antioxidative properties, enabling them to selectively inhibit tumor growth and invasion while preserving healthy cells (Nelson et al. 2016). By altering the redox balance either by increasing ROS or suppressing antioxidant defenses—they trigger programmed cell death in cancer cells—enhancing treatment effectiveness and overcoming resistance to chemotherapy and radiotherapy (Das et al. 2022). Furthermore, their integration with conventional therapies boosts efficacy while minimizing side effects. However, challenges such as refining delivery mechanisms and assessing long-term safety require further exploration. Redox-responsive nanoparticles represent a promising innovation, paving the way for more precise and effective cancer therapies (Deivayanai et al. 2024).

6.11 Targeting TME to Regulate ROS Levels in Cancer Therapy

Regulating ROS in the tumor microenvironment (TME) is a promising strategy in cancer therapy. Targeting antioxidant defenses leaves cancer cells vulnerable to oxidative stress, enhancing tumor suppression (Zhou et al., 2023). Increasing ROS levels selectively induces cancer cell death, while sparing normal cells. Reprogramming immune cells in the TME, such as TAMs and MDSCs, to produce cytotoxic ROS enhances immunological checkpoint treatments (Kennel and Greten 2021). Hypoxia in the TME leads to ROS production through mitochondrial dysfunction and HIF-1α pathways. Disrupting HIF-1α activity impairs tumor survival by altering redox balance (Bae et al. 2024). Personalized therapies targeting ROS dynamics could improve cancer outcomes by leveraging oxidative stress. This redox-based approach is shaping the future of personalized oncology treatments. As research progresses, ROS-targeted strategies continue to unlock new avenues for cancer therapy (Lin et al. 2024).

6.12 Targeted Ferroptosis Inhibits Cancer Development

Ferroptosis, an iron-dependent form of regulated cell death driven by lipid peroxidation, offers a promising strategy for cancer therapy. Cancer cells, especially resistant types, rely on antioxidant defenses like GPX4 to prevent ferroptosis (Zhou et al. 2024). Targeting GPX4 or disrupting iron homeostasis elevates oxidative stress, selectively eliminating tumor cells while sparing healthy tissues. Key therapeutic strategies include GPX4 inhibitors, iron modulation to enhance ROS accumulation, and combination therapies with chemotherapy or immunotherapy (Lei et al. 2022). While these approaches show promise, challenges like selectivity and toxicity remain. Ongoing research aims to optimize ferroptosis-based treatments, offering hope for overcoming treatment resistance and inhibiting cancer progression (Shahid et al. 2024).

6.13 Clinically Available Drugs Targeting Oxidative Stress-Mediated Cancer Progression

Targeting oxidative stress in the course of cancer has enormous therapeutic promise. These medications provide a dual therapeutic strategy by either reducing excessive ROS generation or using it to cause cancer cell death (Jiang et al. 2023). Normalizing oxidative stress levels without negatively impacting healthy cells is an intricate equilibrium that must be maintained. These strategies are being improved by ongoing research and clinical trials, which makes oxidative stress-targeted treatments a viable new avenue for the treatment of cancer. The range of medications available to treat cancer in all of its manifestations will grow along with our comprehension of redox biology (Hegedűs et al. 2018). Different clinical trials are listed in Table 6.2 for different cancer treatments.

Table 6.2 Clinical trials investigating therapeutic intervention targeting reactive oxygen species signaling in cancer treatment

NCT number	Mechanism of action	Status	Conditions	Interventions	Phase
NCT06694129	Entrectinib targets ROS1 rearrangement in NSCLC, offering brain protection and addressing resistance	Not yet recruiting	Non-small cell lung cancer (NSCLC), ROS1 Gene Rearrangement	Drug: Entrectinib (Rozlytrek)	Observational
NCT03088930	Crizotinib targets ALK, ROS1 rearrangements, and MET exon 14 mutations in NSCLC, acting as a tyrosine kinase inhibitor	Completed	Non-small cell lung cancer (NSCLC)	Drug: Crizotinib	Phase 2
NCT06600568	Modulates intracellular ROS levels to selectively eliminate neoplastic T cells, combining apoptosis-inducing drugs and ROS-altering agents	Recruiting	T-cell leukemia/lymphoma, adult	Translation analysis	Observational
NCT03076502	Evaluates ctDNA, PVT1, and ROS as biomarkers for diagnosis, treatment, and recurrence monitoring in gastro-intestinal and hepatobiliary pancre-atic cancer	Not yet recruiting	Gastrointestinal and hepatobiliary pancreatic cancer	Observational study with liq-uid biopsy	Observational
NCT03375437	Molecular screening to detect NTRK1/2/3, ROS1, or ALK gene fusions in soft tissue sarcomas with complex genomic profiles	Active, not recruiting	Soft tissue sarcoma, advanced can-cer, metastatic cancer	Genetic: Blood and tumor samples	Not applicable
NCT01596374	Investigates ROS gene aberrations (mutations, overexpression, rearrangements) in solid tumors and their sensitivity to ROS/ALK inhibitors	Unknown status	Solid tumors, including hepatocel-lular carcinoma, colorectal, gastric, breast, ovarian, cholangiocarcinoma, and NSCLC	Genetic: ROS tyrosine kinase inhibitors	Observational

NCT06081270	Liquid biopsy with next-generation sequencing (NGS) to identify resistance mutations in RTK-positive NSCLC or Ras/MAPK pathway mutations	Recruiting	Non-small cell lung cancer (NSCLC)	Other: Liquid biopsy	Observational
NCT06333899	Lorlatinib targets ROS1 and ALK fusions in high-grade gliomas, aiming to improve disease control and survival rates	Not yet recruiting	High-grade glioma, diffuse intrinsic pontine glioma (DIPG), glioblastoma	Drug: Lorlatinib, Lorlatinib with chemotherapy, Lorlatinib post-radiation	Early phase 1
NCT02207504	Crizotinib acts as an ATP-competitive inhibitor targeting ALK, ROS, and c-Met tyrosine kinases, combined with Enzalutamide, an androgen receptor signaling inhibitor	Completed	Metastatic castration-resistant prostate cancer	Drugs: Crizotinib, Enzalutamide	Phase 1
NCT04459663	JS001 combined with Axitinib targets advanced NSCLC without EGFR mutation, ALK fusion, or ROS fusion, enhancing antitumor activity	Not yet recruiting	Non-small cell lung cancer (NSCLC)	Drugs: Toripalimab injection, Axitinib	Phase 2

6.14 Conclusion

Oxidative stress exhibits an intricate and paradoxical role in cancer progression, functioning as both a promoter and suppressor of tumor development, contingent upon the environment and concentration of ROS. This chapter has elucidated the molecular processes by which oxidative stress impacts cancer. The equilibrium between ROS generation and antioxidant defenses is pivotal in influencing the destiny of cancer cells. Moderate quantities of reactive oxygen species (ROS) might facilitate tumor growth and survival, but severe oxidative stress may trigger apoptosis or senescence. These results present potential opportunities for therapeutic intervention, including pro-oxidant medicines that specifically cause deadly oxidative stress in cancer cells and antioxidant techniques that alleviate tumor-promoting inflammation. In summary, a thorough understanding of the role of oxidative stress in cancer biology is essential for the development of more targeted and effective therapies.

References

Acharya A, Das I, Chandhok D, Saha T (2010) Redox regulation in cancer: a double-edged sword with therapeutic potential. Oxidative Med Cell Longev 3:23–34. https://doi.org/10.4161/oxim.3.1.10095

Afzal S, Abdul Manap AS, Attiq A, Albokhadaim I, Kandeel M, Alhojaily SM (2023) From imbalance to impairment: the central role of reactive oxygen species in oxidative stress-induced disorders and therapeutic exploration. Front Pharmacol 14. https://doi.org/10.3389/fphar.2023.1269581

Aggarwal V, Tuli H, Varol A, Thakral F, Yerer M, Sak K, Varol M, Jain A, Khan M, Sethi G (2019) Role of reactive oxygen species in cancer progression: molecular mechanisms and recent advancements. Biomolecules 9:735. https://doi.org/10.3390/biom9110735

Al-Jamal AN, Al-Hussainy AF, Mohammed BA, Abbas HH, Kadhim IM, Ward ZH, Mahapatra DK, Joseph TM, kianfar E, Thomas S (2025) Photodynamic Therapy (PDT) in drug delivery: Nano-innovations enhancing treatment outcomes. Health Sci Rev 14:100218. https://doi.org/10.1016/j.hsr.2025.100218

Bae T, Hallis SP, Kwak M-K (2024) Hypoxia, oxidative stress, and the interplay of HIFs and NRF2 signaling in cancer. Exp Mol Med 56:501–514. https://doi.org/10.1038/s12276-024-01180-8

Balaban RS, Nemoto S, Finkel T (2005) Mitochondria, oxidants, and aging. Cell 120:483–495. https://doi.org/10.1016/j.cell.2005.02.001

Bauer G (2014) Targeting extracellular ROS signaling of tumor cells. Anticancer Res 34:1467–1482

Chandimali N, Bak SG, Park EH, Lim H-J, Won Y-S, Kim E-K, Park S-I, Lee SJ (2025) Free radicals and their impact on health and antioxidant defenses: a review. Cell Death Discov 11:19. https://doi.org/10.1038/s41420-024-02278-8

Chavda V, Chaurasia B, Garg K, Deora H, Umana GE, Palmisciano P, Scalia G, Lu B (2022) Molecular mechanisms of oxidative stress in stroke and cancer. Brain Disord 5:100029. https://doi.org/10.1016/j.dscb.2021.100029

Das CK, Majumder R, Roy P, Mandal M (2022) The intricacy of ROS in cancer therapy resistance. In: Handbook of oxidative stress in cancer: mechanistic aspects. Springer Nature Singapore, Singapore, pp 1217–1238. https://doi.org/10.1007/978-981-15-9411-3_125

Deivayanai VC, Thamarai P, Karishma S, Saravanan A, Yaashikaa PR, Vickram AS, Hemavathy RV, Kumar RR, Rishikesavan S, Shruthi S (2024) A comprehensive review on advances in nanoparticle-mediated cancer therapeutics: current research and future perspectives. Cancer Pathogen Ther. https://doi.org/10.1016/j.cpt.2024.11.002

Dodson M, Castro-Portuguez R, Zhang DD (2019) NRF2 plays a critical role in mitigating lipid peroxidation and ferroptosis. Redox Biol 23:101107. https://doi.org/10.1016/j.redox.2019.101107

Dudzik T, Domański I, Makuch S (2024) The impact of photodynamic therapy on immune system in cancer—an update. Front Immunol 15. https://doi.org/10.3389/fimmu.2024.1335920

Ezema BO, Eze CN, Ayoka TO, Nnadi CO (2024) Antioxidant-enzyme interaction in non-communicable diseases. J Explor Res Pharmacol 9:262–275. https://doi.org/10.14218/JERP.2024.00020

Forman HJ, Zhang H, Rinna A (2009) Glutathione: overview of its protective roles, measurement, and biosynthesis. Mol Asp Med 30:1–12. https://doi.org/10.1016/j.mam.2008.08.006

Galadari S, Rahman A, Pallichankandy S, Thayyullathil F (2017) Reactive oxygen species and cancer paradox: to promote or to suppress? Free Radic Biol Med 104:144–164. https://doi.org/10.1016/j.freeradbiomed.2017.01.004

Glorieux C, Buc Calderon P (2024) Targeting catalase in cancer. Redox Biol 77:103404. https://doi.org/10.1016/j.redox.2024.103404

Halliwell B, Gutteridge JMC (2015) Free radicals in biology and medicine. Oxford University Press. https://doi.org/10.1093/acprof:oso/9780198717478.001.0001

Hamed A. Hasan AWA-A (2023) Superoxide dismutase activity in breast cancer patients treated with anastrozole. Oncology Radiotherapy

Hayes JD, Dinkova-Kostova AT, Tew KD (2020) Oxidative stress in cancer. Cancer Cell 38:167–197. https://doi.org/10.1016/j.ccell.2020.06.001

Hegedűs C, Kovács K, Polgár Z, Regdon Z, Szabó É, Robaszkiewicz A, Forman HJ, Martner A, Virág L (2018) Redox control of cancer cell destruction. Redox Biol 16:59–74. https://doi.org/10.1016/j.redox.2018.01.015

Herdiana Y, Sriwidodo S, Sofian FF, Wilar G, Diantini A (2023) Nanoparticle-based antioxidants in stress signaling and programmed cell death in breast cancer treatment. Molecules 28. https://doi.org/10.3390/molecules28145305

Hu Y, Rosen DG, Zhou Y, Feng L, Yang G, Liu J, Huang P (2005) Mitochondrial manganese-superoxide dismutase expression in ovarian cancer. J Biol Chem 280:39485–39492. https://doi.org/10.1074/jbc.M503296200

Hu D, Li Y, Li R, Wang M, Zhou K, He C, Wei Q, Qian Z (2024) Recent advances in reactive oxygen species (ROS)-responsive drug delivery systems for photodynamic therapy of cancer. Acta Pharm Sin B 14:5106–5131. https://doi.org/10.1016/j.apsb.2024.10.015

Iqbal MJ, Kabeer A, Abbas Z, Siddiqui HA, Calina D, Sharifi-Rad J, Cho WC (2024) Interplay of oxidative stress, cellular communication and signaling pathways in cancer. Cell Commun Signal 22:7. https://doi.org/10.1186/s12964-023-01398-5

Jelić M, Mandić A, Kladar N, Sudji J, Božin B, Srdjenović B (2018) Lipid peroxidation, Antioxidative defense and level of 8-hydroxy-2-deoxyguanosine in cervical cancer patients. J Med Biochem 37:336–345. https://doi.org/10.1515/jomb-2017-0053

Jena AB, Samal RR, Bhol NK, Duttaroy AK (2023) Cellular red-ox system in health and disease: the latest update. Biomed Pharmacother 162:114606. https://doi.org/10.1016/j.biopha.2023.114606

Jiang J, Wang K, Chen Y, Chen H, Nice EC, Huang C (2017) Redox regulation in tumor cell epithelial–mesenchymal transition: molecular basis and therapeutic strategy. Signal Transduct Target Ther 2:17036. https://doi.org/10.1038/sigtrans.2017.36

Jiang H, Zuo J, Li B, Chen R, Luo K, Xiang X, Lu S, Huang C, Liu L, Tang J, Gao F (2023) Drug-induced oxidative stress in cancer treatments: angel or devil? Redox Biol 63:102754. https://doi.org/10.1016/j.redox.2023.102754

Jomova K, Raptova R, Alomar SY, Alwasel SH, Nepovimova E, Kuca K, Valko M (2023) Reactive oxygen species, toxicity, oxidative stress, and antioxidants: chronic diseases and aging. Arch Toxicol 97:2499–2574. https://doi.org/10.1007/s00204-023-03562-9

Jomova K, Alomar SY, Alwasel SH, Nepovimova E, Kuca K, Valko M (2024) Several lines of antioxidant defense against oxidative stress: antioxidant enzymes, nanomaterials with multiple enzyme-mimicking activities, and low-molecular-weight antioxidants. Arch Toxicol 98:1323–1367. https://doi.org/10.1007/s00204-024-03696-4

Karankar VS, Awasthi S, Srivastava N (2025) Peptide-driven strategies against lung cancer. Life Sci 366–367:123453. https://doi.org/10.1016/j.lfs.2025.123453

Kennel KB, Greten FR (2021) Immune cell - produced ROS and their impact on tumor growth and metastasis. Redox Biol 42:101891. https://doi.org/10.1016/j.redox.2021.101891

Lee JH, Kim C, Lee S-G, Yang WM, Um J-Y, Sethi G, Ahn KS (2018) Ophiopogonin D modulates multiple oncogenic signaling pathways, leading to suppression of proliferation and chemosensitization of human lung cancer cells. Phytomedicine 40:165–175. https://doi.org/10.1016/j.phymed.2018.01.002

Lei G, Zhuang L, Gan B (2022) Targeting ferroptosis as a vulnerability in cancer. Nat Rev Cancer 22:381–396. https://doi.org/10.1038/s41568-022-00459-0

León-Buitimea A, Rodríguez-Fragoso L, Lauer FT, Bowles H, Thompson TA, Burchiel SW (2012) Ethanol-induced oxidative stress is associated with EGF receptor phosphorylation in MCF-10A cells overexpressing CYP2E1. Toxicol Lett 209:161–165. https://doi.org/10.1016/j.toxlet.2011.12.009

Lepara Z, Alić J, Lepara O, Spahović H, Fajkić A (2023) Antioxidant status in patients with bladder cancer regarding cancer stage and grade. Asian J Urol 10:182–188. https://doi.org/10.1016/j.ajur.2021.12.008

Li X, Fang P, Mai J, Choi ET, Wang H, Yang X (2013) Targeting mitochondrial reactive oxygen species as novel therapy for inflammatory diseases and cancers. J Hematol Oncol 6:19. https://doi.org/10.1186/1756-8722-6-19

Li F, Zhang J, Arfuso F, Chinnathambi A, Zayed ME, Alharbi SA, Kumar AP, Ahn KS, Sethi G (2015) NF-κB in cancer therapy. Arch Toxicol 89:711–731. https://doi.org/10.1007/s00204-015-1470-4

Li Y, Zhang X, Wang Z, Li B, Zhu H (2023) Modulation of redox homeostasis: a strategy to overcome cancer drug resistance. Front Pharmacol 14. https://doi.org/10.3389/fphar.2023.1156538

Li S, Yuan H, Li L, Li Q, Lin P, Li K (2025) Oxidative stress and reprogramming of lipid metabolism in cancers. Antioxidants 14:201. https://doi.org/10.3390/antiox14020201

Lin P, Lu Y, Zheng J, Lin Y, Zhao X, Cui L (2024) Strategic disruption of cancer's powerhouse: precise nanomedicine targeting of mitochondrial metabolism. J Nanobiotechnol 22:318. https://doi.org/10.1186/s12951-024-02585-3

Luo M, Zhou L, Huang Z, Li B, Nice EC, Xu J, Huang C (2022) Antioxidant therapy in cancer: rationale and Progress. Antioxidants (Basel) 11. https://doi.org/10.3390/antiox11061128

Maffei F, Angeloni C, Malaguti M, Moraga JMZ, Pasqui F, Poli C, Colecchia A, Festi D, Hrelia P, Hrelia S (2011) Plasma antioxidant enzymes and clastogenic factors as possible biomarkers of colorectal cancer risk. Mutation Research/Fundamental and Molecular Mechanisms of Mutagenesis 714:88–92. https://doi.org/10.1016/j.mrfmmm.2011.06.016

Malafa M, Margenthaler J, Webb B, Neitzel L, Christophersen M (2000) MnSOD expression is increased in metastatic gastric cancer. J Surg Res 88:130–134. https://doi.org/10.1006/jsre.1999.5773

Moloney JN, Cotter TG (2018) ROS signalling in the biology of cancer. Semin Cell Dev Biol 80:50–64. https://doi.org/10.1016/j.semcdb.2017.05.023

Monari M, Trinchero A, Calabrese C, Cattani O, Serrazanetti GP, Foschi J, Fabbri A, Zahlane D, Di Febo G, Tonini V, Cervellera M, Tosi MR, Tugnoli V (2006) Superoxide dismutase in gastric adenocarcinoma: is it a clinical biomarker in the development of cancer? Biomarkers 11:574–584. https://doi.org/10.1080/13547500600899134

Morgan MJ, Liu Z (2011) Crosstalk of reactive oxygen species and NF-κB signaling. Cell Res 21: 103–115. https://doi.org/10.1038/cr.2010.178

Movafagh S, Crook S, Vo K (2015) Regulation of hypoxia-inducible factor-1a by reactive oxygen species : new developments in an old debate. J Cell Biochem 116:696–703. https://doi.org/10.1002/jcb.25074

Nakamura H, Takada K (2021) Reactive oxygen species in cancer: current findings and future directions. Cancer Sci 112:3945–3952. https://doi.org/10.1111/cas.15068

Nathan C, Cunningham-Bussel A (2013) Beyond oxidative stress: an immunologist's guide to reactive oxygen species. Nat Rev Immunol 13:349–361. https://doi.org/10.1038/nri3423

Nelson B, Johnson M, Walker M, Riley K, Sims C (2016) Antioxidant cerium oxide nanoparticles in biology and medicine. Antioxidants 5:15. https://doi.org/10.3390/antiox5020015

Oltra AM, Carbonell F, Tormos C, Iradi A, Sáez GT (2001) Antioxidant enzyme activities and the production of MDA and 8-oxo-dG in chronic lymphocytic lerkenia. Free Radic Biol Med 30: 1286–1292. https://doi.org/10.1016/S0891-5849(01)00521-4

Osama A, Zhang J, Fang J (2023) Harnessing the ROS for cancer treatment. In: Handbook of cancer and immunology. Springer International Publishing, Cham, pp 1–31. https://doi.org/10.1007/978-3-030-80962-1_255-1

Pannala VR, Dash RK (2015) Mechanistic characterization of the thioredoxin system in the removal of hydrogen peroxide. Free Radic Biol Med 78:42–55. https://doi.org/10.1016/j.freeradbiomed.2014.10.508

Perillo B, Di Donato M, Pezone A, Di Zazzo E, Giovannelli P, Galasso G, Castoria G, Migliaccio A (2020) ROS in cancer therapy: the bright side of the moon. Exp Mol Med 52:192–203. https://doi.org/10.1038/s12276-020-0384-2

Prasad S, Gupta SC, Tyagi AK (2017) Reactive oxygen species (ROS) and cancer: role of antioxidative nutraceuticals. Cancer Lett 387:95–105. https://doi.org/10.1016/j.canlet.2016.03.042

Qutub AA, Popel AS (2008) Reactive oxygen species regulate hypoxia-inducible factor 1alpha differentially in cancer and ischemia. Mol Cell Biol 28:5106–5119. https://doi.org/10.1128/MCB.00060-08

Rajput VD, Harish, Singh RK, Verma KK, Sharma L, Quiroz-Figueroa FR, Meena M, Gour VS, Minkina T, Sushkova S, Mandzhieva S (2021) Recent developments in enzymatic antioxidant Defence mechanism in plants with special reference to abiotic stress. Biology (Basel) 10:267. https://doi.org/10.3390/biology10040267

Reczek CR, Birsoy K, Kong H, Martínez-Reyes I, Wang T, Gao P, Sabatini DM, Chandel NS (2017) A CRISPR screen identifies a pathway required for paraquat-induced cell death. Nat Chem Biol 13:1274–1279. https://doi.org/10.1038/nchembio.2499

Redza-Dutordoir M, Averill-Bates DA (2016) Activation of apoptosis signalling pathways by reactive oxygen species. Biochim Biophys Acta 1863:2977–2992. https://doi.org/10.1016/j.bbamcr.2016.09.012

Rhee SG, Yang K-S, Kang SW, Woo HA, Chang T-S (2005) Controlled elimination of intracellular H_2 O2: regulation of peroxiredoxin, catalase, and glutathione peroxidase via post-translational modification. Antioxid Redox Signal 7:619–626. https://doi.org/10.1089/ars.2005.7.619

Ruiz-Ramos R, Lopez-Carrillo L, Rios-Perez AD, De Vizcaya-Ruíz A, Cebrian ME (2009) Sodium arsenite induces ROS generation, DNA oxidative damage, HO-1 and c-Myc proteins, NF-κB activation and cell proliferation in human breast cancer MCF-7 cells. Mutation Research/Genetic Toxicology and Environmental Mutagenesis 674:109–115. https://doi.org/10.1016/j.mrgentox.2008.09.021

Sadani GR, Nadkarni GD (1996) Role of tissue antioxidant defence in thyroid cancers. Cancer Lett 109:231–235. https://doi.org/10.1016/S0304-3835(96)04484-9

Sadikovic B, Al-Romaih K, Squire J, Zielenska M (2008) Cause and consequences of genetic and epigenetic alterations in human cancer. Curr Genomics 9:394–408. https://doi.org/10.2174/138920208785699580

Sanati M, Afshari AR, Kesharwani P, Sukhorukov VN, Sahebkar A (2022) Recent trends in the application of nanoparticles in cancer therapy: the involvement of oxidative stress. J Control Release 348:287–304. https://doi.org/10.1016/j.jconrel.2022.05.035

Sarmiento-Salinas FL, Perez-Gonzalez A, Acosta-Casique A, Ix-Ballote A, Diaz A, Treviño S, Rosas-Murrieta NH, Millán-Perez-Peña L, Maycotte P (2021) Reactive oxygen species: role in carcinogenesis, cancer cell signaling and tumor progression. Life Sci 284:119942. https://doi.org/10.1016/j.lfs.2021.119942

Schägger H, Pfeiffer K (2001) The Ratio of Oxidative Phosphorylation Complexes I–V in Bovine Heart Mitochondria and the Composition of Respiratory Chain Supercomplexes. J Biol Chem 276:37861–37867. https://doi.org/10.1074/jbc.M106474200

Schieber M, Chandel NS (2014) ROS function in redox signaling and oxidative stress. Curr Biol 24: R453–R462. https://doi.org/10.1016/j.cub.2014.03.034

Shah MA, Rogoff HA (2021) Implications of reactive oxygen species on cancer formation and its treatment. Semin Oncol 48:238–245. https://doi.org/10.1053/j.seminoncol.2021.05.002

Shahid W, Iqbal A, Iqbal I, Mehmood A, Jia H (2024) Application of ferroptosis strategy to overcome tumor therapy resistance in breast and different cancer cells. Iran J Basic Med Sci 27:1085–1095. https://doi.org/10.22038/IJBMS.2024.77465.16752

Shao X, Zhang M, Chen Y, Sun S, Yang S, Li Q (2023) Exosome-mediated delivery of superoxide dismutase for anti-aging studies in Caenorhabditis elegans. Int J Pharm 641:123090. https://doi.org/10.1016/j.ijpharm.2023.123090

Shwetha SD, Shastry AH, Arivazhagan A, Santosh V (2016) Manganese superoxide dismutase (MnSOD) is a malignant astrocytoma specific biomarker and associated with adverse prognosis in p53 expressing glioblastoma. Pathol Res Pract 212:17–23. https://doi.org/10.1016/j.prp.2015.11.002

Siegel RL, Giaquinto AN, Jemal A (2024) Cancer statistics, 2024. CA Cancer J Clin 74:12–49. https://doi.org/10.3322/caac.21820

Sies H (2015) Oxidative stress: a concept in redox biology and medicine. Redox Biol 4:180–183. https://doi.org/10.1016/j.redox.2015.01.002

Singh R, Manna PP (2022) Reactive oxygen species in cancer progression and its role in therapeutics. Explor Med:43–57. https://doi.org/10.37349/emed.2022.00073

Snezhkina AV, Kudryavtseva AV, Kardymon OL, Savvateeva MV, Melnikova NV, Krasnov GS, Dmitriev AA (2019) ROS generation and antioxidant defense Systems in Normal and Malignant Cells. Oxidative Med Cell Longev 2019:1–17. https://doi.org/10.1155/2019/6175804

Son Y, Kim S, Chung H-T, Pae H-O (2013) Reactive oxygen species in the activation of MAP kinases:27–48. https://doi.org/10.1016/B978-0-12-405881-1.00002-1

Sporn MB, Liby KT (2012) NRF2 and cancer: the good, the bad and the importance of context. Nat Rev Cancer 12:564–571. https://doi.org/10.1038/nrc3278

Starkov AA (2008) The role of mitochondria in reactive oxygen species metabolism and signaling. Ann N Y Acad Sci 1147:37–52. https://doi.org/10.1196/annals.1427.015

Van Driel BEM, Lyon H, Hoogenraad DCJ, Anten S, Hansen U, Van Noorden CJF (1997) Expression of CuZn- and Mn-superoxide dismutase in human colorectal neoplasms. Free Radic Biol Med 23:435–444. https://doi.org/10.1016/S0891-5849(97)00102-0

Woldeselassie M, Tamene A (2024) Therapeutic controversies over use of antioxidant supplements during cancer treatment: a scoping review. Front Nutr 11. https://doi.org/10.3389/fnut.2024.1480780

Wu Y, Yun D, Zhao Y, Wang Y, Sun R, Yan Q, Zhang S, Lu M, Zhang Z, Lu D, Li Y (2017) Down regulation of RNA binding motif, single-stranded interacting protein 3, along with up regulation of nuclear HIF1A correlates with poor prognosis in patients with gastric cancer. Oncotarget 8: 1262–1277. https://doi.org/10.18632/oncotarget.13605

Xu B, Dong Q, Yu C, Chen H, Zhao Y, Zhang B, Yu P, Chen M (2024) Advances in research on the activity evaluation, mechanism and structure-activity relationships of natural antioxidant peptides. Antioxidants 13:479. https://doi.org/10.3390/antiox13040479

Zalewska-Ziob M, Adamek B, Kasperczyk J, Romuk E, Hudziec E, Chwalińska E, Dobija-Kubica-K, Rogoziński P, Bruliński K (2019) Activity of antioxidant enzymes in the tumor and adjacent noncancerous tissues of non-small-cell lung cancer. Oxidative Med Cell Longev 2019:1–9. https://doi.org/10.1155/2019/2901840

Zapatero A, Morente M, de Vidales CM, Adrados M, Lopez C, Nieto S, González MJA, Arellano R, Conde AC, Vicente FG (2015) HIF1A expression in localized prostate cancer treated with dose escalation radiation therapy. Cancer Biomark 15:41–46. https://doi.org/10.3233/CBM-140439

Zhou X, An B, Lin Y, Ni Y, Zhao X, Liang X (2023) Molecular mechanisms of ROS-modulated cancer chemoresistance and therapeutic strategies. Biomed Pharmacother 165:115036. https://doi.org/10.1016/j.biopha.2023.115036

Zhou Q, Meng Y, Li D, Yao L, Le J, Liu Y, Sun Y, Zeng F, Chen X, Deng G (2024) Ferroptosis in cancer: from molecular mechanisms to therapeutic strategies. Signal Transduct Target Ther 9:55. https://doi.org/10.1038/s41392-024-01769-5

Zińczuk J, Maciejczyk M, Zaręba K, Romaniuk W, Markowski A, Kędra B, Zalewska A, Pryczynicz A, Matowicka-Karna J, Guzińska-Ustymowicz K (2019) Antioxidant barrier, redox status, and oxidative damage to biomolecules in patients with colorectal cancer. Can malondialdehyde and catalase be markers of colorectal cancer advancement? Biomolecules 9:637. https://doi.org/10.3390/biom9100637

Chapter 7
Therapeutic Role of Oxidative Stress in Diabetes

Sayani Saha, Anjali Vijay Karande, Gokul G. Nair, and Nidhi Srivastava

Abbreviations

A-FABP	Adipocyte fatty acid-binding protein
AGEs	Advanced glycation end products
ALA	Alpha-lipoic acid
AOPPs	Advanced oxidation protein products
ARE	Antioxidant response element
BHA	Butylated hydroxyanisole
BHT	Butylated hydroxytoluene
DM	Diabetes mellitus
ELISA	Enzyme-linked immunosorbent assay
ETC	Electron transport chain
FFA	Free fatty acids
GC-MS	Gas chromatography–mass spectrometry
GLUT-4	Glucose transporter type 4
GSH	Glutathione
iNOS	Inducible nitric oxide synthase
IR	Insulin resistance
IRS	Insulin receptor substrates
NAC	N-acetyl-L-cysteine
NADPH	Nicotinamide adenine dinucleotide phosphate (reduced form)
NDGA	Nordihydroguaiaretic acid
NMN	Nicotinamide mononucleotide
Nrf2	Nuclear factor erythroid 2-related factor 2

S. Saha · A. V. Karande · G. G. Nair · N. Srivastava (✉)
Department of Biotechnology, National Institute of Pharmaceutical Education and Research-Raebareli, Lucknow, India
e-mail: nidhi1.srivastava@niperraebareli.edu.in

N. Srivastava, S. A. Saraf (eds.), *Pharmacological Strategies for Prevention and Treatment of Oxidative Stress*, Oxidative Stress in Applied Basic Research and Clinical Practice, https://doi.org/10.1007/978-3-032-19837-2_7

PI3K	Phosphoinositide 3-kinase
RAAS	Renin–angiotensin–aldosterone system
RAGE	Receptor for advanced glycation end products
RNS	Reactive nitrogen species
ROS	Reactive oxygen species
SOD	Superoxide dismutase
T2DM	Type 2 diabetes mellitus
TBHQ	Tertiary butylhydroquinone
UPR	Unfolded protein response

7.1 Introduction

Diabetes mellitus is a chronic metabolic disorder characterized by elevated blood glucose levels due to defects in insulin production, insulin action, or both. The primary types are Type 1, Type 2, and gestational diabetes. Type 1 diabetes is an autoimmune condition where the pancreas produces little or no insulin, typically manifesting in childhood or adolescence. Type 2 diabetes, accounting for the majority of cases, arises from insulin resistance combined with inadequate insulin secretion and is often associated with obesity and physical inactivity. Gestational diabetes occurs during pregnancy and increases the risk of developing Type 2 diabetes later in life (Antar et al. 2023). The global prevalence of diabetes has been rising steadily, with significant increases observed over the past three decades, particularly in low- and middle-income countries. This trend is largely attributed to lifestyle changes, including increased consumption of calorie-dense foods and reduced physical activity (Zakir et al. 2023). Chronic hyperglycemia in diabetes leads to serious complications affecting various organ systems. Microvascular complications include retinopathy, nephropathy, and neuropathy, while macrovascular complications encompass cardiovascular diseases such as coronary artery disease and stroke. Effective management focuses on glycemic control through lifestyle modifications, pharmacotherapy, and regular monitoring to prevent or delay these complications (Khan et al. 2019). Ongoing research aims to enhance understanding of diabetes pathophysiology and develop innovative treatments. Advancements in technology, such as continuous glucose monitoring systems and insulin pumps, have improved diabetes management, offering better quality of life for patients. Public health initiatives continue to emphasize the importance of preventive measures, early diagnosis, and comprehensive management strategies to address the growing diabetes epidemic (Sugandh et al. 2023).

Oxidative stress occurs when the production of reactive oxygen species (ROS) exceeds the capacity of the body's antioxidant defenses, leading to cellular and molecular damage. ROS include either oxygen radicals like superoxide ($^{\bullet}O_2^{-}$), hydroxyl ($HO^{\bullet}$), hydroperoxyl ($HROO^{\bullet -}$), peroxyl ($ROO^{\bullet}$), or non-radical species such as hydrogen peroxide (H_2O_2) or singlet oxygen ($^1O_2$). While ROS are essential byproducts of metabolism involved in cellular signaling and homeostasis, their excessive accumulation can cause damage and contribute to various diseases

(Sugandh et al. 2023; Andrés Juan et al. 2021). Mitochondria are the primary source of ROS during aerobic respiration, particularly when inefficient electron transfer in the electron transport chain (ETC) generates superoxide. Additional ROS sources include enzymes like NADPH oxidases and xanthine oxidase, as well as environmental factors such as ultraviolet radiation, smoking, and pollution (Tirichen et al. 2021). Antioxidant defenses, both enzymatic and non-enzymatic, counteract ROS to maintain redox balance. Enzymatic antioxidants like superoxide dismutase (SOD), catalase, and glutathione peroxidase detoxify ROS, while non-enzymatic antioxidants, including vitamins C and E and dietary polyphenols, also mitigate oxidative damage. Excess ROS levels lead to lipid peroxidation, protein oxidation, and DNA damage, impairing cellular function and stability (Ighodaro and Akinloye 2018). Oxidative stress is implicated in the development of diseases like neurodegeneration, cardiovascular diseases, diabetes, and cancer. However, controlled ROS production is vital for immune defense and signaling. Maintaining a balance between ROS and antioxidants is essential for cellular homeostasis and health (Pizzino et al. 2017).

Oxidative stress plays a crucial role in the pathophysiology of insulin resistance and diabetes mellitus (DM), as well as in the development of various diabetes-related complications. An increasing body of in vivo evidence indicates that elevated levels of free radicals significantly disrupt normal insulin signaling and glucose homeostasis through multiple mechanisms (Yaribeygi et al. 2020). Oxidative stress, defined as an imbalance between free radical production and antioxidant defenses, impairs insulin sensitivity through several molecular pathways. The authors identify five key mechanisms: β-cell dysfunction, where oxidative stress damages pancreatic β-cells, reducing insulin secretion; inflammatory responses, with oxidative stress activating pro-inflammatory pathways that interfere with insulin signaling; downregulation or mislocalization of glucose transporter Type 4 (GLUT-4), leading to decreased glucose uptake in peripheral tissues; mitochondrial dysfunction, resulting in impaired energy metabolism and increased ROS production; and direct impairment of insulin signaling pathways, where oxidative modifications disrupt the insulin signaling cascade. The review emphasizes that these interconnected mechanisms collectively contribute to the pathophysiology of insulin resistance and diabetes. Understanding these pathways is crucial for developing targeted therapeutic strategies to mitigate oxidative stress and improve insulin sensitivity in diabetic patients (Tangvarasittichai 2015).

7.2 Pathophysiological Role of Oxidative Stress in Diabetes

7.2.1 Generation of Reactive Oxygen Species (ROS) in Diabetes

Reactive Oxygen Species (ROS) are produced in the body for regulation of various physiological processes like immune reactions, aging, and cognitive functions. Overproduction or increased concentration of ROS in the body can lead to diabetes

(Newsholme et al. 2016). ROS are usually formed via the mechanisms like β-cell dysfunction, decreased expression of GLUT-4, mitochondrial dysfunction, inflammatory processes, and various insulin signaling pathway defects. In the body, mitochondria and peroxisomes are the two important organelles responsible for ROS production and the higher glucose content can adversely affect these organelles (Eguchi et al. 2021). *β-cell dysfunction*: β-cells are the cells which secrete insulin in the body and are found in Islets of Langerhans in pancreas. Due to the oxidative stress ROS generation from mitochondria increases. The β-cells don't have that much capacity to resist the increasing oxidative stress and thus they become non-functional (Keane et al. 2015). Non-functional β-cells will cause decrease in insulin production leading to diabetes. *Decreased expression of GLUT-4*: GLUT-4 receptor increases the uptake of insulin into cells and thereby assimilation of glucose. When the oxidative stress in the body increases, it may result in the release of ROS which negatively regulates the expression of the receptor and can cause the decreased expression or sometimes localization of GLUT-4 in cells thereby causing diabetes (Donath and Shoelson 2011). *Mitochondrial dysfunction*: Mitochondrial dysfunction leads to imbalance in the ionic state and hance can lead to oxidative stress. Oxidative stress can lead to production of more ROS, and other organelles are also affected. The ROS accumulation affects insulin resistance and also the destruction of β-cells in pancreas and causes diabetes (Tsalamandris et al. 2019). *Inflammatory responses*: Chronic inflammation can lead to insulin resistance and destruction of β-cells and has a major role in diabetes. Inflammation can lead to cytokine release and affects the normal insulin signaling pathway, thus leading to oxidative stress and hence to diabetes.

ROS generation in β-cells is inter-linked and is very much complicated. Because if ROS is there it can cause destruction of β-cells and when this happens, it in turn exaggerates the ROS production (Villalpando-Rodriguez and Gibson 2021). So, the cumulative effect is under secretion and resistance of insulin which leads to diabetes. Antioxidant capacity of β-cells is not that much efficient to resist the overproduction of ROS as they do not have a higher expression level of CAT and GPx1 (Kozlov et al. 2024). Now let's look deeply at the effect of ROS on β-cells.

7.2.2 Oxidative Stress-Induced Complications: Cardiovascular, Renal, and Neurological

As stated before, oxidative stress causes diabetes and related complications which can be cardiovascular, renal, or neurological (Fig. 7.1). As ROS level increases more, the antioxidant ability of the body diminishes slowly and this is the reason for all those complications (Giacco and Brownlee 2010).

Neurological Brain is an organ which needs ROS produced by mitochondria in a normal concentration. But oxidative stress in the body can easily affect the brain, and this condition is the main underlying cause of many neurological diseases like

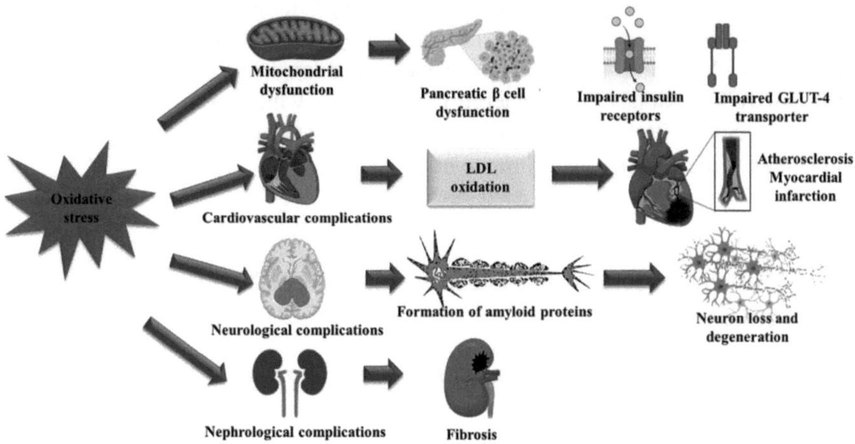

Fig. 7.1 Flow chart illustrating the effect of oxidative stress. It begins with oxidative stress leading to mitochondrial dysfunction, pancreatic β cell dysfunction, impaired insulin receptors, and impaired GLUT-4 transporter. This results in cardiovascular complications, LDL oxidation, and atherosclerosis or myocardial infarction. Neurological complications lead to the formation of amyloid proteins and neuron loss or degeneration. Nephrological complications result in fibrosis. Arrows indicate the progression between stages

Alzheimer's, parkinsonism, Huntington's disease, epilepsy, coma, etc. (Dash et al. 2024). Brain is very much sensitive to ROS because of its high perfusion rate (Watts et al. 2018). The ROS can destroy the neurons especially in the substantia nigrum and can lead to the formation of amyloid plaques and hence cause Alzheimer's disease. In Huntington's disease, the neurons in basal ganglia are destroyed. The neurons or the myelin sheath is destroyed by the oxidative stress (Cheignon et al. 2018).

Cardiovascular Oxidative stress can cause mitochondrial dysfunction, ER dysfunction, endothelial dysfunction, etc. (Picca et al. 2020). Increased ROS generation causes activation of multiple pathways leading to hypertrophy of cardiovascular cells. It can lead to infarction of cardiovascular tissue and may lead to myocardial infarctions or heart attacks (Dubois-Deruy et al. 2020). The oxidative stress leads to oxidation of LDL and causes the deposition on blood vessels, and it can also affect the endothelial cells hence impairing the permeability of blood vessels (Jiang et al. 2022). They also increase vascular smooth muscle proliferation leading to the thickening of blood vessels. Oxidative stress can lead to complications like atherosclerosis, hypertension, and other vascular complications. Sometimes, this can lead to damage in DNA and may lead to mutagenic changes in cells causing cancer (Steven et al. 2019).

Renal Oxidative stress can lead to serious kidney disease like acute/chronic kidney injury, diabetic nephropathy, hypertensive nephropathy, etc. and sometimes permeant damage (Piko et al. 2023). The kidney is also very much affected during oxidative stress as it filters the blood and has a high number of mitochondria. The oxidative stress can lead to glomerular damage by fibrosis; this leads to inefficient filtration of blood and accumulation of waste materials in the body (Gyurászová et al. 2020). The kidney cells undergo inflammation which causes irreversible cell apoptosis, and the ROS also leads to hypertension which causes stress to kidney cells and damages the whole organ (Priante et al. 2019).

7.3 Molecular Pathways Linking Oxidative Stress and Diabetes

7.3.1 Nuclear Factor Erythroid 2-Related Factor 2 (Nrf2) Pathway

Nuclear Factor Erythroid 2-Related Factor 2 (Nrf2) is a transcriptional factor which deals with various enzymes that reduce the complications of oxidative stress. The Nrf2 binds to antioxidant Response Element (ARE) and activates the transcription of antioxidant enzymes and phase II enzymes (Ngo and Duennwald 2022). Normally Kelch-like ECH-associated protein (Keap1) binds with Nrf2 thereby inhibiting from its action. When ROS is present in the cell, the Keap1 undergoes some amino acid modification and gets detached from the Nrf2 and Nrf2 becomes free. This molecule initiates the transcription of antioxidant enzymes thereby neutralizing the ROS effect (Dodson et al. 2022). The Nrf2 is constitutively expressed under normal condition but during oxidative stress condition, more Nrf2 molecules are released (Fig. 7.2). Nrf2 pathway dysregulation can lead to serious disorders (He et al. 2020).

Nrf2 pathway should be in a normal pace. Overactivation or underactivation of this pathway leads to disease condition. Overactivation causes proliferation of cells and causes cancerous cell viability (He et al. 2020). The underactivation of Nrf2 pathway causes complications like diabetes. As the Nrf2 is necessary for the maintenance and proliferation of β-cells in pancreas, its underactivation can cause the destruction or impaired functioning of these cells which leads to decreased release of insulin and leads to diabetes (Rabbani et al. 2019).

7.3.2 Role of Advanced Glycation End Products (AGEs)

Advanced Glycation End Products (AGEs) are formed via Maillard reactions, glycolysis, polyol pathway, and lipid peroxidation. Among these Maillard reactions is the major pathway which produces stable AGEs, and they are known to activate inflammatory response by cytokine release (Khalid et al. 2022). These are toxic to the body as they are formed by non-enzymatic glycation of biomolecules (Peppa et al. 2003). Under hyperglycemic condition, AGEs modify insulin and the insulin becomes non-effective or it affects the release of insulin. The non-effective insulin is

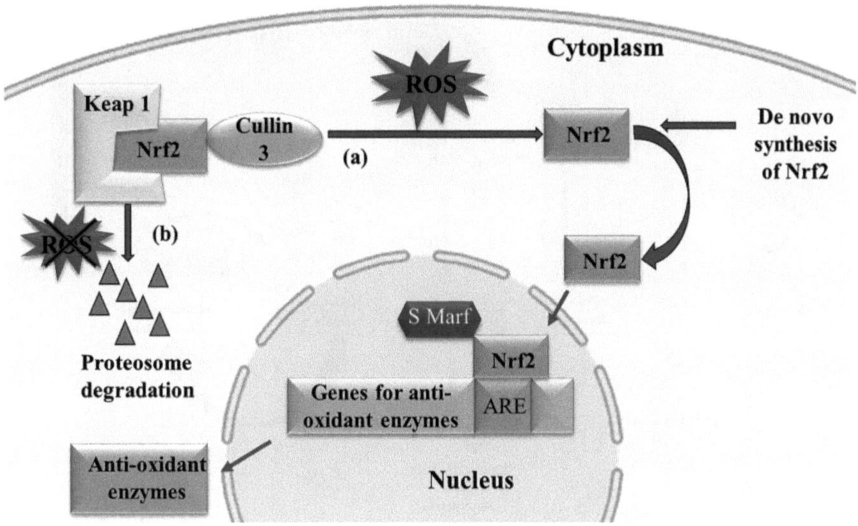

Fig. 7.2 Nrf2 pathway: (**a**) In the presence of ROS, the Nrf2 molecule gets detached from the complex and binds to Antioxidant Response Element (ARE) and initiates the transcription of genes for antioxidant enzymes. (**b**) Nrf2-Keap 1 complex degradation in the absence of ROS

poorly absorbed, and it causes insulin resistance (Nowotny et al. 2015). The AGEs bind to Receptor for Advanced Glycation End products (RAGE) by glycating the albumin and activate various pathways such as JAK/STAT, MAPK, and protein kinase C which alter the insulin signaling pathways. By activating these pathways, RAGE transcription causes the release of NF-κb. The NF-κb leads to inflammation and destruction of pancreatic β-cells hence affecting insulin release (Rhee and Kim 2018). AGE is shown to decrease cytochrome c oxidase activity in mitochondria and hence ATP production leading to impaired secretion of insulin thus causing an increase in glucose level in blood (Reddy et al. 2022). The cytochrome c oxidase enzyme is very much important in the glucose metabolism. Glucose-induced insulin secretion is mediated by cytochrome c oxidase, so cytochrome c oxidase is reduced; it can lead to mitochondrial dysfunction, decreased insulin secretion, β-cells destruction, and again oxidative stress (Aharon-Hananel et al. 2022). It also leads to oxidative stress-induced complications and diabetic induced complications such as cardiovascular diseases, retinopathy, and other neurological disorders (Rungratanawanich et al. 2021) (Fig. 7.3).

7.3.3 Mitochondrial Dysfunction and Oxidative Damage

Mitochondria are known as the powerhouse of the cell as they are the source of ATP for cells. They are also involved in the production of ROS in the body but can get destroyed if the ROS content is very high (Li et al. 2024). There are several enzymes

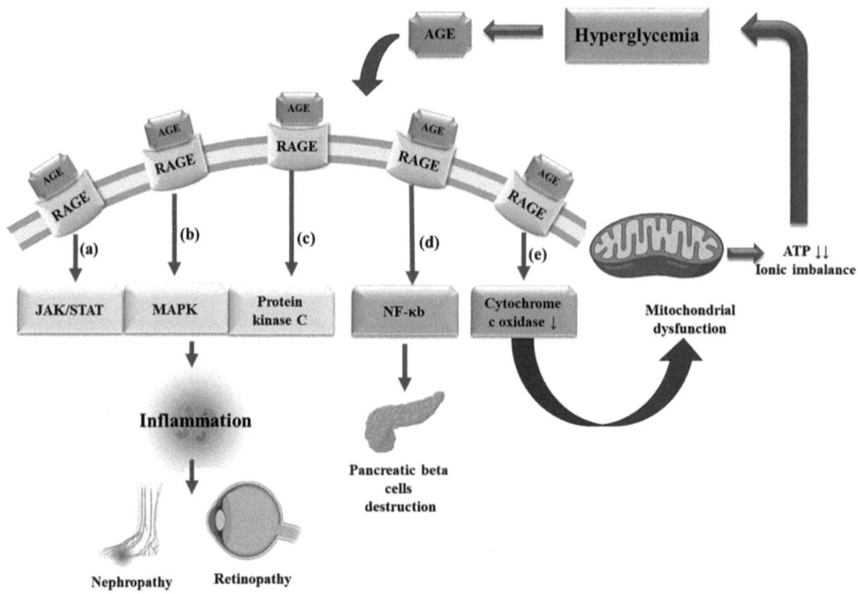

Fig. 7.3 Advanced Glycation End products (AGE) pathway: AGE molecules are formed by the glycation of biomolecules and AGE binds with Receptor for Advanced Glycation End products (RAGE) receptors. (**a**), (**b**), (**c**) represent inflammation pathways via JAK/STAT, MAPK, and protein kinase C. (**d**) represents the destruction of pancreatic beta cells via Nf-kb pathway. (**e**) represents mitochondrial dysfunction via the reduced production of cytochrome c oxidase enzymes leading to the reduced production of ATP and ionic imbalance

to neutralize the ROS produced by mitochondria such as catalase, super oxide dismutase, and glutathione reductase. Mitochondrial dysfunction means the improper functioning of mitochondria leads to ionic imbalance and membrane disruption (Bhatti et al. 2017). Oxidative damage causes increased ROS content in the body, and the antioxidant activity of cells can't cancel it out (Naudi et al. 2012). Oxidative damage causes improper function of NADH dehydrogenase, and cytochrome oxidase enzymes cause serious effect on mitochondrial membrane potential and thus mitochondrial dysfunction (Sivitz and Yorek 2010). The accumulation of ROS leads to the damage of the proteins, lipids, and other components leading to oxidative stress and also can lead to the opening of Mitochondrial Permeability Transition Pore (mPTP) leading to the loss of membrane potential and cell death (Schieber and Chandel 2014).

The membrane dysfunction leads to diabetes by dysregulating the insulin signaling and glucose uptake by cells. The mitochondrial dysfunction leads to accumulation of fatty acids, and this leads to activation of serine threonine kinase activity leading to the inactivation of insulin receptors (Fig. 7.4). Another mechanism is via decreasing activity of P13/Akt pathway (Kim et al. 2008). This decrease in activity affects GLUT-4 transporter leading to impaired uptake of glucose. This leads to accumulation of glucose in blood plasma hence leading to diabetes and insulin resistance (Szendroedi et al. 2009).

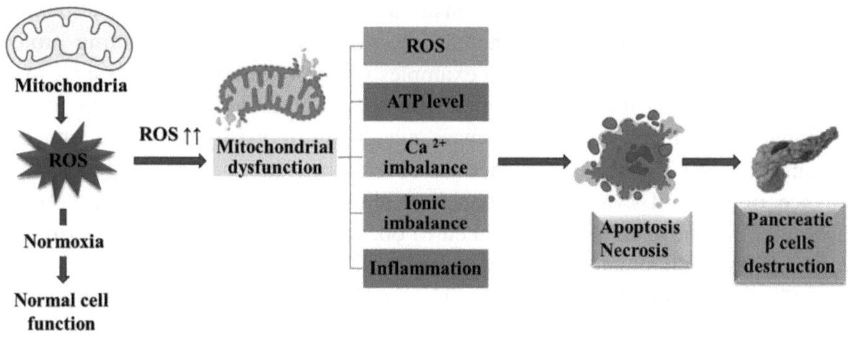

Fig. 7.4 Flow chart illustrating the process of mitochondrial dysfunction leading to pancreatic β-cell destruction. It begins with mitochondria under normoxia, maintaining normal cell function. An increase in reactive oxygen species (ROS) causes mitochondrial dysfunction. This leads to changes such as increased ROS, decreased ATP levels, calcium ion (Ca^2) imbalance, ionic imbalance, and inflammation. These changes result in apoptosis and necrosis, ultimately causing the destruction of pancreatic beta cells

7.4 Insulin Resistance (IR) and Oxidative Stress

Insulin promotes glucose absorption in skeletal muscle and adipose tissue by starting a convoluted series of signaling processes. Phosphoinositide 3-kinase (PI3K) is activated when insulin binds to its receptor and causes downstream tyrosine phosphorylation of protein substrates. This mechanism makes it easier for GLUT4 to move from its intracellular storage locations to the plasma membrane, allowing glucose to enter the cell. A reduced ability of target tissues, including skeletal muscle, the liver, and adipose tissue, to respond to the physiological effects of circulating insulin is known as insulin resistance (IR). Reduced insulin-stimulated glucose uptake in skeletal muscle, poor insulin-mediated hepatic glucose production suppression, and a diminished capacity of insulin to prevent lipolysis in adipose tissue are the outcomes of this disease.

In relation to oxidative stress, this section centers on the manner in which obesity and dietary availability impact inflammatory pathways in three crucial regions: (1) the liver, (2) skeletal muscle and adipose tissue, and (3) the pancreas. Clarifying the molecular processes behind insulin resistance and associated metabolic dysfunctions requires an understanding of these relationships.

7.4.1 Interplay Between Adipose and Skeletal Tissue in Insulin Resistance and Oxidative Stress

Adipocytes, immunological cells (lymphocytes and macrophages), preadipocytes, and endothelial cells are among the several cell types that make up adipose tissue. Adipocytes release proteins like resistin (Jamaluddin et al. 2012) and retinol-binding protein 4 (RBP4) that reduce insulin sensitivity (Palomo 2012) as well as adipokines like leptin and adiponectin that increase it. Adipocyte fatty acid-binding protein (A-FABP) and adiponectin are two examples of adipokines that are thought to be predictive indicators for cardiovascular disease and possible therapeutic targets for its management. Increased body fat mass, a hallmark of obesity, is the outcome of a sustained positive energy balance over several months or years. When energy balance or nutrient availability changes, cellular pathways become responsive. However, insulin resistance (IR) linked to obesity results from the long-term activation of pro-inflammatory pathways in insulin target cells. Increased levels of pro-inflammatory cytokines, including TNF-α, IL-6, and CRP, are closely associated with both diabetes and IR (Wang et al. 2008). In adipose tissue, these peptides—which include hormones, chemokines, and cytokines—have both paracrine and endocrine effects. In insulin-sensitive tissues, TNF-α and IL-6 specifically interfere with insulin signaling, making IR worse (Kalupahana et al. 2012). Oxidative stress and insulin resistance (IR) are intricately connected, with the p66ShcA gene playing a pivotal role as a regulator of apoptotic responses to oxidative stress and a key mediator of IR (Tanaka et al. 2002). P66ShcA not only modulates insulin signaling and promotes fat accumulation but also generates reactive oxygen species (ROS), exacerbating oxidative stress in pathological conditions (Ranieri et al. 2013). Furthermore, it acts as an adaptor molecule, facilitating the assembly of complexes critical to insulin signaling. Oxidative stress activates several serine kinase pathways, including NF-κB, p38 MAPK, and JNK/SAPK, which impair the activity of insulin receptors (IR) and insulin receptor substrates (IRS). Hydrogen peroxide (H_2O_2) has been shown to directly inhibit insulin-stimulated glucose transport (Evans et al. 2002). Additionally, oxidative stress exposure in adipocytes decreases levels of anti-inflammatory adiponectin while increasing pro-inflammatory adipocytokines (Soares et al. 2005). Adiponectin, a protective molecule, mitigates oxidative and nitrosative stress by inhibiting NADPH oxidase and inducible nitric oxide synthase (iNOS) expression, highlighting its critical role in countering oxidative stress-induced IR (Tao et al. 2007). The renin–angiotensin–aldosterone system (RAAS) has also been identified as a contributing factor in mediating oxidative stress within adipocytes (Nguyen Dinh Cat and Touyz 2011). Adipose tissue inflammation, characterized by shifts in immune cell populations and alterations in cytokine profiles, ultimately promotes insulin resistance (IR) in skeletal muscle and the liver, thereby intensifying metabolic dysfunction.

7.4.2 Role of Pancreas in Insulin Resistance and Oxidative Stress

Inflammatory cytokines and oxidative stress produced by immune cells infiltrating pancreatic islets play a significant role in beta-cell destruction, a key feature of Type 2 diabetes. Beta-cell dysfunction and insulin resistance (IR) are central to the progression of this disease. In diabetic conditions, prolonged hyperglycemia leads to pancreatic dysfunction and exacerbates IR, with oxidative stress playing a crucial role in these processes. Beta cells are particularly vulnerable to oxidative stress due to their relatively low expression of antioxidant enzymes like catalase and glutathione peroxidase. In the diabetic state, hyperglycemia and the resulting oxidative stress reduce insulin gene expression and secretion. Markers of oxidative stress, such as 8-hydroxy-2-deoxyguanosine (8-OHdG) and 4-hydroxy-2,3-nonenal (4-HNE), are notably elevated in islets under diabetic conditions. Antioxidant treatments, including probucol (Yamamoto et al. 1983), glutathione peroxidase (Tanaka et al. 2002), N-acetyl-L-cysteine (NAC) (Kaneto et al. 1999), and vitamins C and E, have been shown to prevent the decline in insulin gene expression caused by chronic hyperglycemia.

The activation of the JNK signaling pathway is implicated in the reduction of insulin gene expression induced by oxidative stress, and its suppression helps protect beta cells from oxidative damage. It is thought that JNK activation leads to decreased activity of key pancreatic transcription factors, PDX-1 and Mafa, which are essential for insulin gene transcription (Kaneto et al. 2002). Overexpression of Mafa, in combination with PDX-1, has been shown to improve glucose tolerance in animal models (Kaneto et al. 2005). Beta-cell dysfunction arises from chronic exposure to high glucose or free fatty acids (FFA), or both. Elevated glucose concentrations increase ROS production through enhanced oxidation in the TCA cycle, while high FFA levels in insulin resistance further exacerbate oxidative stress by promoting mitochondrial ROS generation (Poitout and Robertson 2002). Long-term exposure to FFA has been shown to inhibit insulin mRNA and protein synthesis and negatively impact beta-cell function (Maedler et al. 2001).

The pathogenesis of IR is multifactorial, with the endoplasmic reticulum (ER) playing a critical role. The ER is involved in protein folding, maturation, and trafficking, as well as calcium homeostasis (Piperi et al. 2012). ER stress occurs when misfolded or unfolded proteins accumulate, triggering the unfolded protein response (UPR). Protein folding in the ER is closely linked to ROS production, and the interplay between oxidative stress and calcium regulation is critical in understanding the mechanisms underlying insulin resistance and beta-cell dysfunction (Perrin-Sarrado et al. 2007).

7.4.3 The Liver: Insulin Resistance and Oxidative Stress

After a meal, the intestine becomes the primary source of glucose, increasing its availability to the pancreas. This triggers the pancreas to secrete insulin. In one form of insulin resistance (IR), the suppression of hepatic glucose production by insulin is impaired. To compensate, individuals with IR initially maintain normal blood glucose levels through excessive insulin production by the pancreas. However, prolonged hypersecretion of insulin can eventually lead to dysfunction of pancreatic beta cells (Stumvoll et al. 2005). Liver macrophages play a crucial role in maintaining metabolic homeostasis. These cells are present in various tissues, where they respond to metabolic signals by producing pro-inflammatory and/or anti-inflammatory mediators, thereby modulating metabolic processes. While it is well-established that metabolic stressors—such as fatty acids, ceramides, and oxidized LDL/cholesterol—trigger macrophage-driven inflammation, the precise mechanisms by which pro-inflammatory signaling pathways contribute to metabolic diseases, particularly in the context of insulin resistance (IR), remain poorly understood. In the liver, resident macrophages adapt to their local microenvironment, displaying diverse functional and morphological phenotypes. Their capacity to engulf and metabolize lipids links lipid metabolism with inflammation. Moreover, increased lipid accumulation in the liver is recognized as a contributing factor in the development of IR. Increased lipid accumulation in the liver is closely linked to elevated free fatty acid (FFA) flux, which promotes the excessive buildup of fatty acid intermediates, such as ceramide. These lipid intermediates activate intracellular serine kinases, resulting in the inhibition of insulin signaling (Li et al. 2010). The rate of ceramide synthesis is largely determined by the availability of long-chain saturated fats, which are involved in the initial and rate-limiting step of de novo ceramide synthesis (Summers 2006). Ceramide also activates reactive oxygen species (ROS)-generating enzymes, including NADPH oxidase, xanthine oxidase, and nitric oxide (NO) synthase. Furthermore, ceramide has been shown to interact with the mitochondrial electron transport chain, contributing to increased ROS production (Lecour et al. 2006).

7.5 Oxidative Stress and Beta-Cell Dysfunction

Pancreatic beta-cell failure represents the terminal stage in both Type 1 and Type 2 diabetes, driven by several pathological processes, including autoimmune responses, inflammation, insulin resistance, glucotoxicity, lipotoxicity, and beta-cell de- or trans-differentiation (Wajchenberg 2007). Extensive clinical and experimental research has identified non-physiological oxidative stress as a fundamental factor through which these mechanisms impair the structure and function of beta cells, ultimately contributing to the onset of diabetic complications (Robertson 2004). The role of oxidative stress in facilitating various forms of beta-cell death is illustrated in the figure.

Pancreatic beta cells are particularly vulnerable to oxidative stress due to their inherently low antioxidant capacity, which impacts insulin expression and other beta-cell-specific functions (Gerber and Rutter 2017). Comparative analyses of antioxidant enzyme gene expression in human pancreatic islets show that beta cells have significantly lower levels of superoxide-eliminating enzymes, such as CuZn-SOD and Mn-SOD (1.4-fold lower), and hydrogen peroxide-eliminating enzymes, including glutathione peroxidase (GPx1, three-fold lower) and catalase (CAT, 15-fold lower), compared to non-beta cells (Gurgul-Convey et al. 2016). Despite this limitation, recent studies suggest that beta cells can counteract micromolar concentrations of hydrogen peroxide (50 μM) for up to 4 h through a peroxiredoxin/thioredoxin-dependent mechanism, as demonstrated in the human EndoC-βH1 beta cell line (Gurgul-Convey et al. 2016). However, higher concentrations of hydrogen peroxide (100 μM) reduce cell viability, induce DNA damage, and deplete cellular energy, raising concerns about the long-term effectiveness of this antioxidant system under sustained oxidative stress. Low antioxidant capacity also makes beta cells susceptible to reactive oxygen species (ROS)-mediated signaling, which plays a role in glucose-stimulated insulin secretion (GSIS) under physiological conditions (Pi et al. 2010). ROS, primarily generated by the mitochondrial respiratory chain, facilitate GSIS by activating ryanodine receptors (RyR) and increasing intracellular calcium (Llanos et al. 2015).

However, in the context of diabetes pathogenesis, sustained elevations in glycolytic flux caused by impaired glucose clearance lead to excessive ROS and reactive nitrogen species (RNS) production in beta cells. This overproduction can have pathological consequences, including reduced insulin expression and secretion, as well as beta-cell damage. These findings underscore the critical importance of maintaining redox balance and preserving oxidative eustress to ensure proper beta-cell function (Dinić et al. 2016).

7.6 Therapeutic Implications of Oxidative Stress Modulation

7.6.1 Antioxidants as Therapeutic Agents

Antioxidants (Fig. 7.5) are chemicals that regulate the redox state by inhibiting and/or prolonging the oxidation of other substrates, which include food, fat, and oil (Rozoy et al. 2012). These beneficial compounds are very effective and protect

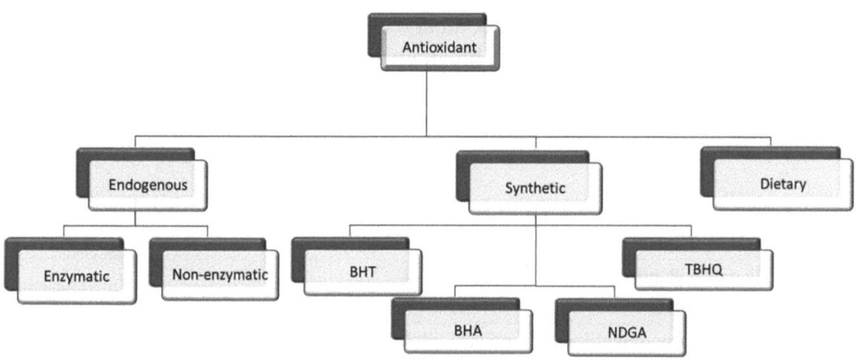

Fig. 7.5 Classification of antioxidant according to origin

against oxidative stress brought on by free radicals through three main mechanisms: ROS scavenging agents help reduce ROS; chelating agents block the production of free radicals by complexing metals; and phenolic agents stop the spread of the reaction by exchanging protons with free radicals (Bazinet and Doyen 2017). Catalytic antioxidants like CAT, SOD, and GPx help reduce harmful substances by breaking down oxygen radicals and other peroxides into water and alcohols (Banerjee and Vats 2014; Marrazzo et al. 2014). Vitamin C can also act as a ROS scavenger by donating an electron to a substrate such as $O^{2\bullet-}$ (Lykkesfeldt et al. 2014). Vitamin C can help remove harmful molecules; certain peptides, carnosine, and anserine, help reduce the production of reactive oxygen species (ROS) by binding to metals such as magnesium, iron, and copper. Trace elements are important for producing reactive oxygen species (ROS), like hydroxyl radicals, which are formed through a reaction involving iron and hydrogen peroxide. Vitamin E, a type of phenolic compound, helps by donating a hydrogen atom from its structure. This process neutralizes harmful lipid peroxide radicals in cell membranes and low-density lipoproteins (LDL) (Jiang 2014). Oxidized vitamin E is stable and does not promote further oxidation. Other antioxidants such as CoQ10, NAC, vitamin D, vitamin B9, and lipoic acid help manage ROS levels differently (Farhangi et al. 2017; Rochette et al. 2014).

7.6.1.1 Endogenous Antioxidants

Endogenous antioxidants, produced by the body, play a crucial role in mitigating oxidative stress. They function by safeguarding, diminishing, or repairing damage to vital biomolecules. There are two primary categories of endogenous antioxidants: enzymatic and non-enzymatic. Major enzymatic antioxidants include catalase, glutathione reductase, glutathione peroxidase, and superoxide dismutase (Table 7.1). Superoxide is converted into hydrogen peroxide and oxygen with the help of superoxide dismutase (SOD), and research indicates that SODs are implicated in various pathological conditions. The glutathione system comprises reduced glutathione and associated enzymes that combat oxidative damage. Catalase is primarily responsible for hydrogen peroxide‘s breakdown into oxygen and water. In contrast, non-enzymatic antioxidants consist of smaller molecules, such as vitamins C and E, which also contribute to the body’s defense against oxidative stress (Santo et al. 2016).

Table 7.1 Types and examples of endogenous antioxidants

Endogenous antioxidants	
Enzymatic antioxidants	Non-enzymatic antioxidants
Superoxide dismutase, glutathione peroxidase, glutathione reductase, catalase	Vitamin C, plant polyphenol, glutathione (GSH), vitamin E

7.6.1.2 Dietary Antioxidants

Naturally occurring antioxidants, which include phytochemicals and vitamins of varying molecular weights synthesized by plants, play a crucial role in health. Important antioxidants have been identified, including alpha-lipoic acid (ALA), uric acid, plant polyphenols, glutathione (GSH), carotenoids, flavonoids, melatonin, polyamines (PAs), curcumin, and vitamins C, E, and A. Dietary antioxidant consumption and protection against diabetic mellitus (DM) are significantly correlated, according to epidemiological research. Additionally, previous studies indicate that antioxidant therapy may protect pancreatic beta cells against apoptosis brought on by oxidative stress, maintaining their functionality and reducing problems associated with diabetes (Kanwugu et al. 2022). Table 7.2 enlists several dietary antioxidants, their sources, and role.

7.6.1.3 Synthetic Antioxidants

Synthetic antioxidants are essential in mitigating oxidative stress linked to diabetes by effectively neutralizing reactive oxygen species (ROS) and minimizing oxidative damage. Compounds such as butylated hydroxytoluene (BHT) and butylated hydroxyanisole (BHA) have demonstrated efficacy in preventing lipid peroxidation, thus safeguarding cell membranes and alleviating oxidative stress in diabetic environments (Banerjee and Vats 2014). Synthetic antioxidants like probucol help prevent fat breakdown and also reduce inflammation and artery hardening in diabetes (Butler et al. 2003). Antioxidants can help reduce oxidative stress and boost the body's defense against it in people with diabetes (Marrazzo et al. 2014). Synthetic compounds, such as edaravone, have demonstrated efficacy in scavenging reactive oxygen species (ROS) and mitigating oxidative damage associated with complications arising from diabetes (Rochette et al. 2014). Phenolic compounds are important parts of synthetic antioxidants. They help reduce oxidative stress, neutralize free radicals, and affect various biological processes related to the harmful effects of free radicals. Some examples of these compounds are Nordihydroguaiaretic acid (NDGA), propyl gallate (a type of gallic acid ester), tertiary butyl hydroquinone (TBHQ), butylated hydroxyl anisole (BHA), and butylated hydroxyl toluene (BHT) (Wolff 1993).

7.6.2 Targeting Mitochondrial Oxidative Stress

Mitochondrial oxidative stress plays a central role in the pathogenesis of diabetes mellitus (DM) and its complications, including retinopathy, nephropathy, and cardiovascular diseases. Excessive production of reactive oxygen species (ROS) by mitochondria, primarily during dysfunctional oxidative phosphorylation, leads to

Table 7.2 List of dietary antioxidants, their sources, and role

Naturally occurring antioxidants	Sources	Role
Vitamin E	Almonds, broccoli, corn oil, eggs, hazelnuts, kiwi fruit, mango, milk, peanut butter, safflower oil, soybean oil, spinach, sunflower oil, tomatoes, wheat germ oil (Rizvi et al. 2014)	Vitamin E enhances the body's ability to utilize glucose more effectively. The substance is key to shielding cells from damage caused by oxidation, mitigating vascular complications associated with hyperglycemia, enhancing beta-cell functionality, and optimizing insulin sensitivity (Koya et al. 1997)
Vitamin C	Blueberries, broccoli, citrus fruits (orange, lemon), grape, kiwi, mango, peppers, strawberries, tomatoes (Lykkesfeldt et al. 2014)	Vitamin C helps protect against complications from diabetes, affecting both small and large blood vessels. It does this by increasing the production of nitric oxide, which is important for widening blood vessels and keeping blood vessel linings healthy (Golbidi et al. 2011; Wollin and Jones 2003)
Vitamin A	Animal sources: Liver, egg yolk (not the white), cheese, butter, and milk (including human breast milk) Plant sources: Sweet potatoes, carrots, maize, mangos, papaya, and various squashes. Other sources: Buriti palm oil and red palm oil (Gilbert 2013)	Vitamin A plays a crucial role in mitigating oxidative stress associated with diabetes, owing to its potent antioxidant properties. It aids in the neutralization of reactive oxygen species (ROS) and promotes the activity of various antioxidant enzymes, thereby contributing to improved metabolic health (Ayeleso et al. 2016)
Carotenoids	Green leafy vegetables: Kale, spinach, broccoli, marigold flower, pumpkin, and carrot. Fruits: Mango, acerola, camu-camu (Myrciaria dubia), rose hip fruits, cherry, guava, watermelon, and Thai papaya. Oils and nuts: Buriti and nuts. Eggs and carotenoids: Egg yolks and beta-carotene. Lycopene sources: Tomato, cherry, guava, watermelon, and Thai papaya (Mezzomo and Ferreira 2016)	Carotenoids are linked to lower levels of fasting plasma glucose and insulin resistance (Ylönen et al. 2003) Carotenoids exhibit significant beneficial effects in diabetes management, primarily by neutralizing oxidative stress, but also through various additional mechanisms. Research indicates that carotenoids can mitigate pancreatic inflammation and enhance insulin sensitivity by modulating immune system responses (Roohbakhsh et al. 2017)

(continued)

Table 7.2 (continued)

Naturally occurring antioxidants	Sources	Role
Flavonoids	Fruits: Apples, grapes, peaches, citrus fruits, and berries. Vegetables: Onions, kale, lettuce, and red pepper. Other sources: Green tea, grape seeds (Panche et al. 2016)	Citrus bioflavonoid supplements help lower oxidative stress and improve diabetes markers in Type 2 diabetes patients (Gupta et al. 2023)
Alpha-lipoic acid	Spinach, broccoli, garden pea, potatoes, carrots, orange and tomato, red meat, yeast (Wollin and Jones 2003)	Alpha-lipoic acid (ALA) has been shown to reduce oxidative stress associated with diabetes mellitus (DM) by enhancing insulin sensitivity, which aids in maintaining glycemic control and mitigating reactive oxygen species (ROS) production linked to hyperglycemia and dyslipidemia. Additionally, ALA exhibits potential in preventing or alleviating pancreatic damage and impaired insulin sensitivity in both Type 1 and Type 2 diabetes (Capece et al. 2022)
Glutathione	Sprouts, garlic, asparagus, cucumber, avocado, cabbage, Brussels, spinach, almonds (Anwar et al. 2018)	Patients diagnosed with Type 2 Diabetes Mellitus (T2DM) who underwent supplementation with glutathione (GSH) exhibited a notable decrease in oxidative stress levels across various blood components (To et al. 2021), reduced oxidative DNA damage (Kalamkar et al. 2022), decreased glycated hemoglobin (HbA1c) (Gaike et al. 2023; Kalamkar et al. 2022) improved insulin sensitivity and increased insulin secretion (Kalamkar et al. 2022; Søndergård et al. 2021)
Polyamines	Cereals, legumes, wheat germs, soybeans, mushrooms, cauliflower, spinach (Larqué et al. 2007)	At physiological pH, polyamines exhibit a strong affinity for DNA and proteins, providing a protective effect against oxidative damage. PAs can help stop glycation because they work as agents that prevent it, and thus they are capable of preventing DM complications (Butkowski and Jelinek 2017)

impaired insulin signaling, beta-cell dysfunction, and chronic inflammation. Targeting mitochondrial oxidative stress represents a promising therapeutic strategy for managing diabetes and its complications (Zhang et al. 2023). Mitochondria-targeted antioxidants, such as mitoquinone (MitoQ) and SkQ1, are specifically designed to accumulate within the mitochondria, neutralizing ROS at their source. These compounds prevent oxidative damage to mitochondrial membranes, proteins, and DNA, thereby preserving mitochondrial function (Modi et al. 2024). Preclinical studies have shown that MitoQ reduces markers of oxidative stress, improves insulin sensitivity, and protects pancreatic beta cells from apoptosis in diabetic models. Pharmacological agents that enhance mitochondrial quality control mechanisms also hold potential in modulating oxidative stress (Sulaimon et al. 2022). Activators of mitophagy, a process that selectively degrades damaged mitochondria, reduce ROS production and restore mitochondrial homeostasis. Compounds such as urolithin A and nicotinamide mononucleotide (NMN) have shown promise in improving mitochondrial function and glucose metabolism in diabetic conditions (Srivastava and Gross 2023).

Inhibiting mitochondrial ROS production directly is another viable approach. Agents targeting mitochondrial electron transport chain dysfunction, such as rotenone analogs or specific inhibitors of Complexes I and III, reduce superoxide generation without impairing energy metabolism. These interventions can help maintain cellular redox balance while preserving ATP production (Zorov et al. 2014). Lifestyle interventions, including exercise and caloric restriction, indirectly target mitochondrial oxidative stress by improving mitochondrial biogenesis and function. These approaches enhance the expression of PGC-1α, a transcriptional coactivator that regulates the formation of new mitochondria and antioxidant enzyme activity (Abu Shelbayeh et al. 2023).

In summary, targeting mitochondrial oxidative stress provides a focused strategy to mitigate diabetes-related oxidative damage. Mitochondria-specific antioxidants, quality control modulators, and lifestyle interventions show promise in preserving mitochondrial function, improving metabolic health, and reducing the burden of diabetes-related complications. Further clinical research is essential to validating these approaches and optimizing their application in diabetes management (Green et al. 2004).

7.7 Markers and Assessment of Oxidative Stress in Diabetes

Oxidative stress arises from a disruption in the equilibrium between the production of reactive oxygen species (ROS) and the synthesis of antioxidants, leading to an increase in ROS generation (Betteridge 2000). The optimal levels of the body's antioxidant reserves are influenced by hyperglycemia, a condition commonly associated with diabetes (Maschirow et al. 2015). Diabetes and its associated hyperglycemia significantly impact the redox state, leading to a reduction in levels of reduced glutathione (GSH) and an elevation in oxidized glutathione (GSSG) (Nwose et al. 2010). 8-OHdG, a marker of oxidative stress in endothelial cells, results from the

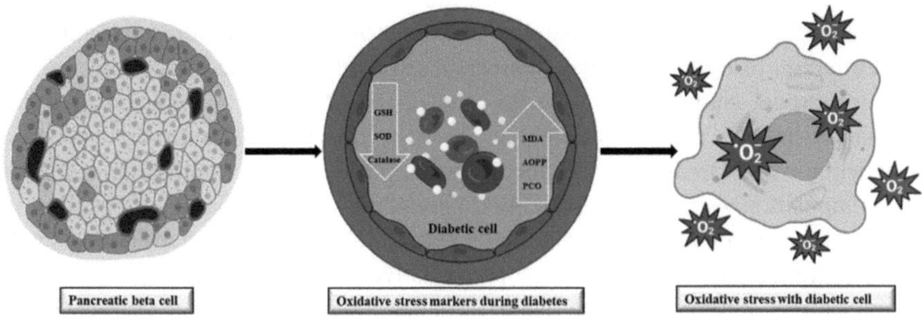

Fig. 7.6 Schematic representation of the status of oxidative stress markers during diabetes. Diagram illustrating the progression of oxidative stress in pancreatic beta cells during diabetes. The first panel shows a healthy pancreatic beta cell with various cellular components. The second panel depicts a diabetic cell with oxidative stress markers, including decreased levels of GSH, SOD, and Catalase, and increased levels of MDA, AOPP, and PCO. The third panel illustrates oxidative stress effects, with reactive oxygen species indicated by red starbursts labeled "O_2". Arrows indicate the progression from a healthy to a stressed state

modification of DNA bases due to the oxidation of deoxyguanosine. Elevated urinary excretion of 8-OHdG has been documented in individuals with prediabetes and Type 2 Diabetes Mellitus (T2DM) when compared to healthy individuals. This biomarker is regarded as a valuable tool for the early detection of both microvascular and macrovascular complications associated with T2DM (Al-Aubaidy and Jelinek 2010). Isoprostanes, stable products formed in vivo through the peroxidation of arachidonic acid as a result of free radical activity, serve as valuable biomarkers for oxidative stress. Research indicates that urinary excretion levels of 8-iso-PGF2α are significantly elevated in diabetic patients compared to healthy controls (Jelinek et al. 2014). Oxidative stress associated with diabetes mellitus leads to numerous detrimental effects on cellular physiology (Fig. 7.6). This phenomenon is especially critical and hazardous for the pancreatic islets, which are characterized by some of the lowest intrinsic antioxidant defenses among various tissues (Fiorentino et al. 2013; Folli et al. 2011).

7.7.1 Biomarkers of Oxidative Stress

Lipid Peroxidation Lipids are recognized as significant targets of reactive oxygen species (ROS). Hydroperoxides exert toxic effects on cellular structures, both directly and through their degradation into highly reactive hydroxyl radicals. Additionally, they can interact with transition metals, such as iron and copper, leading to the formation of stable aldehydes, including malondialdehyde (MDA), which can compromise the integrity of cell membranes (Halliwell and Chirico 1993). Lipid peroxidation generates a range of highly reactive aldehydes, such as malondialdehyde (MDA), acrolein, 4-hydroxynonenal (HNE), 4-oxononenal

(ONE), and isolevuglandins (IsoLGs) (Guo et al. 2012). Research indicates that peroxyl radicals possess the capability to extract hydrogen from lipids, resulting in the formation of hydroperoxides, which subsequently facilitate the continuation of the free-radical chain reaction (Lobo et al. 2010). MDA is recognized as a key biomarker indicative of lipid damage caused by free radicals and oxidative stress (Shodehinde and Oboh 2013). Diabetes has a profound impact on lipid metabolism and composition, especially in patients experiencing vascular complications (Fowler 2008). The elevated levels of malondialdehyde (MDA) observed in individuals with diabetes indicate a potential role of peroxidative damage in the onset of diabetic complications. Furthermore, the rise in lipid peroxidation serves as a marker for the deterioration of both enzymatic and non-enzymatic antioxidant defense mechanisms (Saddala et al. 2013). Oxidized lipids can generate malondialdehyde (MDA) as a decomposition product. The underlying mechanism is believed to involve the formation of prostaglandins, such as endoperoxides, derived from polyunsaturated fatty acids (PUFAs) that contain two or more double bonds (Pandey and Rizvi 2011). Elevated levels of malondialdehyde (MDA) in plasma, serum, and various tissues have been documented in individuals with diabetes (Aldahr and Abd El-Kader 2022).

Protein Oxidation Proteins are essential biomolecules that play a critical role in cellular functions. They are integral to various physiological processes, including cell signaling and transport mechanisms. Additionally, proteins represent a significant target for reactive oxygen species (ROS), which can alter their structure and functionality through modifications. Key side chain targets for protein oxidation include amino acids such as cysteine, methionine, and tyrosine. The formation of carbonyls serves as a byproduct of protein oxidation and is recognized as a potent biomarker for oxidative stress (Suzuki and Miyata 1999). Advanced oxidation protein products (AOPPs) have emerged as a significant biomarker for assessing protein oxidation in the context of oxidative stress. They provide a comprehensive indication of the protein status within cellular and tissue environments (Witko-Sarsat et al. 1996). In the context of chronic oxidative stress, advanced oxidation protein products (AOPPs) are generated through interactions between plasma proteins and chlorinated oxidants. Elevated levels of AOPPs have been observed in individuals with Type 2 diabetes. A significant positive correlation between plasma concentrations of AOPPs and thiobarbituric acid reactive substances (TBARS) during diabetes suggests that both proteins and lipids are similarly affected by reactive oxygen species (ROS) (Savini et al. 2013). The oxidative modification of proteins in individuals with diabetes has significant implications for various physiological functions (Bansal et al. 2013; Piwowar et al. 2007). The elevation of protein carbonyls and advanced oxidation protein products (AOPPs) in diabetic patients highlights the significance of protein conformational alterations in the pathogenesis of diabetic nephropathy (Piarulli et al. 2009). Advanced Oxidation Protein Products (AOPPs), recognized for their pro-inflammatory and pro-oxidative properties, accumulate in aging individuals with diabetes and may significantly contribute to the heightened incidence of endothelial dysfunction and subsequent cardiovascular diseases. AOPPs are characterized by a high concentration of dityrosines, which facilitate crosslinking, the formation of disulfide bridges, and the presence of carbonyl groups. These compounds primarily arise from the action of chlorinated oxidants, including hypochlorous acid and chloramines, which are products of myeloperoxidase activity (Piwowar 2010). Numerous studies have indicated that advanced oxidation protein products (AOPPs) and markers of oxidative stress are elevated in adult individuals diagnosed with Type 2 diabetes, regardless of the presence of microvascular or macrovascular complications (Piwowar et al. 2007; Gradinaru et al. 2013).

Glutathione Level Glutathione (GSH), a tripeptide consisting of γ-L-glutamyl-L-cysteinyl glycine, is present in all mammalian tissues at concentrations of 1–10 mM, with the liver containing the highest levels. It is recognized as the most prevalent non-protein thiol, playing a vital role in safeguarding against oxidative stress (Lu 2013). Glutathione (GSH) plays a key role in keeping protein thiol groups in their reduced form of facilitating amino acid transport, detoxifying reactive radicals, serving as a coenzyme in various enzymatic reactions, and protecting tissues from damage (Tsai et al. 2012). The compound serves as a highly effective antioxidant present in all living cells; it is recognized as a critical biomarker for redox imbalance at the cellular level (Chakravarty and Rizvi 2011). Many studies show that people with diabetes have lower levels of glutathione (GSH) (Calabrese et al. 2012; Rahigude et al. 2012). A reduction in glutathione (GSH) levels may contribute to oxidative DNA damage observed in individuals with Type 2 diabetes (Dinçer et al. 2002). Due to elevated oxidative stress, there was a notable fluctuation in the concentration of glutathione (GSH). Furthermore, the plasma GSH/GSSG ratio exhibited a significant reduction in individuals with Type 2 diabetes when compared to healthy control (Calabrese et al. 2012). Hyperlipidemia, inflammation, and altered antioxidant profiles are common complications associated with diabetes mellitus, primarily resulting from a decreased GSH/GSSG ratio (Das et al. 2012). Abnormal glutathione (GSH) levels are associated with β-cell dysfunction and contribute to the development of long-term complications related to diabetes. This dysregulation is significantly implicated in various disease states (Livingstone and Davis 2007). The enzyme glutathione reductase (GSR) is crucial for cellular redox balance, facilitating the conversion of oxidized glutathione (GSSG) to its reduced form (GSH) while concurrently oxidizing NADPH to NAD+. While GSSG lacks antioxidant capabilities, GSH can be regenerated from GSSG through the action of GSR, utilizing NADPH as a necessary cofactor. However, the efficacy of this GSH regeneration system may be compromised when reactive oxygen species (ROS) are produced in excessive amounts (Morris et al. 2013). Uncontrolled Type 2 diabetes is associated with significantly impaired synthesis of glutathione (GSH), primarily due to a scarcity of available precursors. Supplementation with GSH precursor amino acids has the potential to enhance GSH synthesis, thereby reducing oxidative stress and mitigating oxidant damage in the context of ongoing hyperglycemia (Sekhar et al. 2011). Research indicates that glutathione (GSH) deficiency in individuals with diabetes heightens their vulnerability to melioidosis. It is proposed that the preservation of the GSH redox state could serve as a potential therapeutic strategy to safeguard diabetic patients from certain intracellular bacterial pathogens (Tan et al. 2012).

Catalase Catalase is a crucial antioxidative enzyme found in nearly all living organisms. It plays a significant role in mitigating complications arising from oxidative stress, including conditions such as diabetes (Chelikani et al. 2004). Catalase serves as the primary regulator of hydrogen peroxide metabolism. Hydrogen peroxide is a highly reactive small molecule that is produced naturally as a byproduct of energy metabolism. Elevated levels of hydrogen peroxide can lead to substantial damage to proteins, DNA, RNA, and lipids (Takemoto et al. 2009). Catalase plays a crucial role in the enzymatic conversion of hydrogen peroxide into oxygen and water, effectively neutralizing this potentially harmful compound. Research has indicated a heightened risk of diabetes in individuals with catalase deficiency. This deficiency results in increased oxidative stress within the β-cells, ultimately leading to their dysfunction. Given that β-cells are abundant in mitochondria, these organelles may serve as a significant source of reactive oxygen species (ROS) (Goth and Eaton 2000). Catalase plays a crucial role in safeguarding pancreatic β-cells from oxidative damage induced by hydrogen peroxide (Tiedge et al. 1998). Reduced catalase activity has been observed in individuals diagnosed with schizophrenia and atherosclerosis (Góth and Vitai 1996). The findings align with the hypothesis that prolonged oxidative stress may play a significant role in the onset of various late-onset disorders, including Type 2 diabetes (Góth 2000). The deficiency of catalase leads to an increase in mitochondrial reactive oxygen species (ROS) and fibronectin expression in response to free fatty acids. This effect can be effectively mitigated through the overexpression of catalase or the administration of N-acetylcysteine (Hwang et al. 2012). Reduced catalase activity can lead to methemoglobinemia and hemolytic anemia, potentially resulting from a deficiency in glucose-6-phosphate dehydrogenase or other unidentified factors. This deficiency may also adversely affect heme proteins, induce cellular apoptosis, and, in conjunction with redox-active metal ions, generate highly toxic hydroxyl radicals (Góth et al. 2005; Góth and Bigler 2007).

Superoxide Dismutase Superoxide dismutase (SOD) is an essential antioxidant enzyme that catalyzes the conversion of superoxide anions (O^{2-}) into hydrogen peroxide (H_2O_2) and molecular oxygen (O2). thereby playing a vital role in cellular defense against oxidative stress (Faraci and Didion 2004; Wang et al. 2012). Superoxide dismutase (SOD) serves critical protective functions by mitigating cellular and histological damage induced by reactive oxygen species (ROS). It facilitates the catalytic conversion of superoxide radicals into hydrogen peroxide, which is subsequently processed by other enzymes into molecular oxygen and water (Davari et al. 2013). All mammalian tissues are known to express three distinct forms of superoxide dismutase (SOD): copper–zinc superoxide dismutase (Cu-Zn-SOD), manganese superoxide dismutase (Mn-SOD), and extracellular superoxide dismutase (EC-SOD). Each of these isoforms is encoded by a separate gene, highlighting the complexity and specificity of antioxidant defense mechanisms in mammalian biology (Beyer et al. 1991; Li et al. 2012; Oury et al. 1996). Copper–zinc superoxide dismutase (Cu-Zn-SOD or SOD 1, EC 1.15.1.1) is primarily found in the cytosol, while manganese superoxide dismutase (Mn-SOD or SOD 2, EC 1.15.1.1) is located in the mitochondria. Additionally, extracellular superoxide dismutase (EC-SOD or SOD 3, EC 1.15.1.1) is present in the extracellular space (Guzik et al. 2002; Zelko et al. 2002). Superoxide exhibits a rapid reaction with nitric oxide (NO), leading to a decrease in NO bioactivity and the formation of the oxidative peroxynitrite radical (Guzik et al. 2002). Superoxide dismutase (SOD), a critical antioxidant enzyme, plays a significant role in mitigating elevated superoxide levels in the kidneys during the progression of murine diabetic nephropathy, and, coupled with the downregulation of renal SOD isoforms (SOD1 and SOD3), may contribute significantly to the pathogenesis of the condition (Fujita et al. 2009). The enhancement of superoxide dismutase (SOD) levels or the administration of antioxidant supplements, including SOD mimetics, aimed at mitigating oxidative stress, has demonstrated efficacy in reducing reactive oxygen species (ROS) and elevating antioxidant enzyme activity, thereby contributing to the prevention of diabetes mellitus (Wang et al. 2011). EC-SOD is found in the extracellular matrix of tissues like the pancreas, muscles, and blood vessels. It is the main defender against superoxide radicals outside cells, helping to control oxidative stress (Fattman et al. 2003). Increased levels of EC-SOD led to a six-fold enhancement in the total superoxide dismutase activity within the islets. Consequently, the superoxide radicals released into the extracellular space do not contribute to the destruction of β-cells. The increased levels of superoxide dismutase (SOD) have been demonstrated to mitigate oxidative stress, inhibit the mitochondrial release of cytochrome C, and reduce apoptosis in neuronal cells. Additionally, studies in murine models indicate that SOD may prevent glomerular injury induced by diabetes, highlighting its significant role in the regulation of apoptotic processes (Kowluru et al. 2006). Many studies have shown that the amount of superoxide dismutase (SOD) is lower in the tissues and blood of people with diabetes (Giugliano et al. 1995; He et al. 2011; Shukla et al. 2012). Recent research by Kim (2013) indicates that diabetic skin tissues exhibit a diminished expression of extracellular proteins. The study concludes that extracellular superoxide dismutase (SOD) is associated with the altered metabolic conditions present in diabetic skin, which contributes to an increase in reactive oxygen species (ROS) production (Kim 2013).

7.8 Diagnostic Tools and Methods

Various studies have utilized sophisticated analytical chemistry methods, including gas chromatography–mass spectrometry (GC-MS) and high-performance liquid chromatography (HPLC), to identify markers that reflect oxidative stress levels. With advancements in antibody production technology, it has become feasible to generate specific antibodies targeting stable protein adducts formed through oxidative processes in vivo or in vitro. As a result, the enzyme-linked immunosorbent

Table 7.3 Oxidative stress-related components and processes, diagnostic tools and methods, and limitations and confoundings

Oxidative stress-related components and processes	Diagnostic tools and methods	Limitations and confoundings
Reversible cysteine modifications S-glutathionylation GSH/GSSG SH (Federici et al. 2024)	MS, ELISA, WB Spectrophotometric Methods (Asim et al. 2024; Nasri et al. 2023)	It is necessary to use specialist equipment for a precise quantification (Krishnamurthy et al. 2024)
ROS-regulated transcription factors Nrf-2, NF-kB (Kasai et al. 2020; Patel et al. 2022)	Immunological techniques, RT-PCR Methods (Nasri et al. 2023)	The specificity of antibodies is influenced by varying levels of leukocytes populations (Marcon et al. 2015)
ROS-generating enzymes NOX, MPO, XO, NOS (Kozlov et al. 2024; Montezano and Touyz 2012)	Immunological techniques, WB, PCR, RT-PCR, enzymatic Methods (Nasri et al. 2023)	The specificity of antibodies is influenced by varying levels of leukocytes populations (Marcon et al. 2015)
Antioxidant enzymes SOD, CAT, GPX, GRz (Kozlov et al. 2024)	Immunological techniques WB, PCR, RT-PCR, enzymatic (Nasri et al. 2023)	The specificity of antibodies is influenced by varying levels of leukocytes populations (Marcon et al. 2015)

assay (ELISA), known for its high throughput and versatility, has rapidly gained prominence for analyzing clinical samples. ELISA has become a primary method for assessing oxidative stress in various diseases due to its efficiency and cost-effectiveness. It eliminates the need for expensive equipment and complex sample pretreatments, making it a practical and labor-saving approach. The method is highly reproducible, allows simultaneous measurement of multiple markers, and provides rapid results with minimal sample requirements. These attributes make ELISA exceptionally well-suited for clinical applications. This review aims to highlight recent advancements in the use of ELISA for studying oxidative stress, particularly its potential in investigating diabetes-related oxidative damage. Since directly measuring reactive oxygen species (ROS) is challenging, ELISA-based methods are broadly divided into two categories: detecting evidence of oxidative stress and identifying changes in antioxidant systems in response to ROS. This differentiation underscores ELISA's capability to provide valuable insights into the complex interplay of oxidative and antioxidative mechanisms (Brett et al. 1993). Table 7.3 enlists oxidative stress-related components and processes, diagnostic tools and methods, and limitations and confoundings.

7.9 Conclusion

Diabetes mellitus is a complex metabolic disorder influenced by genetic, environmental, and lifestyle factors. Oxidative stress plays a critical role in the onset and progression of diabetes, contributing to insulin resistance, β-cell dysfunction, and various microvascular and macrovascular complications. The excessive production of reactive oxygen species (ROS) disrupts insulin signaling, promotes inflammation, and damages key metabolic tissues, including the pancreas, liver, adipose tissue, and skeletal muscle. Furthermore, mitochondrial dysfunction and the accumulation of advanced glycation end products (AGEs) exacerbate oxidative damage, perpetuating metabolic dysfunction.

Understanding the molecular pathways linking oxidative stress and diabetes has opened new avenues for therapeutic interventions. Targeting oxidative stress through endogenous and dietary antioxidants, as well as pharmacological agents, offers potential strategies for improving insulin sensitivity and preserving β-cell function. Moreover, advances in continuous glucose monitoring and oxidative stress biomarkers provide valuable tools for early detection and personalized treatment approaches.

Given the rising global prevalence of diabetes, public health initiatives must prioritize preventive measures, including lifestyle modifications such as a balanced diet, regular physical activity, and stress management. Future research should continue exploring novel antioxidant therapies, mitochondrial-targeted interventions, and innovative drug candidates that mitigate oxidative stress-induced metabolic disturbances. A comprehensive approach that integrates oxidative stress management with traditional diabetes care can improve patient outcomes and reduce the burden of diabetes-related complications. Addressing oxidative stress as a central factor in diabetes pathophysiology will be crucial in developing more effective and sustainable treatment strategies for this growing epidemic.

References

Abu Shelbayeh O, Arroum T, Morris S, Busch KB (2023) PGC-1α is a master regulator of mitochondrial lifecycle and ROS stress response. Antioxidants 12:1075. https://doi.org/10.3390/antiox12051075

Aharon-Hananel G, Romero-Afrima L, Saada A, Mantzur C, Raz I, Weksler-Zangen S (2022) Cytochrome c oxidase activity as a metabolic regulator in pancreatic beta-cells. Cells 11:929. https://doi.org/10.3390/cells11060929

Al-Aubaidy HA, Jelinek HF (2010) 8-Hydroxy-2-deoxy-guanosine identifies oxidative DNA damage in a rural prediabetes cohort. Redox Rep 15:155–160. https://doi.org/10.1179/174329210X12650506623681

Aldahr MHS, Abd El-Kader SM (2022) Impact of exercise on renal function, oxidative stress, and systemic inflammation among patients with type 2 diabetic nephropathy. Afr Health Sci 22: 286–295. https://doi.org/10.4314/ahs.v22i3.30

Andrés Juan C, Pérez M, de la Lastra J, Plou FJ, Pérez-Lebeña E, Reinbothe S (2021) Molecular sciences the chemistry of reactive oxygen species (ROS) revisited: outlining their role in biological macromolecules (DNA, lipids and proteins) and induced pathologies. Int J Mol Sci 22:4642

Antar SA, Ashour NA, Sharaky M, Khattab M, Ashour NA, Zaid RT, Roh EJ, Elkamhawy A, Al-Karmalawy AA (2023) Diabetes mellitus: classification, mediators, and complications; a gate to identify potential targets for the development of new effective treatments. Biomed Pharmacother 168:115734. https://doi.org/10.1016/j.biopha.2023.115734

Anwar H, Hussain G, Mustafa I (2018) Antioxidants from natural sources, in: antioxidants in foods and its applications. InTech. https://doi.org/10.5772/intechopen.75961

Asim M, Awan R, Ur Rashid H, Hussain F (2024) Oxidative stress in Type-2 diabetes mellitus. In: Biomarkers of oxidative stress. Springer Nature Switzerland, Cham, pp 103–121. https://doi.org/10.1007/978-3-031-69962-7_5

Ayeleso A, Brooks N, Oguntibeju O, Mukwevho E (2016) Natural antioxidant vitamins: a review of their beneficial roles in management of diabetes mellitus and its complications. Trop J Pharm Res 15:1341. https://doi.org/10.4314/tjpr.v15i6.30

Banerjee M, Vats P (2014) Reactive metabolites and antioxidant gene polymorphisms in Type 2 diabetes mellitus. Redox Biol 2:170–177. https://doi.org/10.1016/j.redox.2013.12.001

Bansal S, Chawla D, Siddarth M, Banerjee BD, Madhu SV, Tripathi AK (2013) A study on serum advanced glycation end products and its association with oxidative stress and paraoxonase activity in type 2 diabetic patients with vascular complications. Clin Biochem 46:109–114. https://doi.org/10.1016/j.clinbiochem.2012.10.019

Bazinet L, Doyen A (2017) Antioxidants, mechanisms, and recovery by membrane processes. Crit Rev Food Sci Nutr 57:677–700. https://doi.org/10.1080/10408398.2014.912609

Betteridge DJ (2000) What is oxidative stress? Metabolism 49:3–8. https://doi.org/10.1016/S0026-0495(00)80077-3

Beyer W, Imlay J, Fridovich I (1991) Superoxide dismutases. Prog Nucleic Acid Res Mol Biol 40: 221–253. https://doi.org/10.1016/S0079-6603(08)60843-0

Bhatti JS, Bhatti GK, Reddy PH (2017) Mitochondrial dysfunction and oxidative stress in metabolic disorders — a step towards mitochondria based therapeutic strategies. Biochim Biophys Acta Mol Basis Dis 1863:1066–1077. https://doi.org/10.1016/j.bbadis.2016.11.010

Brett J, Schmidt AM, Yan SD, Zou YS, Weidman E, Pinsky D, Nowygrod R, Neeper M, Przysiecki C, Shaw A (1993) Survey of the distribution of a newly characterized receptor for advanced glycation end products in tissues. Am J Pathol 143:1699–1712

Butkowski EG, Jelinek HF (2017) Hyperglycaemia, oxidative stress and inflammatory markers. Redox Rep 22:257–264. https://doi.org/10.1080/13510002.2016.1215643

Butler AE, Janson J, Bonner-Weir S, Ritzel R, Rizza RA, Butler PC (2003) β-Cell deficit and increased β-cell apoptosis in humans with type 2 diabetes. Diabetes 52:102–110. https://doi.org/10.2337/diabetes.52.1.102

Calabrese V, Cornelius C, Leso V, Trovato-Salinaro A, Ventimiglia B, Cavallaro M, Scuto M, Rizza S, Zanoli L, Neri S, Castellino P (2012) Oxidative stress, glutathione status, sirtuin and cellular stress response in type 2 diabetes. Biochim Biophys Acta 1822:729–736. https://doi.org/10.1016/j.bbadis.2011.12.003

Capece U, Moffa S, Improta I, Di Giuseppe G, Nista EC, Cefalo CMA, Cinti F, Pontecorvi A, Gasbarrini A, Giaccari A, Mezza T (2022) Alpha-lipoic acid and glucose metabolism: a comprehensive update on biochemical and therapeutic features. Nutrients 15:18. https://doi.org/10.3390/nu15010018

Chakravarty S, Rizvi SI (2011) Day and night GSH and MDA levels in healthy adults and effects of different doses of melatonin on these parameters. Int J Cell Biol 2011:1–5. https://doi.org/10.1155/2011/404591

Cheignon C, Tomas M, Bonnefont-Rousselot D, Faller P, Hureau C, Collin F (2018) Oxidative stress and the amyloid beta peptide in Alzheimer's disease. Redox Biol 14:450–464. https://doi.org/10.1016/j.redox.2017.10.014

Chelikani P, Fita I, Loewen PC (2004) Diversity of structures and properties among catalases. Cell Mol Life Sci 61:192–208. https://doi.org/10.1007/s00018-003-3206-5

Das J, Vasan V, Sil PC (2012) Taurine exerts hypoglycemic effect in alloxan-induced diabetic rats, improves insulin-mediated glucose transport signaling pathway in heart and ameliorates cardiac oxidative stress and apoptosis. Toxicol Appl Pharmacol 258:296–308. https://doi.org/10.1016/j.taap.2011.11.009

Dash UC, Bhol NK, Swain SK, Samal RR, Nayak PK, Raina V, Panda SK, Kerry RG, Duttaroy AK, Jena AB (2024) Oxidative stress and inflammation in the pathogenesis of neurological disorders: mechanisms and implications. Acta Pharm Sin B 15(1):15–34. https://doi.org/10.1016/j.apsb.2024.10.004

Davari S, Talaei SA, Alaei H, salami M (2013) Probiotics treatment improves diabetes-induced impairment of synaptic activity and cognitive function: behavioral and electrophysiological proofs for microbiome–gut–brain axis. Neuroscience 240:287–296. https://doi.org/10.1016/j.neuroscience.2013.02.055

Dinçer Y, Akçay T, Alademir Z, İlkova H (2002) Assessment of DNA base oxidation and glutathione level in patients with type 2 diabetes. Mutat Res Fundam Mol Mech Mutagen 505:75–81. https://doi.org/10.1016/S0027-5107(02)00143-4

Dinić S, Grdović N, Uskoković A, Đorđević M, Mihailović M, Jovanović JA, Poznanović G, Vidaković M (2016) CXCL12 protects pancreatic β-cells from oxidative stress by a Nrf2-induced increase in catalase expression and activity. Proc Jpn Acad Ser B Phys Biol Sci 92: 436–454. https://doi.org/10.2183/pjab.92.436

Dodson M, Shakya A, Anandhan A, Chen J, Garcia JGN, Zhang DD (2022) NRF2 and diabetes: the good, the bad, and the complex. Diabetes 71:2463–2476. https://doi.org/10.2337/db22-0623

Donath MY, Shoelson SE (2011) Type 2 diabetes as an inflammatory disease. Nat Rev Immunol 11: 98–107. https://doi.org/10.1038/nri2925

Dubois-Deruy E, Peugnet V, Turkieh A, Pinet F (2020) Oxidative stress in cardiovascular diseases. Antioxidants 9:864. https://doi.org/10.3390/antiox9090864

Eguchi N, Vaziri ND, Dafoe DC, Ichii H (2021) The role of oxidative stress in pancreatic β cell dysfunction in diabetes. Int J Mol Sci 22:1509. https://doi.org/10.3390/ijms22041509

Evans JL, Goldfine ID, Maddux BA, Grodsky GM (2002) Oxidative stress and stress-activated signaling pathways: a unifying hypothesis of type 2 diabetes. Endocr Rev 23:599–622. https://doi.org/10.1210/er.2001-0039

Faraci FM, Didion SP (2004) Vascular protection. Arterioscler Thromb Vasc Biol 24:1367–1373. https://doi.org/10.1161/01.ATV.0000133604.20182.cf

Farhangi MA, Mesgari-Abbasi M, Hajiluian G, Nameni G, Shahabi P (2017) Adipose tissue inflammation and oxidative stress: the ameliorative effects of vitamin D. Inflammation 40: 1688–1697. https://doi.org/10.1007/s10753-017-0610-9

Fattman CL, Schaefer LM, Oury TD (2003) Extracellular superoxide dismutase in biology and medicine. Free Radic Biol Med 35:236–256. https://doi.org/10.1016/S0891-5849(03)00275-2

Federici L, Masulli M, De Laurenzi V, Allocati N (2024) The role of S-glutathionylation in health and disease: a bird's eye view. Nutrients 16:2753. https://doi.org/10.3390/nu16162753

Fiorentino T, Prioletta A, Zuo P, Folli F (2013) Hyperglycemia-induced oxidative stress and its role in diabetes mellitus related cardiovascular diseases. Curr Pharm Des 19:5695–5703. https://doi.org/10.2174/1381612811319320005

Folli F, Corradi D, Fanti P, Davalli A, Paez A, Giaccari A, Perego C, Muscogiuri G (2011) The role of oxidative stress in the pathogenesis of type 2 diabetes mellitus micro- and macrovascular complications: avenues for a mechanistic-based therapeutic approach. Curr Diabetes Rev 7: 313–324. https://doi.org/10.2174/157339911797415585

Fowler MJ (2008) Microvascular and macrovascular complications of diabetes. Clin Diabetes 26: 77–82. https://doi.org/10.2337/diaclin.26.2.77

Fujita H, Fujishima H, Chida S, Takahashi K, Qi Z, Kanetsuna Y, Breyer MD, Harris RC, Yamada Y, Takahashi T (2009) Reduction of renal superoxide dismutase in progressive diabetic nephropathy. J Am Soc Nephrol 20:1303–1313. https://doi.org/10.1681/ASN.2008080844

Gaike AH, Kalamkar SD, Gajjar V, Divate U, Karandikar-Iyer S, Goel P, Shouche YS, Ghaskadbi SS (2023) Effect of long-term oral glutathione supplementation on gut microbiome of type 2 diabetic individuals. FEMS Microbiol Lett 370:fnad116. https://doi.org/10.1093/femsle/fnad116

Gerber PA, Rutter GA (2017) The role of oxidative stress and hypoxia in pancreatic beta-cell dysfunction in diabetes mellitus. Antioxid Redox Signal 26:501–518. https://doi.org/10.1089/ars.2016.6755

Giacco F, Brownlee M (2010) Oxidative stress and diabetic complications. Circ Res 107:1058–1070. https://doi.org/10.1161/CIRCRESAHA.110.223545

Gilbert C (2013) What is vitamin A and why do we need it? Community Eye Health 26:65

Giugliano D, Ceriello A, Paolisso G (1995) Diabetes mellitus, hypertension, and cardiovascular disease: which role for oxidative stress? Metabolism 44:363–368. https://doi.org/10.1016/0026-0495(95)90167-1

Golbidi S, Badran M, Laher I (2011) Diabetes and alpha lipoic acid. Front Pharmacol 2:69. https://doi.org/10.3389/fphar.2011.00069

Góth L (2000) Lipid and carbohydrate metabolism in acatalasemia. Clin Chem 46:560–576. https://doi.org/10.1093/clinchem/46.4.560g

Góth L, Bigler NW (2007) Catalase deficiency may complicate urate oxidase (rasburicase) therapy. Free Radic Res 41:953–955. https://doi.org/10.1080/10715760701482451

Goth L, Eaton JW (2000) Hereditary catalase deficiencies and increased risk of diabetes. Lancet 356:1820–1821. https://doi.org/10.1016/S0140-6736(00)03238-4

Góth L, Vitai M (1996) Hypocatalasemia in hospital patients. Clin Chem 42:341–342. https://doi.org/10.1093/clinchem/42.2.341

Góth L, Tóth Z, Tarnai I, Bérces M, Török P, Bigler WN (2005) Blood catalase activity in gestational diabetes is decreased but not associated with pregnancy complications. Clin Chem 51:2401–2404. https://doi.org/10.1373/clinchem.2005.055517

Gradinaru D, Borsa C, Ionescu C, Margina D (2013) Advanced oxidative and glycoxidative protein damage markers in the elderly with type 2 diabetes. J Proteome 92:313–322. https://doi.org/10.1016/j.jprot.2013.03.034

Green K, Brand MD, Murphy MP (2004) Prevention of mitochondrial oxidative damage as a therapeutic strategy in diabetes. Diabetes 53:S110–S118. https://doi.org/10.2337/diabetes.53.2007.S110

Guo L, Chen Z, Amarnath V, Davies SS (2012) Identification of novel bioactive aldehyde-modified phosphatidylethanolamines formed by lipid peroxidation. Free Radic Biol Med 53:1226–1238. https://doi.org/10.1016/j.freeradbiomed.2012.07.077

Gupta A, Jamal A, Jamil DA, Al-Aubaidy HA (2023) A systematic review exploring the mechanisms by which citrus bioflavonoid supplementation benefits blood glucose levels and metabolic complications in type 2 diabetes mellitus. Diabetes Metab Syndr Clin Res Rev 17:102884. https://doi.org/10.1016/j.dsx.2023.102884

Gurgul-Convey E, Mehmeti I, Plötz T, Jörns A, Lenzen S (2016) Sensitivity profile of the human EndoC-βH1 beta cell line to proinflammatory cytokines. Diabetologia 59:2125–2133. https://doi.org/10.1007/s00125-016-4060-y

Guzik TJ, West NEJ, Pillai R, Taggart DP, Channon KM (2002) Nitric oxide modulates superoxide release and peroxynitrite formation in human blood vessels. Hypertension 39:1088–1094. https://doi.org/10.1161/01.HYP.0000018041.48432.B5

Gyurászová M, Gurecká R, Bábíčková J, Tóthová Ľ (2020) Oxidative stress in the pathophysiology of kidney disease: implications for noninvasive monitoring and identification of biomarkers. Oxidative Med Cell Longev 2020:1–11. https://doi.org/10.1155/2020/5478708

Halliwell B, Chirico S (1993) Lipid peroxidation: its mechanism, measurement, and significance. Am J Clin Nutr 57:715S–725S. https://doi.org/10.1093/ajcn/57.5.715S

He K, Li X, Chen X, Ye X, Huang J, Jin Y, Li P, Deng Y, Jin Q, Shi Q, Shu H (2011) Evaluation of antidiabetic potential of selected traditional Chinese medicines in STZ-induced diabetic mice. J Ethnopharmacol 137:1135–1142. https://doi.org/10.1016/j.jep.2011.07.033

He F, Ru X, Wen T (2020) NRF2, a transcription factor for stress response and beyond. Int J Mol Sci 21:4777. https://doi.org/10.3390/ijms21134777

Hwang I, Lee J, Huh JY, Park J, Lee HB, Ho Y-S, Ha H (2012) Catalase deficiency accelerates diabetic renal injury through peroxisomal dysfunction. Diabetes 61:728–738. https://doi.org/10.2337/db11-0584

Ighodaro OM, Akinloye OA (2018) First line defence antioxidants-superoxide dismutase (SOD), catalase (CAT) and glutathione peroxidase (GPX): their fundamental role in the entire antioxidant defence grid. Alexandria J Med 54:287–293. https://doi.org/10.1016/j.ajme.2017.09.001

Jamaluddin MS, Weakley SM, Yao Q, Chen C (2012) Resistin: functional roles and therapeutic considerations for cardiovascular disease. Br J Pharmacol 165:622–632. https://doi.org/10.1111/j.1476-5381.2011.01369.x

Jelinek H, Jamil D, Al-Aubaidy H (2014) Impaired fasting glucose & 8-iso-prostaglandin F2α in diabetes disease progression. Br J Med Med Res 4:5229–5237. https://doi.org/10.9734/BJMMR/2014/11147

Jiang Q (2014) Natural forms of vitamin E: metabolism, antioxidant, and anti-inflammatory activities and their role in disease prevention and therapy. Free Radic Biol Med 72:76–90. https://doi.org/10.1016/j.freeradbiomed.2014.03.035

Jiang H, Zhou Y, Nabavi SM, Sahebkar A, Little PJ, Xu S, Weng J, Ge J (2022) Mechanisms of oxidized LDL-mediated endothelial dysfunction and its consequences for the development of atherosclerosis. Front Cardiovasc Med 9:925923. https://doi.org/10.3389/fcvm.2022.925923

Kalamkar S, Acharya J, Kolappurath Madathil A, Gajjar V, Divate U, Karandikar-Iyer S, Goel P, Ghaskadbi S (2022) Randomized clinical trial of how long-term glutathione supplementation offers protection from oxidative damage and improves HbA1c in elderly type 2 diabetic patients. Antioxidants 11:1026. https://doi.org/10.3390/antiox11051026

Kalupahana NS, Moustaid-Moussa N, Claycombe KJ (2012) Immunity as a link between obesity and insulin resistance. Mol Asp Med 33:26–34. https://doi.org/10.1016/j.mam.2011.10.011

Kaneto H, Kajimoto Y, Miyagawa J, Matsuoka T, Fujitani Y, Umayahara Y, Hanafusa T, Matsuzawa Y, Yamasaki Y, Hori M (1999) Beneficial effects of antioxidants in diabetes: possible protection of pancreatic beta-cells against glucose toxicity. Diabetes 48:2398–2406. https://doi.org/10.2337/diabetes.48.12.2398

Kaneto H, Xu G, Fujii N, Kim S, Bonner-Weir S, Weir GC (2002) Involvement of c-Jun N-terminal kinase in oxidative stress-mediated suppression of insulin gene expression. J Biol Chem 277: 30010–30018. https://doi.org/10.1074/jbc.M202066200

Kaneto H, Matsuoka T, Nakatani Y, Miyatsuka T, Matsuhisa M, Hori M, Yamasaki Y (2005) A crucial role of MafA as a novel therapeutic target for diabetes. J Biol Chem 280:15047–15052. https://doi.org/10.1074/jbc.M412013200

Kanwugu ON, Glukhareva TV, Danilova IG, Kovaleva EG (2022) Natural antioxidants in diabetes treatment and management: prospects of astaxanthin. Crit Rev Food Sci Nutr 62:5005–5028. https://doi.org/10.1080/10408398.2021.1881434

Kasai S, Shimizu S, Tatara Y, Mimura J, Itoh K (2020) Regulation of Nrf2 by mitochondrial reactive oxygen species in physiology and pathology. Biomolecules 10:320. https://doi.org/10.3390/biom10020320

Keane KN, Cruzat VF, Carlessi R, de Bittencourt PIH, Newsholme P (2015) Molecular events linking oxidative stress and inflammation to insulin resistance and β-cell dysfunction. Oxidative Med Cell Longev 2015:1–15. https://doi.org/10.1155/2015/181643

Khalid M, Petroianu G, Adem A (2022) Advanced glycation end products and diabetes mellitus: mechanisms and perspectives. Biomolecules 12:542. https://doi.org/10.3390/biom12040542

Khan MAB, Hashim MJ, King JK, Govender RD, Mustafa H, Al Kaabi J (2019) Epidemiology of type 2 diabetes—global burden of disease and forecasted trends. J Epidemiol Glob Health 10: 107. https://doi.org/10.2991/jegh.k.191028.001

Kim CH (2013) Expression of extracellular superoxide dismutase protein in diabetes. Arch Plast Surg 40:517–521. https://doi.org/10.5999/aps.2013.40.5.517

Kim J, Wei Y, Sowers JR (2008) Role of mitochondrial dysfunction in insulin resistance. Circ Res 102:401–414. https://doi.org/10.1161/CIRCRESAHA.107.165472

Kowluru RA, Atasi L, Ho Y-S (2006) Role of mitochondrial superoxide dismutase in the development of diabetic retinopathy. Invest Opthalmol Vis Sci 47:1594. https://doi.org/10.1167/iovs.05-1276

Koya D, Lee IK, Ishii H, Kanoh H, King GL (1997) Prevention of glomerular dysfunction in diabetic rats by treatment with d-alpha-tocopherol. J Am Soc Nephrol 8:426–435. https://doi.org/10.1681/ASN.V83426

Kozlov AV, Javadov S, Sommer N (2024) Cellular ROS and antioxidants: physiological and pathological role. Antioxidants 13:602. https://doi.org/10.3390/antiox13050602

Krishnamurthy HK, Pereira M, Rajavelu I, Jayaraman V, Krishna K, Wang T, Bei K, Rajasekaran JJ (2024) Oxidative stress: fundamentals and advances in quantification techniques. Front Chem 12:1470458. https://doi.org/10.3389/fchem.2024.1470458

Larqué E, Sabater-Molina M, Zamora S (2007) Biological significance of dietary polyamines. Nutrition 23:87–95. https://doi.org/10.1016/j.nut.2006.09.006

Lecour S, van der Merwe E, Opie LH, Sack MN (2006) Ceramide attenuates hypoxic cell death via reactive oxygen species signaling. J Cardiovasc Pharmacol 47:158–163. https://doi.org/10.1097/01.fjc.0000198520.28674.41

Li X, Becker KA, Zhang Y (2010) Ceramide in redox signaling and cardiovascular diseases. Cell Physiol Biochem 26:41–48. https://doi.org/10.1159/000315104

Li H, Xia N, Förstermann U (2012) Cardiovascular effects and molecular targets of resveratrol. Nitric Oxide 26:102–110. https://doi.org/10.1016/j.niox.2011.12.006

Li Y, Zhang H, Yu C, Dong X, Yang F, Wang M, Wen Z, Su M, Li B, Yang L (2024) New insights into mitochondria in health and diseases. Int J Mol Sci 25:9975. https://doi.org/10.3390/ijms25189975

Livingstone C, Davis J (2007) Review: targeting therapeutics against glutathione depletion in diabetes and its complications. Br J Diabetes Vasc Dis 7:258–265. https://doi.org/10.1177/14746514070070060201

Llanos P, Contreras-Ferrat A, Barrientos G, Valencia M, Mears D, Hidalgo C (2015) Glucose-dependent insulin secretion in pancreatic β-cell islets from male rats requires Ca2+ release via ROS-stimulated ryanodine receptors. PLoS One 10:e0129238. https://doi.org/10.1371/journal.pone.0129238

Lobo V, Patil A, Phatak A, Chandra N (2010) Free radicals, antioxidants and functional foods: impact on human health. Pharmacogn Rev 4:118–126. https://doi.org/10.4103/0973-7847.70902

Lu SC (2013) Glutathione synthesis. Biochim Biophys Acta Gen Subj 1830:3143–3153. https://doi.org/10.1016/j.bbagen.2012.09.008

Lykkesfeldt J, Michels AJ, Frei B (2014) Vitamin C. Adv Nutr 5:16–18. https://doi.org/10.3945/an.113.005157

Maedler K, Spinas GA, Dyntar D, Moritz W, Kaiser N, Donath MY (2001) Distinct effects of saturated and monounsaturated fatty acids on β-cell turnover and function. Diabetes 50:69–76. https://doi.org/10.2337/diabetes.50.1.69

Marcon E, Jain H, Bhattacharya A, Guo H, Phanse S, Pu S, Byram G, Collins BC, Dowdell E, Fenner M, Guo X, Hutchinson A, Kennedy JJ, Krastins B, Larsen B, Lin Z-Y, Lopez MF, Loppnau P, Miersch S, Nguyen T, Olsen JB, Paduch M, Ravichandran M, Seitova A, Vadali G, Vogelsang MS, Whiteaker JR, Zhong G, Zhong N, Zhao L, Aebersold R, Arrowsmith CH, Emili A, Frappier L, Gingras A-C, Gstaiger M, Paulovich AG, Koide S, Kossiakoff AA, Sidhu SS, Wodak SJ, Gräslund S, Greenblatt JF, Edwards AM (2015) Assessment of a method to characterize antibody selectivity and specificity for use in immunoprecipitation. Nat Methods 12:725–731. https://doi.org/10.1038/nmeth.3472

Marrazzo G, Barbagallo I, Galvano F, Malaguarnera M, Gazzolo D, Frigiola A, D'Orazio N, Li Volti G (2014) Role of dietary and endogenous antioxidants in diabetes. Crit Rev Food Sci Nutr 54:1599–1616. https://doi.org/10.1080/10408398.2011.644874

Maschirow L, Khalaf K, Al-Aubaidy HA, Jelinek HF (2015) Inflammation, coagulation, endothelial dysfunction and oxidative stress in prediabetes — biomarkers as a possible tool for early disease detection for rural screening. Clin Biochem 48:581–585. https://doi.org/10.1016/j.clinbiochem.2015.02.015

Mezzomo N, Ferreira SRS (2016) Carotenoids functionality, sources, and processing by supercritical technology: a review. J Chem 2016:1–16. https://doi.org/10.1155/2016/3164312

Modi HR, Musyaju S, Ratcliffe M, Shear DA, Scultetus AH, Pandya JD (2024) Mitochondria-targeted antioxidant therapeutics for traumatic brain injury. Antioxidants 13:303. https://doi.org/10.3390/antiox13030303

Montezano AC, Touyz RM (2012) Reactive oxygen species and endothelial function—role of nitric oxide synthase uncoupling and Nox family nicotinamide adenine dinucleotide phosphate oxidases. Basic Clin Pharmacol Toxicol 110:87–94. https://doi.org/10.1111/j.1742-7843.2011.00785.x

Morris D, Khurasany M, Nguyen T, Kim J, Guilford F, Mehta R, Gray D, Saviola B, Venketaraman V (2013) Glutathione and infection. Biochim Biophys Acta Gen Subj 1830:3329–3349. https://doi.org/10.1016/j.bbagen.2012.10.012

Nasri M, Adibhesami G, Mahdavifard S, Babaeenezhad E, Ahmadvand H (2023) Exogenous glutamine ameliorates diabetic nephropathy in a rat model of type 2 diabetes mellitus through its antioxidant and anti-inflammatory activities. Arch Physiol Biochem 129:363–372. https://doi.org/10.1080/13813455.2020.1828478

Naudi A, Jove M, Ayala V, Cassanye A, Serrano J, Gonzalo H, Boada J, Prat J, Portero-Otin M, Pamplona R (2012) Cellular dysfunction in diabetes as maladaptive response to mitochondrial oxidative stress. Exp Diabetes Res 2012:1–14. https://doi.org/10.1155/2012/696215

Newsholme P, Cruzat VF, Keane KN, Carlessi R, de Bittencourt PIH (2016) Molecular mechanisms of ROS production and oxidative stress in diabetes. Biochem J 473:4527–4550. https://doi.org/10.1042/BCJ20160503C

Ngo V, Duennwald ML (2022) Nrf2 and oxidative stress: a general overview of mechanisms and implications in human disease. Antioxidants 11:2345. https://doi.org/10.3390/antiox11122345

Nguyen Dinh Cat A, Touyz RM (2011) A new look at the renin–angiotensin system—focusing on the vascular system. Peptides 32:2141–2150. https://doi.org/10.1016/j.peptides.2011.09.010

Nowotny K, Jung T, Höhn A, Weber D, Grune T (2015) Advanced glycation end products and oxidative stress in type 2 diabetes mellitus. Biomolecules 5:194–222. https://doi.org/10.3390/biom5010194

Nwose EU, Richards RS, McDonald S, Jelinek HF, Kerr PG, Tinley P (2010) Assessment of diabetic macrovascular complications: a prediabetes model. Br J Biomed Sci 67:59–66. https://doi.org/10.1080/09674845.2010.11730292

Oury TD, Day BJ, Crapo JD (1996) Extracellular superoxide dismutase: a regulator of nitric oxide bioavailability. Lab Investig 75:617–636

Palomo I (2012) Gene expression of adipose tissue, endothelial cells and platelets in subjects with metabolic syndrome (review). Mol Med Rep 5(5):1135–1140. https://doi.org/10.3892/mmr.2012.785

Panche AN, Diwan AD, Chandra SR (2016) Flavonoids: an overview. J Nutr Sci 5:e47. https://doi.org/10.1017/jns.2016.41

Pandey KB, Rizvi SI (2011) Biomarkers of oxidative stress in red blood cells. Biomed Pap Med Fac Univ Palacky Olomouc Czech Repub 155:131–136. https://doi.org/10.5507/bp.2011.027

Patel S, Khan H, Majumdar A (2022) Crosstalk between Sirtuins and Nrf2: SIRT1 activators as emerging treatment for diabetic neuropathy. Metab Brain Dis 37:2181–2195. https://doi.org/10.1007/s11011-022-00956-z

Peppa M, Uribarri J, Vlassara H (2003) Glucose, advanced glycation end products, and diabetes complications: what is new and what works. Clin Diabetes 21:186–187. https://doi.org/10.2337/diaclin.21.4.186

Perrin-Sarrado C, Bouchot O, Vergely C, Rochette L (2007) Release of secondary free radicals during post-ischaemic reperfusion is not influenced by extracellular calcium levels in isolated rat hearts. Mol Cell Biochem 297:199–207. https://doi.org/10.1007/s11010-006-9347-8

Pi J, Zhang Q, Fu J, Woods CG, Hou Y, Corkey BE, Collins S, Andersen ME (2010) ROS signaling, oxidative stress and Nrf2 in pancreatic beta-cell function. Toxicol Appl Pharmacol 244:77–83. https://doi.org/10.1016/j.taap.2009.05.025

Piarulli F, Sartore G, Ceriello A, Ragazzi E, Reitano R, Nollino L, Cosma C, Fedele D, Lapolla A (2009) Relationship between glyco-oxidation, antioxidant status and microalbuminuria in type 2 diabetic patients. Diabetologia 52:1419–1425. https://doi.org/10.1007/s00125-009-1367-y

Picca A, Calvani R, Coelho-Junior HJ, Landi F, Bernabei R, Marzetti E (2020) Mitochondrial dysfunction, oxidative stress, and neuroinflammation: intertwined roads to neurodegeneration. Antioxidants 9:647. https://doi.org/10.3390/antiox9080647

Piko N, Bevc S, Hojs R, Ekart R (2023) The role of oxidative stress in kidney injury. Antioxidants 12:1772. https://doi.org/10.3390/antiox12091772

Piperi C, Adamopoulos C, Dalagiorgou G, Diamanti-Kandarakis E, Papavassiliou AG (2012) Crosstalk between advanced glycation and endoplasmic reticulum stress: emerging therapeutic targeting for metabolic diseases. J Clin Endocrinol Metab 97:2231–2242. https://doi.org/10.1210/jc.2011-3408

Piwowar A (2010) Advanced oxidation protein products. Part I. Mechanism of the formation, characteristics and property. Pol Merkur Lekarski 28:166–169

Piwowar A, Knapik-Kordecka M, Warwas M (2007) AOPP and its relations with selected markers of oxidative/antioxidative system in type 2 diabetes mellitus. Diabetes Res Clin Pract 77:188–192. https://doi.org/10.1016/j.diabres.2006.12.007

Pizzino G, Irrera N, Cucinotta M, Pallio G, Mannino F, Arcoraci V, Squadrito F, Altavilla D, Bitto A (2017) Oxidative stress: harms and benefits for human health. Oxidative Med Cell Longev 2017:8416763. https://doi.org/10.1155/2017/8416763

Poitout V, Robertson RP (2002) Minireview: secondary β-cell failure in type 2 diabetes—a convergence of glucotoxicity and lipotoxicity. Endocrinology 143:339–342. https://doi.org/10.1210/endo.143.2.8623

Priante G, Gianesello L, Ceol M, Del Prete D, Anglani F (2019) Cell death in the kidney. Int J Mol Sci 20:3598. https://doi.org/10.3390/ijms20143598

Rabbani PS, Soares MA, Hameedi SG, Kadle RL, Mubasher A, Kowzun M, Ceradini DJ (2019) Dysregulation of Nrf2/Keap1 redox pathway in diabetes affects multipotency of stromal cells. Diabetes 68:141–155. https://doi.org/10.2337/db18-0232

Rahigude A, Bhutada P, Kaulaskar S, Aswar M, Otari K (2012) Participation of antioxidant and cholinergic system in protective effect of naringenin against type-2 diabetes-induced memory dysfunction in rats. Neuroscience 226:62–72. https://doi.org/10.1016/j.neuroscience.2012.09.026

Ranieri SC, Fusco S, Pani G (2013) p66ShcA:linking mammalian longevity with obesity-induced insulin resistance. Vitam Horm 91:219–241. https://doi.org/10.1016/B978-0-12-407766-9.00009-2

Reddy VP, Aryal P, Darkwah EK (2022) Advanced glycation end products in health and disease. Microorganisms 10:1848. https://doi.org/10.3390/microorganisms10091848

Rhee SY, Kim YS (2018) The role of advanced glycation end products in diabetic vascular complications. Diabetes Metab J 42:188. https://doi.org/10.4093/dmj.2017.0105

Rizvi S, Raza ST, Ahmed F, Ahmad A, Abbas S, Mahdi F (2014) The role of vitamin E in human health and some diseases. Sultan Qaboos Univ Med J 14:e157–e165

Robertson RP (2004) Chronic oxidative stress as a central mechanism for glucose toxicity in pancreatic islet beta cells in diabetes. J Biol Chem 279:42351–42354. https://doi.org/10.1074/jbc.R400019200

Rochette L, Zeller M, Cottin Y, Vergely C (2014) Diabetes, oxidative stress and therapeutic strategies. Biochim Biophys Acta Gen Subj 1840:2709–2729. https://doi.org/10.1016/j.bbagen.2014.05.017

Roohbakhsh A, Karimi G, Iranshahi M (2017) Carotenoids in the treatment of diabetes mellitus and its complications: a mechanistic review. Biomed Pharmacother 91:31–42. https://doi.org/10.1016/j.biopha.2017.04.057

Rozoy E, Simard S, Liu Y, Kitts D, Lessard J, Bazinet L (2012) The use of cyclic voltammetry to study the oxidation of l-5-methyltetrahydrofolate and its preservation by ascorbic acid. Food Chem 132:1429–1435. https://doi.org/10.1016/j.foodchem.2011.11.132

Rungratanawanich W, Qu Y, Wang X, Essa MM, Song B-J (2021) Advanced glycation end products (AGEs) and other adducts in aging-related diseases and alcohol-mediated tissue injury. Exp Mol Med 53:168–188. https://doi.org/10.1038/s12276-021-00561-7

Saddala RR, Thopireddy L, Ganapathi N, Kesireddy SR (2013) Regulation of cardiac oxidative stress and lipid peroxidation in streptozotocin-induced diabetic rats treated with aqueous extract of Pimpinella tirupatiensis tuberous root. Exp Toxicol Pathol 65:15–19. https://doi.org/10.1016/j.etp.2011.05.003

Santo A, Zhu H, Li YR (2016) Free radicals: from health to disease. React Oxyg Species 2:245–263. 10.20455/ros.2016.847

Savini I, Catani M, Evangelista D, Gasperi V, Avigliano L (2013) Obesity-associated oxidative stress: strategies finalized to improve redox state. Int J Mol Sci 14:10497–10538. https://doi.org/10.3390/ijms140510497

Schieber M, Chandel NS (2014) ROS function in redox signaling and oxidative stress. Curr Biol 24: R453–R462. https://doi.org/10.1016/j.cub.2014.03.034

Sekhar RV, McKay SV, Patel SG, Guthikonda AP, Reddy VT, Balasubramanyam A, Jahoor F (2011) Glutathione synthesis is diminished in patients with uncontrolled diabetes and restored by dietary supplementation with cysteine and glycine. Diabetes Care 34:162–167. https://doi.org/10.2337/dc10-1006

Shodehinde SA, Oboh G (2013) Antioxidant properties of aqueous extracts of unripe Musa paradisiaca on sodium nitroprusside induced lipid peroxidation in rat pancreas in vitro. Asian Pac J Trop Biomed 3:449–457. https://doi.org/10.1016/S2221-1691(13)60095-7

Shukla K, Dikshit P, Tyagi MK, Shukla R, Gambhir JK (2012) Ameliorative effect of Withania coagulans on dyslipidemia and oxidative stress in nicotinamide–streptozotocin induced diabetes mellitus. Food Chem Toxicol 50:3595–3599. https://doi.org/10.1016/j.fct.2012.07.026

Sivitz WI, Yorek MA (2010) Mitochondrial dysfunction in diabetes: from molecular mechanisms to functional significance and therapeutic opportunities. Antioxid Redox Signal 12:537–577. https://doi.org/10.1089/ars.2009.2531

Soares AF, Guichardant M, Cozzone D, Bernoud-Hubac N, Bouzaïdi-Tiali N, Lagarde M, Géloën A (2005) Effects of oxidative stress on adiponectin secretion and lactate production in 3T3-L1 adipocytes. Free Radic Biol Med 38:882–889. https://doi.org/10.1016/j.freeradbiomed.2004.12.010

Søndergård SD, Cintin I, Kuhlman AB, Morville TH, Bergmann ML, Kjær LK, Poulsen HE, Giustarini D, Rossi R, Dela F, Helge JW, Larsen S (2021) The effects of 3 weeks of oral glutathione supplementation on whole body insulin sensitivity in obese males with and without type 2 diabetes: a randomized trial. Appl Physiol Nutr Metab 46:1133–1142. https://doi.org/10.1139/apnm-2020-1099

Srivastava V, Gross E (2023) Mitophagy-promoting agents and their ability to promote healthy-aging. Biochem Soc Trans 51:1811–1846. https://doi.org/10.1042/BST20221363

Steven S, Frenis K, Oelze M, Kalinovic S, Kuntic M, Bayo Jimenez MT, Vujacic-Mirski K, Helmstädter J, Kröller-Schön S, Münzel T, Daiber A (2019) Vascular inflammation and oxidative stress: major triggers for cardiovascular disease. Oxid Med Cell Longev 2019:1–26. https://doi.org/10.1155/2019/7092151

Stumvoll M, Goldstein BJ, van Haeften TW (2005) Type 2 diabetes: principles of pathogenesis and therapy. Lancet 365:1333–1346. https://doi.org/10.1016/S0140-6736(05)61032-X

Sugandh F, Chandio M, Raveena F, Kumar L, Karishma F, Khuwaja S, Memon UA, Bai K, Kashif M, Varrassi G, Khatri M, Kumar S (2023) Advances in the management of diabetes mellitus: a focus on personalized medicine. Cureus 15(8):e43697. https://doi.org/10.7759/cureus.43697

Sulaimon LA, Afolabi LO, Adisa RA, Ayankojo AG, Afolabi MO, Adewolu AM, Wan X (2022) Pharmacological significance of MitoQ in ameliorating mitochondria-related diseases. Adv Redox Res 5:100037. https://doi.org/10.1016/j.arres.2022.100037

Summers SA (2006) Ceramides in insulin resistance and lipotoxicity. Prog Lipid Res 45:42–72. https://doi.org/10.1016/j.plipres.2005.11.002

Suzuki D, Miyata T (1999) Carbonyl stress in the pathogenesis of diabetic nephropathy. Intern Med 38:309–314. https://doi.org/10.2169/internalmedicine.38.309
Szendroedi J, Schmid AI, Meyerspeer M, Cervin C, Kacerovsky M, Smekal G, Gräser-Lang S, Groop L, Roden M (2009) Impaired mitochondrial function and insulin resistance of skeletal muscle in mitochondrial diabetes. Diabetes Care 32:677–679. https://doi.org/10.2337/dc08-2078
Takemoto K, Tanaka M, Iwata H, Nishihara R, Ishihara K, Wang D-H, Ogino K, Taniuchi K, Masuoka N (2009) Low catalase activity in blood is associated with the diabetes caused by alloxan. Clin Chim Acta 407:43–46. https://doi.org/10.1016/j.cca.2009.06.028
Tan KS, Lee KO, Low KC, Gamage AM, Liu Y, Tan G-YG, Koh HQV, Alonso S, Gan Y-H (2012) Glutathione deficiency in type 2 diabetes impairs cytokine responses and control of intracellular bacteria. J Clin Invest 122:2289–2300. https://doi.org/10.1172/JCI57817
Tanaka Y, Tran POT, Harmon J, Robertson RP (2002) A role for glutathione peroxidase in protecting pancreatic β cells against oxidative stress in a model of glucose toxicity. Proc Natl Acad Sci 99:12363–12368. https://doi.org/10.1073/pnas.192445199
Tangvarasittichai S (2015) Oxidative stress, insulin resistance, dyslipidemia and type 2 diabetes mellitus. World J Diabetes 6:456. https://doi.org/10.4239/wjd.v6.i3.456
Tao L, Gao E, Jiao X, Yuan Y, Li S, Christopher TA, Lopez BL, Koch W, Chan L, Goldstein BJ, Ma XL (2007) Adiponectin cardioprotection after myocardial ischemia/reperfusion involves the reduction of oxidative/nitrative stress. Circulation 115:1408–1416. https://doi.org/10.1161/CIRCULATIONAHA.106.666941
Tiedge M, Lortz S, Munday R, Lenzen S (1998) Complementary action of antioxidant enzymes in the protection of bioengineered insulin-producing RINm5F cells against the toxicity of reactive oxygen species. Diabetes 47:1578–1585. https://doi.org/10.2337/diabetes.47.10.1578
Tirichen H, Yaigoub H, Xu W, Wu C, Li R, Li Y (2021) Mitochondrial reactive oxygen species and their contribution in chronic kidney disease progression through oxidative stress. Front Physiol 12:627837. https://doi.org/10.3389/fphys.2021.627837
To K, Cao R, Yegiazaryan A, Owens J, Nguyen T, Sasaninia K, Vaughn C, Singh M, Truong E, Medina A, Avitia E, Villegas J, Pham C, Sathananthan A, Venketaraman V (2021) Effects of oral liposomal glutathione in altering the immune responses against mycobacterium tuberculosis and the mycobacterium bovis BCG strain in individuals with type 2 diabetes. Front Cell Infect Microbiol 11:657775. https://doi.org/10.3389/fcimb.2021.657775
Tsai C-J, Hsieh C-J, Tung S-C, Kuo M-C, Shen F-C (2012) Acute blood glucose fluctuations can decrease blood glutathione and adiponectin levels in patients with type 2 diabetes. Diabetes Res Clin Pract 98:257–263. https://doi.org/10.1016/j.diabres.2012.09.013
Tsalamandris S, Antonopoulos AS, Oikonomou E, Papamikroulis G-A, Vogiatzi G, Papaioannou S, Deftereos S, Tousoulis D (2019) The role of inflammation in diabetes: current concepts and future perspectives. Eur Cardiol Rev 14:50–59. https://doi.org/10.15420/ecr.2018.33.1
Villalpando-Rodriguez GE, Gibson SB (2021) Reactive oxygen species (ROS) regulates different types of cell death by acting as a rheostat. Oxidative Med Cell Longev 2021:9912436. https://doi.org/10.1155/2021/9912436
Wajchenberg BL (2007) β-Cell failure in diabetes and preservation by clinical treatment. Endocr Rev 28:187–218. https://doi.org/10.1210/10.1210/er.2006-0038
Wang P, Mariman E, Renes J, Keijer J (2008) The secretory function of adipocytes in the physiology of white adipose tissue. J Cell Physiol 216:3–13. https://doi.org/10.1002/jcp.21386
Wang C, Li S, Shang D-J, Wang X-L, You Z-L, Li H-B (2011) Antihyperglycemic and neuroprotective effects of one novel Cu–Zn SOD mimetic. Bioorg Med Chem Lett 21:4320–4324. https://doi.org/10.1016/j.bmcl.2011.05.051
Wang X, Tao L, Hai CX (2012) Redox-regulating role of insulin: the essence of insulin effect. Mol Cell Endocrinol 349:111–127. https://doi.org/10.1016/j.mce.2011.08.019
Watts ME, Pocock R, Claudianos C (2018) Brain energy and oxygen metabolism: emerging role in normal function and disease. Front Mol Neurosci 11:216. https://doi.org/10.3389/fnmol.2018.00216

Witko-Sarsat V, Friedlander M, Capeillère-Blandin C, Nguyen-Khoa T, Nguyen AT, Zingraff J, Jungers P, Descamps-Latscha B (1996) Advanced oxidation protein products as a novel marker of oxidative stress in uremia. Kidney Int 49:1304–1313. https://doi.org/10.1038/ki.1996.186

Wolff SP (1993) Diabetes mellitus and free radicals. Br Med Bull 49:642–652. https://doi.org/10.1093/oxfordjournals.bmb.a072637

Wollin SD, Jones PJH (2003) α-lipoic acid and cardiovascular disease. J Nutr 133:3327–3330. https://doi.org/10.1093/jn/133.11.3327

Yamamoto A, Matsuzawa Y, Kishino B, Hayashi R, Hirobe K, Kikkawa T (1983) Effects of probucol on homozygous cases of familial hypercholesterolemia. Atherosclerosis 48:157–166. https://doi.org/10.1016/0021-9150(83)90102-8

Yaribeygi H, Sathyapalan T, Atkin SL, Sahebkar A (2020) Molecular mechanisms linking oxidative stress and diabetes mellitus. Oxidative Med Cell Longev 2020:1–13. https://doi.org/10.1155/2020/8609213

Ylönen K, Alfthan G, Groop L, Saloranta C, Aro A, Virtanen SM (2003) Dietary intakes and plasma concentrations of carotenoids and tocopherols in relation to glucose metabolism in subjects at high risk of type 2 diabetes: the Botnia Dietary Study. Am J Clin Nutr 77:1434–1441. https://doi.org/10.1093/ajcn/77.6.1434

Zakir M, Ahuja N, Surksha MA, Sachdev R, Kalariya Y, Nasir M, Kashif M, Shahzeen F, Tayyab A, Khan MS, moazzam, Junejo M, Manoj Kumar F, Varrassi G, Kumar S, Khatri M, Mohamad T (2023) Cardiovascular complications of diabetes: from microvascular to macrovascular pathways. Cureus 15(9):e45835. https://doi.org/10.7759/cureus.45835

Zelko IN, Mariani TJ, Folz RJ (2002) Superoxide dismutase multigene family: a comparison of the CuZn-SOD (SOD1), Mn-SOD (SOD2), and EC-SOD (SOD3) gene structures, evolution, and expression. Free Radic Biol Med 33:337–349. https://doi.org/10.1016/S0891-5849(02)00905-X

Zhang Z, Huang Q, Zhao D, Lian F, Li X, Qi W (2023) The impact of oxidative stress-induced mitochondrial dysfunction on diabetic microvascular complications. Front Endocrinol (Lausanne) 14:1112363. https://doi.org/10.3389/fendo.2023.1112363

Zorov DB, Juhaszova M, Sollott SJ (2014) Mitochondrial reactive oxygen species (ROS) and ROS-induced ROS release. Physiol Rev 94:909–950. https://doi.org/10.1152/physrev.00026.2013

Chapter 8
Mechanistic Action of Oxidative Stress in Depression

Animesh Singh and Shubha Shukla

Abbreviations

ATP	Adenosine Triphosphate
BDNF	Brain-Derived Neurotrophic Factor
COX	Cyclooxygenase
DNA	Deoxyribonucleic Acid
HPA axis	Hypothalamic–Pituitary–Adrenal Axis
IL-18	Interleukin-18
IL-1β	Interleukin-1 Beta
JAK/STAT	Janus Kinase/Signal Transducer and Activator of Transcription
MAO	Monoamine Oxidase
MAPK	Mitogen-Activated Protein Kinase
NAC	N-Acetylcysteine
NLRP3	NOD-Like Receptor Family Pyrin Domain Containing 3
ROS	Reactive Oxygen Species
SERT	Serotonin Reuptake Transporter
SOD	Superoxide Dismutase
SPD	Sensory Processing Disorder
SSRI	Selective Serotonin Reuptake Inhibitors
TLR	Toll-Like Receptor
TNF-α	Tumor Necrosis Factor-Alpha
WHO	World Health Organization

A. Singh · S. Shukla (✉)
Division of Neurosciences and Ageing biology, CSIR-Central Drug Research Institute, Lucknow, Uttar Pradesh, India

Academy of Scientific and Innovative Research (AcSIR), Ghaziabad, India
e-mail: shubha.shukla.cdri@csir.res.in

N. Srivastava, S. A. Saraf (eds.), *Pharmacological Strategies for Prevention and Treatment of Oxidative Stress*, Oxidative Stress in Applied Basic Research and Clinical Practice, https://doi.org/10.1007/978-3-032-19837-2_8

8.1 Introduction

Depression is a psychiatric disorder, which can be characterized by consistent feeling of sadness, hopelessness, and lack of interest in performing daily activities (Vahia 2013). It affects individuals of different age groups across the globe and remains a leading cause of disability among them. It is estimated that more than 280 million people worldwide suffer from depression according to the latest report of the World Health Organization (WHO), highlighting its significant impact on public health. Depression in itself is a socio-economic burden on society directly affecting the cost on health care and indirectly costing productivity and disability (Greenberg et al. 2015).

Previously, neurotransmitter loss involving Serotonin, norepinephrine, and dopamine was seen as a major cause of depression (Schildkraut 1965). However, with recent advancements in technology, depression is now recognized as a disorder occurring due to complex genetic, environmental, and biological interplay (Krishnan and Nestler 2008). Out of all these factors, oxidative stress has emerged as a major player in pathogenesis of depression, offering novel insights into its molecular mechanistic pathway (Maes et al. 2011).

Imbalance between the body's antioxidant defense and reactive oxygen species (ROS) production leads to oxidative stress (Sies 2015). Reactive oxygen species production leads to damage of various cellular components including DNA, lipids, and proteins (Valko et al. 2007). The brain is generally highly vulnerable to oxidative stress due to its higher oxygen consumption, abundant lipid content, and relatively low levels of antioxidant enzymes (Halliwell 2006). This vulnerability becomes the key contributor to the neurobiological changes which occur in depression (Salim 2014).

Mitochondria are known as the power house of the cell and plays a pivotal role in energy production through oxidative phosphorylation; its dysfunction can trigger various cascades of cellular pathways which can lead to depression (Wallace 2013). Oxidative stress can cause neural energy deficits, impaired neurogenesis, and increased neurodegeneration which is associated with depression, as it impairs mitochondrial function, leading to reduction in ATP, increased ROS production, and release of pro-apoptotic factors (Andreazza et al. 2008). The causes and consequences of oxidative stress are shown in Fig. 8.1.

Other than mitochondrial dysfunction, oxidative stress also causes neurotransmitter disruption, which is crucial for mood regulation (Ng et al. 2008). Oxidative stress generally alters the synthesis, release, and reuptake of neurotransmitters such as serotonin, dopamine, and glutamate (Bhatt et al. 2020). These neurotransmitters are vital for regulating mood, and their dysregulation is strongly linked to depression symptoms (Belmaker and Agam 2008). Serotonin, often known as "happiness hormone," is particularly susceptible to oxidative damage, which can exacerbate depressive symptoms.

Oxidative stress also activates proinflammatory signaling pathways such as NF-κB, MAPK, NLRP3 inflammasome, JAK/STAT, and TLR signaling, triggering the release of proinflammatory cytokines like tumor necrosis factor-alpha (TNF-α),

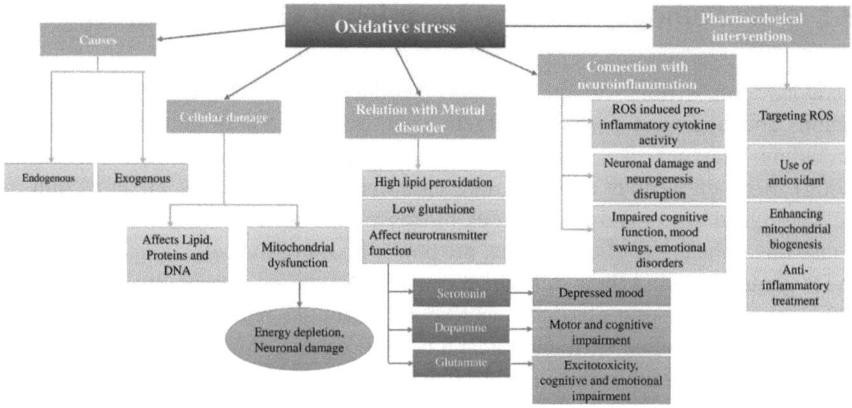

Fig. 8.1 Illustration depicting the cause and consequence of oxidative stress and its relation to mental disorders, neuroinflammation, and pharmacological approaches

interleukin-1 beta (IL-1β), and interleukin-18 (IL-18), leading to a state of chronic low-grade inflammation in the brain (Miller and Raison 2016). The neuroinflammatory environment further exacerbates neuronal damage which contributes to the dysregulation of neurotransmitter system (Dantzer et al. 2008). There have been numerous studies establishing the link between depression and inflammation by studying the elevated levels of various neuroinflammatory markers in individuals with depression (Dowlati et al. 2010).

Studying the role of oxidative stress in pathogenesis of depression by targeting oxidative stress has become a promising therapeutic strategy. Antioxidants, which target and neutralize ROS and restore the balance between oxidative and antioxidant forces, have been used as a potential treatment for depression (Scapagnini et al. 2012). Additionally, other neuroprotective drugs which can enhance mitochondrial function, reduce inflammation, and modulate neurotransmitter systems are being investigated for the potential to eradicate the effects of oxidative stress in depression (Berk et al. 2011).

This chapter provides a detailed overview of molecular mechanisms by which oxidative stress contributes to the pathogenesis of depression. It describes how oxidative stress disrupts mitochondrial dysfunction, neurotransmitter regulation, and synaptic plasticity, leading to the development and progression of depressive symptoms. This chapter also describes the role of neuroprotective agents and their role in restoring neuronal health. By examining the mechanism of occurrence of oxidative stress in depression, this chapter describes new therapeutic strategies that can improve the outcomes for individuals suffering from this debilitating disorder.

Overall, understanding the role of oxidative stress in depression is crucial for developing better and effective treatments. As research continue to uncover complexities and the interplay among oxidative stress, mitochondrial dysfunction, neurotransmitter dysregulation, and inflammation, new therapeutic approaches may emerge as a promising hope for improved mental health outcomes.

8.2 The Pathophysiological Intrusion of Oxidative Stress

Oxidative stress refers to the excessive production of ROS, such as superoxide anions and hydrogen peroxide, targeting the body's natural antioxidant defenses including enzymes like superoxide dismutase (SOD) and catalase (Valko et al. 2006). Both exogenous variables like environmental pollutants, inflammation, or psychological stress and endogenous processes like mitochondrial respiration can cause this imbalance. Mitochondria, the power house of the cell, are responsible for ATP production, and their imbalance can seriously affect the brain processes like neurogenesis, synaptic plasticity, and neurotransmitter regulation, which are vital processes for maintaining mental health of an organism (Mattson et al. 2008). The resulting energy deficits in neurons lead to decreased plasticity, thereby reducing the brain's ability to adapt and process mood regulation, ultimately contributing to depressive behavior (Manji et al. 2003).

Oxidative stress causes a major impact on neurotransmitter systems, specifically on glutamate, serotonin, and dopamine systems, which are essential for regulating mood and processing emotions (Michel et al. 2012). Anhedonia (lack of pleasure) and mood dysfunction are the classic signs of depression caused by oxidative damage interfering with synthesis, release, and reuptake of above-mentioned neurotransmitters (Dantzer et al. 2011). People suffering with depression are reported to have lower levels of serotonin which is known to elevate depressed mood and hopelessness (Belmaker and Agam 2008). Likewise, disrupted glutamate signaling further disrupts cognitive and emotional balance, and oxidative damage disrupts dopamine pathway resulting in motivational deficiencies (Sanacora et al. 2012).

Oxidative stress, alongside its impact on neurotransmitters, also activates inflammatory processes that accelerate brain damage. When ROS are overproduced, they trigger NF-κB, which initiates the release of inflammatory molecules called cytokines. These cytokines create a harmful inflammatory environment in the brain leading to further oxidative damage, thus creating a harmful cycle (Miller and Raison 2016). High levels of cytokines are often seen in patients of depression linking it to inflammation-driven oxidative stress damage. This inflammation-driven oxidative stress damages neurons and decreases the formation of new brain cells, especially in areas like hippocampus, which regulates memory and emotions. Together, inflammation and oxidative stress work together in worsening the damage and impairing normal brain functions (Miller and Raison 2016).

The structural integrity of cells, including neurons, is another victim to chronic oxidative stress. DNA, lipids, and proteins are the major targets of oxidation across a cell, leading to cellular damage. Oxidative damage to membrane lipids has an impact on the transmission of neuronal signal between cells, which leads to severe emotional and cognitive dysfunctions; in addition, oxidation of proteins linked with synaptic function could potentially compromise neuronal connections and cause the reduced synaptic plasticity in sadness (Yao et al. 2001; Butterfield et al. 1997). Additionally, damaged neurons exhibit higher susceptibility to the program that regulates cell death and accelerates neurodegeneration in response to chronic depression (Salim 2014).

The interventions targeting ROS and mitochondrial dysfunction have been deemed as an effective approach because oxidative stress is considered the key factor in the progression of depression. Because of their capacity to reduce ROS, antioxidants have the potential for meditating oxidative homeostasis and some of the cytotoxic changes, believed to occur in depression (Berk et al. 2008). For instance, N-acetylcysteine, coenzyme Q10, and vitamin E have been explored as potential agents which may help in the therapy against oxidative stress and to enhance mitochondrial biogenesis (Dean et al. 2011). By relenishing ATP levels, drugs designed to increase mitochondrial energy supply, reduce symptoms related to depression (Scaini et al. 2016). Due to the suppression of proinflammatory pathways and cytokine production, innovative treatments are also directed to target the inflammatory aspect of oxidative stress (Maes et al. 2011). All of these treatment approaches aim at interrupting the vicious cycle of oxidative stress and inflammation, so that neurons are not subjected to further damage.

Therefore, oxidative stress is extremely damaging not only by causing the breakdown of fundamental cell physiology and psychological functions, but also causes inflammation and neuronal destruction. Aiming at these mechanisms by antioxidant therapies, inflammation modulation, and mitochondrial protection, there are possibilities of a better treatment of depression in future (Moylan et al. 2014).

8.3 Oxidative Stress in Mental Illness

There are growing evidences about the role of oxidative stress in the development and manifestation of many mental disorders. Oxidative stress is defined at the cellular level as occurrence of a relative increase in production of ROS in relation to the antioxidant ability of the body (Halliwell and Gutteridge 2015). Oxidative stress is the condition where ROS concentration increases beyond the body's capability of neutralizing their toxic effects and hence negatively alters lipids, proteins, and DNA of the cell (Sies 2015). The damage is more in neurons because of their high metabolic rates, accumulated oxidants, and low ability to regenerate (Valko et al. 2007). This damage is further amplified by mitochondrial dysfunction, which is associated with oxidative stress and results in energy depletion as well as neuronal dysfunction, changes that are responsible for many psychiatric disorders.

There are many reports on the oxidative stress and depression, schizophrenia, bipolar disorder, and other mental illnesses where raised oxidative stress markers and low antioxidant protection have been documented (Ng et al. 2008). Research indicates that patients with depression exhibit increased levels of by-products from lipd peroxidation, whereas, a concomitant decrease in glutathione, an important antioxidant enzyme (Gawryluk et al. 2011). This oxidative stress interferes with the normal functioning of neurotransmitters like serotonin, dopamine, and glutamate that are vital in controlling mood, cognition, and stability (Michel et al. 2012).

Dysfunction in these systems has been attributed to the reason as to why patients with mental illnesses exhibit poor cognitive functioning, mood swings, and other emotional disorders.

Oxidative stress is closely connected to neuroinflammation which is becoming a matter of concern in most mental issues. ROS play a significant part in causing inflammation, and it induces the signaling of transcription factors for the activity of proinflammatory cytokines such as NF-κB which creates a neuroinflammatory state, which in turn harms neurons and disrupts neurogenesis in the brain, including the hippocampus and prefrontal cortex, which are important in processes including memory formation and regulation of emotions. This inflammation not only increases oxidative stress, but also continuously causes progressive injury to neurons as well as cognitive disabilities resulting in mental illness (Maes et al. 2011; Miller and Raison 2016).

Because of the interconnectivity between oxidative stress, inflammation, and neurodegeneration, antioxidants are considered to be the primary candidate for pharmacological management. Antioxidants, whether natural or prescribed, can help to ward off oxidative stress during mental illness. Antioxidants such as vitamins C and E, coenzyme Q10, and polyphenol are shown to possess the ability of neuroprotection by removing free radicals and thus minimizing oxidative stress. Furthermore, new treatments designed to treat mitochondrial disorders also have the potential to mend the cell's ability to produce energy and further protect neurons against oxidative stress, giving potential treatments for mental disorders associated with oxidative stress (Scaini et al. 2016).

Therefore, oxidative stress is a key factor in the etiology and development of mental illnesses (Salim 2017). How it influences the state of neuronal transmission and inflammation in the brain is attributed to the development of psychiatric disorders. Knowledge of these mechanisms paves the way to novel approaches in using therapies, especially those which are directed at diminishing oxidative injury and enhancing brain function. It also creates a chance to reduce the load of oxidative stress by applying antioxidants and mitochondrial targeted treatments to provide new hope to the patients suffering from mental health issues (Berk et al. 2011; Rezin et al. 2009).

8.4 Association of Oxidative Stress with Depression

Free radical plays a central role in the development as well as the manifestation of depression (Miller and Raison 2016). This pathophysiological process can be described in terms of an increased generation of reactive oxygen species (ROS) with a decreased capacity for their detoxification by antioxidants (Sies 2015). In depression, there is oxidative injury through the mechanism of oxidative stress that leads to neuronal dysfunction and cellular injury and disruption of almost all important pathways in the brain. Various studies have pointed toward oxidative stress as being associated with the primary depressive symptoms with the brain being a high metabolic organ and is most susceptible to oxidative stress.

Psychological stress that leads to oxidative stress affects several biological systems which are involved in mood and emotional functioning. It harms neurons and disrupts important processes of the brain, including transmission of signals between neurons and neuronal plasticity along with the ratio of proinflammatory and anti-inflammatory mediators. This disruption in cellular homeostasis depletes the body's capacity to resynchronize and alter states in response to new information, decreasing the brain's ability to cope with the stressor, and lessening the chance of recovering from depression (Duman and Aghajanian 2012). It is, therefore, important to establish the connection between oxidative stress and depression so as to have specific molecular targets for the treatment (Table 8.1).

Table 8.1 Effect of oxidative stress on key biological systems during depression

Key biological system	Effect of oxidative stress	Mechanism
Neuroendocrine dysregulation	Disruption of the hypothalamic–pituitary–adrenal (HPA) axis, leading to chronic cortisol release and hormonal imbalance	Damage to glucocorticoid receptors impairs feedback regulation, exacerbating cortisol release and oxidative damage (Pariante and Lightman 2008)
Neuroinflammatory imbalance	Increased inflammation due to ROS, promoting chronic neuroinflammation and worsening depression	Activation of transcription factors (e.g., NF-κB) leads to the release of proinflammatory cytokines (TNF-α, IL-6), causing neurodegeneration (Halliwell and Gutteridge 2015)
Serotonergic system dysfunction	Impairment of serotonin signaling, reduction in serotonin availability, leading to mood disorders	ROS oxidize serotonin and its receptors, affect serotonin transporter (SERT), and disrupt synthesis enzymes (Anderson and Maes 2014)
Neuronal plasticity impairment	Reduced neurogenesis and synaptic function, resulting in cognitive deficits and emotional instability	Mitochondrial dysfunction and reduced neurotrophic factors (BDNF) impair synaptic connections and energy production in neurons (Duman and Aghajanian 2012)
Genetic/epigenetic modifications	Oxidative stress causes DNA damage and alters gene expression, leading to vulnerability to depression	ROS induce mutations and epigenetic changes like DNA methylation and histone modification, affecting genes regulating mood and plasticity (Uddin et al. 2010)

8.4.1 Neuroendocrine Dysregulation

HPA axis plays a pivotal role in stress response in the body, and the Sensory Processing Disorder (SPD) is characterized by dysregulation of this axis. Oxidative stress in the long run therefore has a central role to play in this disruption of the normal functioning of this axis by augmenting cortisol levels through sustained activation. Cortisol stress hormone increases oxidative stress in neurons impairing their functionality in terms of signaling (McEwen and Gianaros 2010). The oxidative damage also decreases the effectiveness of glucocorticoid receptors, fails at regulating the HPA axis, and thereby promotes a state of chronic stress (Miller and Raison 2016). In this manner, the mentioned chronic neuroendocrine abnormalities that have developed due to oxidative stress are directly associated with the emergence and duration of depressive disorders.

Further, oxidative stress affects other hormonal systems including the thyroid and gonadal axis that are somehow affiliated to mood regulation. This chronic disruption of the hormonal balance has the cascade effect of depressing the supply of mood chemicals in the form of neurotransmitters such as serotonin and dopamine (Miller and Raison 2016). This generates a biochemical milieu that leads the person into developing depressive behaviors.

8.4.2 Neuroinflammatory Complex Imbalance

Oxidative stress is largely understood as inflammation resulting from injury or stress with neuroinflammation being a process that happens in the brain (Dantzer et al. 2008). The crosstalk between oxidative stress and inflammation has been referred as "neuroinflammatory stress," where due to increased oxidative stress, proinflammatory cytokines interleukin-6 (IL-6), interleukin-1 beta (IL-1β), and tumor necrosis factor (TNF-a) are produced in an exaggerated manner. This inflammatory action results in synaptic dysconnectivity, neuronal loss, and impaired neurogenesis which are characteristics of depression (Miller et al. 2009).

Also, ROS compromise the integrity of the blood-brain barrier and permit peripheral immune cells and cytokines to infiltrate the brain exacerbating inflammation (Vahia 2013). Oxidative stress causes activation of microglial cells which are part of the immune system in the central nervous system. These cells when chronically activated secrete even higher ROS and cytokines escalating the oxidative–inflammatory spiral. This kind of uncontrolled immune response in the brain is now considered to be a key factor in depression development because such inflammatory changes cause pathophysiological disturbances in brain function and behavior (Dantzer et al. 2008).

8.4.3 Serotonergic System Dysfunction

Oxidative stress is widely known to influence the regulation of mood through having an impact on the neurotransmission system via the serotonergic system. Serotonin is the other neurotransmitter that is important in mood regulation, sleep patterns, and emotional homeostasis and is highly prone to oxidative stress. ROS can potentially alter the serotonin and its receptors and can lower their functionality (Miller and Raison 2016). In addition, oxidative stress is an important modulator of the Serotonin Reuptake Transporter (SERT) that specifically controls reuptake and recycling of serotonin in synapses. Low levels of serotonin available to interact with postsynaptic receptors are the reason for low mood, fatigue, and cognitive deficit observed in depression.

Moreover, oxidative stress affects Serotonin synthesis by inhibiting enzymes involved in the synthesis of Serotonin including tryptophan hydroxylase. This enzyme is also dysregulated, and the breakdown of serotonin by this enzyme in addition to monoamine oxidase (MAO) results in depletion of serotonin, hence playing a role in depression. As such, the serotonergic system is instantly singled out as the fighting tool for the therapeutic interventions, which are to fight both the oxidative stress and apparently the depression.

8.4.4 Impaired Neuronal Plasticity

Neuronal plasticity, which is also known as synaptic plasticity, is the brain's capacity to change and modify the connections between neurons. This process is critical for learning, memory, and emotional regulation processes that are considered to be affected in depression (Miller et al. 2009). It is clear that oxidative stress interferes with immediate neuronal plasticity because of synapse and dendrite and neuronal cell damage. This damage is mainly demonstrated to occur in the hippocampus, an area in the brain involved in the processes of learning and memory formation and in the regulation of mood and emotions, and where new neurons are produced less under conditions of oxidative stress (Miller and Raison 2016).

Due to oxidative stress, mitochondria lose their efficiency to produce enough energy; therefore, neurons cannot support synaptic circuits necessary for cognition and mood stability. Due to oxidative stress, there is a decrease of neurotrophic factors such as the brain-derived neurotrophic factor (BDNF) that helps in impeding the depotentiation of depressive behaviors and synaptic plasticity (Citri and Malenka 2008).

8.4.5 Genetic or Epigenetic Damage

ROS not only invades the cellular aspects of the body but interferes with the genetic and epigenetic makeup, where changes to the gene's settings are for life. In fact, ROS can itself bring damage to the DNA which in turn leads to faulty gene expression that affects cerebral function (Lu et al. 2014). In addition, oxidative stress affects epigenetic regulation, DNA methylation, and histone modification and also plays a key role in controlling the gene accessibility for transcription (Sies 2015).

In depression, oxidative stress might cause gene hypermethylation with the genes controlling mood and neuroplasticity in the human body. Epigenetic repression of these important genes interferes with proper cellular processes of the brain that makes the neuron more susceptible to stress and neurodegenerative diseases. It is important to make these considerations because several genetic and epigenetic changes caused by oxidative stress may increase the vulnerability of people to develop depression if exposed to chronic stress or possessing a vulnerable genetic profile for mood disorders (Choi and Friso 2010).

8.5 Oxidative Stress as a Pharmacological Target in the Treatment of Depression

It has been proposed that oxidative stress has pivotal functions in depression. It is associated with neuronal dysfunction as well as decreased neurogenesis and development observed during depression establishment. Mitochondrial damage due to oxidative stress results in lack of energy and loss of normal balance within the cell which in turn could disturb synaptic plasticity needed for mood regulation as well as neurotransmitter synthesis and receptor function. Pharmacological interventions capable of attenuating mitochondrial oxidative stress may have therapeutic benefits by preventing impairment in mitochondrial function, maintaining neuronal bioenergetic state and redox balance (Miller and Raison 2016; Choi and Friso 2010; Lee and Giuliani 2019).

The following are some of the pharmacological interventions that target the breakdown of oxidative stress in depression: Some of the interventions include the augmentation of antioxidants in the body, or the administration of exogenous antioxidant agents. Direct ROS scavengers like the antioxidant precursors of the compounds include N-acetylcysteine (NAC) that has been found to alleviate depressive symptoms. Selenium and Vitamin E are also viewed as possessing antioxidant effects in the case of fighting the neurotoxic action in the brain. Moreover, medications for depression such as Selective Serotonin Reuptake Inhibitors (SSRIs) and tricyclic antidepressant apart from controlling the neurotransmitters have been observed to have antioxidant properties (Calsolaro and Edison 2016).

Mitochondrial health may also be targeted, establishing another emerging avenue in drug development. Oxidative stress is both produced and received within mitochondria making it a key component that can be addressed. Any pharmacological compound that had additive, synergistic, or protective effects on mitochondrial actions, or any protective effects against ROS, may help in the improvement of energy production in the affected cells, decrease apoptosis, and increase neurogenesis, leading to reversal of depressive symptoms. Newer antidepressants such as the mitochondria-targeted antioxidant like MitoQ or any agents which can induce mitochondrial genesis can be other valuable treatments that can directly combat the oxidative diameter of depression and at the same time enhance brain functioning.

Another actor in depression, neuroinflammation, is intimately connected with oxidative stress. Activation of caspases stimulates the releasing of proinflammatory cytokines leading to cumulative inflammation and oxidative damage. Hence, renewed interest is being placed on anti-inflammatory drugs with antioxidant potentials. For instance, omega-3 fatty acids, apart from having the anti-inflammatory effects, also are associated with lowered oxidative stress, and thus are two-pronged in the fight against depression. In the same way, drugs that block COX-2 such as anti-inflammatory agents may help to reduce oxidative stress because they cut down ROS emissions (Miller and Raison 2016; Michel et al. 2012; Schildkraut 1995).

Therefore, oxidative stress is becoming accepted as a core pathological hallmark of depressive disorders, which could well explain why it has emerged as a clear and very attractive pharmacological therapeutic target. The application of the antioxidant-based therapies is quite promising, as these therapies focus not only on the repair of the oxidative damage but also on the recovery of the neurons and the brain's ability to adapt to different conditions. Further studies regarding both ordinary antioxidants and novel existent drugs that regulate oxidative stress will be important for future developments in the topic of depression (Table 8.2).

Table 8.2 Pharmacological approaches and their mechanism of action

Pharmacological approach	Drugs	Mechanism of action
Direct ROS scavengers	N-acetylcysteine (NAC)	Increases glutathione levels, scavenges free radicals, reduces oxidative damage (Dean et al. 2011)
Endogenous antioxidants	Selenium, vitamin E	Enhances the body's antioxidant defenses, reduces neurotoxicity, protects against oxidative stress (Dean et al. 2011)
Antidepressants with antioxidant properties	SSRIs (e.g., fluoxetine), tricyclics (e.g., amitriptyline)	Primarily modulates neurotransmitters (serotonin), but also reduces oxidative stress (Kenis and Maes 2002)
Mitochondria-targeted antioxidants	MitoQ	Targets mitochondrial oxidative stress, improves mitochondrial function and energy production (Smith and Murphy 2010)
Anti-inflammatory antioxidants	Omega-3 fatty acids, COX-2 inhibitors	Reduces neuroinflammation, lowers oxidative stress, mitigates ROS production (Müller and Schwarz 2008)
Mitochondrial biogenesis inducers	Agents that stimulate mitochondrial genesis	Promotes mitochondrial health, improves neurogenesis, restores neuronal function (Wu et al. 1999)

References

Anderson G, Maes M (2014) Oxidative/nitrosative stress and immuno-inflammatory pathways in depression: treatment implications. Curr Pharm Des 20(23):3812–3847. https://doi.org/10.2174/13816128113196660738

Andreazza AC, Kauer-Sant'anna M, Frey BN, Bond DJ, Kapczinski F, Young LT, Yatham LN (2008) Oxidative stress markers in bipolar disorder: a meta-analysis. J Affect Disord 111(2–3): 135–144. https://doi.org/10.1016/j.jad.2008.04.013

Belmaker RH, Agam G (2008) Major depressive disorder. N Engl J Med 358(1):55–68. https://doi.org/10.1056/NEJMra073096

Berk M et al (2008) Glutathione: a novel treatment target in psychiatry. Trends Pharmacol Sci 29(7):346–351. https://doi.org/10.1016/j.tips.2008.05.001

Berk M et al (2011) Pathways underlying neuroprogression in bipolar disorder: focus on inflammation, oxidative stress and neurotrophic factors. Neurosci Biobehav Rev 35(3):804–817. https://doi.org/10.1016/j.neubiorev.2010.10.001

Bhatt S, Pandey DK, Patil CR, Nagappa AN (2020) Oxidative stress in depression and other comorbid disorders. In: Maurya P, Dua K (eds) Role of oxidative stress in pathophysiology of diseases. Springer, Singapore, pp 149–162. https://doi.org/10.1007/978-981-15-1568-2_9

Butterfield DA et al (1997) Free radical oxidation of brain proteins in accelerated senescence and its modulation by N-tert-butyl-alpha-phenylnitrone. Proc Natl Acad Sci 94(2):674–678. https://doi.org/10.1073/pnas.94.2.674

Calsolaro V, Edison P (2016) Neuroinflammation in Alzheimer's disease: current evidence and future directions. Alzheimers Dement 12(6):719–732. https://doi.org/10.1016/j.jalz.2016.02.010

Choi SW, Friso S (2010) Epigenetics: a new bridge between nutrition and health. Adv Nutr 1(1): 8–16. https://doi.org/10.3945/an.110.1004

Citri A, Malenka RC (2008) Synaptic plasticity: multiple forms, functions, and mechanisms. Neuropsychopharmacology 33(1):18–41. https://doi.org/10.1038/sj.npp.1301559

Dantzer R, O'Connor JC, Freund GG, Johnson RW, Kelley KW (2008) From inflammation to sickness and depression: when the immune system subjugates the brain. Nat Rev Neurosci 9(1): 46–56. https://doi.org/10.1038/nrn2297

Dantzer R, O'Connor JC, Lawson MA, Kelley KW (2011) Inflammation-associated depression: from serotonin to kynurenine. Psychoneuroendocrinology 36(3):426–436. https://doi.org/10.1016/j.psyneuen.2010.09.012

Dean O, Giorlando F, Berk M (2011) N-acetylcysteine in psychiatry: current therapeutic evidence and potential mechanisms of action. J Psychiatry Neurosci 36(2):78–86. https://doi.org/10.1503/jpn.100057

Dowlati Y, Herrmann N, Swardfager W, Liu H, Sham L, Reim EK, Lanctôt KL (2010) A meta-analysis of cytokines in major depression. Biol Psychiatry 67(5):446–457. https://doi.org/10.1016/j.biopsych.2009.09.033

Duman RS, Aghajanian GK (2012) Synaptic dysfunction in depression: potential therapeutic targets. Science 338(6103):68–72. https://doi.org/10.1126/science.1222939

Gawryluk JW et al (2011) Decreased levels of glutathione, the major brain antioxidant, in post-mortem prefrontal cortex from patients with psychiatric disorders. Int J Neuropsychopharmacol 14(1):123–130. https://doi.org/10.1017/S1461145710000805

Greenberg PE, Fournier AA, Sisitsky T, Pike CT, Kessler RC (2015) The economic burden of adults with major depressive disorder in the United States (2005 and 2010). J Clin Psychiatry 76(2):155–162. https://doi.org/10.4088/JCP.14m09298

Halliwell B (2006) Oxidative stress and neurodegeneration: where are we now? J Neurochem 97(6): 1634–1658. https://doi.org/10.1111/j.1471-4159.2006.03907.x

Halliwell B, Gutteridge JMC (2015) Free radicals in biology and medicine, 5th edn. Oxford University Press, Oxford. https://doi.org/10.1093/acprof:oso/9780198717478.001.0001

Kenis G, Maes M (2002) Effects of antidepressants on the production of cytokines. Int J Neuropsychopharmacol 5(4):401–412. https://doi.org/10.1017/S1461145702003164

Krishnan V, Nestler EJ (2008) The molecular neurobiology of depression. Nature 455(7215): 894–902. https://doi.org/10.1038/nature07455

Lee CH, Giuliani F (2019) The role of inflammation in depression and fatigue. Front Immunol 10: 1696. https://doi.org/10.3389/fimmu.2019.01696

Lu B, Nagappan G, Lu YB (2014) BDNF and synaptic plasticity, cognitive function, and dysfunction. In: Neurotrophic factors. Springer, Berlin, pp 223–250. https://doi.org/10.1007/978-3-642-45106-5_13

Maes M, Leonard BE, Myint AM, Kubera M, Verkerk R (2011) The new '5-HT' hypothesis of depression: cell-mediated immune activation induces indoleamine 2,3-dioxygenase, which leads to lower plasma tryptophan and an increased synthesis of detrimental tryptophan catabolites (TRYCATs), both of which contribute to the onset of depression. Prog Neuro-Psychopharmacol Biol Psychiatry 35(3):702–721. https://doi.org/10.1016/j.pnpbp.2010.12.017

Maes M, Galecki P, Chang YS, Berk M (2011a) A review on the oxidative and nitrosative stress (O&NS) pathways in major depression and their possible contribution to the (neuro)-degenerative processes in that illness. Prog Neuro-Psychopharmacol Biol Psychiatry 35(3): 676–692. https://doi.org/10.1016/j.pnpbp.2010.05.004

Manji HK et al (2003) Enhancing neuronal plasticity and cellular resilience to develop novel, improved therapeutics for difficult-to-treat depression. Biol Psychiatry 53(8):707–742. https://doi.org/10.1016/S0006-3223(03)00117-3

Mattson MP, Gleichmann M, Cheng A (2008) Mitochondria in neuroplasticity and neurological disorders. Neuron 60(5):748–766. https://doi.org/10.1016/j.neuron.2008.10.010

McEwen BS, Gianaros PJ (2010) Central role of the brain in stress and adaptation: links to socioeconomic status, health, and disease. Ann N Y Acad Sci 1186:190–222. https://doi.org/10.1111/j.1749-6632.2009.05331.x

Michel TM, Pülschen D, Thome J (2012) The role of oxidative stress in depressive disorders. Curr Pharm Des 18(36):5890–5899. https://doi.org/10.2174/138161212803523554

Miller AH, Raison CL (2016) The role of inflammation in depression: from evolutionary imperative to modern treatment target. Nat Rev Immunol 16(1):22–34. https://doi.org/10.1038/nri.2015.5

Miller AH, Maletic V, Raison CL (2009) Inflammation and its discontents: the role of cytokines in the pathophysiology of major depression. Biol Psychiatry 65(9):732–741. https://doi.org/10.1016/j.biopsych.2008.11.029

Moylan S et al (2014) Oxidative & nitrosative stress in depression: why so much stress? Neurosci Biobehav Rev 45:46–62. https://doi.org/10.1016/j.neubiorev.2014.05.007

Müller N, Schwarz MJ (2008) COX-2 inhibition in schizophrenia and major depression. Curr Pharm Des 14(14):1452–1465. https://doi.org/10.2174/138161208784480243

Ng F, Berk M, Dean O, Bush AI (2008) Oxidative stress in psychiatric disorders: evidence base and therapeutic implications. Int J Neuropsychopharmacol 11(6):851–876. https://doi.org/10.1017/S1461145707008401

Pariante CM, Lightman SL (2008) The HPA axis in major depression: classical theories and new developments. Trends Neurosci 31(9):464–468. https://doi.org/10.1016/j.tins.2008.06.006

Rezin GT et al (2009) Mitochondrial dysfunction and psychiatric disorders. Neurochem Res 34(6): 1021–1029. https://doi.org/10.1007/s11064-008-9865-8

Salim S (2014) Oxidative stress and psychological disorders. Curr Neuropharmacol 12(2):140–147. https://doi.org/10.2174/1570159X11666131120230309

Salim S (2017) Oxidative stress and the central nervous system. J Pharmacol Exp Ther 360(1): 201–205. https://doi.org/10.1124/jpet.116.237503

Sanacora G, Treccani G, Popoli M (2012) Towards a glutamate hypothesis of depression: an emerging frontier of neuropsychopharmacology for mood disorders. Neuropharmacology 62(1):63–77. https://doi.org/10.1016/j.neuropharm.2011.07.036

Scaini G, Rezin GT, Carvalho AF, Streck EL, Berk M, Quevedo J (2016) Mitochondrial dysfunction in bipolar disorder: evidence, pathophysiology and translational implications. Neurosci Biobehav Rev 68:694–713. https://doi.org/10.1016/j.neubiorev.2016.06.040

Scapagnini G, Davinelli S, Drago F, De Lorenzo A, Oriani G (2012) Antioxidants as antidepressants: fact or fiction? CNS Drugs 26(6):477–490. https://doi.org/10.2165/11633190-000000000-00000

Schildkraut JJ (1965) The catecholamine hypothesis of affective disorders: a review of supporting evidence. Am J Psychiatry 122(5):509–522. https://doi.org/10.1176/ajp.122.5.509

Schildkraut JJ (1995) The catecholamine hypothesis of affective disorders: a review of supporting evidence. J Neuropsychiatry Clin Neurosci 7(4):524–533. https://doi.org/10.1176/jnp.7.4.524

Sies H (2015) Oxidative stress: a concept in redox biology and medicine. Redox Biol 4:180–183. https://doi.org/10.1016/j.redox.2015.01.002

Smith RA, Murphy MP (2010) Animal and human studies with the mitochondria-targeted antioxidant MitoQ. Ann N Y Acad Sci 1201:96–103. https://doi.org/10.1111/j.1749-6632.2010.05627.x

Uddin M et al (2010) Epigenetic and immune function profiles associated with posttraumatic stress disorder. Proc Natl Acad Sci 107(20):9470–9475. https://doi.org/10.1073/pnas.0910794107

Vahia VN (2013) Diagnostic and statistical manual of mental disorders 5: a quick glance. Indian J Psychiatry 55(3):220–223

Valko M et al (2006) Free radicals, metals and antioxidants in oxidative stress-induced cancer. Chem Biol Interact 160(1):1–40. https://doi.org/10.1016/j.cbi.2005.12.009

Valko M, Leibfritz D, Moncol J, Cronin MT, Mazur M, Telser J (2007) Free radicals and antioxidants in normal physiological functions and human disease. Int J Biochem Cell Biol 39(1):44–84. https://doi.org/10.1016/j.biocel.2006.07.001

Wallace DC (2013) A mitochondrial bioenergetic etiology of disease. J Clin Invest 123(4):1405–1412. https://doi.org/10.1172/JCI61398

Wu Z et al (1999) Mechanisms controlling mitochondrial biogenesis and respiration through the thermogenic coactivator PGC-1. Cell 98(1):115–124. https://doi.org/10.1016/S0092-8674(00)80611-X

Yao JK, Reddy RD, van Kammen DP (2001) Oxidative damage and schizophrenia: an overview of the evidence and its therapeutic implications. CNS Drugs 15(4):287–310. https://doi.org/10.2165/00023210-200115040-00004

Chapter 9
Understanding Oxidative Stress in Ischemia-Reperfusion Injury

Alpana Singh, Shivangi Gupta, Suniti Das, Santoshi Gupta, Rishabh Chaudhary, and Vikas Mishra

Abbreviations

$^1O_2$	Singlet oxygen
4-HNE	4-hydroxy-2-nonenal
8-OH-dA	8-oxo-7,8-dihydro-deoxyadenosine
8-OH-dG	8-oxo-7,8-dihydro-deoxyguanosine
ALA	α-linolenic acid
ALDH2	Aldehyde dehydrogenase 2
AMPK	AMP-activated protein kinase
AP-1	Activator protein 1
ATF6	Activating transcription factor 6
BBB	Blood–brain barrier
Ca2+	Calcium ion
CAT	Catalase
CNS	Central nervous system
COX-2	Cyclooxygenase-2
CRF	Corticotrophin releasing factor
CyA	Cyclosporine A
CyD	Cyclophilin D

A. Singh
Department of Pharmacy, Dr. M.C. Saxena Group of Colleges, Dubagga, Lucknow, India

S. Gupta · S. Das · S. Gupta · V. Mishra (✉)
Department of Pharmaceutical Sciences, Babasaheb Bhimrao Ambedkar University, Lucknow, Uttar Pradesh, India
e-mail: dr.vikasmishra@bbau.ac.in

R. Chaudhary
Department of Pharmacy, School of Medical and Allied Sciences, Galgotias University, Greater Noida, India

N. Srivastava, S. A. Saraf (eds.), *Pharmacological Strategies for Prevention and Treatment of Oxidative Stress*, Oxidative Stress in Applied Basic Research and Clinical Practice, https://doi.org/10.1007/978-3-032-19837-2_9

DAMPs	Damage-associated molecular patterns
DCFDA	2′,7′-dichlorodihydrofluorescein diacetate
DHA	Docosahexaenoic acid
DHE	Dihydroethidium
DMECs	Dermal microvascular endothelial cells
DNA	Deoxyribonucleic acid
DNPH	2,4-dinitrophenylhydrazine
DUOX1/2	Dual oxidases1/2
ECM	Extracellular matrix
ELISA	Enzyme-linked immunosorbent assay
EOR	Endoplasmic reticulum overload response
EPA	Eicosapentaenoic acid
ER	Endoplasmic reticulum
ERAD	Endoplasmic-reticulum-associated protein degradation
ERK1/2	Extracellular signal-regulated kinases 1 and 2
ET-1	Endothelin-1
ETC	Electron transport chain
FADD	Fas-associated death domain
FGF-2	Fibroblast growth factor 2
FRAP	Ferric-reducing antioxidant power
GCs	Guanylyl cyclases
Gi proteins	Inhibitory G-proteins
GLP-1	Glucagon-like peptide 1
GPCRs	G-protein coupled receptors
GPx	Glutathione peroxidase
GPX4	Glutathione peroxidase 4
H_2O_2	Hydrogen peroxide
HI	Hypoxia-ischemia
HPLC	High-performance liquid chromatography
I/R	Ischemia-reperfusion
ICAM-1	Intercellular adhesion molecule-1
IGF-1	Insulin-like growth factor-1
IL-1β	Interleukin-1β
IL-6	Interleukin-6
IRE1	Inositol-requiring enzyme 1
IRI	Ischemia-reperfusion injury
JAK/STAT	Janus kinase/signal transducers and activators of transcription
JNK	c-Jun N-terminal kinase
LIF	Leukemia inhibitory factor
MAPK	Mitogen-activated protein kinase
MDA	Malondialdehyde
MEK	Mitogen extracellular kinase
MI	Myocardial infarction
MitoSOX	Mitochondrial Superoxide Indicators

mK_{ATP}	Mitochondrial ATP-sensitive K+ channel
mPTP	Mitochondrial permeability transition pore
mtDNA	Mitochondrial DNA
mTOR	Mammalian target of rapamycin
Na^+/K^+-ATPase	Sodium–potassium adenosine triphosphatase
NAC	N-acetylcysteine
NADPH	Nicotinamide adenine dinucleotide phosphate
NF-κB	Nuclear factor-kappa B
NGAL	Neutrophil gelatinase-associated lipocalin
NO	Nitric oxide
NOx	Nitrogen oxide
NPs	Natriuretic peptides
Nrf2/HO-1	Nuclear factor erythroid 2-related factor 2/heme oxygenase 1
NSAID	Non-steroidal anti-inflammatory drug
$O_2^{-}\bullet$	Superoxide anion
OH	Hydroxyl radical
$ONOO^-$	Peroxynitrite
PCs	Carbonylated proteins
PDE-5	Phosphodiesterase-5
PDK	Pyruvate dehydrogenase kinase
PEG	Polyethylene glycol
PERK	Protein kinase R-like endoplasmic reticulum kinase
PGE2	Prostaglandin E2
PKA	Protein kinase A
PM	Particulate matter
PRRs	Pattern recognition receptors
PUFAs	Polyunsaturated fatty acids
RET	Reverse electron transport
RISK	Reperfusion Injury Salvage Kinase
RNA	Ribonucleic acid
RNS	Reactive nitrogen species
RO•	Alkoxyl radical
ROO•	Peroxyl radical
ROS	Reactive oxygen species
RSS	Reactive sulfur species
RTKs	Receptor tyrosine kinases
S1QELs	Suppressors of site I Q Electron Leak
SAFE	Survivor Activating Factor Enhancement
sGCs	Soluble guanylyl cyclases
SO_2	Sulfur dioxide
SOD	Superoxide dismutase
SREBP	Sterol regulatory element binding protein
TAC	Total antioxidant capacity
TBARS	Thiobarbituric acid reactive substance

TEAC	Trolox equivalent antioxidant capacity
TGF-β1	Transforming growth factor-β1
TNF-α	Tumor necrosis factor-alpha
TTFA	2-thenoyltrifluoroacetone
UPR	Unfolded protein response
UV	Ultraviolet
VCAM-1	Vascular cell adhesion molecule-1
vWF	Von Willebrand factor
XBP1	X-box binding protein 1

9.1 Introduction

Ischemia-reperfusion (I/R) injury represents a serious pathological condition that occurs when blood flow to a tissue is temporarily obstructed (ischemia) and afterwards restored (reperfusion) (Zhang et al. 2024). Although it is widely acknowledged that restoring blood flow is vital to prevent irreversible tissue damage, it is also known that reperfusion can inevitably worsen the injury. The underlying mechanisms driving IRI are intricate and multifactorial, encompassing a complex interplay between inflammatory responses, oxidative damage, and cell death pathways (Sánchez-Hernández et al. 2020). During ischemia, a significant decrease in blood flow leads to hypoxia, causing a significant disruption in cellular metabolism. Oxidative phosphorylation in the mitochondria is severely affected when there is a lack of oxygen following a substantial decrease in adenosine diphosphate (ATP) production. ATP plays a crucial role in supporting a wide variety of cellular activities, such as maintaining ion gradient flow across the plasma membrane. The depletion of this substance causes a disturbance in the functioning of ion pumps, like the Na^+/K^+-ATPase and Ca^{2+}-ATPase, resulting in an increase in the levels of sodium and calcium inside the cells (Akhigbe et al. 2024). The imbalance of ions leads to cellular edema, compromised membrane integrity, and, ultimately, cellular dysfunction and injury. The absence of oxygen also results in the buildup of metabolic by-products, including lactate, which contributes to the acidification of the intracellular microenvironment. The acidic pH plays a role in the activation of different proteolytic enzymes, such as proteases and phospholipases. These enzymes break down cellular proteins and membrane lipids, leading to further tissue damage. Following reperfusion, the restoration of blood flow and reintroduction of oxygen give rise to a rapid increase in the generation of reactive oxygen species (ROS) (Bagheri et al. 2016). Resuscitation of oxygen-deficient tissues provides substrates for enzymes that produce ROS, such as NADPH oxidases and xanthine oxidase. Furthermore, the sudden increase in oxygen levels can result in the generation of hydrogen peroxide and superoxide radicals. An excessive release of ROS affects the antioxidant defense mechanisms, leading to the oxidative damage to lipids, proteins, and DNA. The impairment of cellular function caused by this damage exacerbates the injury (Firdous et al. 2024; Zhao et al. 2023). At the same time, reperfusion activates pathways related to inflammation. The release of damage-associated molecular patterns (DAMPs) and activation of pattern recognition receptors (PRRs) result in the production of pro-inflammatory molecules. The inflammatory response is able to attract immune cells, like neutrophils and macrophages, to the area of injury (Algoet et al. 2023). These cells continue to release ROS, proteases, and inflammatory mediators, which further contribute to tissue damage and worsen the initial injury. Moreover, cell death in IRI is influenced by various pathways, such as apoptosis, necrosis, and necroptosis. The interaction between oxidative stress and inflammation triggers these mechanisms of cell death, resulting in permanent tissue damage (Dugbartey 2024; Luo et al. 2024). Considering the intricate nature of IRI, therapeutic approaches have been directed toward various elements of the injury process. One area of interest involves the reduction of oxidative stress by utilizing antioxidants, which strive to counteract ROS and safeguard against oxidative harm. There is growing potential in the field of nutraceuticals, particularly in the use of polyphenols, as well as novel pharmacological agents that can effectively target enzymes responsible for generating ROS (Elsayed Abouzed et al. 2024). Preclinical studies have provided encouraging results in this regard (Sedighi et al. 2019). In addition, researchers are investigating therapies that can adjust mitochondrial function, decrease calcium overload, and block inflammatory pathways as possible interventions to prevent IRI. This chapter aims to provide a detailed overview of the molecular mechanisms underlying oxidative stress during IRI, focusing on the subsequent activation of inflammatory cascades and cell death pathways that contribute to tissue damage. A detailed investigation of current therapeutic strategies, such as nutraceuticals and novel pharmacological agents, to mitigate oxidative stress, modulate inflammation, and prevent cell death, is presented.

9.2 Reactive Oxygen Species (ROS)

Due to their role in multiple cellular and disease processes, reactive species—such as reactive nitrogen species (RNS), reactive oxygen species (ROS), and reactive sulfur species (RSS)—have attracted significant interest in various fields of biology and medicine (Sies and Jones 2020). ROS are generated as a result of regular metabolic processes that involve oxygen. During these reactions, an oxygen molecule acquires an electron, leading to the formation of superoxide anion, which is a type of oxygen radical. Superoxide ion serves as a precursor for the generation of different ROS, including hydrogen peroxide, hydroxyl radical, and singlet oxygen (Fig. 9.1) (Juan et al. 2021). These molecules are involved in important processes such as cell signaling, maintaining balance within the body, and supporting the immune system. Every RO species possesses a distinct level of reactivity that plays a critical role in determining its molecular effects and biological implications. The various ways by which ROS can work are largely determined by their wide range in reactivity. It can act as a signaling molecule, leading to reversible modifications, or it can cause

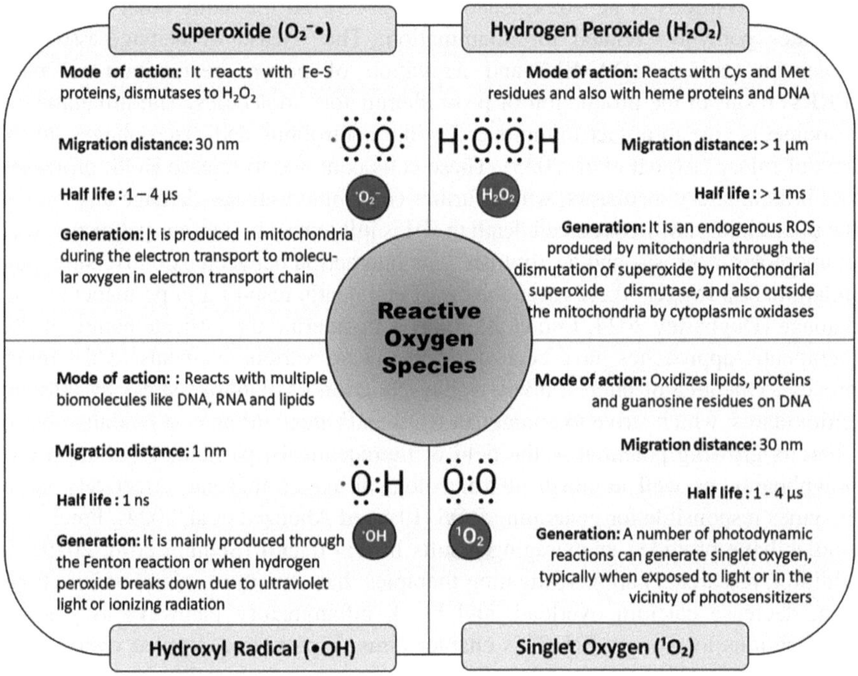

Fig. 9.1 A schematic overview of different reactive oxygen species and their characteristic features

irreversible damage to various important biomolecules like DNA, proteins, and lipids (Patel et al. 2018; Sies and Jones 2020). ROS play a vital role in various physiological processes, such as redox regulation, cellular signaling, and immunological responses. They are not just metabolic by-products, but rather important participants in these processes. ROS play a crucial role as secondary messengers in important signaling pathways, impacting transcription factors such as NF-κB (nuclear factor-kappa B) and AP-1 (activator protein 1). These factors are eventually responsible for controlling multiple genes related to cellular apoptosis, immune and inflammatory responses (Ray et al. 2012). Moreover, ROS play a crucial role in the immune system, specifically in phagocytic cells. They help in destroying pathogens through oxidative burst. This process produces significant amounts of ROS to eliminate invading microbes. In addition to their role in pathogen clearance, ROS also have an impact on immune signaling and the release of cytokines. However, a healthy state of cellular homeostasis depends on a delicate balance between the generation of ROS and antioxidant defense systems. Any perturbation in this equilibrium can lead to oxidative stress (Sies 2020). The heightened level of oxidative stress can lead to significant harm to cellular components such as lipids, proteins, and DNA, which in turn plays a role in the development of different diseases and disorders. Prolonged oxidative stress condition is linked to the emergence and advancement of various conditions. For instance, it contributes to genetic mutations and tumor growth in cancer, causes damage to neurons in neurodegenerative disorders like Alzheimer's and Parkinson's disease, and worsens inflammation and tissue damage in chronic diseases such as rheumatoid arthritis and psoriasis (Pizzino et al. 2017). Thus, it is critical to identify the intricate relationship between ROS as signaling molecules and their potential to induce oxidative damage in order to fully comprehend their impact on both human well-being and illnesses.

9.2.1 Types of ROS

1. *Superoxide Anion (O_2^{-}•)*
   Superoxide exhibits a moderate level of reactivity and acts as a precursor to species that are even more reactive. This process occurs primarily in the mitochondria during electron transport and is then transformed into hydrogen peroxide by superoxide dismutase (SOD). While superoxide alone may not usually result in substantial harm to macromolecules, its function as a precursor to more reactive species places it as a crucial factor in oxidative stress (Hayyan et al. 2016).
2. *Hydrogen Peroxide (H_2O_2)*
   Hydrogen peroxide exhibits a lower reactivity when compared to other RO species. However, it possesses a higher stability and the ability to easily diffuse across cell membranes. Thus, it is an important signaling molecule that is essential for controlling a wide range of biological activities, including cell division, cell proliferation, and apoptosis. Moreover, the reactivity of the substance is greatly enhanced when it comes into contact with transition metals such as copper or iron. This leads to its involvement in the Fenton reaction, resulting in the production of the highly reactive hydroxyl radical (Collin 2019).
3. *Hydroxyl Radical (•OH)*
   The hydroxyl radical is extremely reactive and poses a significant threat to various molecules, including cellular/molecular proteins, nucleic acids, and lipids. Because of its notable reactivity, the hydroxyl radical has a very brief half-life and primarily affects the area where it is generated. It is mainly produced through the Fenton reaction or when hydrogen peroxide breaks down due to ultraviolet (UV) light or ionizing radiation. This radical is associated with significant oxidative damage, which plays a role in mutagenesis, lipid peroxidation, and protein oxidation. It is also linked to several diseases, including cancer and neurodegeneration (Tvrdá and Benko 2020).
4. *Singlet Oxygen ($^1O_2$)*
   Singlet oxygen is a highly energized form of molecular oxygen, with one of its electrons elevated to a higher energy state. It exhibits a strong reactivity toward electron-rich biomolecules like lipids and proteins. A number of photodynamic reactions can generate singlet oxygen, typically when exposed to light or in the vicinity of photosensitizers. The reactivity of this substance can result in lipid peroxidation, protein oxidation, and oxidative harm to cellular membranes (Murotomi et al. 2020).

5. *Peroxyl Radicals (ROO•) and Alkoxyl Radicals (RO•)*
   These radicals are commonly generated during lipid peroxidation and involved in the expansion of oxidative chains of events. Peroxyl radicals (ROO•) interact with other lipid molecules, continuing the lipid peroxidation process, which may lead to disruption of cellular integrity and membrane damage. On the contrary, alkoxyl radicals (RO•) exhibit a remarkable level of reactivity, enabling them to directly target lipids, proteins, and DNA (Martemucci et al. 2022).
6. *Peroxynitrite ($ONOO^-$)*
   Peroxynitrite is a molecule that is formed when nitric oxide (NO) and superoxide anion react with each other. It is known for its high reactivity. It exhibits dual properties, causing harm to cellular components by acting as an oxidant and a nitrating agent. Peroxynitrite has the potential to induce lipid peroxidation, protein nitration, and DNA damage, which can ultimately lead to cellular dysfunction and the development of numerous diseases (Bartesaghi and Radi 2018).

9.2.2 Generation of ROS

Free radicals are atoms or molecules that possess unpaired electrons in their outermost electron shell, which gives them a high level of reactivity. In biological systems, the most common types of free radicals and reactive molecules are those that come from oxygen, called ROS, and those that come from nitrogen, known as RNS. These species are formed mainly through electron transfer processes, which involve the addition or removal of electrons (Halliwell and Gutteridge 2015). In biological systems, the production of ROS arises from a variety of sources, which can be classified into exogenous and endogenous origins (Fig. 9.2). Exogenous factors, like pollution, UV radiation, and toxins, can introduce ROS into cells from the environment. Endogenous sources arise from within the organism and encompass metabolic progressions such as mitochondrial respiration, inflammation, and enzymatic activities that generate ROS as by-products. Both sources play a role in the maintenance of the cellular redox state and have an impact on oxidative stress and damage.

A. *Exogenous Sources of ROS*

   1. *UV Radiation*
      UV radiation from sunlight is a notable external source of ROS. UV light, specifically UV-B and UV-C, has the ability to penetrate the skin and other tissues, causing the stimulation of oxygen molecules and the ROS generation for example singlet oxygen and hydroxyl radicals. ROS can result in DNA damage, which can lead to mutations and contribute to the development or progression of various diseases (De Almeida et al. 2022).
   2. *Pollution*
      Air pollutants such as nitrogen oxides (NOx), sulfur dioxide (SO_2), and particulate matter (PM) can interact with biological tissues, potentially leading to the generation of ROS. Pollutants have the ability to trigger oxidative

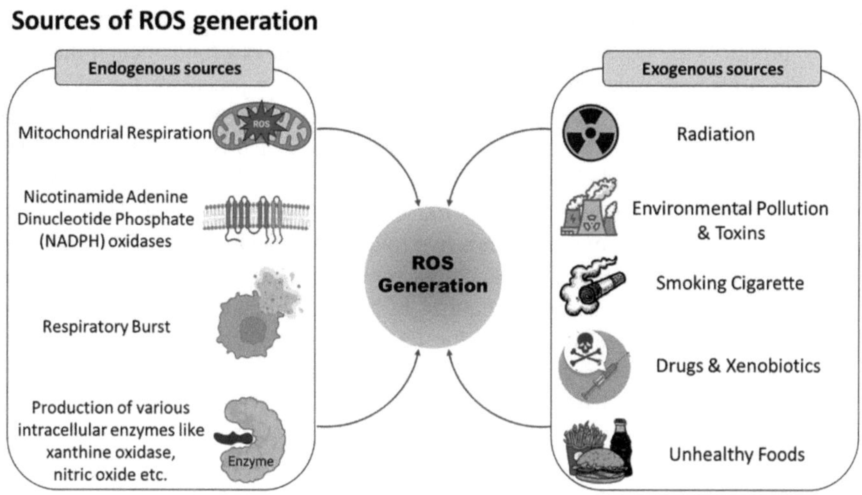

Fig. 9.2 A schematic overview of distinct generative sources of reactive oxygen species

stress by facilitating the production of superoxide anions and other RO species via different means, including inflammation and direct chemical interactions with cellular components. Long-term exposure to such pollutants are associated with the generation of various health-related complications, including respiratory diseases and cardiovascular conditions (Leni et al. 2020).

3. *Toxins*

   Production of ROS in cells can also be triggered via heightened environmental toxins such as heavy metals (like lead and mercury), pesticides, and industrial chemicals. These toxins have the ability to disturb cellular processes and cause oxidative stress by producing ROS through redox cycling or by interfering with mitochondrial function (Poljšak and Fink 2014).

B. *Endogenous Sources of ROS*

1. *Mitochondrial Respiration*

   Mitochondrial respiration is a significant contributor to the ROS generation. This process takes place in the inner membrane of the mitochondria, where the electron transport chain (ETC) plays a crucial role in transferring electrons through a series of protein complexes. These complexes, including Complex I (NADH oxidoreductase), Complex II (succinate dehydrogenase), Complex III (ubiquinol c oxidoreductase), and Complex IV (cytochrome c oxidase) are responsible for facilitating this electron transfer. Electrons are transferred efficiently through these complexes, leading to the conversion of molecular oxygen (O_2) into water (H_2O) at Complex IV, facilitated by cytochrome c oxidase activity. However, in the course of this process, a fraction of electrons may react with molecular oxygen before reaching Complex IV, leading to an incomplete reduction of oxygen (Sarniak et al. 2016). This aberrant response produces the O_2^{-}•, a form of ROS. Superoxide serves a fundamental function as a primary reactive oxygen species and acts as a precursor for additional reactive species via secondary reactions. For example, the enzyme SOD can reduce superoxide to form hydrogen peroxide, which is a more stable ROS. Hydrogen peroxide has the ability to take part in the Fenton or Haber–Weiss reactions when transition metals like iron or copper are present. This leads to the generation of extremely reactive hydroxyl radical species. The oxidative regulation of ROS within the cell is strengthened by such series of reactive processes. Furthermore, a number of factors, including cellular substrate accessibility, the efficiency of the ETC complexes, and the redox status of the cell, affect the generation of ROS through mitochondrial respiration (Sarniak et al. 2016).

2. *NADPH Oxidase*

   Nicotinamide Adenine Dinucleotide Phosphate (NADPH) oxidases, often known as NOX, are intricate enzyme systems that are found in the plasma membrane or the membranes of the phagosomes of cells. NADPH oxidases are made up of a number of different subunits, some of which are catalytic subunits (NOX proteins) and then there are regulatory subunits. When it comes to the production of ROS, the catalytic subunits, which include NOX1, 2, 3, 4, 5, and DUOX1/2 (dual oxidases1/2), are extremely important. There is a significant contribution made by regulatory subunits to the process of fine-tuning the activity of these catalytic subunits (Magnani and Mattevi 2019). The NADPH oxidase complex is composed of a number of different components, some of which are membrane-bound while others are cytosolic. Cytochrome b558 is the membrane-bound component of the NADPH oxidase complex and is composed of $gp91^{PHOX}$ (also known as NOX2) and $p22^{PHOX}$. In contrast, $p47^{PHOX}$, $p67^{PHOX}$, and Rac1 or Rac2 (Ras-related C3 botulinum toxin substrate 1 or 2) are cytosolic components that translocate to the membrane upon activation. The NADPH oxidase complex is activated when the cytosolic components travel to the membrane and join with cytochrome b558. This occurs when the cytosolic components are triggered by the identification of a pathogen or by inflammatory signals. Through the utilization of this complex mechanism, electrons can be transferred from NADPH to molecular oxygen in an effective manner, which ultimately results in the production of superoxide anions (Babior 2004).

3. *Xanthine Oxidase*

   One of the most important aspects of purine metabolism is the breakdown of purine bases, which requires xanthine oxidase as a necessary component. When it comes to the oxidation of hypoxanthine and xanthine, which are both key stages in the process of purine nucleotide breakdown, this enzyme is absolutely necessary. This process ultimately results in the production of uric acid, which is subsequently eliminated from the body through the urine. The enzymatic reaction catalyzed by xanthine oxidase occurs in two primary steps: (1) the conversion of hypoxanthine to xanthine which involves the addition of two oxygen atoms by xanthine oxidase, resulting in the formation of a peroxide species as an intermediate; (2) the oxidation of Xanthine to Uric Acid involves the transfer of additional oxygen atoms. During each of these stages, xanthine oxidase generates RO species as a by-product. Specifically, hydrogen peroxide and superoxide anions are formed as a result of this process. It is the enzyme's capacity to oxidize hypoxanthine and xanthine that is responsible for the generation of ROS. Hydrogen peroxide is a secondary product that is created from the dismutation of superoxide, whereas superoxide is produced through the process of electron transfer (Cantu-Medellin and Kelley 2013).

4. *Cytochrome P450*
   Cytochrome P450 enzymes play a vital role in the oxidative metabolism of drugs, steroids, fatty acids, and xenobiotics. They are a diverse family of heme-containing monooxygenases. These enzymes are primarily found in the liver's endoplasmic reticulum. They play a crucial role in incorporating one atom of molecular oxygen into substrates and simultaneously reducing the other atom to water. The catalytic cycle of cytochrome P450 enzymes encompasses various crucial stages. At first, the substrate attaches to the active site of the enzyme, interacting with the heme group. The transfer of electrons from NADPH to cytochrome P450 occurs through a series of intermediate steps that involve the assistance of cytochrome P450 reductase (Veith and Moorthy 2018). This process facilitates the reduction of the heme iron within the enzyme. The heme iron interacts with molecular oxygen, resulting in the development of extremely reactive oxygen species, like a peroxo or hydroperoxo complex. This reactive intermediate reacts with the substrate, adding an oxygen atom and transforming it into a metabolite with higher polarity. Meanwhile, the other O_2 atom is usually reduced to H_2O. At various stages of this catalytic process, cytochrome P450 enzymes can generate ROS as secondary products. Specifically, the process of reducing molecular oxygen can result in the formation of superoxide anions. On the other hand, hydrogen peroxide can be produced either directly upon oxygen reduction or as a secondary product resulting from the dismutation of superoxide (Veith and Moorthy 2018).

9.2.3 *Oxidative Stress Due to Reactive Oxygen Species*

Helmut Sies was the first to define "oxidative stress" as a state in which the balance between the production of ROS and the cellular antioxidant defense systems is disrupted, which may result in damage to the functioning of biological systems (Jones and Radi 2014). In normal physiological settings, ROS are frequently generated as a result of cellular metabolism, and their quantities are adequately controlled by a complex network of antioxidant mechanisms. Nevertheless, if the production of ROS surpasses the limits of the antioxidant defenses, oxidative stress takes place.

This may occur when there is an increase in ROS generation or a decrease in antioxidant activity (Lushchak 2011). At present, the more precise and scientifically nuanced definition of oxidative stress can be expressed as follows: oxidative stress is a condition characterized by increased steady-state levels of ROS, either temporarily or permanently, that disrupts the delicate equilibrium of cellular and metabolic signaling networks, particularly those influenced by redox processes. This includes dysregulation that triggers oxidative alterations in essential macromolecules, which could damage their structural integrity and functionality. When the cell's antioxidant defenses fail to adequately counteract these oxidative changes, they can trigger a series of cellular dysfunctions, gradually leading to programmed cell death processes like apoptosis or uncontrolled cell death through necrosis. The given description emphasizes the intricate and divergent characteristics of oxidative stress, thereby highlighting its significance not just in causing harm to cells but also in the wider framework of pathological mechanisms (Forman and Zhang 2021; Lushchak and Storey 2021). Oxidative stress is an intricate and multifaceted phenomenon. The impact of this phenomenon on an organism is influenced by several factors, including the specific type of oxidant species involved, the location and intensity of its production, the composition and activity levels of various antioxidant defenses, and the efficiency of cellular repair mechanisms. The interplay between these factors determines whether oxidative stress results in reversible damage, triggers adaptive responses, or leads to irreversible cellular injury and disease progression. A wide variety of pathogenic processes, including inflammation, ischemia/reperfusion injury, and infection, are significantly influenced by oxidative stress. It plays a role in the development and advancement of several chronic and acute disorders, including diabetes, inflammatory bowel disease, renal dysfunction, and cardiovascular disorders, as well as cancer (Olsson et al. 2012).

Under normal biological settings, the levels of ROS are regulated within a certain range called the steady-state level, which represents a balance between ROS generation and antioxidant defenses. When external factors, such as exposure to an oxidant, disturb this equilibrium, ROS levels may temporarily rise (Rauf et al. 2024). If the antioxidant defense systems are adequately robust, they can promptly mitigate the excess ROS, returning levels to their steady state within minutes. This temporary increase in ROS is known as "acute oxidative stress." It is believed that the reaction to acute oxidative stress does not necessitate the activation of gene expression to counteract reactive oxygen species (ROS). However, in many cases, living organisms are unable to readily or rapidly adapt to even a single administration of an oxidative stress trigger, resulting in prolonged elevated levels of ROS over extended periods, such as hours, days, and even longer. This type of stress is referred to as "chronic oxidative stress" and typically involves the activation of several genes implicated with stress (Lushchak and Storey 2021). Chronic oxidative stress can be categorized into two distinct forms. The first occurs when ROS levels are persistently elevated or slightly increased beyond the normal steady-state range, remaining consistently higher than their baseline levels. The second form involves the stabilization of ROS levels at a new, elevated quasi-stationary state. This sustained elevation of ROS may not return to the original steady state, reflecting a more permanent shift in the cellular environment. This latter case could be indicative of a new physiological state of the organism, where the elevated ROS levels become a part of the body's adjusted homeostasis, potentially influencing various cellular and systemic functions (Jomova et al. 2023; Lushchak and Storey 2021). Due to their very low total antioxidant capacity, several biological systems, including the immune, endocrine, cardiovascular, and central nervous system (CNS), are highly vulnerable to oxidative damage. The ROS produced within these tissues can cause significant oxidative damage to several vital molecules, including lipids, nucleic acids, and proteins (Di Meo et al. 2016; Snezhkina et al. 2020). In the CNS, for example, the high metabolic demand and abundant polyunsaturated fatty acids make neuronal cells especially prone to lipid peroxidation, which can lead to neurodegenerative disorders like Parkinson's and Alzheimer-like diseases (Singh et al. 2019; Solleiro-Villavicencio and Rivas-Arancibia 2018). Similarly, in the cardiovascular system, oxidative damage to endothelial cells, lipoproteins, and other components can contribute to the development of atherosclerosis, hypertension, and other cardiovascular pathologies. ROS can interfere with immune cell signaling and function in the immune system, which may compromise the body's ability to effectively respond to threats (Panth et al. 2016). The endocrine system is also susceptible to oxidative stress, which can lead to dysfunction in hormone-secreting glands, such as in diabetes mellitus and thyroid disorders (Vitale et al. 2013). The vulnerability of these systems demonstrates the vital significance of maintaining redox homeostasis in safeguarding cellular well-being and halting the advancement of diseases in various physiological areas.

ROS can oxidize polyunsaturated fatty acids (PUFAs) with greater efficiency than any other targets. Oxygen-derived free radicals, such as alkylperoxyl radical, hydroxyl radical, and superoxide anion, are particularly powerful for triggering lipid peroxidation, a process linked to the etiology of many disorders. Once lipid peroxidation begins, it triggers a chain reaction of free radical generation that continues until termination products are formed. As a result, lipid peroxidation end-products, such as malondialdehyde (MDA), 4-hydroxy-2-nonenal (4-HNE), and F2-isoprostanes, accumulate in biological systems (Perluigi et al. 2012; Tsikas 2017). These compounds serve as biomarkers of oxidative stress and are indicative of ongoing cellular damage. MDA, for instance, can be metabolized by various enzymes, notably aldehyde dehydrogenase in mitochondria, or it can form covalent adducts with proteins and nucleic acids. These adducts lead to the formation of DNA-protein crosslinks and other modifications, contributing to the disruption of cellular functions and the promotion of disease processes, including cancer, neurodegeneration, and cardiovascular conditions (Marnett 1999). In relation to protein damage, the presence of free radicals produced during oxidative stress can cause alterations in polypeptide chains, resulting in the production of carbonylated proteins (PCs). A non-reversible post-translational alteration, protein carbonylation is regarded as a crucial indicator of oxidative stress. This alteration commonly induces the incorporation of reactive carbonyl groups, such as aldehydes, ketones, and lactams, into the protein structure, therefore potentially modifying the function and stability of the protein. The buildup of carbonylated proteins might hinder biological functions, as these proteins that have been damaged by oxidation may accumulate or lose their ability to function as enzymes. Carbonylation, being irreversible, frequently marks the protein for catabolism by the proteasome (Akagawa 2021). Nevertheless, if the intricate mechanisms of cellular repair and degradation are overloaded it can result in cellular malfunction and facilitate pre-determined cell death pathways, such as apoptosis or necrosis, hence worsening tissue damage and the advancement of the disease (Fedorova et al. 2014). Oxidative stress-induced damage to DNA and RNA can give rise to mutations and genetic abnormalities, all of which culminate in cellular destruction and contribute to numerous pathologies. Free radicals constitute one of the stressors that accelerate the degradation of nucleic acids, thereby increasing the probability of DNA and RNA impairment (Kong and Lin 2010; Poetsch 2020). Breaks in double-stranded DNA within the nucleus and oxidation of the free bases in the deoxynucleotide triphosphate pool, especially in mitochondria, are examples of the various manifestations of this damage (Ichikawa et al. 2008). Oxidative DNA damage occurs through reactions involving pyrimidine and purine bases, deoxyribose, or the phosphodiester backbone of DNA. In this context, the hydroxyl radical exhibits noteworthy potency. It has the ability to create adducts with the double bonds of DNA bases or extract hydrogen atoms from deoxyribose residues. This process can result in strand breaks or modifications to the DNA bases (Sharma et al. 2012). One important reaction occurs when hydroxyl radicals are added to the double bond of purines at the 7,8-position, resulting in the formation of 8-oxo-7,8-dihydro-deoxyguanosine (8-OH-dG) from 8-OH-dGTP. The elimination of the diphosphate group from 8-OH-dGTP and subsequent breakdown of 8-OH-dGMP, or the repair of damaged DNA, additionally aids in the breakdown of nucleic acids (Kasai et al. 2008). Out of the various oxidative modifications, 8-OH-dG is more commonly found than 8-oxo-7,8-dihydro-deoxyadenosine (8-OH-dA) and is frequently investigated as an indicator of endogenous oxidative DNA damage. Additional by-products of oxidation include 8-oxo-2′-deoxyguanosine (derived from 8-OH-dG), 8-hydroxyguanine (from DNA or RNA), and 8-hydroxyguanosine (from RNA) (Cadet et al. 1999). These molecules often serve as biomarkers to evaluate the level of damage caused by oxidative stress to nucleic acids and their potential involvement in the development and advancement of various diseases.

Although not considered direct indicators of oxidative stress, the activity of important antioxidant enzymes responsible for eliminating ROS can provide valuable insight into oxidative stress equilibrium. The diminished activity of these enzymes may suggest an imbalance in the generation and removal of ROS. Superoxide dismutase (SOD), catalase (CAT), and glutathione peroxidase (GPx) are the primary enzymes involved in neutralizing ROS (Fukai and Ushio-Fukai 2011). SOD is essential for defense toward superoxide radicals, facilitating their transformation into hydrogen peroxide. This detoxification step is crucial, as superoxide radicals have the potential to cause extensive damage to cellular structures. The hydrogen peroxide produced is then further reduced to H_2O molecule by CAT, GPx, or peroxiredoxins. These reactions are vital as they effectively neutralize hydrogen peroxide without generating any additional free radicals. CAT, in particular, has demonstrated remarkable activity in tissues that produce increased levels of hydrogen peroxide, such as the liver and red blood cells. CAT is primarily found in peroxisomes, where it effectively transforms hydrogen peroxide into water and oxygen molecules (Goyal and Basak 2010). GPx and peroxiredoxins, in contrast, function in distinct cellular compartments and play a crucial role in reducing peroxides, such as hydrogen peroxide and organic hydroperoxides, to water or corresponding alcohols. The activity of these isoforms varies depending on their location within the cell, such as the cytoplasm, mitochondria, and other cellular regions. This compartmentalization guarantees a comprehensive defense system against ROS in different cellular environments. Working in unison, these enzymes play a crucial role in maintaining the delicate balance of redox reactions (Molavian et al. 2015; Perkins et al. 2015). Together, all these enzymes maintain redox balance, protecting cells from oxidative damage and preserving cellular function.

9.3 Ischemia-Reperfusion (I/R) Injury

Ischemia refers to a condition where there is an inadequate supply of oxygen and nutrients to tissues, resulting from diminished or obstructed blood flow (Fig. 9.3). This can lead to serious and potentially life-threatening outcomes, particularly when it impacts essential organs such as the heart and brain (Frangogiannis 2015). The process of tissue reperfusion, which involves restoring blood flow to areas of ischemic tissue, is a key strategy employed to avert irreversible damage during ischemic incidents like myocardial infarction (MI) or stroke. Nonetheless, although reperfusion is crucial for preserving viable tissue, it can abruptly exacerbate the injury through a mechanism referred to as ischemia-reperfusion (I/R) injury. This occurrence is observed in various organs such as the heart, brain, liver, kidneys, and intestines, potentially resulting in further complications, disability, and, in extreme cases, mortality (Yapca et al. 2013). I/R injury encompasses various pathological processes that together lead to tissue damage. Important mechanisms include cell death pathways such as apoptosis, necrosis, and ferroptosis, wherein oxidative stress and mitochondrial dysfunction initiate both regulated and unregulated cell death. The generation of excessive ROS during reperfusion leads to lipid peroxidation, protein modification, and DNA damage, which further intensifies the injury. The inflammatory response plays a crucial role, with immune cells, especially neutrophils and macrophages, infiltrating the injured tissue and releasing cytokines, proteases, and reactive oxygen species, thereby intensifying inflammation and contributing to tissue damage. I/R results in the deterioration of the blood–brain barrier (BBB) and interferes with extracellular matrix (ECM) remodeling, impacting tissue integrity. In the heart, it can induce cardiomyocyte hypertrophy and result in fibrosis, compromising organ function and leading to long-term complications such as heart failure (Zhang et al. 2024). In the section below, we provide a comprehensive overview of the key pathological mechanisms associated with I/R injuries (Fig. 9.4).

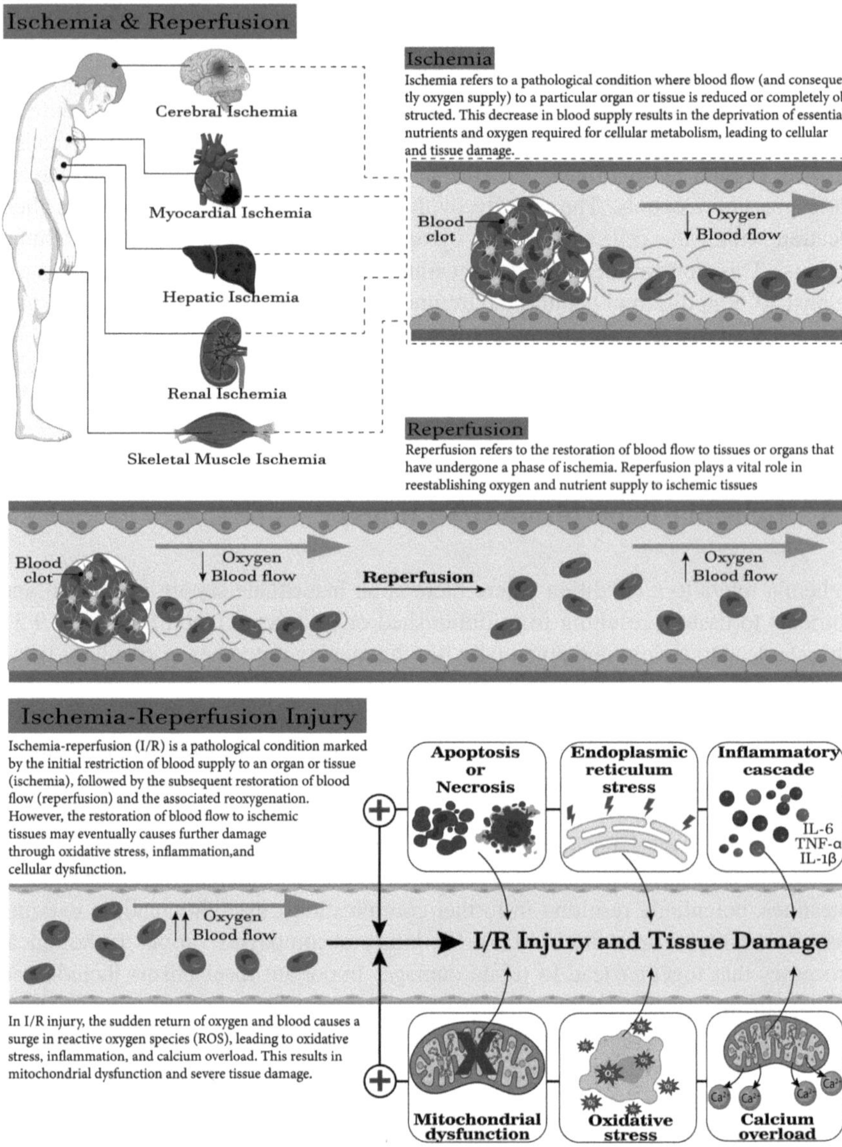

Fig. 9.3 A schematic overview of ischemia, reperfusion, and ischemia-reperfusion injury

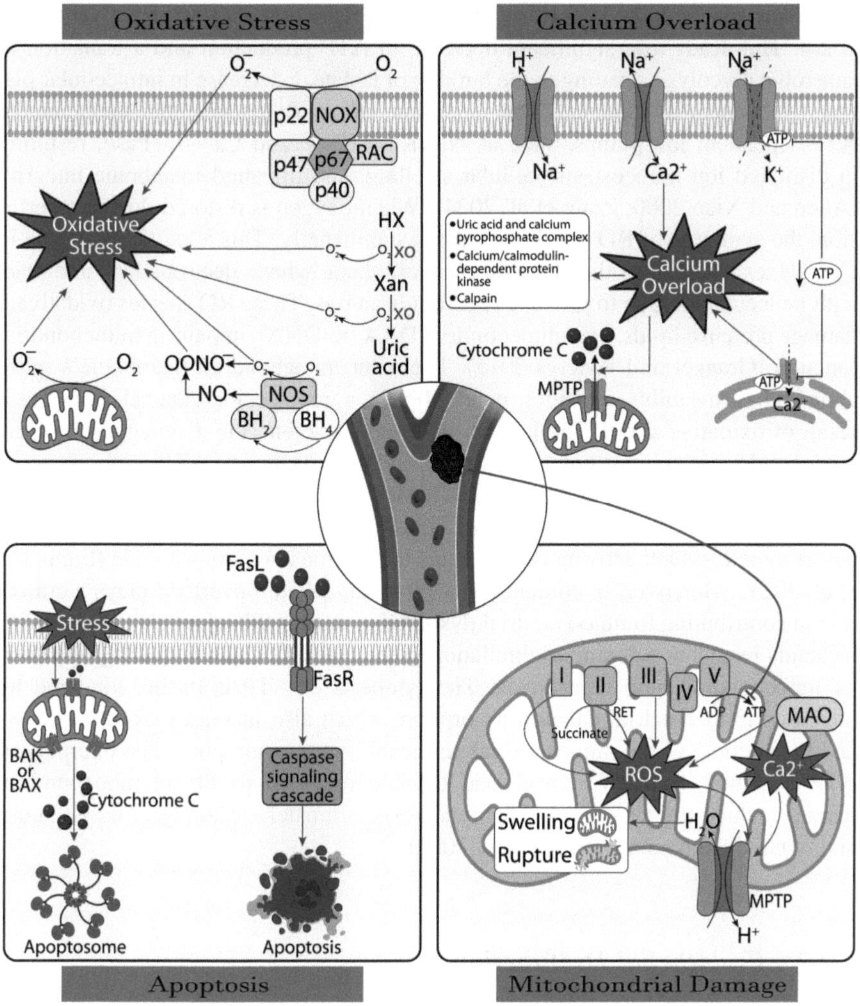

Fig. 9.4 A schematic overview of different key pathological mechanisms associated with ischemia-reperfusion injury

9.4 Key Pathological Mechanisms Associated with I/R Injury

9.4.1 Mitochondrial Dysfunction and ATP Production

Mitochondrial dysfunction during I/R injury plays a pivotal part in cellular energy failure and damage to the tissues (Zong et al. 2024). Mitochondria serve as an essential spot for oxidative phosphorylation, facilitating the generation of ATP via the ETC. During ischemia, absence of oxygen supply interrupts the ETC, particularly at Complex IV, obstructing the final reduction of O_2 to H_2O (Naito et al. 2020). This leads to a significant decrease in ATP production and a transition to anaerobic glycolysis, leading to the buildup of lactate, a decrease in intracellular pH, and the onset of acidosis. The depletion of ATP hinders the operation of ATP-dependent ion pumps, such as Na^+/K^+-ATPase and Ca^{2+}-ATPase, resulting in disrupted ion homeostasis, cellular swelling, and impaired membrane integrity (Allen and Xiao 2000; Zong et al. 2024). When oxygen is restored during reperfusion, the synthesis of RO species increases significantly. This is particularly true in Complexes I and III of the electron transport chain, where electrons leak and react with molecular oxygen to produce superoxide anions. These RO species oxidatively damage proteins, lipids, and mitochondrial DNA (mtDNA), impairing mitochondrial function (Granger and Kvietys 2015). The inner mitochondrial membrane's mitochondrial permeability transition pore (MPTP), a non-specific channel, opens as a result of oxidative stress. The loss of the proton gradient, the disintegration of the mitochondrial membrane potential ($\Delta\psi m$), and the cessation of ATP synthesis occur upon the start of the MPTP. This process leads to mitochondrial swelling, the rupture of the outer mitochondrial membrane, and the release of pro-apoptotic factors like cytochrome c, which activate caspases and trigger the apoptotic cascade (Endlicher et al. 2023). Moreover, in conjunction with ROS, calcium overload plays a critical role in contributing to mitochondrial dysfunction during I/R injury (Xu et al. 2020). Ischemia results in calcium accumulation in the cytosol, which is quickly absorbed by mitochondria upon reperfusion. The synthesis of ATP is further hindered by increased mitochondrial calcium absorption, which also increases oxidative stress and the opening of the mitochondrial permeability transition pore. The interplay of ATP depletion, oxidative stress, and calcium overload results in mitochondrial damage, which triggers apoptosis and necrosis, ultimately resulting in tissue injury and organ dysfunction (Bertero et al. 2024).

9.4.2 Endothelial Dysfunction

Endothelial dysfunction is pivotal in I/R injury and greatly influences the progression of tissue damage. The inner wall of blood vessels is lined with vascular endothelium, which is highly susceptible to oxidative damage and inflammation spurred on by I/R injury (Theofilis et al. 2021). The production of nitric oxide (NO) is reduced during ischemia due to endothelial cell dysfunction caused by the reduced oxygen supply and lack of nutrients. NO serves as a powerful vasodilator crucial for sustaining vascular tone, averting platelet aggregation, and obstructing leukocyte adhesion. A reduction in NO bioavailability during ischemia leads to vasoconstriction and heightened vascular resistance (Singhal et al. 2014). Upon reperfusion, the abrupt reintroduction of oxygen supply initiates an excessive generation of ROS, which subsequently diminishes NO by interacting with it to produce peroxynitrite, a highly reactive nitrogen species (Bagheri et al. 2016). This process intensifies oxidative stress, leading to a direct harm to endothelial cells and further diminishing NO bioavailability. The disparity between vasodilators, such as NO, and vasoconstrictors, like endothelin-1 (ET-1), results in compromised vascular relaxation, endothelial dysfunction, and heightened vascular permeability. An inflammatory response is also promoted by endothelial dysfunction. Adhesion molecules, such as vascular cell adhesion molecule-1 (VCAM-1) and intercellular adhesion molecule-1 (ICAM-1) that are presented by the active endothelium, are essential for drawing neutrophils and other immune cells to the site of injury. The attached neutrophils release RO species, proteolytic enzymes, and pro-inflammatory cytokines, leading to additional endothelial injury and compromising the integrity of the microvasculature. This results in heightened vascular permeability, which plays a role in edema and the "no-reflow" phenomenon, where microvascular blood flow is compromised even after successful reperfusion of large vessels (Bouleti et al. 2015). Moreover, endothelial cells generate pro-thrombotic factors like von Willebrand factor (vWF) in response to I/R injury, facilitating platelet aggregation and microvascular thrombosis, thereby worsening ischemic injury (Yau et al. 2015). The microthrombi that form lead to additional endothelial dysfunction and restrict oxygen delivery to the impacted tissue, thereby exacerbating the ischemic damage. In short, endothelial dysfunction during ischemia/reperfusion injury disturbs vascular balance, increases inflammation, and leads to tissue ischemia, resulting in worsened organ damage and unfavorable clinical results (Yang et al. 2016).

9.4.3 Oxidative Stress

Oxidative stress plays an integral role in I/R injury, marked by a disparity between the production of RO species and the ability of antioxidant defenses to counteract them (D'Oria et al. 2020). Ischemia leads to lower oxygen levels, hindering mitochondrial oxidative phosphorylation and causing an accumulation of reduced intermediates in the ETC. Following reperfusion, the rapid inflow of oxygen results in a marked increase in the generation of ROS, including superoxide anions, hydrogen peroxide, and hydroxyl radicals. ROS cause harm to biological constituents through lipid peroxidation, protein oxidation, and alterations to DNA (Su et al. 2019). Lipid peroxidation compromises membrane integrity, whereas oxidative modifications of proteins hinder enzyme function and disrupt signaling pathways (Valgimigli 2023). ROS also cause mitochondrial dysfunction by activating the MPTP, resulting in a loss of membrane potential, halting ATP synthesis, and the release of pro-apoptotic factors (Brillo et al. 2021). The resultant cellular damage exacerbates inflammation, as ROS generation trigger the nuclear factor-kappa B (NF-κB) and other transcriptional factors that augment the synthesis of adhesion molecules and pro-inflammatory cytokines, thereby worsening tissue injury (Morgan and Liu 2011). ROS have a substantial effect on endothelial cells by reducing the bioavailability of NO, which causes endothelial dysfunction, increased vascular permeability, and the development of microthrombi. Taken together, the severe oxidative stress that results from I/R injury upsets the equilibrium between cells, causes cell death, and exacerbates tissue injury (Zhang et al. 2024).

9.4.4 Cell Death

Cellular mortality in I/R injury happens through various pathways, such as apoptosis, necrosis, and ferroptosis, with each playing a role in tissue damage through distinct yet interconnected mechanisms (Dugbartey 2024; Gottlieb 2011). The interplay among mitochondrial malfunction, oxidative stress, and calcium overload is essential to these mechanisms of cell death, which lead to organ failure and exacerbation of injury. Mitochondrial failure and oxidative stress during I/R produce apoptosis, which is a controlled type of cell death pathway (Malko et al. 2021). The process starts with the opening of MPTP, which results in the release of cytochrome c into the cytosol and a loss of mitochondrial membrane potential. Consequently, this causes the apoptosome to become active, which in turn causes caspase 3 and 9 to become active. This results in the execution of a regulated cell death program that is characterized by chromatin condensation and DNA fragmentation. Necrosis, on the other hand, is an unregulated type of cell death that occurs due to significant energy depletion and the rupture of membranes (Chen et al. 2019). During ischemia, the depletion of ATP disrupts ion pumps, leading to swelling and damage to the cellular membrane. After reperfusion, the influx of oxygen and calcium intensifies oxidative damage, resulting in cellular lysis and the release of intracellular contents, which triggers inflammation. Ferroptosis, a relatively new form of regulated cell death, is triggered by iron-dependent lipid peroxidation (Stockwell et al. 2017). The generation of excessive RO species during reperfusion accelerates the oxidation of polyunsaturated fatty acids in cell membranes, resulting in ferroptosis in cells with impaired glutathione peroxidase 4 (GPX4) activity (Yu et al. 2021a). The activation of the caspase-1 and inflammasome is a characteristic of pyroptosis, a unique form of inflammatory cell death mechanism (Yu et al. 2021b). This results in the development of gasdermin D pores inside the cell membrane, ultimately leading to cell lysis and the subsequent release of distinct pro-inflammatory cytokines. The interaction among these cell death pathways intensifies tissue damage and plays a significant role in the pathology seen in I/R injury.

9.4.5 Autophagy

Autophagy is an inherent mechanism through which the body removes and reuses damaged or non-functional cells to promote repair and maintain cellular life (Wu et al. 2018b). The membrane-bound vesicles are used in the highly controlled and vital cellular process of autophagy, which comprises the recycling and breakdown of damaged organelles, aberrant biological macromolecules, and proteins. This procedure is essential for preserving cellular homeostasis, particularly in stressful situations like starvation or cellular injury. In the process of autophagy, cellular components are encapsulated by double-membrane structures known as autophagosomes (Decuypere et al. 2015). Autophagosomes subsequently merge with lysosomes to create autolysosomes, in which their contents undergo degradation by hydrolytic enzymes. The degraded components, which include fatty acids, amino acids, and nucleotides, are subsequently transported back into the cytoplasm for recycling in cellular metabolism and repair processes. Autophagy is essential for eliminating damaged mitochondria (mitophagy), dysfunctional proteins, and various cellular debris, thus averting the buildup of harmful substances that may lead to diseases like neurodegeneration, cancer, and cardiovascular issues. Three primary forms of autophagy exist: macroautophagy, characterized by the degradation of large cellular structures; microautophagy, which entails the direct engulfment of cytoplasmic material by lysosomes; and chaperone-mediated autophagy, a more selective mechanism that targets specific proteins for degradation (Parzych and Klionsky 2014). The disruption of autophagy is linked with a range of pathological conditions, including I/R injury. In I/R injury, autophagy exhibits a multifaceted and complex role, serving as both a protective mechanism and, in certain instances, leading to cellular damage. During the ischemic phase, autophagy is primarily viewed as a mechanism that protects cells. Through the selective degradation of damaged mitochondria, known as mitophagy, autophagy effectively diminishes mitochondrial ROS production, safeguards against cell death, and maintains ATP levels. This reduces the harmful impacts of oxidative stress, calcium overload, and inflammation that usually arise in ischemic conditions (Shi et al. 2019). Autophagy facilitates the elimination of misfolded proteins and other harmful aggregates that build up during ischemia, thereby enhancing cellular survival. Nonetheless, during the reperfusion phase, an overactive or erroneously regulated autophagy may lead to detrimental effects. The abrupt introduction of oxygen during reperfusion results in the formation of ROS, an excess of calcium, and cellular swelling, potentially worsening mitochondrial dysfunction and initiating an excessive autophagic response. In these instances, autophagy may become overly pronounced or lead to "autophagic cell death," resulting in the breakdown of vital cellular components and worsening tissue damage. The dual nature of autophagy in I/R injury underscores the delicate equilibrium necessary for its regulation (Lin et al. 2018). Insufficient autophagy may result in the inability to eliminate damaged components, whereas excessive autophagy could cause the undue breakdown of essential cellular structures, ultimately leading to cell death. Some essential molecular pathways, including the AMP-activated protein kinase (AMPK) and the mechanistic target of rapamycin (mTOR), play a crucial role in regulating autophagy during I/R injury (Mihaylova and Shaw 2011). AMPK initiates autophagy during ischemia by detecting low energy levels and inhibiting mTOR, which typically restrains autophagy in nutrient-abundant environments. Regulating these pathways via pharmacological agents, including rapamycin (mTOR inhibitor) or AMPK activators, presents a promising therapeutic strategy to enhance autophagy in the condition of I/R injury. Researchers are focused on optimizing autophagy to bolster its protective functions during ischemia, while also mitigating its negative impacts during reperfusion.

9.4.6 Endoplasmic Reticulum Stress

The endoplasmic reticulum (ER) is essential for regulating various cellular activities, including intracellular calcium levels, enabling lipid metabolism, and synthesizing numerous proteins. It is a central hub for protein folding and processing in order to ensure that newly synthesized proteins achieve their proper three-dimensional structure prior to being sent to their final destinations (Schwarz and Blower 2016). Furthermore, the ER regulates the storage and release of calcium, which is crucial for numerous cellular signaling pathways (Carreras-Sureda et al. 2018). It also participates in the biosynthesis of lipids and sterols, significantly contributing to membrane formation and energy metabolism (Basseri and Austin 2012). When cellular stressors like oxidative stress, hypoxia, or nutrient deprivation disrupt ER function, the ability to correctly fold proteins becomes disrupted. Consequently, unfolded or misfolded proteins build up in the ER lumen, resulting in a condition referred to as ER stress (Kalogeris et al. 2012). ER stress initiates three unique responses that are essential for reestablishing cellular equilibrium: the unfolded protein response (UPR), the endoplasmic reticulum overload response (EOR), and the sterol regulatory element binding protein (SREBP) pathway. The UPR stands out as the most crucial mechanism for mitigating the stress resulting from the buildup of unfolded or misfolded proteins in the ER lumen (Wang et al. 2022). In the ischemic phase, hypoxia diminishes ATP production, disrupting the energy-dependent mechanisms of protein folding and ER-associated degradation (ERAD) (Ren et al. 2021). The following reperfusion phase amplifies this stress by introducing reactive oxygen species and pro-inflammatory cytokines, including tumor necrosis factor-alpha (TNF-α), interleukin-6 (IL-6), and interleukin-1β (IL-1β), which cause additional harm to cellular structures and heighten oxidative stress. ROS specifically alter ER proteins and disturb the redox environment essential for disulfide bond formation during protein folding, resulting in the buildup of misfolded proteins within the ER lumen (Wang and Kaufman 2016). Consequently, ER stress is triggered by the activation of three key sensors of the UPR: inositol-requiring enzyme 1 (IRE1), protein kinase R-like endoplasmic reticulum kinase (PERK), and activating transcription factor 6 (ATF6) (Hetz 2012). These sensors work to reestablish ER homeostasis by decreasing protein synthesis (through PERK), improving protein folding capabilities (via ATF6 and X-box binding protein 1, XBP1, which is a downstream effector of IRE1), and facilitating the breakdown of misfolded proteins (ERAD). However, in events where ER stress becomes severe or extended, particularly in the context of I/R injury, the UPR switches from its protective function to the initiation of apoptotic pathways (Yang et al. 2024). This shift is primarily facilitated by the IRE1-mediated activation of c-Jun N-terminal kinase (JNK) and CHOP (C/EBP homologous protein), which serves as a pro-apoptotic transcription factor. Pro-inflammatory cytokines further intensify ER stress via their signaling pathways, notably by activating NF-κB and JAK/STAT pathways, which boost the transcription of UPR-associated genes and worsen the inflammatory response

(Li et al. 2020). This initiates a destructive cycle, in which ER stress triggers the release of more cytokines, intensifying both systemic and local inflammation. Furthermore, ER stress plays a role in the generation of ROS, establishing a feedback-loop between ER dysfunction and oxidative stress, which continues to inflict cellular damage in the impacted tissues (Burgos-Morón et al. 2019).

9.5 Detection of I/R Injury Caused by ROS

Detecting injury induced by ROS encompasses various methods that can reveal oxidative harm to cellular elements, such as lipids, proteins, and DNA (Dikalov and Harrison 2014). These methods facilitate the measurement of ROS-mediated damage and evaluate the effectiveness of possible therapeutic approaches. The assessment of ROS levels can be carried out through an array of fluorescent and chemiluminescent probes. Fluorescent probes like dihydroethidium (DHE) are utilized for the detection of superoxide anions (Kumar and Gullapalli 2024). When reacting with O_2^-, DHE undergoes oxidation to produce a fluorescent product that can be observed and measured through fluorescence microscopy or flow cytometry. In a similar manner, 2′,7′-dichlorodihydrofluorescein diacetate (DCFDA) serves to assess overall ROS levels, such as hydrogen peroxide, by transforming into a fluorescent form when exposed to ROS (Kim and Xue 2020). MitoSOX Red is designed to specifically target mitochondrial superoxide, enabling researchers to evaluate oxidative stress in mitochondria (Wang and Zou 2018). Chemiluminescent assays utilizing probes such as luminol or lucigenin emit light when they react with ROS, offering a measurable assessment of oxidative activity (Sharma et al. 2018). Lipid peroxidation serves as a crucial marker of oxidative damage, typically assessed by quantifying malondialdehyde (MDA), a significant by-product of this process (Tsikas 2017). The thiobarbituric acid reactive substances (TBARS) assay entails the reaction of MDA with thiobarbituric acid, resulting in the formation of a colored complex that can be measured using spectrophotometry. Another notable indicator is 4-hydroxynonenal (4-HNE), which arises from lipid peroxidation. Detection of 4-HNE can be accomplished through methods like enzyme-linked immunosorbent assay (ELISA), immunohistochemistry, and mass spectrometry, each offering valuable insights into the degree of lipid damage and its biological significance (Schaur et al. 2015). Protein oxidation is evaluated through the measurement of protein carbonyls, which arise when ROS alter amino acid side chains. The protein carbonyl assay employs derivatization with 2,4-dinitrophenylhydrazine (DNPH), succeeded by spectrophotometric analysis of the resultant carbonyl-DNPH adducts (Weber et al. 2015). Furthermore, Western blotting enables the detection of specific oxidative modifications to proteins through the use of antibodies targeting oxidized protein forms, facilitating the identification and quantification of damaged proteins that play roles in various cellular functions (Kehm et al. 2021). The Comet assay, also known as single-cell gel electrophoresis, is employed to assess DNA damage by quantifying DNA fragmentation at the individual cell level. In this assay, cells embedded in agarose undergo electrophoresis, resulting in a "comet" with a tail that reflects the degree of DNA damage (Collins 2004). Another important indicator of oxidative DNA damage is 8-hydroxydeoxyguanosine (8-OHdG) (Al-Taie et al. 2021). This lesion, which signifies oxidative damage to guanine, can be quantified through methods such as ELISA, immunohistochemistry, and high-performance liquid chromatography (HPLC), offering a precise assessment of oxidative stress at the genomic level. The assessment of antioxidant capacity and defense mechanisms is carried out to better understand the equilibrium between the production of ROS and their scavenging. The total antioxidant capacity (TAC) assay assesses the comprehensive capability of a sample to counteract ROS, employing techniques such as the Trolox equivalent antioxidant capacity (TEAC) assay or the ferric-reducing antioxidant power (FRAP) assay (Silvestrini et al. 2023). Glutathione (GSH) levels, recognized as a key intracellular antioxidant, are assessed through colorimetric assays or HPLC, offering valuable information about the cellular antioxidant defense status (Forman et al. 2009). These methods together provide an in-depth perspective on ROS-mediated damage and the oxidative stress response, enabling investigators to evaluate the extent of injury and investigate potential therapeutic approaches to reduce oxidative damage in I/R like diverse pathological conditions.

9.6 Therapeutic Interventions in I/R Injury: A Pharmacological Basis

Therapeutic interventions for I/R injury aim to reduce cellular damage and restore function by addressing the multiple mechanisms involved in this intricate condition. Pharmacological therapies play a crucial role in this context. The underlying mechanisms of I/R injury have been the topic of extensive research for decades; however, a definitive understanding on the precise intrinsic biochemical pathways activated during this process remains elusive. Here, we systematically classify pharmacological agents based on the current understanding of I/R injury signaling mechanisms and their respective molecular targets. A considerable number of potential target sites are focused within several well-defined pathways, such as the Survivor Activating Factor Enhancement (SAFE) pathway, the Reperfusion Injury Salvage Kinase (RISK) pathway, and the cyclic guanosine monophosphate/Protein Kinase G (cGMP/PKG) signaling cascade (Fig. 9.5) (Hadebe et al. 2018; Numata and Takimoto 2022; Rossello and Yellon 2018). Furthermore, an interplay of various essential pathways—like those regulating inflammatory responses, metabolic functions, intrinsic/mitochondrial elements, and nuclear DNA control—significantly contributes to the pathophysiology of I/R injury.

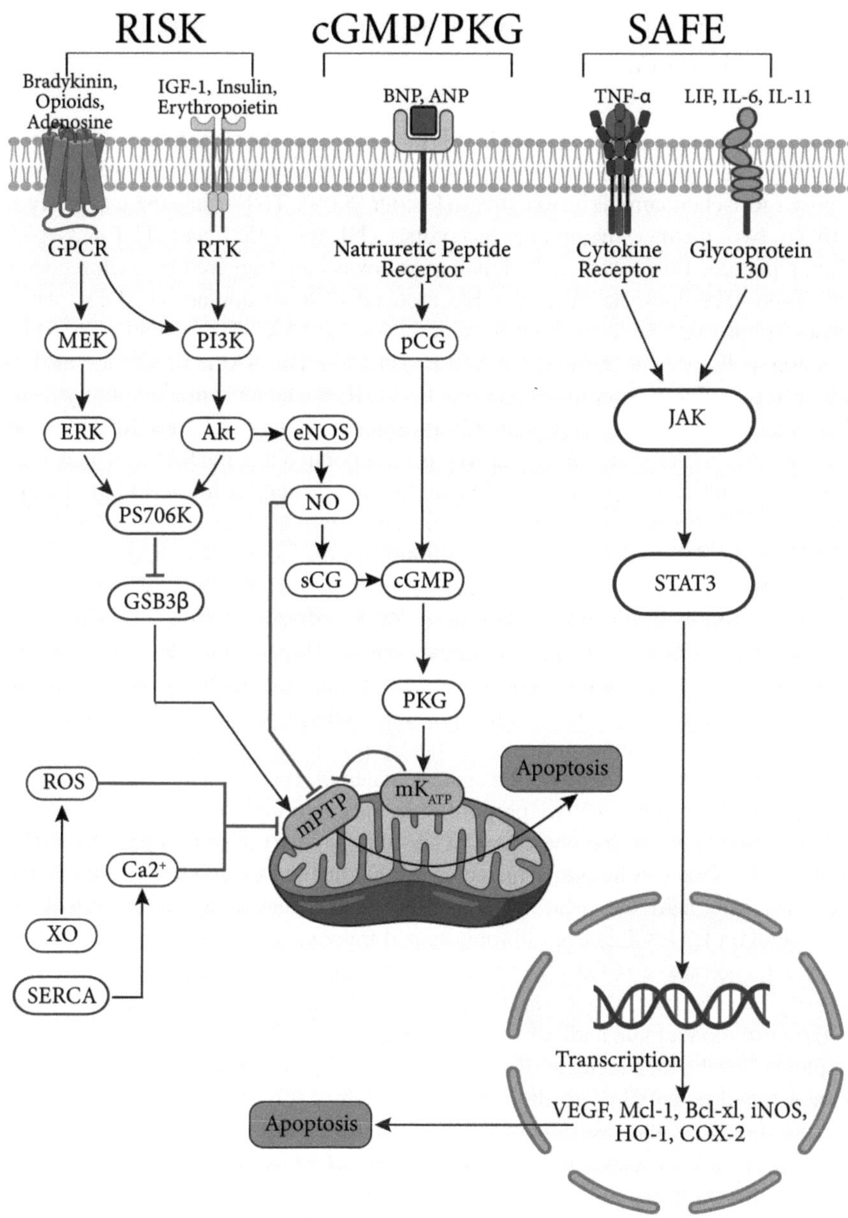

Fig. 9.5 A schematic overview of different signaling pathways that are relevant for the development of I/R-related pharmacological therapies

9.6.1 *Targeting Survivor Activating Factor Enhancement (SAFE) Pathway*

Sandrine Lecour first identified the SAFE pathway in 2009; she brought attention to the intriguing paradox of TNF-α's protective but detrimental effects on myocytes exposed to ischemia/reperfusion stress (Lecour 2009). TNF-α has the capability to activate two distinct membrane receptors, TNF-R1 (p55) and TNF-R2 (p75) (Tartaglia et al. 1991). Different signaling pathways are triggered by each receptor. The Type 1 receptor activates the Fas-associated death domain (FADD), which causes apoptosis. The Type 2 receptor activates the JAK/STAT-3 pathway, which increases cell survival through the SAFE pathway. The SAFE route may also be activated by inflammatory molecules like IL-11, IL-6, and leukemia inhibitory factor (LIF) that activate the glycoprotein 130 receptor (Lecour and James 2011). As per that, the JAK/STAT-3 signaling pathway is incorporated into the SAFE, which, once activated, inhibits the opening of mPTP and increases nuclear transcription of genes linked to cell differentiation, proliferation, and survival, such as Bcl-x(L), Bcl-2, VEGF, cyclin D1, and numerous other transcription factors. The SAFE route can significantly lessen myocardial damage after reperfusion while also increasing STAT-3 activity. It functions apart from RISK, offering options for alternative approaches. A study conducted by Shibata and colleagues provided evidence that TNF-α is accountable for the reduction of adiponectin (metabolic cytokine) (Shibata et al. 2005). This adipokine, which promotes energy expenditure by activating mitochondrial uncoupling proteins, is mostly produced by adipose tissue, and is associated with increased protection against ischemia (Bauche et al. 2007). Gao et al. demonstrated cardioprotective properties in rats exposed to 30-min myocardial ischemia by administering etanercept (TNF-α antagonist) prior to reperfusion (Gao et al. 2015). This can be partly linked to the heightened expression of adiponectin that may safeguard the heart by activating AMPK, encouraging the release of prostaglandin E2 (PGE2) from fibroblasts and myocytes, and promoting the release of cyclooxygenase-2 (COX-2), which in turn suppress the synthesis of TNF-α in monocytic cells. Remarkable reactivity has also been seen by infliximab, a humanized mouse monoclonal antibody that is another anti-TNF-α medication. Frequently employed for the management of several autoimmune disorders, infliximab has also demonstrated potential as an alternative to I/R injury (Lu et al. 2016). After administering this drug, the researchers successfully decreased lipid peroxidation in the kidneys of rats by elevating GSH levels and SOD activity. Furthermore, they observed the recovery of histological abnormalities (Tasdemir et al. 2012). Furthermore, Nagata et al. discovered that administering infliximab before ischemia had an equivalent protective impact to preventive splenectomy in rat kidney transplantation. This treatment effectively decreased the infiltration of renal macrophages and monocytes, and consequently reduced the release of inflammatory cytokines (Nagata et al. 2016). Propofol, along with the gaseous anesthetic agents sevoflurane and isoflurane, exert protective effects through activation of the SAFE pathway, thereby providing protection against I/R injury (Agarwal et al. 2014). Another study found that the concurrent use of sevoflurane or isoflurane enhances the safeguarding activity of ischemia pre-conditioning on the liver. Nevertheless, the researchers discovered that the exclusive use of either medication without pre-conditioning did not provide any substantial protection against I/R injury. This finding supports the notion of a beneficial synergy among different processes. The results were partially ascribed to the 1.7-fold and 2.3-fold elevation in the expression of the anti-apoptotic gene Bcl-2 in the sevoflurane/IP and isoflurane/IP groups, respectively (Jeong et al. 2017). Propofol, another widely used anesthetic, was also found to exert a similar anti-inflammatory impact on the rat kidney by increasing the expression of bone morphogenetic protein 2 (BMP-2) (Yang et al. 2013). Subsequently, Xu et al. improved the concept that both propofol and sevoflurane can diminish the level of oxidative stress in the liver, while also modifying the expression of pro- and anti-apoptotic genes, thereby inhibiting apoptosis. The research revealed that the anti-inflammatory properties of sevoflurane and propofol can be linked to the suppression of the transcriptional factor NF-κB. This suppression leads to a decrease in the expression of pro-inflammatory cytokines and an increase in the expression of the anti-inflammatory cytokine in the rat liver (Xu et al. 2016). Notably, although both substances offered equivalent defense, sevoflurane significantly diminished TNF-α leakage while propofol resulted in reduced IL-1 release. Another example of the activation of the SAFE pathway and the increase of nuclear anti-apoptotic expression was presented by Katayama et al. Erythromycin was used to pre-condition the rat brain, which results in the achievement of this goal (Katayama et al. 2014). This well-recognized antibiotic can enhance neuronal survival following I/R injury by decreasing the size of the infarct and the degree of edema. This makes it a potentially effective neuroprotective treatment for acute ischemic stroke. The suggested mechanism is characterized by antioxidant and anti-inflammatory effectiveness, accompanied by a temporary upregulation of Bcl-2 expression. Furthermore, the preventive effects of erythromycin were confirmed in a pig model by reducing the degree of apoptosis in the neocortex after the ischemia phase, 12 h before the generated acute hypothermic circulatory arrest (Koutsogiannidis and Johnson 2017).

9.6.2 Targeting Reperfusion Injury Salvage Kinase (RISK) Pathway

The RISK pathway is closely linked to a complex system of protein kinases that promote survival (and inhibit apoptosis) and are essential in pathogenesis of I/R injury. The system is comprised of two distinct signaling cascades: the PI3K-Akt pathway, which involves phosphoinositide-3 kinase/protein kinase B, and the MAPK/MEK1/ERK1/2 pathway, which involves mitogen-activated extracellular signal transduction. Both cascades are triggered when reperfusion begins and have been shown to provide protective effects by inhibiting apoptotic signaling and decreasing cell necrosis (Rossello and Yellon 2018). Two major categories of membrane receptors, G-protein coupled receptors (GPCRs) and receptor tyrosine kinases (RTKs), are involved in the activation of the RISK pathway. G-protein-coupled receptors (GPCRs) bind to ligands such as adenosine receptors, bradykinin receptor, and opioids receptor. On the other hand, RTKs are activated by signals from cytokines, erythropoietin, insulin-like growth factor-1 (IGF-1), insulin, and certain other growth factors (Ghaboura et al. 2011; Hausenloy et al. 2011; Hausenloy and Yellon 2004). Although RISK has two clearly defined pathways, its overall regulation always results in the suppression of the opening of the mPTP and the subsequent promoter of survival (anti-apoptotic) action. A multitude of pharmacological investigations have analyzed the RISK pathways by implementing various interventions with ERK and PI3K inhibitors at different concentrations. Subsequently, numerous studies have highlighted the significance of these findings (Cao et al. 2010; He et al. 2022; Walters et al. 2012). Preclinical studies have verified that inducing at least partial activation of RISK upon reperfusion after ischemia does provide protection and thereby greatly decreases the severity of infarction (Hausenloy and Yellon 2007). Furthermore, it is noteworthy to recognize that RISK has been demonstrated to play a role in the protective non-pharmacological strategy for ischemic pre- and post-conditioning. This aspect is especially advantageous for studying important biochemical processes, with the goal of rationally designing and/or discovering mimetic medications (Hausenloy and Yellon 2007). While several medications have the potential to target the RISK pathway, it was urocortin that initially became linked to it. Urocortin was primarily derived from rat tissue and then from the human brain. It is composed of a 40 amino acid α-helix shape; it is a stress-related mediator in mammals and is associated with the hypothalamic hormone corticotrophin-releasing factor (CRF) (Zhao et al. 1998). It was demonstrated that urocortin interacts with its cardiac GPCRs, resulting in enhanced phosphorylation of ERK1 and ERK2. This activation safeguards the heart from severe reperfusion injury (Schulman et al. 2002). The authors confirmed these findings by demonstrating that the protective effects of urocortin were abolished when co-administered with the ERK 1/2 inhibitor PD98059. Baxter et al. reported similar findings, noting that the protective effects of transforming growth factor-β1 (TGF-β1) were negated when the same ERK 1/2 inhibitor (PD98059) was employed (Baxter et al. 2001). Several beneficial features of polyethylene glycols (PEG) have been demonstrated to provide protective aids in myocytes involved in post-hypoxia reperfusion. In addition to their common usage as oncotic agents, PEGs can reduce interstitial recombination injury (IRI) by enhancing the protection of the mitochondria and cytoskeleton in preservation solutions, rinse solutions, pre-conditioning, and supercooling. Furthermore, evidence indicated that PEGs can enhance the pro-survival pathway via phosphorylating AKT and activating two crucial cytoprotective proteins, AMPK and eNOS (Bejaoui et al. 2015; Malhotra et al. 2011; Pasut et al. 2016). More recently, PEGs of different molecular sizes have been effectively employed as pre-conditioning agents by intravenous administration, significantly improving the transplantation survivability of liver implants. The observed result can be ascribed to both the physical and chemical characteristics of PEGs, including their comparatively large molecular mass, low toxicity, remarkable flexibility, hydrophilicity, enhanced hydrodynamic volume, and strong protein-rejecting qualities (Abuchowski et al. 1977). Lopez and colleagues showed that including a high weight polyethylene glycol (PEG35) in preservation solutions during cold ischemia time can enhance defenses against hypoxia by preserving the structural strength of the endothelial glycocalyx and so preserving its capacity to facilitate NO production (Lopez et al. 2018). Without uncertainty, the number of medications and mechanisms that activate, to some extent, the RISK pathway and induce cardio protection is vast and continuously growing. Their impact may manifest either before or after the onset of arterial ischemia and may or may not be facilitated by receptors. IGF-1, insulin, fibroblast growth factor 2 (FGF-2), leptin, erythropoietin, apelin, visfatin, and other cytokines are important activators of the RTK receptor. Compounds that exhibit IRI protection and bind to the GPCR receptor include glucagon-like peptide 1 (GLP-1), opioids, adenomedullin, bradykinin, urocortin, and adenosine. Although it is not possible to provide a comprehensive description of all of them in this chapter, we suggest that readers who have a specific interest in this topic refer to a work authored by Hausenloy and Yellon (2007).

9.6.3 Targeting Cyclic Guanosine Monophosphate/Protein Kinase G (cGMP/PKG) Pathway

The cGMP/PKG pathway regulates the synthesis of cGMP through guanylyl cyclases, which then triggers cGMP-dependent protein kinase (PKG) and cAMP-dependent protein kinase (PKA). These protein kinases are crucial for implementing survival-promoting effects, such as the activation of mitochondrial ATP-sensitive K^+ channels (mK_{ATP}) and the regulation of $Ca2^+$ levels in the sarcoplasmic reticulum (Burley et al. 2007; Feyzizadeh and Badalzadeh 2017). Guanylyl cyclases (GCs) are categorized into two distinct groups. The first category consists of particulate guanylyl cyclases (pGCs), which are proteins located in the cell membrane. These proteins are activated by extracellular natriuretic peptides (NPs), such as atrial and brain natriuretic peptides, upon binding to their specialized membrane receptors. Furthermore, more than one GPCR is implicated in both the pre-conditioning and post-conditioning of ischemia, and it is possible for these receptors to be activated autonomously (Burley et al. 2007). The second category consists of soluble guanylyl cyclases (sGCs) located in the cytoplasm. These enzymes are triggered by NO and other nitrites. Nitrites can originate from systemic sources or be locally synthesized by endothelial nitric oxide synthase (eNOS) within cells (Costa et al. 2006). It has been demonstrated that pharmacological interventions involving endogenous NO stimulating drugs and exogenous cGMP/PKG modulators, such as donated natriuretic factors, NO, and phosphodiesterase inhibitors, effectively protect the heart and liver by inducing both mitochondrial and cytosolic processes (Cohen et al. 2010; Duranski et al. 2005; Shiva et al. 2007). To obtain a more comprehensive analysis of the cGMP/PKG pathway, check the review by Inserte and Garcia-Dorado (Inserte and Garcia-Dorado 2015). Administering phosphodiesterase-5 (PDE-5) inhibitors, such as vardenafil or sildenafil, either as a bolus or continuous infusion at doses lower than those used for erectile dysfunction, has shown cardioprotective effects during reperfusion. This protection is mediated by the opening of mitochondrial K_{ATP} channels in the heart, which results in a reduced infarct size in vivo in rat and rabbit models (Ebner et al. 2013; Salloum et al. 2007). Additionally, other PDE-5 inhibitors such as ordonafil and tadalafil also demonstrate similar cardioprotective effects against I/R injury in animal models. This suggests that the observed protective effect may be a characteristic shared across the entire class of PDE-5 inhibitors (Khatib 2019). Inhibiting PDE-5 leads to increased levels of cGMP, which subsequently activates anti-apoptotic mechanisms by promoting the opening of mK_{ATP} channels. Similar effects and outcomes have been observed in sGC activator BAY 60–2770 (Lee et al. 2017). For over 20 years, exogenous adenosine was recognized for enhancing the protective properties of ischemic pre-conditioning, notably through the increased involvement of Akt/eNOs (Peralta et al. 1997). Nonetheless, the understanding of the pharmaceutical dynamics of I/R injury protection through adenosine has significantly advanced, facilitating broader applications. A study by Boros et al. demonstrated that adenosine mitigates myocardial damage caused by I/R injury via reducing immune cell infiltration, a process mediated through the suppression of pro-inflammatory cytokines (Boros et al. 2016). According to a study by Paez et al., the early activation of adenosine (A1) receptors in the mitochondria during reperfusion was the primary cause of the reduction in myocardial infarct size observed in rats through remote ischemic pre-conditioning. This activation then led to the involvement of the RISK pathway and Akt/eNOS phosphorylation (Paez et al. 2019). Cao et al. employed the A_{2A} antagonist ZM-241385 and the agonist CGS-21680 to assess the apoptotic index in dermal microvascular endothelial cells (DMECs) subjected to ischemic post-conditioning (Cao et al. 2019). The study revealed that ischemic post-conditioning or stimulation of A_{2A} receptors significantly reduced I/R injury to DMECs, while blocking these receptors had the opposite effect, exacerbating the condition. It is noted that the activation of A_{2A} receptors can reduce I/R injury in a pig model of extracorporeal cardiopulmonary resuscitation by decreasing the elevated lactate levels following reperfusion, along with the indicators of inflammation and stress in the liver and kidneys (Mehaffey et al. 2019). Recently, Mohamed et al. have demonstrated that central A_{2A} antagonism can effectively reduce inflammatory signals, thereby suppressing apoptotic pathways in the rat brain (Mohamed et al. 2012, 2016). Huerter et al. validated these protective and anti-inflammatory properties in the mouse lung subjected to I/R injury (Huerter et al. 2016). These results reinforce the concept that both A_{2A} and A_{2B} receptors are activated locally in response to hypoxic conditions, contributing to the protection and preservation of myocardial tissue during I/R injury (Ahmad et al. 2009). Moreover, the existing body of research indicates that A_3 agonism offers ischemic tolerance and demonstrates both efficacy and safety. Through the usage of the A_3 agonist CP-532,903, Wan et al. successfully safeguarded mouse myocardium from ischemia-reperfusion injury, a finding that was negated when implementing in the A_3 deficient knockout mice specifically within the cardiomyocytes (Wan et al. 2019). They propose that the protective mechanistic pathway is associated with the stimulation of K_{ATP} channels by inhibitory G-proteins (Gi proteins). Additional evidence supporting this claim is provided by the use of a highly specific A_3 receptor agonist, CF102, which exhibited hepatoprotective benefits via suppressing the NF-κB signaling pathway. The findings indicate that activation of A_3 receptors not only decreases the damage caused by I/R in cardiac tissue, but may also provide protective advantages to other organs via regulating inflammatory and apoptotic pathways (Ohana et al. 2016).

9.6.4 Targeting Mitochondria

It is well-established that I/R injury obeys the principle that "almost all routes lead to the opening of the mPTP." Throughout this chapter, we have discussed various drugs that influence mPTP opening indirectly by targeting upstream pathways. In this section, however, we will focus on examples of drugs that bypass these intermediary pathways and directly inhibit mPTP opening, addressing the core issue at its source. These direct mPTP inhibitors represent a more targeted approach to preventing cellular damage during reperfusion. The majority of research on I/R injury protection that focuses on preventing the mPTP opening has been carried out since the 1990s by inhibiting cyclophilin D (CyD), which is a crucial and well-characterized regulator of the mPTP and a major target for preventing necrotic cell death (Schinzel et al. 2005). CyD was named for its strong binding affinity to cyclosporine A (CyA), an immunosuppressant drug that became the prototype for CyD inhibitors. Numerous studies have reported successful attenuation of I/R injury in various organs, including the brain, heart, and liver following treatment with CyA (Bernardi and Di Lisa 2015; Fakharnia et al. 2017; Naoumov 2014). However, in recent decades, there has been a steady decline in reports supporting CyA's use, largely due to deleterious results from recent clinical studies. In these studies, CyA failed to exhibit significant effectiveness in reducing myocardial infarction-related damage, leading to a reevaluation of its potential as a treatment for I/R injury (Ghaffari et al. 2013; Ottani et al. 2016). As a result, novel approaches are being investigated to either enhance the therapeutic application of CyA or to assess CyA homologs as substitutes. Ikeda and colleagues had previously developed a delivery method that utilizes poly-lactic/glycolic acid (PLGA) nanoparticles to specifically target CyA to the mitochondria during the process of reperfusion. As shown in a mouse model of ischemia-reperfusion injury, this method increases the concentration of CyA at the site of action, therefore enhancing its cardioprotective properties in vivo (Ikeda et al. 2016). Conversely, CyA analogs such as NIM811, sanglifehrin A, and alisporivir (Debio-025) are demonstrating promise in mitigating I/R injury in the brain, liver, and myocardium of animal models. Additionally, while not a CyA analog, nitroglycerin, a traditional vasodilator, has recently been found to inhibit cyclophilin D (CyD) through S-nitrosation. This action helps reduce myocardial infarction driven by I/R in rabbit models (Bibli et al. 2019). In the past few years, significant focus has shifted toward discovering new compounds that directly inhibit mPTP opening without targeting CyD. This strategy promises to unlock innovative therapeutic options and could potentially offer enhanced efficacy when used in combination with CyA-based treatments. Such an approach aims to provide more effective and targeted interventions for I/R injury. To date, several compounds, including cinnamic anilides, N-phenylbenzamides, TRO40303, and isoxazoles have demonstrated effective mPTP inhibition with varying degrees of protective effects in animal models (Fancelli et al. 2014; Roy et al. 2015; Schaller et al. 2010). However, they still need to undergo trial studies to evaluate their safety and clinical efficacy. Moreover, hydroxytyrosol, a phenolic compound present in olive oil, has demonstrated notable cardioprotective properties against I/R injury by blocking the opening of mPTP. Nevertheless, it is not clear if this protective effect depends on CyD or operates through other mechanisms (Jia et al. 2018). The activation of mK_{ATP} during insulin resistance is a beneficial defensive mechanism, as it alters the threshold for mPTP activation, thereby prolonging the initiation of the apoptotic pathway. Another type of protection from I/R injury can be exerted by non-mitochondrial K_{ATP} by boosting blood flow, and which helps to ensure a surplus of nutrients before ischemia occurs. Shimizy et al. successfully prevented I/R injury in the rat kidney by targeting the proximal tubules with cromakalim (non-selective K_{ATP} agonist and vasodilator) and nicorandil (selective mK_{ATP} agonist and NO donor) (Shimizu et al. 2011). Both cromakalim and nicorandil demonstrated efficacy against cerebral I/R injury in diabetic rats by reducing caspase-3 levels. The treatment with nicorandil also notably decreased the volume of cerebral infarction (Pithadia et al. 2017). Additionally, diazoxide has demonstrated promising results in enhancing ischemic tolerance and providing protection against I/R injury across various organs, including the spinal cord, heart, smooth muscle, pancreas, gut, liver, and endothelium (Coetzee 2013; Fernandes de Mattos Dourado et al. 2018; Yamanaka et al. 2019). Although diazoxide predominantly acts as a selective agonist of mK_{ATP}, its pre-treatment may additionally activate the RISK and SAFE pathways. This is achieved through the enhancement of β common receptor subunit expression of the erythropoietin receptor and the optimization of the STAT3 signaling pathway, respectively (Yamanaka et al. 2018, 2019). Existing literature highlights a significant pharmacological similarity between respiratory Complex II inhibitors and mK_{ATP} channel agonists. Notably, compounds such as atpenin A5 and malonate can activate mK_{ATP} channels, offering cardioprotection in rat models following reperfusion (Wojtovich and Brookes 2008, 2009). The uncoupling of Complexes II and III has been shown to enhance the activation of the mK_{ATP} channel, which reduces hyperoxygenation and ROS generation after reperfusion in the rat heart (Liu et al. 2010). All of these mechanisms are influenced by acute oxidative stress, particularly through the production of ROS (St-Pierre et al. 2002). One of the most notable cellular processes affected is the mitochondrial ETC, especially during reperfusion when reverse electron transport occurs and succinate accumulates—a state known as "oxidant-induced reduction" (Chouchani et al. 2014). Thus, upon reperfusion, reverse electron transport could be prevented by using direct modulators of the respiratory chain (Murphy 2009; Pell et al. 2016). This approach would help control too much ROS production, thereby reducing mPTP activation and subsequent cell death. It has been noted that various sites of abnormal ROS formation within mitochondrial respiratory complexes are linked to distinct abnormalities. Evidence indicates that Complexes I, II, and III play significant roles in amplifying I/R injury. These complexes contribute to increased ROS generation, which exacerbates tissue damage during reperfusion (Orr et al. 2015). Gadicherla et al. demonstrated that the anti-anginal medication ranolazine alleviates damage to Complex I following myocardial infarction and reperfusion, thereby reducing tissue damage. Notably, ranolazine does not directly interact with Complex I. Instead, its protective effect is due to its ability to decrease the exchange of sodium and calcium ions in the cytosol. This reduction in cytosolic ion exchange ultimately leads to a decrease in the harmful overload of mitochondrial calcium, mitigating the associated damage (Gadicherla et al. 2012). An interdisciplinary study conducted by Orr et al. (2015), Brand et al. (2016), and their colleagues has successfully identified specific categories of small molecules that can selectively attach to the IQ site in Complex I (S1QELs) or to the Qo site in Complex III (S3QELs). This binding prevents the leakage of electrons and reduces the production of ROS during reverse electron transport (RET), without impacting oxidative phosphorylation. As a result, the I/R injury in the mouse heart is attenuated. Although the respiratory Complex II is not directly responsible for generating ROS during the period of reperfusion, it does influence the efficacy of Complexes I and III by accumulating succinate (Chouchani et al. 2014). Reduced inner mitochondrial membrane potential and electron supply resulting from mild inhibition of Complex II likely diminish ROS generation caused by RET (Dröse et al. 2011). Dröse and colleagues demonstrated that inhibiting Complex II with agents such as diazoxide, atpenin A5, 2-thenoyltrifluoroacetone (TTFA), and malonate, during reperfusion can either reduce or enhance the production of ROS generated by RET in Complexes I and III (Dröse et al. 2011). By utilizing this approach, they achieved cardioprotection without depending on the presence of K^+. This finding validates the independent role of Complex II in providing protective effects under the experimental conditions, highlighting its significance in modulating ROS production and mitigating I/R injury. Citric acid cycle intermediates such as oxaloacetate and malate, along with the succinate-competitor malonate, are well-established inhibitors of Complex II (Gutman and Silman 1975; Zeylemaker and Slater 1967). As a result, they have been extensively studied for their potential to mitigate I/R injury. For example, treating isolated mouse hearts with malonate either before ischemia or at the onset of reperfusion has demonstrated a prominent decrease in infarction size (Chouchani et al. 2014; Valls-Lacalle et al. 2016). This is accomplished via preventing mPTP opening and reducing the generation of ROS. Additionally, positive findings have been noted in a pig model of transient coronary artery occlusion (Valls-Lacalle et al. 2018). Intracoronary administration of malonate during early reperfusion effectively prevents excessive accumulation of ROS and limits the extent of myocardial infarcts. Researchers have reported that this therapy significantly reduces systolic shortening in the at-risk region of the heart while having no impact on the incidence of reperfusion arrhythmias. Similarly, in a mouse myocardial ischemia model, treatment with dimethyl malonate—a more membrane-permeable derivative of the Complex II inhibitor malonate—resulted in reduced accumulation of ischemic succinate and subsequent decrease in localized ROS generation. This indicates that dimethyl malonate effectively mitigates oxidative stress associated with ischemia (Boylston et al. 2015; Pell et al. 2016). Guo et al. reported that the anti-cancer agent lonidamine selectively inhibits the succinate-ubiquinone reductase activity of respiratory Complex II, while not significantly obstructing its succinate dehydrogenase function. This results in a decrease in malate and fumarate levels, leading to the buildup of succinate without the occurrence of hypoxia (Guo et al. 2016). Though the above-mentioned study did not specifically investigate the potential of lonidamine for I/R injury protection, the agent's ability to induce succinate accumulation could be of significant interest. This property might be useful for future research aimed at creating chemical ischemia-like states to better understand and develop interventions for I/R injury.

9.6.5 Protection Via Alternative Mechanistic Pathways

Diverse anti-apoptotic and anti-inflammatory medications exert their effects on multiple biochemical pathways besides the typically established intracellular mechanisms. Some of the mechanisms by which these drugs can alleviate inflammation include lowering leukocyte activation and adhesion, controlling nitric oxide levels, stimulating autophagy instead of apoptosis, and inhibiting inflammasome pathways. Through these diverse actions, they can directly diminish or suppress inflammatory responses. Among a wide range of well-accepted and safe therapeutic drugs, a few have shown especially encouraging results in reducing ischemia-reperfusion damage and associated inflammatory reactions. Methylprednisolone, a corticosteroid, is widely known for its strong immunosuppressive and anti-inflammatory properties, which significantly decrease tissue damage induced by inflammation during reperfusion (Abu-Amara et al. 2010). Ulinastatin, a serine protease inhibitor, has been shown to modulate inflammatory responses, protect against oxidative stress, and reduce organ injury in various I/R models (Li et al. 2017). Edaravone, a free radical scavenger, has proven to be effective in reducing oxidative stress and protecting tissues from reperfusion-induced damage (Ito et al. 2005). Aripiprazole, typically used as an antipsychotic, has exhibited anti-inflammatory and neuroprotective effects in experimental I/R injury studies (Li et al. 2017). Flurbiprofen axetil, a non-steroidal anti-inflammatory drug (NSAID) derivative, has been highlighted for its capacity to reduce postoperative pain and inflammation (Wu et al. 2018a). Neutrophil gelatinase-associated lipocalin (NGAL), both endogenous and exogenous forms, has emerged as a key player in protecting against kidney damage via decreasing inflammation and oxidative stress (Cannavo et al. 2018; Sweitzer and Airhart 2018). Furthermore, ghrelin, a gastrointestinal hormone with anti-inflammatory and anti-apoptotic properties, has shown potential in reducing the severity of I/R injury by modulating inflammatory cascades and preserving cellular function in various organs (Wang et al. 2017). The protective benefits against I/R damage by antioxidants have been consistently shown by regulating several inflammatory pathways. The techniques by which they operate are varied but can be broadly classified into two primary functions. First, antioxidants stimulate protective signaling pathways, typically by inhibiting mitophagy, a mechanism that, when overly triggered, leads to cell death and tissue damage during immune-mediated inflammatory response. Furthermore, antioxidants counteract the harmful effects of free radicals and ROS, which are produced prominently during reperfusion and play a crucial role in oxidative stress and resulting cellular damage. By inhibiting the buildup of ROS, these chemicals ensure the preservation of mitochondrial integrity, decrease inflammation, and alleviate cell death, thereby improving tissue survival and recovery during I/R injury. Several compounds have demonstrated significant potential in reducing I/R damage by means of their antioxidant, anti-inflammatory, or other protective properties. The enzyme aldehyde dehydrogenase 2 (ALDH2) is essential for the detoxification of reactive aldehydes produced during oxidative stress, therefore providing defense against cells damage (Budas et al. 2009; Chen et al. 2008; Panisello-Roselló et al. 2018). Through the inhibition of pro-inflammatory cytokines, candesartan, an angiotensin II receptor blocker, exerts antioxidant and anti-inflammatory actions (Soliman et al. 2015). In addition to its conventional antibacterial effects, minocycline exhibits neuroprotective characteristics by decreasing microglial activation and suppressing ROS generation (Soliman et al. 2015). Melatonin, a well-known antioxidant, neutralizes free radicals and boosts the function of antioxidant enzymes, while external NADPH facilitates the restoration of reduced glutathione, a crucial component in cellular protection against oxidative stress (Weng et al. 2018; Zhou et al. 2018). A highly effective intracellular antioxidant, glutathione counteracts free radicals and preserves redox equilibrium (Crack et al. 2001; Suyavaran et al. 2015). Similarly, N-acetylcysteine (NAC) restores glutathione levels and decreases oxidative stress response (Kalimeris et al. 2016). Mangafodipir and calmangafodipir, two compounds derived from manganese, have shown the ability to imitate the activity of SOD, thereby decreasing production of superoxide anion (Ben Mosbah et al. 2012; Karlsson et al. 2015). The enzyme superoxide dismutase plays a crucial role in the catalytic breakdown of superoxide radicals (Marcelino et al. 2017; Pinto et al. 2016). Through the preservation of mitochondrial function, diltiazem, a calcium channel blocker, has demonstrated protective effects against I/R injury (Chen et al. 2017). Idebenone and rasagiline are neuroprotective drugs. Idebenone functions as a synthetic antioxidant that enhances mitochondrial electron transport and decreases oxidative stress (Wieland et al. 1995). On the other hand, rasagiline hinders the degradation of dopamine and also demonstrates neuroprotective effects in models of I/R injury (Lei et al. 2018). Another well-recognized strategy to address I/R injury is to improve glucose metabolism, as well documented in the literature. Its primary objective is to enhance glucose metabolism and promote cellular glucose uptake before the ischemia "starvation" phase. The primary objective of this approach is to redirect the cell's energy generation from fatty acid β-oxidation, which requires more oxygen, to the more highly efficient and less dependent glucose oxidation pathway (Kantor et al. 2000; Lopaschuk 2017). In doing so, cells are more effectively prepared to produce energy under conditions of oxygen deprivation, therefore maintaining cellular function and minimizing harm during periods of reduced blood flow and subsequent restoration of blood flow. Drugs within this classification may differ in their particular processes but eventually converge on this biological change in metabolism. For example, substances such as glucagon-like peptide-1 (GLP-1) agonists improve the responsiveness of insulin and the absorption of glucose, whereas pyruvate dehydrogenase kinase (PDK) inhibitors function by facilitating the oxidation of glucose instead of fatty acids. Dichloroacetate and similar chemicals function by directly inhibiting PDK, therefore promoting improved glucose utilization. These techniques serve the dual purpose of maintaining cellular energy levels during ischemia and preventing the buildup of detrimental metabolites that can worsen tissue damage upon reperfusion. Metabolic adaptability is especially vital for energy-intensive organs such as the heart and brain during conditions of ischemia-induced stress.

9.6.6 Nutraceuticals: A Complementary Therapeutic Approach

Nutraceuticals are bioactive substances obtained from food or dietary supplements that provide health advantages beyond their fundamental nutritional value. The term "Nutraceutical" combines "nutrition" and "pharmaceutical," indicating its two-fold function in promoting health and possibly preventing or curing illness. These substances are commonly present in common foods such as fruits, vegetables, cereals, and herbs, and can also be ingested in concentrated quantities via supplements. Phytochemicals, including flavonoids, carotenoids, and polyphenols, are molecules obtained from plants that possess antioxidant characteristics. These features enable them to counteract ROS and mitigate oxidative stress, therefore preventing cellular damage and a range of harmful conditions. Functional foods, which are fortified with supplementary bioactive constituents such as probiotics, fiber, and omega-3 fatty acids, offer augmented health advantages and can potentially improve ailments such as gastrointestinal, neurological, and cardiovascular health. In order to enhance nutritional intake and rectify deficiencies, individuals use dietary supplements such as vitamins, minerals, amino acids, and other chemicals. Inhibition of pro-inflammatory cytokines, enhancement of antioxidant defenses, and support of immune system function are among the mechanisms by which nutraceuticals exert their effects. The modulation of various biological processes by nutraceuticals can contribute to the prevention of chronic diseases, promotion of overall health, and supplementation of traditional medical therapies. Polyphenols are a wide variety of plant-derived chemical substances, which are primarily distinguished by the presence of multiple hydroxyl groups on the aromatic rings. These compounds are secondary metabolites that play multiple physiological activities. In general, they are involved in the defense mechanism against the harmful effects of UV radiation or the aggression of pathogens (Manach et al. 2004). Polyphenols exhibit a wide range of biological activities, and research has demonstrated the potential benefits of dietary polyphenols found in fruits and vegetables for the treatment and prevention of several illnesses, such as cancer, diabetes, cardiovascular, neurodegenerative, and many other diseases.

Resveratrol, chemically known as 2,5,4′-trihydroxystilbene, is a naturally occurring stilbene polyphenol that can be found in a wide variety of plant species and foods, including grapes, peanuts, pomegranates, and some berries (Tian and Liu 2020). Resveratrol is an extensively studied polyphenol particularly known for its antioxidant, anti-inflammatory, and anti-cancer properties (Rauf et al. 2017). In the context of newborn hypoxia-ischemia (HI), a number of research have investigated the neuroprotective characteristics of resveratrol when it is provided intraperitoneally to pups prior to an HI insult or when it is given to mothers as a dietary supplementation (Arteaga et al. 2015; Dumont et al. 2020; Gao et al. 2018; Isac et al. 2017; Revuelta et al. 2016; West et al. 2007). When administered intraperitoneally to mouse pups 10 min or 24 h before to HI, West et al. showed that resveratrol at a dosage of 20 mg/kg prevented tissue loss in the hippocampus and striatum and decreased necrotic and apoptotic cell death in a dose-dependent manner (West et al. 2007). Moreover, using the same treatment protocol, Arteaga et al. demonstrated that resveratrol may mitigate the morphological damage caused by HI. This was accomplished by lowering the extent of infarcts, myelin loss, and the loss of cells, particularly in cortex and hippocampal regions. Similar studies showed that this resveratrol pre-treatment not only decreased astrogliosis and morphological damage in the inferior colliculus but also improved the auditory brainstem's functional responsiveness, which had been impaired by HI (Arteaga et al. 2015). It was postulated that resveratrol's ability to prevent HI could include maintaining the integrity of the mitochondrial inner membrane and transmembrane potential in addition to the decrease in ROS generation (Arteaga et al. 2015). In a similar way, another study demonstrated that intraperitoneal injection of resveratrol prior to the production of HI in P14 rat pups led to a reduction in the infarct area as well as HI-induced cerebral edema. A decrease in the expression of pro-inflammatory cytokines and NF-κB p65 subunit was seen as a result of the attenuation of the HI inflammatory response by resveratrol consumption. Furthermore, resveratrol diminished oxidative stress by augmenting the activity of the antioxidative enzymes GPx, CAT, and SOD. The investigators proposed that resveratrol stimulates the nuclear factor erythroid 2-related factor 2/heme oxygenase 1 (Nrf2/HO-1)-dependent signaling to generate endogenous antioxidants, which may enhance its neural protective activities (Gao et al. 2018).

Quercetin, chemically known as 3,3′,4′,5,7-pentahydroxyflavone, is a flavanol that can be found in large quantities in the majority of fruits, plants, and vegetables. Some of the most important foods that contain quercetin in the human diet include onions, broccoli, apples, leeks, kales, tea, and red wine. It is possible for quercetin to reach levels in the human diet that are as high as 3–38 mg/day (Hertog et al. 1993,

1995). A number of studies have demonstrated that quercetin possesses neuroprotective characteristics in the context of oxygen–glucose deprivation injury or focal ischemia (Huang et al. 2012; Le et al. 2020; Ossola et al. 2009; Rivera et al. 2004; Shah et al. 2024; Wang et al. 2020). Quercetin has also been found to have cardioprotective and antioxidant properties. Quercetin-filled nanosomes, which were developed by Blasina et al. (2015), were intravenously administered to piggies that had undergone bilateral transient carotid ligation, followed by 40 min of hypoxia. The administration of quercetin nanosomes (10 mg/kg) resulted in an improvement in electroencephalographic amplitude records. However, there were no histopathological alterations in brain lesions detected when compared to untreated groups. Piggies that received nanosomes were able to stabilize their blood pressure and recover spontaneous breathing 8 h after hypoxia, and they demonstrated improved walking and suckling abilities 3 days after hypoxia (Blasina et al. 2015). Moreover, the Rice–Vannucci paradigm was utilized by Wu et al. (2019) who gave quercetin orally at a dose of 40 mg/kg/day for a period of 1 week prior to the induction of hormone induction in P7 rats. Twenty-four hours after HI, it was demonstrated that quercetin had the ability to decrease the rate of apoptosis in cortical cells by modulating the expression of proteins in the apoptotic pathway. Furthermore, it was found to attenuate microgliosis and astrogliosis in cortical cells. Furthermore, it partly reduced the neuronal inflammation that was caused by HI via decreasing the expression of pro-inflammatory molecules. This was observed in other polyphenols as well (Gao et al. 2018; Isac et al. 2017; Li et al. 2016).

Fatty acids belonging to the omega-3 (Ω-3) family are classified as long-chain polyunsaturated fatty acids (PUFAs). These fatty acids are distinguished by the presence of a double bond in their chemical structure, which is located three atoms away from the terminal methyl group. There are a number of other Ω-3 fatty acids; however, the three most important Ω-3 fatty acids are eicosapentaenoic acid (EPA), α-linolenic acid (ALA), and docosahexaenoic acid (DHA) (Scorletti and Byrne 2013). Multiple studies have shown that Ω-3 fatty acids, such as ALA, EPA, and DHA, are packed with therapeutic potential and helped in a variety of focal ischemia models (Arteaga et al. 2017a; Belayev et al. 2009; Blondeau et al. 2009; Bourourou et al. 2016; Heurteaux et al. 2006; Nguemeni et al. 2010; Okabe et al. 2011; Pan et al. 2009; Ueda et al. 2013). In particular, DHA have demonstrated promising neuroprotective characteristics in a number of studies conducted with HI experimental models. These studies were conducted both when Ω-3 fatty acids were administered intraperitoneally to P7–P10 pups prior to an HI insult or as dietary supplements to the mother during pregnancy (Arteaga et al. 2017b; Berman et al. 2009, 2010; de Barros Mucci et al. 2015; Decker et al. 2016; Williams et al. 2013). Zhang and colleagues have conducted a series of studies that demonstrated the neuroprotective characteristics of EPA and DHA in neonatal HI (Zhang et al. 2010, 2015). The authors provided dams with a fish-oil-enriched diet from E2 to P14. DHA and EPA dietary supplementation has been shown to diminish brain tissue loss and edema, while also inhibiting the inflammatory reactions associated with HI injury by lowering the activated microglia and various inflammatory marker levels. Furthermore, it offers protection against cell death induced by HI conditions by facilitating the development of membrane phosphatidylserine and triggering the pro-survival PI3K/Akt signaling. Moreover, various other studies have demonstrated that when injected intraperitoneally shortly prior to HI damage, omega-3 fatty acids exhibit positive benefits. As per the findings of Williams et al., the administration of two doses of fish oil triglyceride emulsions, which are abundant in DHA and EPA, promptly following the closure of the common carotid artery during the hypoxic phase, resulted in a reduction in the volume of the brain tumor (Williams et al. 2013). In the setting of *Escherichia coli* lipopolysaccharide-induced systemic inflammation, Berman et al. demonstrated that pre-treatment with 1 mg/kg DHA, which was administered via intraperitoneal route 2.5 h prior to HI, reduced the amount of brain impairment and enhanced functional activities subsequent to neonatal HI (Berman et al. 2009). Altogether, there is an increasing amount of literature that supports the neuroprotective characteristics of such nutraceuticals, which are responsible for minimizing the damage to tissue and cells that is produced by neonatal hypoxia and alleviating the neurological and cardiac abnormalities that are due to such diseases.

9.7 Concluding Remarks

I/R injury represents a detrimental pathophysiological event in numerous medical situations, such as stroke, myocardial infarction, organ transplantation, and surgeries involving temporary blood flow occlusion. The intricate relationship between ischemia and reperfusion triggers an excessive generation of ROS, leading to heightened oxidative stress and exacerbating tissue damage or injury. The damage mediated by ROS not only leads to cellular injury but also initiates severe inflammatory responses, thereby worsening the recovery process even further as well as prolonging tissue injury. Moreover, addressing oxidative stress in I/R injury continues to be a significant challenge. Therefore, understanding the underlying processes of ROS generation and the pathways associated with oxidative stress during I/R injury is essential for establishing an effective treatment strategy which can minimize tissue damage in affected individuals. Strategies for intervention focused on reducing I/R injury have drawn considerable interest, with both pharmacological and nutraceutical treatments demonstrating potential effectiveness. Pharmacological therapies are designed to specifically target the generation of ROS, neutralize free radicals, and modulate oxidative stress pathways to safeguard tissues during the process of reperfusion. Nutraceuticals, derived from natural sources and other bioactive substances, have shown promise in mitigating oxidative damage associated with I/R injury. These compounds, abundant in antioxidants such as polyphenols, flavonoids, vitamins, and minerals, possess the ability to neutralize free radicals and diminish the overall oxidative burden on tissues. In addition to their antioxidant properties, numerous drugs demonstrate anti-inflammatory, anti-apoptotic, and cytoprotective effects, making them vital adjuncts in the prevention and management of I/R injury. The introduction of nutraceuticals into treatment plans has the potential to supplement standard pharmacological therapies, providing an integrated approach for safeguarding tissues against the harmful impacts of ischemia-reperfusion occurrences. Thus, ongoing research into such interventional approaches is of paramount significance since it may provide novel insights into the optimization of clinical outcomes for patients who experience I/R-related complications.

References

Abu-Amara M, Gurusamy K, Hori S, Glantzounis G, Fuller B, Davidson BR (2010) Systematic review of randomized controlled trials of pharmacological interventions to reduce ischaemia-reperfusion injury in elective liver resection with vascular occlusion. HPB 12:4–14. https://doi.org/10.1111/j.1477-2574.2009.00120.x

Abuchowski A, van Es T, Palczuk NC, Davis FF (1977) Alteration of immunological properties of bovine serum albumin by covalent attachment of polyethylene glycol. J Biol Chem 252:3578–3581. https://doi.org/10.1016/s0021-9258(17)40291-2

Agarwal B, Dash RK, Stowe DF, Bosnjak ZJ, Camara AKS (2014) Isoflurane modulates cardiac mitochondrial bioenergetics by selectively attenuating respiratory complexes. Biochim Biophys Acta Bioenerg 1837:354–365. https://doi.org/10.1016/j.bbabio.2013.11.006

Ahmad A, Ahmad S, Glover L, Miller SM, Shannon JM, Guo X, Franklin WA, Bridges JP, Schaack JB, Colgan SP, White CW (2009) Adenosine A2A receptor is a unique angiogenic target of HIF-2α in pulmonary endothelial cells. Proc Natl Acad Sci USA 106:10684–10689. https://doi.org/10.1073/pnas.0901326106

Akagawa M (2021) Protein carbonylation: molecular mechanisms, biological implications, and analytical approaches. Free Radic Res 55:307–320. https://doi.org/10.1080/10715762.2020.1851027

Akhigbe RE, Odetayo AF, Akhigbe TM, Hamed MA, Ashonibare PJ (2024) Pathophysiology and management of testicular ischemia/reperfusion injury: lessons from animal models. Heliyon 10: e27760. https://doi.org/10.1016/j.heliyon.2024.e27760

Algoet M, Janssens S, Himmelreich U, Gsell W, Pusovnik M, Van den Eynde J, Oosterlinck W (2023) Myocardial ischemia-reperfusion injury and the influence of inflammation. Trends Cardiovasc Med 33:357–366. https://doi.org/10.1016/j.tcm.2022.02.005

Allen D, Xiao XH (2000) Activity of the Na+/H+ exchanger contributes to cardiac damage following ischaemia and reperfusion. Clin Exp Pharmacol Physiol 27:727–733. https://doi.org/10.1046/j.1440-1681.2000.03329.x

Al-Taie A, Sancar M, Izzettin FV (2021) 8-Hydroxydeoxyguanosine: a valuable predictor of oxidative DNA damage in cancer and diabetes mellitus. In: Preedy VR, Patel VB (eds) Cancer: oxidative stress and dietary antioxidants, 2nd edn. Academic, San Diego, pp 179–187. https://doi.org/10.1016/B978-0-12-819547-5.00017-1

Arteaga O, Revuelta M, Urigüen L, Álvarez A, Montalvo H, Hilario E (2015) Pretreatment with resveratrol prevents neuronal injury and cognitive deficits induced by perinatal hypoxia-ischemia in rats. PLoS One 10:e0142424. https://doi.org/10.1371/journal.pone.0142424

Arteaga O, Álvarez A, Revuelta M, Santaolalla F, Urtasun A, Hilario E (2017a) Role of antioxidants in neonatal hypoxic–ischemic brain injury: new therapeutic approaches. Int J Mol Sci 18(2):265. https://doi.org/10.3390/ijms18020265

Arteaga O, Revuelta M, Urigüen L, Martínez-Millán L, Hilario E, Álvarez A (2017b) Docosahexaenoic acid reduces cerebral damage and ameliorates long-term cognitive impairments caused by neonatal hypoxia–ischemia in rats. Mol Neurobiol 54:7137–7155. https://doi.org/10.1007/s12035-016-0221-8

Babior BM (2004) NADPH oxidase. Curr Opin Immunol 16:42–47. https://doi.org/10.1016/j.coi.2003.12.001

Bagheri F, Khori V, Alizadeh AM, Khalighfard S, Khodayari S, Khodayari H (2016) Reactive oxygen species-mediated cardiac-reperfusion injury: mechanisms and therapies. Life Sci 165: 43–55. https://doi.org/10.1016/j.lfs.2016.09.013

Bartesaghi S, Radi R (2018) Fundamentals on the biochemistry of peroxynitrite and protein tyrosine nitration. Redox Biol 14:618–625. https://doi.org/10.1016/j.redox.2017.09.009

Basseri S, Austin RC (2012) Endoplasmic reticulum stress and lipid metabolism: mechanisms and therapeutic potential. Biochem Res Int 2012:841362. https://doi.org/10.1155/2012/841362

Bauche IB, El Mkadem SA, Pottier AM, Senou M, Many MC, Rezsohazy R, Penicaud L, Maeda N, Funahashi T, Brichard SM (2007) Overexpression of adiponectin targeted to adipose tissue in transgenic mice: impaired adipocyte differentiation. Endocrinology 148:1539–1549. https://doi.org/10.1210/en.2006-0838

Baxter GF, Mocanu MM, Brar BK, Latchman DS, Yellon DM (2001) Cardioprotective effects of transforming growth factor-β1 during early reoxygenation or reperfusion are mediated by p42/p44 MAPK. J Cardiovasc Pharmacol 38:930–939. https://doi.org/10.1097/00005344-200112000-00015

Bejaoui M, Pantazi E, Folch-Puy E, Panisello A, Calvo M, Pasut G, Rimola A, Navasa M, Adam R, Roselló-Catafau J (2015) Protective effect of intravenous high molecular weight polyethylene glycol on fatty liver preservation. Biomed Res Int 2015:794287. https://doi.org/10.1155/2015/794287

Belayev L, Khoutorova L, Atkins KD, Bazan NG (2009) Robust docosahexaenoic acid-mediated neuroprotection in a rat model of transient, focal cerebral ischemia. Stroke 40:3121–3126. https://doi.org/10.1161/STROKEAHA.109.555979

Ben Mosbah I, Mouchel Y, Pajaud J, Ribault C, Lucas C, Laurent A, Boudjema K, Morel F, Corlu A, Compagnon P (2012) Pretreatment with mangafodipir improves liver graft tolerance to ischemia/reperfusion injury in rat. PLoS One 7:e50235. https://doi.org/10.1371/journal.pone.0050235

Berman DR, Mozurkewich E, Liu YQ, Barks J (2009) Docosahexaenoic acid pretreatment confers neuroprotection in a rat model of perinatal cerebral hypoxia-ischemia. Am J Obstet Gynecol 200:305.e1–305.e6. https://doi.org/10.1016/j.ajog.2009.01.020

Berman DR, Liu YQ, Barks J, Mozurkewich E (2010) Docosahexaenoic acid confers neuroprotection in a rat model of perinatal hypoxia-ischemia potentiated by Escherichia coli lipopolysaccharide-induced systemic inflammation. Am J Obstet Gynecol 202:469.e1–469.e6. https://doi.org/10.1016/j.ajog.2010.01.076

Bernardi P, Di Lisa F (2015) The mitochondrial permeability transition pore: molecular nature and role as a target in cardioprotection. J Mol Cell Cardiol 78:100–106. https://doi.org/10.1016/j.yjmcc.2014.09.023

Bertero E, Popoiu TA, Maack C (2024) Mitochondrial calcium in cardiac ischemia/reperfusion injury and cardioprotection. Basic Res Cardiol 119:569–585. https://doi.org/10.1007/s00395-024-01060-2

Bibli SI, Papapetropoulos A, Iliodromitis EK, Daiber A, Randriamboavonjy V, Steven S, Brouckaert P, Chatzianastasiou A, Kypreos KE, Hausenloy DJ, Fleming I, Andreadou I (2019) Nitroglycerine limits infarct size through S-nitrosation of cyclophilin D: a novel mechanism for an old drug. Cardiovasc Res 115:625–636. https://doi.org/10.1093/cvr/cvy222

Blasina F, Vaamonde L, Silvera F, Tedesco AC, Dajas F (2015) Intravenous nanosomes of quercetin improve brain function and hemodynamic instability after severe hypoxia in newborn piglets. Neurochem Int 89:149–156. https://doi.org/10.1016/j.neuint.2015.08.007

Blondeau N, Nguemeni C, Debruyne DN, Piens M, Wu X, Pan H, Hu XZ, Gandin C, Lipsky RH, Plumier JC, Marini AMS, Heurteaux C (2009) Subchronic alpha-linolenic acid treatment enhances brain plasticity and exerts an antidepressant effect: a versatile potential therapy for stroke. Neuropsychopharmacology 34:2548–2559. https://doi.org/10.1038/npp.2009.84

Boros D, Thompson J, Larson DF (2016) Adenosine regulation of the immune response initiated by ischemia reperfusion injury. Perfusion 31:103–110. https://doi.org/10.1177/0267659115586579

Bouleti C, Mewton N, Germain S (2015) The no-reflow phenomenon: state of the art. Arch Cardiovasc Dis 108:661–674. https://doi.org/10.1016/j.acvd.2015.09.006

Bourourou M, Heurteaux C, Blondeau N (2016) Alpha-linolenic acid given as enteral or parenteral nutritional intervention against sensorimotor and cognitive deficits in a mouse model of ischemic stroke. Neuropharmacology 108:60–72. https://doi.org/10.1016/j.neuropharm.2016.04.040

Boylston JA, Sun J, Chen Y, Gucek M, Sack MN, Murphy E (2015) Characterization of the cardiac succinylome and its role in ischemia-reperfusion injury. J Mol Cell Cardiol 88:73–81. https://doi.org/10.1016/j.yjmcc.2015.09.005

Brand MD, Goncalves RLS, Orr AL, Vargas L, Gerencser AA, Borch Jensen M, Wang YT, Melov S, Turk CN, Matzen JT, Dardov VJ, Petrassi HM, Meeusen SL, Perevoshchikova IV, Jasper H, Brookes PS, Ainscow EK (2016) Suppressors of superoxide-H2O2 production at site IQ of mitochondrial complex I protect against stem cell hyperplasia and ischemia-reperfusion injury. Cell Metab 24:582–592. https://doi.org/10.1016/j.cmet.2016.08.012

Brillo V, Chieregato L, Leanza L, Muccioli S, Costa R (2021) Mitochondrial dynamics, ROS, and cell signaling: a blended overview. Life 11(4):332. https://doi.org/10.3390/life11040332

Budas GR, Disatnik MH, Mochly-Rosen D (2009) Aldehyde dehydrogenase 2 in cardiac protection: a new therapeutic target? Trends Cardiovasc Med 19:158–164. https://doi.org/10.1016/j.tcm.2009.09.003

Burgos-Morón E, Abad-Jiménez Z, de Marañón AM, Iannantuoni F, Escribano-López I, López-Domènech S, Salom C, Jover A, Mora V, Roldan I, Solá E, Rocha M, Víctor VM (2019) Relationship between oxidative stress, ER stress, and inflammation in type 2 diabetes: the battle continues. J Clin Med 8(9):1385. https://doi.org/10.3390/jcm8091385

Burley DS, Ferdinandy P, Baxter GF (2007) Cyclic GMP and protein kinase-G in myocardial ischaemia-reperfusion: opportunities and obstacles for survival signaling. Br J Pharmacol 152:855–869. https://doi.org/10.1038/sj.bjp.0707409

Cadet J, Delatour T, Douki T, Gasparutto D, Pouget JP, Ravanat JL, Sauvaigo S (1999) Hydroxyl radicals and DNA base damage. Mutat Res 424:9–21. https://doi.org/10.1016/S0027-5107(99)00004-4

Cannavo A, Komici K, Bencivenga L, D'amico ML, Gambino G, Liccardo D, Ferrara N, Rengo G (2018) GRK2 as a therapeutic target for heart failure. Expert Opin Ther Targets 22:75–83. https://doi.org/10.1080/14728222.2018.1406925

Cantu-Medellin N, Kelley EE (2013) Xanthine oxidoreductase-catalyzed reactive species generation: a process in critical need of reevaluation. Redox Biol 1:353–358. https://doi.org/10.1016/j.redox.2013.05.002

Cao CM, Zhang Y, Weisleder N, Ferrante C, Wang X, Lv F, Zhang Y, Song R, Hwang M, Jin L, Guo J, Peng W, Li G, Nishi M, Takeshima H, Ma J, Xiao RP (2010) MG53 constitutes a primary determinant of cardiac ischemic preconditioning. Circulation 121:2565–2574. https://doi.org/10.1161/CIRCULATIONAHA.110.954628

Cao J, Lin H, Li W, Dong Z, Shi Y, Zhang X, Xiao R (2019) Ischemia postconditioning protects dermal microvascular endothelial cells of rabbit epigastric skin flaps against apoptosis via adenosine A2a receptors. J Plast Surg Hand Surg 53:76–82. https://doi.org/10.1080/2000656X.2018.1550417

Carreras-Sureda A, Pihán P, Hetz C (2018) Calcium signaling at the endoplasmic reticulum: fine-tuning stress responses. Cell Calcium 70:24–31. https://doi.org/10.1016/j.ceca.2017.08.004

Chen CH, Budas GR, Churchill EN, Disatnik MH, Hurley TD, Mochly-Rosen D (2008) Activation of aldehyde dehydrogenase-2 reduces ischemic damage to the heart. Science 321:1493–1495. https://doi.org/10.1126/science.1158554

Chen C, Lu W, Wu G, Lv L, Chen W, Huang L, Wu X, Xu N, Wu Y (2017) Cardioprotective effects of combined therapy with diltiazem and superoxide dismutase on myocardial ischemia-reperfusion injury in rats. Life Sci 183:50–59. https://doi.org/10.1016/j.lfs.2017.06.024

Chen J, Kos R, Garssen J, Redegeld F (2019) Molecular insights into the mechanism of necroptosis: the necrosome as a potential therapeutic target. Cells 8(12):1486. https://doi.org/10.3390/cells8121486

Chouchani ET, Pell VR, Gaude E, Aksentijević D, Sundier SY, Robb EL, Logan A, Nadtochiy SM, Ord ENJ, Smith AC, Eyassu F, Shirley R, Hu CH, Dare AJ, James AM, Rogatti S, Hartley RC, Eaton S, Costa ASH, Brookes PS, Davidson SM, Duchen MR, Saeb-Parsy K, Shattock MJ, Robinson AJ, Work LM, Frezza C, Krieg T, Murphy MP (2014) Ischaemic accumulation of succinate controls reperfusion injury through mitochondrial ROS. Nature 515:431–435. https://doi.org/10.1038/nature13909

Coetzee WA (2013) Multiplicity of effectors of the cardioprotective agent, diazoxide. Pharmacol Ther 140:167–175. https://doi.org/10.1016/j.pharmthera.2013.06.007

Cohen MV, Yang XM, Liu Y, Solenkova NV, Downey JM (2010) Cardioprotective PKG-independent NO signaling at reperfusion. Am J Physiol Heart Circ Physiol 299:H2028–H2036. https://doi.org/10.1152/ajpheart.00527.2010

Collin F (2019) Chemical basis of reactive oxygen species reactivity and involvement in neurodegenerative diseases. Int J Mol Sci 20(10):2407. https://doi.org/10.3390/ijms20102407

Collins AR (2004) The comet assay for DNA damage and repair: principles, applications, and limitations. Mol Biotechnol 26:249–261. https://doi.org/10.1385/MB:26:3:249

Costa MÁ, Elesgaray R, Balaszczuk AM, Arranz C (2006) Role of NPR-C natriuretic receptor in nitric oxide system activation induced by atrial natriuretic peptide. Regul Pept 135:63–68. https://doi.org/10.1016/j.regpep.2006.04.002

Crack PJ, Taylor JM, Flentjar NJ, De Haan J, Hertzog P, Iannello RC, Kola I (2001) Increased infarct size and exacerbated apoptosis in the glutathione peroxidase-1 (Gpx-1) knockout mouse brain in response to ischemia/reperfusion injury. J Neurochem 78:1389–1399. https://doi.org/10.1046/j.1471-4159.2001.00535.x

D'Oria R, Schipani R, Leonardini A, Natalicchio A, Perrini S, Cignarelli A, Laviola L, Giorgino F (2020) The role of oxidative stress in cardiac disease: from physiological response to injury factor. Oxidative Med Cell Longev 2020:5732956. https://doi.org/10.1155/2020/5732956

De Almeida AJPO, De Oliveira JCPL, Da Silva Pontes LV, De Souza Júnior JF, Gonçalves TAF, Dantas SH, De Almeida Feitosa MS, Silva AO, De Medeiros IA (2022) ROS: basic concepts, sources, cellular signaling, and its implications in aging pathways. Oxidative Med Cell Longev 2022:1225578. https://doi.org/10.1155/2022/1225578

de Barros Mucci D, Fernandes FS, dos Santos Souza A, de Carvalho Sardinha FL, Soares-Mota M, Tavares do Carmo M d G (2015) Flaxseed mitigates brain mass loss, improving motor hyperactivity and spatial memory, in a rodent model of neonatal hypoxic-ischemic encephalopathy. Prostaglandins Leukot Essent Fatty Acids 97:13–19. https://doi.org/10.1016/j.plefa.2015.03.001

de Mattos Dourado SF, Barbeiro DF, Koike MK, Barbeiro HV, Pinheiro da Silva F, César Machado MC (2018) Diazoxide reduces local and remote organ damage in a rat model of intestinal ischemia reperfusion. J Surg Res 225:118–124. https://doi.org/10.1016/j.jss.2018.01.009

Decker MJ, Jones K, Keating GL, Damato EG, Darrah R (2016) Maternal dietary supplementation with omega-3 polyunsaturated fatty acids confers neuroprotection to the newborn against hypoxia-induced dopamine dysfunction. Sleep Sci 9:94–99. https://doi.org/10.1016/j.slsci.2016.05.007

Decuypere JP, Ceulemans LJ, Agostinis P, Monbaliu D, Naesens M, Pirenne J, Jochmans I (2015) Autophagy and the kidney: implications for ischemia-reperfusion injury and therapy. Am J Kidney Dis 66:699–709. https://doi.org/10.1053/j.ajkd.2015.05.021

Di Meo S, Reed TT, Venditti P, Victor VM (2016) Role of ROS and RNS sources in physiological and pathological conditions. Oxidative Med Cell Longev 2016:1245049. https://doi.org/10.1155/2016/1245049

Dikalov SI, Harrison DG (2014) Methods for detection of mitochondrial and cellular reactive oxygen species. Antioxid Redox Signal 20:372–382. https://doi.org/10.1089/ars.2012.4886

Dröse S, Bleier L, Brandt U (2011) A common mechanism links differently acting complex II inhibitors to cardioprotection: modulation of mitochondrial reactive oxygen species production. Mol Pharmacol 79:814–822. https://doi.org/10.1124/mol.110.070342

Dugbartey GJ (2024) Cellular and molecular mechanisms of cell damage and cell death in ischemia–reperfusion injury in organ transplantation. Mol Biol Rep 51:473. https://doi.org/10.1007/s11033-024-09261-7

Dumont U, Sanchez S, Olivier B, Chateil JF, Deffieux D, Quideau S, Pellerin L, Beauvieux MC, Bouzier-Sore AK, Roumes H (2020) Maternal alcoholism and neonatal hypoxia-ischemia: neuroprotection by stilbenoid polyphenols. Brain Res 1738:146798. https://doi.org/10.1016/j.brainres.2020.146798

Duranski MR, Greer JJM, Dejam A, Jaganmohan S, Hogg N, Langston W, Patel RP, Yet SF, Wang X, Kevil CG, Gladwin MT, Lefer DJ (2005) Cytoprotective effects of nitrite during in vivo ischemia-reperfusion of the heart and liver. J Clin Invest 115:1232–1240. https://doi.org/10.1172/JCI22493

Ebner B, Ebner A, Reetz A, Böhme S, Schauer A, Strasser RH, Weinbrenner C (2013) Pharmacological postconditioning by bolus injection of phosphodiesterase-5 inhibitors vardenafil and sildenafil. Mol Cell Biochem 379:43–49. https://doi.org/10.1007/s11010-013-1625-7

Elsayed Abouzed DE, Ezelarab HAA, Selim HMRM, Elsayed MMA, El Hamd MA, Aboelez MO (2024) Multimodal modulation of hepatic ischemia/reperfusion-induced injury by phytochemical agents: a mechanistic evaluation of hepatoprotective potential and safety profiles. Int Immunopharmacol 138:112445. https://doi.org/10.1016/j.intimp.2024.112445

Endlicher R, Drahota Z, Štefková K, Červinková Z, Kučera O (2023) The mitochondrial permeability transition pore—current knowledge of its structure, function, and regulation, and optimized methods for evaluating its functional state. Cells 12(9):1273. https://doi.org/10.3390/cells12091273

Fakharnia F, Khodagholi F, Dargahi L, Ahmadiani A (2017) Prevention of cyclophilin D-mediated mPTP opening using cyclosporine-a alleviates the elevation of necroptosis, autophagy and apoptosis-related markers following global cerebral ischemia-reperfusion. J Mol Neurosci 61: 52–60. https://doi.org/10.1007/s12031-016-0843-3

Fancelli D, Abate A, Amici R, Bernardi P, Ballarini M, Cappa A, Carenzi G, Colombo A, Contursi C, Di Lisa F, Dondio G, Gagliardi S, Milanesi E, Minucci S, Pain G, Pelicci PG, Saccani A, Storto M, Thaler F, Varasi M, Villa M, Plyte S (2014) Cinnamic anilides as new mitochondrial permeability transition pore inhibitors endowed with ischemia-reperfusion injury protective effect in vivo. J Med Chem 57:5333–5347. https://doi.org/10.1021/jm500547c

Fedorova M, Bollineni RC, Hoffmann R (2014) Protein carbonylation as a major hallmark of oxidative damage: update of analytical strategies. Mass Spectrom Rev 33:79–97. https://doi.org/10.1002/mas.21381

Feyzizadeh S, Badalzadeh R (2017) Application of ischemic postconditioning's algorithms in tissues protection: response to methodological gaps in preclinical and clinical studies. J Cell Mol Med 21:2257–2267. https://doi.org/10.1111/jcmm.13159

Firdous SM, Khan SA, Maity A (2024) Oxidative stress–mediated neuroinflammation in Alzheimer's disease. Naunyn Schmiedebergs Arch Pharmacol 397(11):8189–8209. https://doi.org/10.1007/s00210-024-03188-3

Forman HJ, Zhang H (2021) Targeting oxidative stress in disease: promise and limitations of antioxidant therapy. Nat Rev Drug Discov 20:689–709. https://doi.org/10.1038/s41573-021-00233-1

Forman HJ, Zhang H, Rinna A (2009) Glutathione: overview of its protective roles, measurement, and biosynthesis. Mol Asp Med 30:1–12. https://doi.org/10.1016/j.mam.2008.08.006

Frangogiannis NG (2015) Pathophysiology of myocardial infarction. Compr Physiol 5:1841–1875. https://doi.org/10.1002/cphy.c150006

Fukai T, Ushio-Fukai M (2011) Superoxide dismutases: role in redox signaling, vascular function, and diseases. Antioxid Redox Signal 15:1583–1606. https://doi.org/10.1089/ars.2011.3999

Gadicherla AK, Stowe DF, Antholine WE, Yang M, Camara AKS (2012) Damage to mitochondrial complex I during cardiac ischemia reperfusion injury is reduced indirectly by anti-anginal drug ranolazine. Biochim Biophys Acta Bioenerg 1817:419–429. https://doi.org/10.1016/j.bbabio.2011.11.021

Gao C, Liu Y, Yu Q, Yang Q, Li B, Sun L, Yan W, Cai X, Gao E, Xiong L, Wang H, Tao L (2015) TNF-α antagonism ameliorates myocardial ischemia-reperfusion injury in mice by upregulating adiponectin. Am J Physiol Heart Circ Physiol 308:H1583–H1591. https://doi.org/10.1152/ajpheart.00346.2014

Gao Y, Fu R, Wang J, Yang X, Wen L, Feng J (2018) Resveratrol mitigates the oxidative stress mediated by hypoxic-ischemic brain injury in neonatal rats via nrf2/ho-1 pathway. Pharm Biol 56:440–449. https://doi.org/10.1080/13880209.2018.1502326

Ghaboura N, Tamareille S, Ducluzeau PH, Grimaud L, Loufrani L, Croué A, Tourmen Y, Henrion D, Furber A, Prunier F (2011) Diabetes mellitus abrogates erythropoietin-induced cardioprotection against ischemic-reperfusion injury by alteration of the RISK/GSK-3β signaling. Basic Res Cardiol 106:147–162. https://doi.org/10.1007/s00395-010-0130-3

Ghaffari S, Kazemi B, Toluey M, Sepehrvand N (2013) The effect of prethrombolytic cyclosporine-a injection on clinical outcome of acute anterior st-elevation myocardial infarction. Cardiovasc Ther 31:e34–e39. https://doi.org/10.1111/1755-5922.12010

Gottlieb RA (2011) Cell death pathways in acute ischemia/reperfusion injury. J Cardiovasc Pharmacol Ther 16:233–238. https://doi.org/10.1177/1074248411409581

Goyal MM, Basak A (2010) Human catalase: looking for complete identity. Protein Cell 1:888–897. https://doi.org/10.1007/s13238-010-0113-z

Granger DN, Kvietys PR (2015) Reperfusion injury and reactive oxygen species: the evolution of a concept. Redox Biol 6:524–551. https://doi.org/10.1016/j.redox.2015.08.020

Guo L, Shestov AA, Worth AJ, Nath K, Nelson DS, Leeper DB, Glickson JD, Blair IA (2016) Inhibition of mitochondrial complex II by the anticancer agent lonidamine. J Biol Chem 291: 42–57. https://doi.org/10.1074/jbc.M115.697516

Gutman M, Silman N (1975) The steady state activity of succinate dehydrogenase in the presence of opposing effectors—I. The effect of l malate and CoQH2 on the enzymic activity. Mol Cell Biochem 7:51–58. https://doi.org/10.1007/BF01732163

Hadebe N, Cour M, Lecour S (2018) The SAFE pathway for cardioprotection: is this a promising target? Basic Res Cardiol 113:9. https://doi.org/10.1007/S00395-018-0670-5

Halliwell B, Gutteridge JMC (2015) Free radicals in biology and medicine. Oxford university press. https://doi.org/10.1093/acprof:oso/9780198717478.001.0001

Hausenloy DJ, Yellon DM (2004) New directions for protecting the heart against ischaemia-reperfusion injury: targeting the Reperfusion Injury Salvage Kinase (RISK)-pathway. Cardiovasc Res 61:448–460. https://doi.org/10.1016/j.cardiores.2003.09.024

Hausenloy DJ, Yellon DM (2007) Reperfusion injury salvage kinase signalling: taking a RISK for cardioprotection. Heart Fail Rev 12:217–234. https://doi.org/10.1007/s10741-007-9026-1

Hausenloy DJ, Lecour S, Yellon DM (2011) Reperfusion injury salvage kinase and survivor activating factor enhancement prosurvival signaling pathways in ischemic postconditioning: two sides of the same coin. Antioxid Redox Signal 14:893–907. https://doi.org/10.1089/ars.2010.3360

Hayyan M, Hashim MA, Alnashef IM (2016) Superoxide ion: generation and chemical implications. Chem Rev 116:3029–3085. https://doi.org/10.1021/acs.chemrev.5b00407

He J, Liu D, Zhao L, Zhou D, Rong J, Zhang L, Xia Z (2022) Myocardial ischemia/reperfusion injury: mechanisms of injury and implications for management (review). Exp Ther Med 23:430. https://doi.org/10.3892/etm.2022.11357

Hertog MGL, Hollman PCH, van de Putte B (1993) Content of potentially anticarcinogenic flavonoids of tea infusions, wines, and fruit juices. J Agric Food Chem 41:1242–1246. https://doi.org/10.1021/jf00032a015

Hertog MGL, Kromhout D, Aravanis C, Blackburn H, Buzina R, Fidanza F, Giampaoli S, Jansen A, Menotti A, Nedeljkovic S, Pekkarinen M, Simic BS, Toshima H, Feskens EJM, Hollman PCH, Katan MB (1995) Flavonoid intake and long-term risk of coronary heart disease and cancer in the seven countries study. Arch Intern Med 155:381–386. https://doi.org/10.1001/archinte.1995.00430040053006

Hetz C (2012) The unfolded protein response: controlling cell fate decisions under ER stress and beyond. Nat Rev Mol Cell Biol 13:89–102. https://doi.org/10.1038/nrm3270

Heurteaux C, Laigle C, Blondeau N, Jarretou G, Lazdunski M (2006) Alpha-linolenic acid and riluzole treatment confer cerebral protection and improve survival after focal brain ischemia. Neuroscience 137:241–251. https://doi.org/10.1016/j.neuroscience.2005.08.083

Huang JJ, Liu X, Wang XQ, Yang LH, Qi DS, Yao RQ (2012) Effects of quercetin on the learning and memory ability of neonatal rats with hypoxic-ischemic brain damage. Chinese J Contemp Pediatr 14:454–457

Huerter ME, Sharma AK, Zhao Y, Charles EJ, Kron IL, Laubach VE (2016) Attenuation of pulmonary ischemia-reperfusion injury by adenosine A2B receptor antagonism. Ann Thorac Surg 102:385–393. https://doi.org/10.1016/j.athoracsur.2016.02.060

Ichikawa J, Tsuchimoto D, Oka S, Ohno M, Furuichi M, Sakumi K, Nakabeppu Y (2008) Oxidation of mitochondrial deoxynucleotide pools by exposure to sodium nitroprusside induces cell death. DNA Repair (Amst) 7:418–430. https://doi.org/10.1016/j.dnarep.2007.11.007

Ikeda G, Matoba T, Nakano Y, Nagaoka K, Ishikita A, Nakano K, Funamoto D, Sunagawa K, Egashira K (2016) Nanoparticle-mediated targeting of cyclosporine a enhances cardioprotection against ischemia-reperfusion injury through inhibition of mitochondrial permeability transition pore opening. Sci Rep 6:20467. https://doi.org/10.1038/srep20467

Inserte J, Garcia-Dorado D (2015) The cGMP/PKG pathway as a common mediator of cardioprotection: translatability and mechanism. Br J Pharmacol 172:1996–2009. https://doi.org/10.1111/bph.12959

Isac S, Panaitescu AM, Spataru A, Iesanu M, Totan A, Udriste A, Cucu N, Peltecu G, Zagrean L, Zagrean AM (2017) Trans-resveratrol enriched maternal diet protects the immature hippocampus from perinatal asphyxia in rats. Neurosci Lett 653:308–313. https://doi.org/10.1016/j.neulet.2017.06.003

Ito K, Ozasa H, Horikawa S (2005) Edaravone protects against lung injury induced by intestinal ischemia/reperfusion in rat. Free Radic Biol Med 38:369–374. https://doi.org/10.1016/j.freeradbiomed.2004.10.029

Jeong JS, Kim D, Kim KY, Ryu S, Han S, Shin BS, Kim GS, Gwak MS, Ko JS (2017) Ischemic preconditioning produces comparable protection against hepatic ischemia/reperfusion injury under isoflurane and sevoflurane anesthesia in rats. Transplant Proc 49:2188–2193. https://doi.org/10.1016/j.transproceed.2017.07.002

Jia P, Liu C, Wu N, Jia DL, Sun YX (2018) Agomelatine protects against myocardial ischemia reperfusion injury by inhibiting mitochondrial permeability transition pore opening. Am J Transl Res 10:1310–1323. https://doi.org/10.3892/etm.2018.7016

Jomova K, Raptova R, Alomar SY, Alwasel SH, Nepovimova E, Kuca K, Valko M (2023) Reactive oxygen species, toxicity, oxidative stress, and antioxidants: chronic diseases and aging. Arch Toxicol 97:2499–2574. https://doi.org/10.1007/s00204-023-03562-9

Jones DP, Radi R (2014) Redox pioneer: professor Helmut Sies. Antioxid Redox Signal 21:2459–2468. https://doi.org/10.1089/ars.2014.6037

Juan CA, de la Lastra JMP, Plou FJ, Pérez-Lebeña E (2021) The chemistry of reactive oxygen species (ROS) revisited: outlining their role in biological macromolecules (dna, lipids and proteins) and induced pathologies. Int J Mol Sci 22(9):4642. https://doi.org/10.3390/ijms22094642

Kalimeris K, Briassoulis P, Ntzouvani MSA, Nomikos T, Papaparaskeva K, Politi A, Batistaki C, Kostopanagiotou G (2016) N-acetylcysteine ameliorates liver injury in a rat model of intestinal ischemia reperfusion. J Surg Res 206:263–272. https://doi.org/10.1016/j.jss.2016.08.049

Kalogeris T, Baines CP, Krenz M, Korthuis RJ (2012) Cell biology of ischemia/reperfusion injury. Int Rev Cell Mol Biol 298:229–317. https://doi.org/10.1016/B978-0-12-394309-5.00006-7

Kantor PF, Lucien A, Kozak R, Lopaschuk GD (2000) The antianginal drug trimetazidine shifts cardiac energy metabolism from fatty acid oxidation to glucose oxidation by inhibiting mitochondrial long- chain 3-ketoacyl coenzyme A thiolase. Circ Res 86:580–588. https://doi.org/10.1161/01.RES.86.5.580

Karlsson JOG, Ignarro LJ, Lundström I, Jynge P, Almén T (2015) Calmangafodipir [Ca4Mn(DPDP)5], mangafodipir (MnDPDP) and MnPLED with special reference to their SOD mimetic and therapeutic properties. Drug Discov Today 20:411–421. https://doi.org/10.1016/j.drudis.2014.11.008

Kasai H, Kawai K, Li Y (2008) Analysis of 8-OH-dG and 8-OH-Gua as biomarkers of oxidative stress. Genes Environ 30:33–40. https://doi.org/10.3123/jemsge.30.33

Katayama Y, Inaba T, Nito C, Ueda M, Katsura K (2014) Neuroprotective effects of erythromycin on cerebral ischemia reperfusion-injury and cell viability after oxygen-glucose deprivation in cultured neuronal cells. Brain Res 1588:159–167. https://doi.org/10.1016/j.brainres.2014.09.016

Kehm R, Baldensperger T, Raupbach J, Höhn A (2021) Protein oxidation—formation mechanisms, detection and relevance as biomarkers in human diseases. Redox Biol 42:101901. https://doi.org/10.1016/j.redox.2021.101901

Khatib SY (2019) The protective effects of phosphodiesterase 5 inhibitors "Sildenafil & Ordonafil" on ischemia-reperfusion injury. FASEB J 33:833–837. https://doi.org/10.1096/fasebj.2019.33.1_supplement.833.7

Kim H, Xue X (2020) Detection of total reactive oxygen species in adherent cells by 2′,7′--dichlorodihydrofluorescein diacetate staining. J Vis Exp 2020:1–5. https://doi.org/10.3791/60682

Kong Q, Lin CLG (2010) Oxidative damage to RNA: mechanisms, consequences, and diseases. Cell Mol Life Sci 67:1817–1829. https://doi.org/10.1007/s00018-010-0277-y

Koutsogiannidis CPC, Johnson EO (2017) Pharmacological neuroprotection in cardiac surgery: effectiveness of pharmacologic-preconditioning with erythromycin. Curr Vasc Pharmacol 15:329–335. https://doi.org/10.2174/1570161115666171010120953

Kumar R, Gullapalli RR (2024) High throughput screening assessment of reactive oxygen species (ROS) generation using Dihydroethidium (DHE) fluorescence dye. J Vis Exp. https://doi.org/10.3791/66238

Le K, Song Z, Deng J, Peng X, Zhang J, Wang L, Zhou L, Bi H, Liao Z, Feng Z (2020) Quercetin alleviates neonatal hypoxic-ischemic brain injury by inhibiting microglia-derived oxidative stress and TLR4-mediated inflammation. Inflamm Res 69:1201–1213. https://doi.org/10.1007/s00011-020-01402-5

Lecour S (2009) Activation of the protective Survivor Activating Factor Enhancement (SAFE) pathway against reperfusion injury: does it go beyond the RISK pathway? J Mol Cell Cardiol 47:32–40. https://doi.org/10.1016/j.yjmcc.2009.03.019

Lecour S, James RW (2011) When are pro-inflammatory cytokines SAFE in heart failure? Eur Heart J 32:680–685. https://doi.org/10.1093/eurheartj/ehq484

Lee KH, Lee SR, Cho H, Woo JS, Kang JH, Jeong YM, Cheng XW, Kim WS, Kim W (2017) Cardioprotective effects of PKG activation by soluble GC activator, BAY 60-2770, in ischemia-reperfusion-injured rat hearts. PLoS One 12:e0180207. https://doi.org/10.1371/journal.pone.0180207

Lei D, Shao Z, Zhou X, Yuan H (2018) Synergistic neuroprotective effect of rasagiline and idebenone against retinal ischemia-reperfusion injury via the Lin28-let-7- Dicer pathway. Oncotarget 9:12137–12153. https://doi.org/10.18632/oncotarget.24343

Leni Z, Künzi L, Geiser M (2020) Air pollution causing oxidative stress. Curr Opin Toxicol 20–21:1–8. https://doi.org/10.1016/j.cotox.2020.02.006

Li D, Song T, Yang L, Wang X, Yang C, Jiang Y (2016) Neuroprotective actions of pterostilbene on hypoxic-ischemic brain damage in neonatal rats through upregulation of heme oxygenase-1. Int J Dev Neurosci 54:22–31. https://doi.org/10.1016/j.ijdevneu.2016.08.005

Li X, Su L, Zhang X, Zhang C, Wang L, Li Y, Zhang Y, He T, Zhu X, Cui L (2017) Ulinastatin downregulates TLR4 and NF-kB expression and protects mouse brains against ischemia/reperfusion injury. Neurol Res 39:367–373. https://doi.org/10.1080/01616412.2017.1286541

Li W, Cao T, Luo C, Cai J, Zhou X, Xiao X, Liu S (2020) Crosstalk between ER stress, NLRP3 inflammasome, and inflammation. Appl Microbiol Biotechnol 104:6129–6140. https://doi.org/10.1007/s00253-020-10614-y

Lin XL, Xiao WJ, Le Xiao L, Liu MH (2018) Molecular mechanisms of autophagy in cardiac ischemia/reperfusion injury (review). Mol Med Rep 18:675–683. https://doi.org/10.3892/mmr.2018.9028

Liu B, Zhu X, Chen CL, Hu K, Swartz HM, Chen YR, He G (2010) Opening of the mitoKATP channel and decoupling of mitochondrial complex II and III contribute to the suppression of myocardial reperfusion hyperoxygenation. Mol Cell Biochem 337:25–38. https://doi.org/10.1007/s11010-009-0283-2

Lopaschuk GD (2017) Metabolic modulators in heart disease: past, present, and future. Can J Cardiol 33:838–849. https://doi.org/10.1016/j.cjca.2016.12.013

Lopez A, Panisello-Rosello A, Castro-Benitez C, Adam R (2018) Glycocalyx preservation and NO production in fatty livers—the protective role of high molecular polyethylene glycol in cold ischemia injury. Int J Mol Sci 19(8):2375. https://doi.org/10.3390/ijms19082375

Lu L, Zhou H, Ni M, Wang X, Busuttil R, Kupiec-Weglinski J, Zhai Y (2016) Innate immune regulations and liver ischemia-reperfusion injury. Transplantation 100:2601–2610. https://doi.org/10.1097/TP.0000000000001411

Luo S, Luo R, Deng G, Huang F, Lei Z (2024) Programmed cell death, from liver Ischemia–Reperfusion injury perspective: an overview. Heliyon 10:e32480. https://doi.org/10.1016/j.heliyon.2024.e32480

Lushchak VI (2011) Environmentally induced oxidative stress in aquatic animals. Aquat Toxicol 101:13–30. https://doi.org/10.1016/j.aquatox.2010.10.006

Lushchak VI, Storey KB (2021) Oxidative stress concept updated: definitions, classifications, and regulatory pathways implicated. EXCLI J 20:956–967. https://doi.org/10.17179/excli2021-3596

Magnani F, Mattevi A (2019) Structure and mechanisms of ROS generation by NADPH oxidases. Curr Opin Struct Biol 59:91–97. https://doi.org/10.1016/j.sbi.2019.03.001

Malhotra R, Valuckaite V, Staron ML, Theccanat T, D'Souza KM, Alverdy JC, Akhter SA (2011) High-molecular-weight polyethylene glycol protects cardiac myocytes from hypoxia- and reoxygenation-induced cell death and preserves ventricular function. Am J Physiol Heart Circ Physiol 300:H1733–H1742. https://doi.org/10.1152/ajpheart.01054.2010

Malko P, Ding R, Jiang LH (2021) TRPM2 channel in oxidative stress-induced mitochondrial dysfunction and apoptotic cell death. Adv Protein Chem Struct Biol 125:51–72. https://doi.org/10.1016/bs.apcsb.2020.12.001

Manach C, Scalbert A, Morand C, Rémésy C, Jiménez L (2004) Polyphenols: food sources and bioavailability. Am J Clin Nutr 79:727–747. https://doi.org/10.1093/ajcn/79.5.727

Marcelino P, Marinho HS, Campos MC, Neves AR, Real C, Fontes FS, Carvalho A, Feio G, Martins MBF, Corvo ML (2017) Therapeutic activity of superoxide dismutase-containing enzymosomes on rat liver ischaemia-reperfusion injury followed by magnetic resonance microscopy. Eur J Pharm Sci 109:464–471. https://doi.org/10.1016/j.ejps.2017.09.008

Marnett LJ (1999) Lipid peroxidation—DNA damage by malondialdehyde. Mutat Res 424:83–95. https://doi.org/10.1016/S0027-5107(99)00010-X

Martemucci G, Costagliola C, Mariano M, D'andrea L, Napolitano P, D'Alessandro AG (2022) Free radical properties, source and targets, antioxidant consumption and health. Oxygen 2:48–78. https://doi.org/10.3390/oxygen2020006

Mehaffey JH, Money D, Charles EJ, Schubert S, Piñeros AF, Wu D, Bontha SV, Hawkins R, Teman NR, Laubach VE, Mas VR, Tribble CG, Maluf DG, Sharma AK, Yang Z, Kron IL, Roeser ME (2019) Adenosine 2A receptor activation attenuates ischemia reperfusion injury during extracorporeal cardiopulmonary resuscitation. Ann Surg 269:1176–1183. https://doi.org/10.1097/SLA.0000000000002685

Mihaylova MM, Shaw RJ (2011) The AMPK signalling pathway coordinates cell growth, autophagy and metabolism. Nat Cell Biol 13:1016–1023. https://doi.org/10.1038/ncb2329

Mohamed RA, Agha AM, Nassar NN (2012) SCH58261 the selective adenosine a 2A receptor blocker modulates ischemia reperfusion injury following bilateral carotid occlusion: role of inflammatory mediators. Neurochem Res 37:538–547. https://doi.org/10.1007/s11064-011-0640-x

Mohamed RA, Agha AM, Abdel-Rahman AA, Nassar NN (2016) Role of adenosine A2A receptor in cerebral ischemia reperfusion injury: signaling to phosphorylated extracellular signal-regulated protein kinase (pERK1/2). Neuroscience 314:145–159. https://doi.org/10.1016/j.neuroscience.2015.11.059

Molavian H, Madani Tonekaboni A, Kohandel M, Sivaloganathan S (2015) The synergetic coupling among the cellular antioxidants glutathione peroxidase/peroxiredoxin and other antioxidants and its effect on the concentration of H2O2. Sci Rep 5(1):13620. https://doi.org/10.1038/srep13620

Morgan MJ, Liu ZG (2011) Crosstalk of reactive oxygen species and NF-κB signaling. Cell Res 21:103–115. https://doi.org/10.1038/cr.2010.178

Murotomi K, Umeno A, Sugino S, Yoshida Y (2020) Quantitative kinetics of intracellular singlet oxygen generation using a fluorescence probe. Sci Rep 10:10616. https://doi.org/10.1038/s41598-020-67155-7

Murphy MP (2009) How mitochondria produce reactive oxygen species. Biochem J 417:1–13. https://doi.org/10.1042/BJ20081386

Nagata Y, Fujimoto M, Nakamura K, Isoyama N, Matsumura M, Fujikawa K, Uchiyama K, Takaki E, Takii R, Nakai A, Matsuyama H (2016) Anti-TNF-α agent infliximab and splenectomy are protective against renal ischemia-reperfusion injury. Transplantation 100:1675–1682. https://doi.org/10.1097/TP.0000000000001222

Naito H, Nojima T, Fujisaki N, Tsukahara K, Yamamoto H, Yamada T, Aokage T, Yumoto T, Osako T, Nakao A (2020) Therapeutic strategies for ischemia reperfusion injury in emergency medicine. Acute Med Surg 7:e501. https://doi.org/10.1002/ams2.501

Naoumov NV (2014) Cyclophilin inhibition as potential therapy for liver diseases. J Hepatol 61:1166–1174. https://doi.org/10.1016/j.jhep.2014.07.008

Nguemeni C, Delplanque B, Rovère C, Simon-Rousseau N, Gandin C, Agnani G, Nahon JL, Heurteaux C, Blondeau N (2010) Dietary supplementation of alpha-linolenic acid in an enriched rapeseed oil diet protects from stroke. Pharmacol Res 61:226–233. https://doi.org/10.1016/j.phrs.2009.12.007

Numata G, Takimoto E (2022) Cyclic GMP and PKG signaling in heart failure. Front Pharmacol 13:792798. https://doi.org/10.3389/fphar.2022.792798

Ohana G, Cohen S, Rathwolfson L, Fishman P (2016) A3 adenosine receptor agonist, CF102, protects against hepatic ischemia/reperfusion injury following partial hepatectomy. Mol Med Rep 14:4335–4341. https://doi.org/10.3892/mmr.2016.5746

Okabe N, Nakamura T, Toyoshima T, Miyamoto O, Lu F, Itano T (2011) Eicosapentaenoic acid prevents memory impairment after ischemia by inhibiting inflammatory response and oxidative damage. J Stroke Cerebrovasc Dis 20:188–195. https://doi.org/10.1016/j.jstrokecerebrovasdis.2009.11.016

Olsson MG, Allhorn M, Bülow L, Hansson SR, Ley D, Olsson ML, Schmidtchen A, Åkerström B (2012) Pathological conditions involving extracellular hemoglobin: molecular mechanisms, clinical significance, and novel therapeutic opportunities for α1-microglobulin. Antioxid Redox Signal 17:813–846. https://doi.org/10.1089/ars.2011.4282

Orr AL, Vargas L, Turk CN, Baaten JE, Matzen JT, Dardov VJ, Attle SJ, Li J, Quackenbush DC, Goncalves RLS, Perevoshchikova IV, Petrassi HM, Meeusen SL, Ainscow EK, Brand MD (2015) Suppressors of superoxide production from mitochondrial complex III. Nat Chem Biol 11:834–836. https://doi.org/10.1038/nchembio.1910

Ossola B, Kääriäinen TM, Männistö PT (2009) The multiple faces of quercetin in neuroprotection. Expert Opin Drug Saf 8:397–409. https://doi.org/10.1517/14740330903026944

Ottani F, Latini R, Staszewsky L, La Vecchia L, Locuratolo N, Sicuro M, Masson S, Barlera S, Milani V, Lombardi M, Costalunga A, Mollichelli N, Santarelli A, De Cesare N, Sganzerla P, Boi A, Maggioni AP, Limbruno U (2016) Cyclosporine a in reperfused myocardial infarction the multicenter, controlled, open-label CYCLE trial. J Am Coll Cardiol 67:365–374. https://doi.org/10.1016/j.jacc.2015.10.081

Paez DT, Garces M, Calabró V, Bin EP, D'Annunzio V, Del Mauro J, Marchini T, Höcht C, Evelson P, Gelpi RJ, Donato M (2019) Adenosine A(1) receptors and mitochondria: targets of remote ischemic preconditioning. Am J Physiol Heart Circ Physiol 316:H743–H750. https://doi.org/10.1152/ajpheart.00071.2018

Pan HC, Kao TK, Ou YC, Yang DY, Yen YJ, Wang CC, Chuang YH, Liao SL, Raung SL, Wu CW, Chiang AN, Chen CJ (2009) Protective effect of docosahexaenoic acid against brain injury in ischemic rats. J Nutr Biochem 20:715–725. https://doi.org/10.1016/j.jnutbio.2008.06.014

Panisello-Roselló A, Lopez A, Folch-Puy E, Carbonell T, Rolo A, Palmeira C, Adam R, Net M, Roselló-Catafau J (2018) Role of aldehyde dehydrogenase 2 in ischemia reperfusion injury: an update. World J Gastroenterol 24:2984–2994. https://doi.org/10.3748/wjg.v24.i27.2984

Panth N, Paudel KR, Parajuli K (2016) Reactive oxygen species: a key hallmark of cardiovascular disease. Adv Med 2016:1–12. https://doi.org/10.1155/2016/9152732

Parzych KR, Klionsky DJ (2014) An overview of autophagy: morphology, mechanism, and regulation. Antioxid Redox Signal 20:460–473. https://doi.org/10.1089/ars.2013.5371

Pasut G, Panisello A, Folch-Puy E, Lopez A, Castro-Benítez C, Calvo M, Carbonell T, García-Gil A, Adam R, Roselló-Catafau J (2016) Polyethylene glycols: an effective strategy for limiting liver ischemia reperfusion injury. World J Gastroenterol 22:6501–6508. https://doi.org/10.3748/wjg.v22.i28.6501

Patel R, Rinker L, Peng J, Chilian WM (2018) Reactive oxygen species: the good and the bad. In: Filip C, Albu E (eds) Reactive oxygen species (ROS) in living cells. IntechOpen, Rijeka. https://doi.org/10.5772/intechopen.71547

Pell VR, Chouchani ET, Frezza C, Murphy MP, Krieg T (2016) Succinate metabolism: a new therapeutic target for myocardial reperfusion injury. Cardiovasc Res 111:134–141. https://doi.org/10.1093/cvr/cvw100

Peralta C, Hotter G, Closa D, Gelpí E, Bulbena O, Roselló-Catafau J (1997) Protective effect of preconditioning on the injury associated to hepatic ischemia-reperfusion in the rat: role of nitric oxide and adenosine. Hepatology 25:934–937. https://doi.org/10.1002/hep.510250424

Perkins A, Nelson KJ, Parsonage D, Poole LB, Karplus PA (2015) Peroxiredoxins: guardians against oxidative stress and modulators of peroxide signaling. Trends Biochem Sci 40:435–445. https://doi.org/10.1016/j.tibs.2015.05.001

Perluigi M, Coccia R, Allan Butterfield D (2012) 4-hydroxy-2-nonenal, a reactive product of lipid peroxidation, and neurodegenerative diseases: a toxic combination illuminated by redox proteomics studies. Antioxid Redox Signal 17:1590–1609. https://doi.org/10.1089/ars.2011.4406

Pinto A, Immohr MB, Jahn A, Jenke A, Boeken U, Lichtenberg A, Akhyari P (2016) The extracellular isoform of superoxide dismutase has a significant impact on cardiovascular ischaemia and reperfusion injury during cardiopulmonary bypass. Eur J Cardiothorac Surg 50:1035–1044. https://doi.org/10.1093/ejcts/ezw216

Pithadia AB, Panchal SS, Patel DJ (2017) Neuroprotective effects of potassium channel openers on cerebral ischemia–reperfusion injury in diabetic rats. Bull Fac Pharm Cairo Univ 55:95–100. https://doi.org/10.1016/j.bfopcu.2016.09.002

Pizzino G, Irrera N, Cucinotta M, Pallio G, Mannino F, Arcoraci V, Squadrito F, Altavilla D, Bitto A (2017) Oxidative stress: harms and benefits for human health. Oxidative Med Cell Longev 2017:8416763. https://doi.org/10.1155/2017/8416763

Poetsch AR (2020) The genomics of oxidative DNA damage, repair, and resulting mutagenesis. Comput Struct Biotechnol J 18:207–219. https://doi.org/10.1016/j.csbj.2019.12.013

Poljšak B, Fink R (2014) The protective role of antioxidants in the defence against ROS/RNS-mediated environmental pollution. Oxidative Med Cell Longev 2014:671539. https://doi.org/10.1155/2014/671539

Rauf A, Imran M, Sulera HAR, Ahmad B, Peters DG, Mubarak MS (2017) A comprehensive review of the health perspectives of resveratrol. Food Funct 8:4284–4305. https://doi.org/10.1039/c7fo01300k

Rauf A, Khalil AA, Awadallah S, Khan SA, Abu-Izneid T, Kamran M, Hemeg HA, Mubarak MS, Khalid A, Wilairatana P (2024) Reactive oxygen species in biological systems: pathways, associated diseases, and potential inhibitors—a review. Food Sci Nutr 12:675–693. https://doi.org/10.1002/fsn3.3784

Ray PD, Huang BW, Tsuji Y (2012) Reactive oxygen species (ROS) homeostasis and redox regulation in cellular signaling. Cell Signal 24:981–990. https://doi.org/10.1016/j.cellsig.2012.01.008

Ren J, Bi Y, Sowers JR, Hetz C, Zhang Y (2021) Endoplasmic reticulum stress and unfolded protein response in cardiovascular diseases. Nat Rev Cardiol 18:499–521. https://doi.org/10.1038/s41569-021-00511-w

Revuelta M, Arteaga O, Montalvo H, Alvarez A, Hilario E, Martinez-Ibargüen A (2016) Antioxidant treatments recover the alteration of auditory-evoked potentials and reduce morphological damage in the inferior colliculus after perinatal asphyxia in rat. Brain Pathol 26:186–198. https://doi.org/10.1111/bpa.12272

Rivera F, Urbanavicius J, Gervaz E, Morquio A, Dajas F (2004) Some aspects of the in vivo neuroprotective capacity of flavonoids: bioavailability and structure-activity relationship. Neurotox Res 6:543–553. https://doi.org/10.1007/BF03033450

Rossello X, Yellon DM (2018) The RISK pathway and beyond. Basic Res Cardiol 113:2. https://doi.org/10.1007/s00395-017-0662-x

Roy S, Šileikyte J, Schiavone M, Neuenswander B, Argenton F, Aubé J, Hedrick MP, Chung TDY, Forte MA, Bernardi P, Schoenen FJ (2015) Discovery, synthesis, and optimization of diarylisoxazole-3-carboxamides as potent inhibitors of the mitochondrial permeability transition pore. ChemMedChem 10:1655–1671. https://doi.org/10.1002/cmdc.201500284

Salloum FN, Takenoshita Y, Ockaili RA, Daoud VP, Chou E, Yoshida K i, Kukreja RC (2007) Sildenafil and vardenafil but not nitroglycerin limit myocardial infarction through opening of mitochondrial KATP channels when administered at reperfusion following ischemia in rabbits. J Mol Cell Cardiol 42:453–458. https://doi.org/10.1016/j.yjmcc.2006.10.015

Sánchez-Hernández CD, Torres-Alarcón LA, González-Cortés A, Peón AN (2020) Ischemia/reperfusion injury: pathophysiology, current clinical management, and potential preventive approaches. Mediat Inflamm 2020:8405370. https://doi.org/10.1155/2020/8405370

Sarniak A, Lipińska J, Tytman K, Lipińska S (2016) Endogenous mechanisms of reactive oxygen species (ROS) generation. Postepy Hig Med Dosw (Online) 70:1150–1165. https://doi.org/10.5604/17322693.1224259

Schaller S, Paradis S, Ngoh GA, Assaly R, Buisson B, Drouot C, Ostuni MA, Lacapere JJ, Bassissi F, Bordet T, Berdeaux A, Jones SP, Morin D, Pruss RM (2010) TRO40303, a new cardioprotective compound, inhibits mitochondrial permeability transition. J Pharmacol Exp Ther 333:696–706. https://doi.org/10.1124/jpet.110.167486

Schaur RJ, Siems W, Bresgen N, Eckl PM (2015) 4-Hydroxy-Nonenal—a bioactive lipid peroxidation product. Biomolecules 5:2247–2337. https://doi.org/10.3390/biom5042247

Schinzel AC, Takeuchi O, Huang Z, Fisher JK, Zhou Z, Rubens J, Hetz C, Danial NN, Moskowitz MA, Korsmeyer SJ (2005) Cyclophilin D is a component of mitochondrial permeability transition and mediates neuronal cell death after focal cerebral ischemia. Proc Natl Acad Sci USA 102:12005–12010. https://doi.org/10.1073/pnas.0505294102

Schulman D, Latchman DS, Yellon DM (2002) Urocortin protects the heart from reperfusion injury via upregulation of p42/p44 MAPK signaling pathway. Am J Physiol Heart Circ Physiol 283: H1481–H1488. https://doi.org/10.1152/ajpheart.01089.2001

Schwarz DS, Blower MD (2016) The endoplasmic reticulum: structure, function and response to cellular signaling. Cell Mol Life Sci 73:79–94. https://doi.org/10.1007/s00018-015-2052-6

Scorletti E, Byrne CD (2013) Omega-3 fatty acids, hepatic lipid metabolism, and nonalcoholic fatty liver disease. Annu Rev Nutr 33:231–248. https://doi.org/10.1146/annurev-nutr-071812-161230

Sedighi M, Sewell RDE, Nazari A, Abbaszadeh S, Cheraghi M, Amini A, Heydari Z, Rafieian-Kopaei M (2019) A review on the most important medicinal plants effective in cardiac ischemia-reperfusion injury. Curr Pharm Des 25:352–358. https://doi.org/10.2174/1381612825666190329144016

Shah FA, Albaqami F, Alattar A, Alshaman R, Zaitone SA, Gabr AM, Abdel-Moneim AMH, El dosoky M, Koh PO (2024) Quercetin attenuated ischemic stroke induced neurodegeneration by modulating glutamatergic and synaptic signaling pathways. Heliyon 10:e28016. https://doi.org/10.1016/j.heliyon.2024.e28016

Sharma P, Jha AB, Dubey RS, Pessarakli M (2012) Reactive oxygen species, oxidative damage, and antioxidative defense mechanism in plants under stressful conditions. J Bot 2012:1–26. https://doi.org/10.1155/2012/217037

Sharma R, Selvam MKP, Agarwal A (2018) Reactive oxygen species methodology using chemiluminescence assay. In: Henkel R, Samanta L, Agarwal Antioxidants and Impact of the Oxidative Status in Male Reproduction, A.B.T.-O (eds) Oxidants, antioxidants, and impact of the oxidative status in male reproduction. Academic, pp 183–193. https://doi.org/10.1016/B978-0-12-812501-4.00017-1

Shi B, Ma M, Zheng Y, Pan Y, Lin X (2019) mTOR and Beclin1: two key autophagy-related molecules and their roles in myocardial ischemia/reperfusion injury. J Cell Physiol 234:12562–12568. https://doi.org/10.1002/jcp.28125

Shibata R, Sato K, Pimentel DR, Takemura Y, Kihara S, Ohashi K, Funahashi T, Ouchi N, Walsh K (2005) Adiponectin protects against myocardial ischemia-reperfusion injury through AMPK- and COX-2-dependent mechanisms. Nat Med 11:1096–1103. https://doi.org/10.1038/nm1295

Shimizu S, Saito M, Kinoshita Y, Ohmasa F, Dimitriadis F, Shomori K, Hayashi A, Satoh K (2011) Nicorandil ameliorates ischaemia-reperfusion injury in the rat kidney. Br J Pharmacol 163:272–282. https://doi.org/10.1111/j.1476-5381.2011.01231.x

Shiva S, Sack MN, Greer JJ, Duranski M, Ringwood LA, Burwell L, Wang X, MacArthur PH, Shoja A, Raghavachari N, Calvert JW, Brookes PS, Lefer DJ, Gladwin MT (2007) Nitrite augments tolerance to ischemia/reperfusion injury via the modulation of mitochondrial electron transfer. J Exp Med 204:2089–2102. https://doi.org/10.1084/jem.20070198

Sies H (2020) Oxidative stress: concept and some practical aspects. Antioxidants 9:1–6. https://doi.org/10.3390/antiox9090852

Sies H, Jones DP (2020) Reactive oxygen species (ROS) as pleiotropic physiological signalling agents. Nat Rev Mol Cell Biol 21:363–383. https://doi.org/10.1038/s41580-020-0230-3

Silvestrini A, Meucci E, Ricerca BM, Mancini A (2023) Total antioxidant capacity: biochemical aspects and clinical significance. Int J Mol Sci 24(13):10978. https://doi.org/10.3390/ijms241310978

Singh A, Kukreti R, Saso L, Kukreti S (2019) Oxidative stress: a key modulator in neurodegenerative diseases. Molecules 24(8):1583. https://doi.org/10.3390/molecules24081583

Singhal AK, Symons JD, Boudina S, Jaishy B, Shiu Y-TE (2014) Role of endothelial cells in myocardial ischemia-reperfusion injury. Vasc Dis Prev 7:1–14. https://doi.org/10.2174/1874120701007010001

Snezhkina AV, Kudryavtseva AV, Kardymon OL, Savvateeva MV, Melnikova NV, Krasnov GS, Dmitriev AA (2020) ROS generation and antioxidant defense systems in normal and malignant cells. Oxidative Med Cell Longev 2019:6175804. https://doi.org/10.1155/2019/6175804

Soliman S, Ishrat T, Fouda AY, Patel A, Pillai B, Fagan SC (2015) Sequential therapy with minocycline and candesartan improves long-term recovery after experimental stroke. Transl Stroke Res 6:309–322. https://doi.org/10.1007/s12975-015-0408-8

Solleiro-Villavicencio H, Rivas-Arancibia S (2018) Effect of chronic oxidative stress on neuroinflammatory response mediated by CD4+T cells in neurodegenerative diseases. Front Cell Neurosci 12:114. https://doi.org/10.3389/fncel.2018.00114

Stockwell BR, Friedmann Angeli JP, Bayir H, Bush AI, Conrad M, Dixon SJ, Fulda S, Gascón S, Hatzios SK, Kagan VE, Noel K, Jiang X, Linkermann A, Murphy ME, Overholtzer M, Oyagi A, Pagnussat GC, Park J, Ran Q, Rosenfeld CS, Salnikow K, Tang D, Torti FM, Torti SV, Toyokuni S, Woerpel KA, Zhang DD (2017) Ferroptosis: a regulated cell death nexus linking metabolism, redox biology, and disease. Cell 171:273–285. https://doi.org/10.1016/j.cell.2017.09.021

St-Pierre J, Buckingham JA, Roebuck SJ, Brand MD (2002) Topology of superoxide production from different sites in the mitochondrial electron transport chain. J Biol Chem 277:44784–44790. https://doi.org/10.1074/jbc.M207217200

Su LJ, Zhang JH, Gomez H, Murugan R, Hong X, Xu D, Jiang F, Peng ZY (2019) Reactive oxygen species-induced lipid peroxidation in apoptosis, autophagy, and ferroptosis. Oxidative Med Cell Longev 2019:5080843. https://doi.org/10.1155/2019/5080843

Suyavaran A, Ramamurthy C, Mareeswaran R, Subastri A, Lokeswara Rao P, Thirunavukkarasu C (2015) TNF-α suppression by glutathione preconditioning attenuates hepatic ischemia reperfusion injury in young and aged rats. Inflamm Res 64:71–81. https://doi.org/10.1007/s00011-014-0785-6

Sweitzer NK, Airhart S (2018) New therapeutic target in heart failure: achieving and maintaining normokalemia. Circulation 137(13):1331–1333. https://doi.org/10.1161/CIRCULATIONAHA.118.033854

Tartaglia LA, Weber RF, Figari IS, Reynolds C, Palladino MA, Goeddel DV (1991) The two different receptors for tumor necrosis factor mediate distinct cellular responses. Proc Natl Acad Sci USA 88:9292–9296. https://doi.org/10.1073/pnas.88.20.9292

Tasdemir C, Tasdemir S, Vardi N, Ates B, Parlakpinar H, Kati B, Karaaslan MG, Acet A (2012) Protective effect of infliximab on ischemia/reperfusion-induced damage in rat kidney. Ren Fail 34:1144–1149. https://doi.org/10.3109/0886022X.2012.717490

Theofilis P, Sagris M, Oikonomou E, Antonopoulos AS, Siasos G, Tsioufis C, Tousoulis D (2021) Inflammatory mechanisms contributing to endothelial dysfunction. Biomedicine 9(7):781. https://doi.org/10.3390/biomedicines9070781

Tian B, Liu J (2020) Resveratrol: a review of plant sources, synthesis, stability, modification and food application. J Sci Food Agric 100:1392–1404. https://doi.org/10.1002/jsfa.10152

Tsikas D (2017) Assessment of lipid peroxidation by measuring malondialdehyde (MDA) and relatives in biological samples: analytical and biological challenges. Anal Biochem 524:13–30. https://doi.org/10.1016/j.ab.2016.10.021

Tvrdá E, Benko F (2020) Free radicals: what they are and what they do. In: Preedy VR (ed) Pathology: oxidative stress and dietary antioxidants. Academic, pp 3–13. https://doi.org/10.1016/B978-0-12-815972-9.00001-9

Ueda M, Inaba T, Nito C, Kamiya N, Katayama Y (2013) Therapeutic impact of eicosapentaenoic acid on ischemic brain damage following transient focal cerebral ischemia in rats. Brain Res 1519:95–104. https://doi.org/10.1016/j.brainres.2013.04.046

Valgimigli L (2023) Lipid peroxidation and antioxidant protection. Biomolecules 13(9):1291. https://doi.org/10.3390/biom13091291

Valls-Lacalle L, Barba I, Miró-Casas E, Alburquerque-Béjar JJ, Ruiz-Meana M, Fuertes-Agudo M, Rodríguez-Sinovas A, García-Dorado D (2016) Succinate dehydrogenase inhibition with malonate during reperfusion reduces infarct size by preventing mitochondrial permeability transition. Cardiovasc Res 109:374–384. https://doi.org/10.1093/cvr/cvv279

Valls-Lacalle L, Barba I, Miró-Casas E, Ruiz-Meana M, Rodríguez-Sinovas A, García-Dorado D (2018) Selective inhibition of succinate dehydrogenase in reperfused myocardium with intracoronary malonate reduces infarct size. Sci Rep 8:2442. https://doi.org/10.1038/s41598-018-20866-4

Veith A, Moorthy B (2018) Role of cytochrome P450s in the generation and metabolism of reactive oxygen species. Curr Opin Toxicol 7:44–51. https://doi.org/10.1016/j.cotox.2017.10.003

Vitale G, Salvioli S, Franceschi C (2013) Oxidative stress and the ageing endocrine system. Nat Rev Endocrinol 9:228–240. https://doi.org/10.1038/nrendo.2013.29

Walters AM, Porter GA, Brookes PS (2012) Mitochondria as a drug target in ischemic heart disease and cardiomyopathy. Circ Res 111:1222–1236. https://doi.org/10.1161/CIRCRESAHA.112.265660

Wan TC, Tampo A, Kwok WM, Auchampach JA (2019) Ability of CP-532,903 to protect mouse hearts from ischemia/reperfusion injury is dependent on expression of A 3 adenosine receptors in cardiomyoyctes. Biochem Pharmacol 163:21–31. https://doi.org/10.1016/j.bcp.2019.01.022

Wang M, Kaufman RJ (2016) Protein misfolding in the endoplasmic reticulum as a conduit to human disease. Nature 529:326–335. https://doi.org/10.1038/nature17041

Wang Q, Zou MH (2018) Measurement of reactive oxygen species (ROS) and mitochondrial ROS in AMPK knockout mice blood vessels. Methods Mol Biol 1732:507–517. https://doi.org/10.1007/978-1-4939-7598-3_32

Wang Q, Lin P, Li P, Feng L, Ren Q, Xie X, Xu J (2017) Ghrelin protects the heart against ischemia/reperfusion injury via inhibition of TLR4/NLRP3 inflammasome pathway. Life Sci 186:50–58. https://doi.org/10.1016/j.lfs.2017.08.004

Wang YY, Chang CY, Lin SY, Der Wang J, Wu CC, Chen WY, Kuan YH, Liao SL, Wang WY, Chen CJ (2020) Quercetin protects against cerebral ischemia/reperfusion and oxygen glucose deprivation/reoxygenation neurotoxicity. J Nutr Biochem 83:108436. https://doi.org/10.1016/j.jnutbio.2020.108436

Wang L, Liu Y, Zhang X, Ye Y, Xiong X, Zhang S, Gu L, Jian Z, Wang H (2022) Endoplasmic reticulum stress and the unfolded protein response in cerebral ischemia/reperfusion injury. Front Cell Neurosci 16:864426. https://doi.org/10.3389/fncel.2022.864426

Weber D, Davies MJ, Grune T (2015) Determination of protein carbonyls in plasma, cell extracts, tissue homogenates, isolated proteins: focus on sample preparation and derivatization conditions. Redox Biol 5:367–380. https://doi.org/10.1016/j.redox.2015.06.005

Weng XF, Li ST, Song Q, Zhu Q, Song DD, Qin ZH, Xie Y (2018) Protective effect of nicotinamide adenine dinucleotide phosphate on renal ischemia-reperfusion injury. Kidney Blood Press Res 43:651–663. https://doi.org/10.1159/000489620

West T, Atzeva M, Holtzman DM (2007) Pomegranate polyphenols and resveratrol protect the neonatal brain against hypoxic-ischemic injury. Dev Neurosci 29:363–372. https://doi.org/10.1159/000105477

Wieland E, Schütz E, Armstrong VW, Küthe F, Heller C, Oellerich M (1995) Idebenone protects hepatic microsomes against oxygen radical-mediated damage in organ preservation solutions. Transplantation 60:444–451. https://doi.org/10.1097/00007890-199509000-00007

Williams JJ, Mayurasakorn K, Vannucci SJ, Mastropietro C, Bazan NG, Ten VS, Deckelbaum RJ (2013) N-3 fatty acid rich triglyceride emulsions are neuroprotective after cerebral hypoxic-ischemic injury in neonatal mice. PLoS One 8:e56233. https://doi.org/10.1371/journal.pone.0056233

Wojtovich AP, Brookes PS (2008) The endogenous mitochondrial complex II inhibitor malonate regulates mitochondrial ATP-sensitive potassium channels: implications for ischemic preconditioning. Biochim Biophys Acta Bioenerg 1777:882–889. https://doi.org/10.1016/j.bbabio.2008.03.025

Wojtovich AP, Brookes PS (2009) The complex II inhibitor atpenin A5 protects against cardiac ischemia-reperfusion injury via activation of mitochondrial KATP channels. Basic Res Cardiol 104:121–129. https://doi.org/10.1007/s00395-009-0001-y

Wu H, Tang C, Tai LW, Yao W, Guo P, Hong J, Yang X, Li X, Jin Z, Ke J, Wang Y (2018a) Flurbiprofen axetil attenuates cerebral ischemia/reperfusion injury by reducing inflammation in a rat model of transient global cerebral ischemia/reperfusion. Biosci Rep 38(4):BSR20171562. https://doi.org/10.1042/BSR20171562

Wu MY, Yiang GT, Liao WT, Tsai APY, Cheng YL, Cheng PW, Li CY, Li CJ (2018b) Current mechanistic concepts in ischemia and reperfusion injury. Cell Physiol Biochem 46:1650–1667. https://doi.org/10.1159/000489241

Wu M, Liu F, Guo Q (2019) Quercetin attenuates hypoxia-ischemia induced brain injury in neonatal rats by inhibiting TLR4/NF-κB signaling pathway. Int Immunopharmacol 74: 105704. https://doi.org/10.1016/j.intimp.2019.105704

Xu Z, Yu J, Wu J, Qi F, Wang H, Wang Z, Wang Z (2016) The effects of two anesthetics, propofol and sevoflurane, on liver ischemia/reperfusion injury. Cell Physiol Biochem 38:1631–1642. https://doi.org/10.1159/000443103

Xu HX, Cui SM, Zhang YM, Ren J (2020) Mitochondrial Ca2+ regulation in the etiology of heart failure: physiological and pathophysiological implications. Acta Pharmacol Sin 41:1301–1309. https://doi.org/10.1038/s41401-020-0476-5

Yamanaka K, Eldeiry M, Aftab M, Mares J, Ryan TJ, Meng X, Weyant MJ, Cleveland JC, Fullerton DA, Reece TB (2018) Optimized induction of beta common receptor enhances the neuroprotective function of erythropoietin in spinal cord ischemic injury. J Thorac Cardiovasc Surg 155:2505–2516. https://doi.org/10.1016/j.jtcvs.2017.12.132

Yamanaka K, Eldeiry M, Aftab M, Ryan TJ, Roda G, Meng X, Weyant MJ, Cleveland JC, Fullerton DA, Reece TB (2019) Pretreatment with diazoxide attenuates spinal cord ischemia-reperfusion injury through signaling transducer and activator of transcription 3 pathway. Ann Thorac Surg 107:733–739. https://doi.org/10.1016/j.athoracsur.2018.09.031

Yang S, Chou WP, Pei L (2013) Effects of propofol on renal ischemia/reperfusion injury in rats. Exp Ther Med 6:1177–1183. https://doi.org/10.3892/etm.2013.1305

Yang Q, He GW, Underwood MJ, Yu CM (2016) Cellular and molecular mechanisms of endothelial ischemia/reperfusion injury: perspectives and implications for postischemic myocardial protection. Am J Transl Res 8:765–777

Yang Y, Lu D, Wang M, Liu G, Feng Y, Ren Y, Sun X, Chen Z, Wang Z (2024) Endoplasmic reticulum stress and the unfolded protein response: emerging regulators in progression of traumatic brain injury. Cell Death Dis 15:156. https://doi.org/10.1038/s41419-024-06515-x

Yapca OE, Borekci B, Suleyman H (2013) Ischemia-reperfusion damage. Eurasian J Med 45:126–127. https://doi.org/10.5152/eajm.2013.24

Yau JW, Teoh H, Verma S (2015) Endothelial cell control of thrombosis. BMC Cardiovasc Disord 15:130. https://doi.org/10.1186/s12872-015-0124-z

Yu Y, Yan Y, Niu F, Wang Y, Chen X, Su G, Liu Y, Zhao X, Qian L, Liu P, Xiong Y (2021a) Ferroptosis: a cell death connecting oxidative stress, inflammation and cardiovascular diseases. Cell Death Discov 7:193. https://doi.org/10.1038/s41420-021-00579-w

Yu P, Zhang X, Liu N, Tang L, Peng C, Chen X (2021b) Pyroptosis: mechanisms and diseases. Signal Transduct Target Ther 6:128. https://doi.org/10.1038/s41392-021-00507-5

Zeylemaker WP, Slater EC (1967) The inhibition of succinate dehydrogenase by oxaloacetate. Biochim Biophys Acta 132:210–212. https://doi.org/10.1016/0005-2744(67)90214-8

Zhang W, Hu X, Yang W, Gao Y, Chen J (2010) Omega-3 polyunsaturated fatty acid supplementation confers long-term neuroprotection against neonatal hypoxic-ischemic brain injury through anti-inflammatory actions. Stroke 41:2341–2347. https://doi.org/10.1161/STROKEAHA.110.586081

Zhang W, Liu J, Hu X, Li P, Leak RK, Gao Y, Chen J (2015) N-3 polyunsaturated fatty acids reduce neonatal hypoxic/ischemic brain injury by promoting phosphatidylserine formation and akt signaling. Stroke 46:2943–2950. https://doi.org/10.1161/STROKEAHA.115.010815

Zhang M, Liu Q, Meng H, Duan H, Liu X, Wu J, Gao F, Wang S, Tan R, Yuan J (2024) Ischemia-reperfusion injury: molecular mechanisms and therapeutic targets. Signal Transduct Target Ther 9:12. https://doi.org/10.1038/s41392-023-01688-x

Zhao L, Donaldson CJ, Smith GW, Vale WW (1998) The structures of the mouse and human urocortin genes (Ucn and UCN). Genomics 50:23–33. https://doi.org/10.1006/geno.1998.5292

Zhao K, Chen X, Bian Y, Zhou Z, Wei X, Zhang J (2023) Broadening horizons: the role of ferroptosis in myocardial ischemia-reperfusion injury. Naunyn Schmiedeberg's Arch Pharmacol 396:2269–2286. https://doi.org/10.1007/s00210-023-02506-5

Zhou H, Ma Q, Zhu P, Ren J, Reiter RJ, Chen Y (2018) Protective role of melatonin in cardiac ischemia-reperfusion injury: from pathogenesis to targeted therapy. J Pineal Res 64(3):e12471. https://doi.org/10.1111/jpi.12471

Zong Y, Li H, Liao P, Chen L, Pan Y, Zheng Y, Zhang C, Liu D, Zheng M, Gao J (2024) Mitochondrial dysfunction: mechanisms and advances in therapy. Signal Transduct Target Ther 9:124. https://doi.org/10.1038/s41392-024-01839-8

Chapter 10
Oxidative Stress in Neurodegeneration: Insights into Alzheimer's, Parkinson's, Huntington's, and Multiple Sclerosis

Sandeep Bolla, Shashank Reddy Pasika, Hari P. Kandagatla, and Deepika Godugu

Abbreviations

Aβ	Amyloid-Beta
AD	Alzheimer's disease
ALA	Alpha-Lipoic Acid
AMPK	AMP-activated protein kinase
APP	Amyloid Precursor Protein
AOPP	Advanced Oxidation Protein Products
ATP	Adenosine Triphosphate
BACE1	Beta-Site APP-Cleaving Enzyme 1
BBB	Blood-Brain Barrier
CAT	Catalase
CNS	Central Nervous System
CoQ10	Coenzyme Q10
COPD	Chronic Obstructive Pulmonary Disease
Cu^{2*}	Copper Ion
DMTs	Disease-Modifying Therapies
DNA	Deoxyribonucleic Acid

S. Bolla
Department of Biometrics, Fortrea, Durham, NC, USA

S. R. Pasika
Department of Biotechnology, National Institute of Pharmaceutical Education and Research-Raebareli (NIPER-R), Raebareli, India

College of Pharmacy and Health Sciences, St John's University, Queens, NY, USA

H. P. Kandagatla · D. Godugu (✉)
College of Pharmacy and Health Sciences, St John's University, Queens, NY, USA

N. Srivastava, S. A. Saraf (eds.), *Pharmacological Strategies for Prevention and Treatment of Oxidative Stress*, Oxidative Stress in Applied Basic Research and Clinical Practice, https://doi.org/10.1007/978-3-032-19837-2_10

DJ-1	Parkinsonism Associated Deglycase
ETC	Electron Transport Chain
Fe^{2*}	Ferrous Ion
GABA	Gamma-Aminobutyric Acid
GAPDH	Glyceraldehyde 3-Phosphate Dehydrogenase
GPx	Glutathione Peroxidase
GSH	Reduced Glutathione
GSSG	Oxidized Glutathione
H_20_2	Hydrogen Peroxide
HAP-1	Huntingtin-Associated Protein 1
HD	Huntington's Disease
Htt	Huntingtin
IFN-γ	Interferon-gamma
IL-17	Interleukin-17
IL-6	Interleukin-6
IV	Intravenous
MAO	Monoamine Oxidase
mHTT	Mutant Huntingtin
MHC	Major Histocompatibility Complex
MRI	Magnetic Resonance Imaging
MS	Multiple Sclerosis
mtDNA	Mitochondrial DNA
NAC	N-Acetylcysteine
NADPH	Nicotinamide Adenine Dinucleotide Phosphate
NF-κB	Nuclear Factor Kappa-light-chain-enhancer of Activated B cells
NOX	NADPH Oxidase
Nrf2	Nuclear Factor Erythroid 2-Related Factor 2
O_2^-	Superoxide Anion
OH^-	Hydroxyl Radical
OS	Oxidative stress
PD	Parkinson's Disease
PEG	Polyethylene Glycol
PGC-1α	Peroxisome Proliferator-Activated Receptor Gamma Coactivator 1-alpha
Prx	Peroxiredoxins
PUFAs	Polyunsaturated Fatty Acids
RNAi	RNA interference
ROS	Reactive oxygen species
SOD	Superoxide Dismutase
SOD1	Cu-Zn Superoxide Dismutase
SOD2	Mn Superoxide Dismutase
SOD3	Extracellular Superoxide Dismutase
TLR	Toll-Like Receptor
TNF-α	Tumor Necrosis Factor-alpha

Trx	Thioredoxin
UV	Ultraviolet
4-HNE	4-Hydroxynonenal
8-oxoG	8-oxoguanine

10.1 Introduction

Oxidative stress (OS) is a condition characterized by an imbalance between the production of reactive oxygen species (ROS) and the ability of the body to neutralize these harmful compounds using antioxidant defense systems (Chaudhary et al. 2023; Fujii et al. 2011). ROS are highly reactive molecules containing oxygen, including free radicals like superoxide anion (O_2^-) and hydroxyl radical (OH^-), as well as non-radical species like hydrogen peroxide (H_2O_2) (Brand 2010; Phaniendra et al. 2015). While ROS play vital roles in physiological processes, excessive production can lead to cellular damage, contributing to various diseases, aging, and even death (Brand 2010; Chaudhary et al. 2023; Phaniendra et al. 2015; Pizzino et al. 2017; Taniyama and Griendling 2003; Valko et al. 2007).

Oxidative stress is implicated in the pathogenesis of numerous chronic conditions, including cardiovascular diseases (Dai et al. 2011), neurodegenerative diseases, diabetes (Anderson et al. 2009), and cancer (Chaudhary et al. 2023; Naito and Yoshikawa 2002; Pizzino et al. 2017; Riley 1994; Smith et al. 2016). Understanding the origins of reactive oxygen species (ROS), the body's protective systems against oxidative damage, and the cellular elements most susceptible to ROS-induced injury is crucial for developing therapeutic approaches to address disorders associated with oxidative stress.

10.2 Sources of ROS That Cause Oxidative Stress

The sources of ROS that cause oxidative stress are two types: endogenous sources and exogenous sources (Fig. 10.1, Table 10.1) (Bhattacharyya et al. 2014; Jomova et al. 2023).

10.2.1 Endogenous Sources

The primary sources of ROS in the body are metabolic processes that occur in the mitochondria (Aparicio-Trejo et al. 2018), peroxisomes (Lismont et al. 2019), and other organelles (Jomova et al. 2023) as follows.

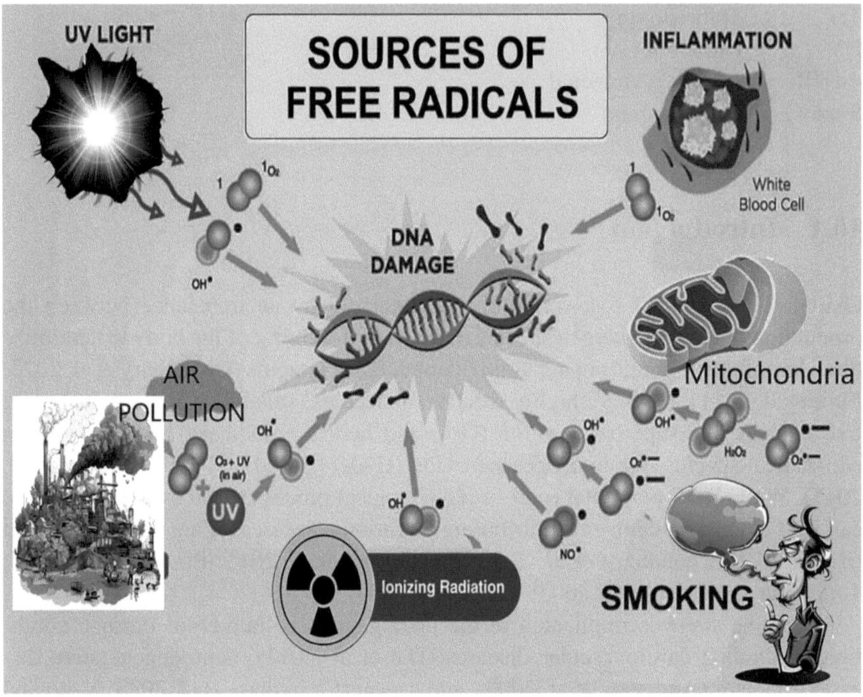

Fig. 10.1 Different types of sources of free radicals are endogenous sources and exogenous sources

Table 10.1 Key sources, types of ROS, and impact of oxidative stress

Source of ROS	Type of ROS Produced	Primary Mechanism	Impact of ROS
Mitochondrial respiration	Superoxide (O_2^-), hydrogen Peroxide (H_2O_2)	Electron leakage during oxidative phosphorylation	Damage to mitochondrial DNA, proteins, and lipids
Peroxisomes	Hydrogen Peroxide (H_2O_2)	Fatty acid β-oxidation	Lipid peroxidation and cellular membrane damage
NADPH oxidase	Superoxide (O_2^-)	ROS generation as a part of the immune response	Oxidative burst leading to inflammation and tissue injury
Iron accumulation	Hydroxyl Radicals (OH^-)	Fenton reaction between iron and H_2O_2	DNA strand breaks, protein oxidation
Pollution and environmental toxins	Hydroxyl Radicals (OH^-), nitrogen species (NO^-)	Exposure to UV, air pollutants, and chemicals	Cellular mutations, chronic diseases like COPD and cancer

10.2.1.1 Mitochondria

The mitochondrial respiratory chain is a major endogenous source of ROS (Zorova et al. 2018). During oxidative phosphorylation, electrons are transferred along the electron transport chain (ETC) to generate ATP (Aparicio-Trejo et al. 2018; Jomova et al. 2023). However, approximately 1–3% of oxygen is incompletely reduced, leading to the generation of superoxide radicals (O_2^-). Superoxide dismutase (SOD) quickly converts these radicals into hydrogen peroxide (H_2O_2), which can then be neutralized by other antioxidant systems (Aparicio-Trejo et al. 2018; Zorova et al. 2018).

10.2.1.2 Peroxisomes

These organelles contribute to ROS production via fatty acid oxidation. Specifically, during β-oxidation, acyl-CoA oxidase generates hydrogen peroxide as a by-product (Jomova et al. 2023; Lismont et al. 2019). If not detoxified, the buildup of H_2O_2 can lead to oxidative stress (Lismont et al. 2019; Schrader and Fahimi 2006; Valko et al. 2007).

10.2.1.3 NADPH Oxidases

These membrane-bound enzymes generate ROS as part of the body's immune response. Activated immune cells, such as neutrophils and macrophages, produce ROS to kill invading pathogens (Mortezaee 2018; Valko et al. 2007). However, excessive ROS release during chronic inflammation can contribute to oxidative damage (Bedard and Krause 2007; Schalk et al. 1996).

10.2.1.4 Enzymatic Reactions

Other enzymatic processes, such as xanthine oxidase activity, cytochrome P450 reactions, and the Fenton reaction (which involves iron and hydrogen peroxide), also produce ROS. In the Fenton reaction, Fe^{2+} reacts with H_2O_2 to generate the highly reactive hydroxyl radical (OH^-), which is one of the most damaging ROS (Jomova et al. 2023; Taniyama and Griendling 2003; Valko et al. 2007).

10.2.2 Exogenous Sources

In addition to endogenous production, ROS can be generated from external environmental factors (Gangwar et al. 2020; Jomova et al. 2023) as follows.

10.2.2.1 Pollution

Air pollutants, such as ozone (O_3), nitrogen dioxide (NO_2), and particulate matter, can generate ROS in the respiratory system, contributing to lung diseases like asthma, chronic obstructive pulmonary disease (COPD), and cancer (Gangwar et al. 2020; Jomova et al. 2023; Mazuryk et al. 2020).

10.2.2.2 Radiation

Ultraviolet (UV) and ionizing radiation (e.g., X-rays) are potent generators of ROS. UV radiation can excite oxygen molecules in the skin, leading to the formation of ROS that can damage DNA and proteins (Jia et al. 2010; Jomova et al. 2023; Riley 1994).

10.2.2.3 Chemicals and Toxins

Exposure to chemicals like tobacco smoke, pesticides, and heavy metals (e.g., lead, mercury, cadmium) can elevate ROS levels, leading to cellular and tissue damage (Yang et al. 2008). Cigarette smoke, for example, contains high levels of free radicals and ROS that overwhelm the body's antioxidant defenses (Jomova et al. 2023; Mazuryk et al. 2020; Yang et al. 2008).

10.2.2.4 Drugs

Certain medications, such as chemotherapy agents and some antibiotics, can induce oxidative stress by promoting the production of ROS (Nuszkiewicz et al. 2020). While this can be beneficial in killing cancer cells or bacteria, it can also result in collateral damage to healthy tissues (Bhattacharyya et al. 2014; Conklin 2004; Jomova et al. 2023).

10.3 Antioxidant Defense Systems

The human body has evolved several antioxidant defense mechanisms to counterbalance the harmful effects of ROS (Valko et al. 2007; Willcox et al. 2004). These systems, which neutralize ROS and repair oxidative damage, are made up of both enzymatic and non-enzymatic antioxidants. Table 10.2 explain the enzymatic and non-enzymatic antioxidant systems.

Table 10.2 Enzymatic and non-enzymatic antioxidant systems

Antioxidant system	Type of antioxidant	Function	Key role in oxidative stress defense
Superoxide dismutase (SOD)	Enzymatic	Converts superoxide radicals (O_2^-) into hydrogen peroxide	Prevents accumulation of superoxide in mitochondria
Catalase	Enzymatic	Converts hydrogen peroxide into oxygen and water	Protects cells from hydrogen peroxide-induced damage
Glutathione peroxidase (GPx)	Enzymatic	Lowers lipid and hydrogen peroxide levels	Inhibits lipid peroxidation, crucial for membrane integrity
Glutathione (GSH)	Non-enzymatic	Scavenges ROS, regenerates other antioxidants	Neutralizes free radicals directly and supports GPx
Vitamin C and vitamin E	Non-enzymatic	Scavenges ROS in aqueous and lipid environments	Protects DNA, lipids, and proteins from oxidative damage

10.3.1 Enzymatic Antioxidants

Enzymatic antioxidants work by catalyzing reactions that convert ROS into less harmful molecules (Al-Gubory et al. 2012; Lustgarten et al. 2011; Nozik-Grayck et al. 2005; Shults et al. 2002). Key enzymes involved in antioxidant defense include the following.

10.3.1.1 Superoxide Dismutase (SOD)

SOD catalyzes the dismutation of superoxide radicals (O_2^-) into hydrogen peroxide (H_2O_2) and molecular oxygen (Nozik-Grayck et al. 2005; Van Camp et al. 1997). There are three types of SOD in humans: copper-zinc SOD (SOD1) found in the cytosol, manganese SOD (SOD2) located in the mitochondria, and extracellular SOD (SOD3) present in the extracellular space (Bhattacharyya et al. 2014; Nozik-Grayck et al. 2005; Zelko et al. 2002).

10.3.1.2 Catalase

Catalase is an enzyme located primarily in peroxisomes that decomposes hydrogen peroxide (H_2O_2) into water and oxygen (Nandi et al. 2019; Schrader and Fahimi 2006). This enzyme plays a crucial role in protecting cells from oxidative damage caused by excessive levels of H_2O_2 (Bhattacharyya et al. 2014; Nandi et al. 2019).

10.3.1.3 Glutathione Peroxidase (GPx)

GPx catalyzes the reduction of hydrogen peroxide (H_2O_2) and lipid hydroperoxides to water and lipid alcohols, respectively, using glutathione (GSH) as an electron donor (Bompart et al. 1990; Lubos et al. 2011). It is one of the most important antioxidant enzymes in preventing lipid peroxidation and maintaining cellular integrity (Bompart et al. 1990; Brigelius-Flohé and Maiorino 2013; Taniyama and Griendling 2003).

10.3.1.4 Thioredoxin (Trx) and Peroxiredoxins (Prx)

Thioredoxin is a small protein that acts as an antioxidant by reducing oxidized proteins. Peroxiredoxins (Prx) are enzymes that reduce peroxides, especially hydrogen peroxide, using thioredoxin as a cofactor (Kondo et al. 2004; Lustgarten et al. 2011).

10.3.2 Non-enzymatic Antioxidants

Non-enzymatic antioxidants are small molecules that scavenge free radicals and terminate oxidative chain reactions (Mills 1988).

10.3.2.1 Glutathione (GSH)

Glutathione is a tripeptide composed of glutamine, cysteine, and glycine. It exists in both reduced (GSH) and oxidized (GSSG) forms (Couto et al. 2016). GSH acts as a direct antioxidant by donating electrons to neutralize ROS and as a cofactor for GPx and other antioxidant enzymes (Couto et al. 2016; Lushchak 2012; Valko et al. 2007). In addition, GSH plays a role in antioxidant defense by regenerating the radicals of vitamins C and E to their active, non-radical forms (Jones 2002).

10.3.2.2 Vitamin C (Ascorbic Acid)

Vitamin C is a water-soluble antioxidant that scavenges ROS in the aqueous compartments of cells (Kaźmierczak-Barańska et al. 2020; Stohs and Bagchi 1995). It also regenerates vitamin E from its oxidized form, thereby enhancing its antioxidant capacity (Traber and Atkinson 2007).

10.3.2.3 Vitamin E (Tocopherols)

Vitamin E is a fat-soluble antioxidant that protects cell membranes from lipid peroxidation (Traber and Atkinson 2007). It scavenges lipid radicals and prevents the propagation of lipid peroxidation in polyunsaturated fatty acids (Mills 1988; Traber and Atkinson 2007).

10.3.2.4 Carotenoids and Flavonoids

These plant-based compounds possess antioxidant properties and are found in fruits and vegetables. Carotenoids like β-carotene scavenge singlet oxygen, while flavonoids neutralize free radicals and inhibit oxidative enzymes (Mortensen et al. 2001; Simunkova et al. 2021).

10.3.2.5 Coenzyme Q10

This fat-soluble molecule, present in the mitochondrial membrane, acts as an antioxidant by preventing lipid peroxidation and regenerating vitamin E. It also participates in the electron transport chain (Lee et al. 2017; Stohs and Bagchi 1995).

10.4 Mechanisms Involved in Oxidative Damage

When the production of ROS overwhelms the body's antioxidant defenses, oxidative damage occurs, affecting lipids (Barrera 2012; Valko et al. 2007), proteins (Suzuki et al. 2010), and DNA (Gonzalez-Hunt et al. 2018). These macromolecules are particularly vulnerable to oxidative modification, leading to cellular dysfunction and disease.

10.4.1 Lipid Peroxidation

Lipid peroxidation refers to the oxidative degradation of lipids, particularly polyunsaturated fatty acids (PUFAs), in cell membranes (Barrera 2012; Valko et al. 2007). This process is initiated by ROS such as hydroxyl radicals, which abstract hydrogen atoms from the lipid chains, forming lipid radicals. These lipid radicals react with molecular oxygen to form lipid peroxyl radicals, which can further propagate the chain reaction, leading to the formation of lipid hydroperoxides (Barrera 2012).

Lipid peroxidation alters membrane fluidity and permeability, disrupting cellular homeostasis. The breakdown products of lipid peroxidation, such as malondialdehyde (MDA) and 4-hydroxynonenal (4-HNE), are highly reactive and can form adducts with proteins and DNA, exacerbating oxidative damage (Pizzimenti et al. 2010; Ramana et al. 2014).

10.4.2 Protein Oxidation

ROS can oxidize amino acid residues in proteins, leading to the formation of carbonyl groups (oxidized side chains) and disulfide bonds (between cysteine residues) (Dalle-Donne et al. 2003). Protein oxidation can result in the loss of enzyme activity, altered protein structure, and impaired cellular signaling (Gonos et al. 2018). Carbonylation is the addition of carbonyl groups to proteins (carbonylation) which is a hallmark of oxidative stress. These oxidized proteins are often degraded by the proteasome, but if the degradation system is overwhelmed, they can accumulate and form aggregates that impair cellular function (Gonos et al. 2018; Suzuki et al. 2010). Advanced Oxidation Protein Products (AOPP): Proteins modified by oxidative stress form AOPP, which are considered biomarkers of oxidative damage, particularly in inflammatory conditions like chronic kidney disease (Gargouri et al. 2019; Gryszczyńska et al. 2017).

10.4.3 DNA Damage

Oxidative stress can cause significant damage to DNA, leading to mutations, strand breaks, and base modifications (Martins et al. 2021; Smith et al. 2016). The hydroxyl radical (OH^-) is particularly efficient at damaging DNA by attacking purine and pyrimidine bases as well as the deoxyribose backbone (Gonzalez-Hunt et al. 2018; Martins et al. 2021). 8-Oxoguanine (8-oxoG): A modified form of guanine that can result in G-to-T transversion mutations during DNA replication, 8-oxoguanine (8-oxoG) is one of the most prevalent oxidative DNA lesions. Accumulation of 8-oxoG is associated with aging and cancer development (Hazra et al. 1998; Li et al. 2022). Single- and Double-Strand Breaks: ROS can induce single-strand breaks (SSBs) (Martins et al. 2021).

10.4.4 Significance of Oxidative Stress in Neurodegenerative Diseases with Epidemiology

Neurodegenerative disorders are noncommunicable diseases with problems relating to the gradual destruction of neurons and organ's capabilities that have become a challenge on a global scale (Lamptey et al. 2022). For many of the patients with these disorders, Alzheimer's disease (AD), Parkinson's disease (PD), Huntington's disease (HD), and multiple sclerosis (MS) are highly correlated with oxidative stress (Uttara et al. 2009). Each of these proves to have unique syndromic and pathological manifestation, but they all suffer in unison from oxidative damage that enhances their neuronal malfunctions and ultimately renders the organ inoperable. The current book chapter aims to elucidate the critical role that oxidative stress plays in the pathophysiology of AD, PD, HD, and MS. Specifically, it will

1. Examine the molecular mechanisms via which oxidative stress causes damage to neurons in each disease.

2. Examine the interplay between oxidative stress and disease-specific pathological features, such as protein aggregates and genetic mutations.
3. Assess the evidence from preclinical and clinical studies linking oxidative stress to disease progression.
4. Discuss possible treatment approaches that target oxidative stress to lessen neurodegeneration.

By synthesizing current knowledge, this chapter seeks to provide a comprehensive understanding of how oxidative stress serves as both a cause and consequence of these neurodegenerative disorders, offering insights into future research directions and clinical applications.

10.5 Alzheimer's Disease (AD)

10.5.1 Pathophysiology of Alzheimer's Disease

AD is a complex neurodegenerative disorder characterized by progressive cognitive decline and memory loss. It is the most common cause of dementia ("Alzheimer's Disease Fact Sheet," 2023; "Is Alzheimer's Disease Hereditary?," n.d.). AD is a biological process that begins with the appearance of a buildup of proteins in the form of amyloid plaques and neurofibrillary tangles in the brain. This causes brain cells to die over time and the brain to shrink ("Alzheimer's disease—Symptoms and causes," n.d.; "Alzheimer's Disease Fact Sheet," 2023). In the US, 6.9 million people with an age range of 65 and older suffer from AD. This could reach 13.8 million by 2060 ("2024 Alzheimer's disease facts and figures," 2024). Forgetting recent discussions or experiences is one of the early signs of AD. Further symptoms include impaired capacity to do daily chores and causes significant memory loss over time ("10 Early Signs and Symptoms of Alzheimer's & Dementia | alz.org," n.d.).

The formations of amyloid-beta (Aβ) peptides and neurofibrillary tangles of the microtubule-binding protein tau are hallmarks in AD pathology (Rukmangadachar and Bollu 2025). Oxidative stress plays a critical role in pathogenesis of AD and is increased with aging, characterized by an imbalance between reactive oxygen species (ROS) and the cellular antioxidant defense mechanisms in the brain. This imbalance leads to neuronal damage, specifically amyloid precursor protein to Aβ (APP) processing (Ishola et al. 2023; Roy et al. 2023). APP is metabolized by beta-secretase (BACE1) and gamma-secretase (Willem et al. 2004). Oxidative stress upregulates the formation of BACE1 through nuclear factor kappa B (NF-κB) pathway activation, which ultimately leads to increased formation of Aβ peptides. Further oxidative stress promotes Aβ peptide aggregation, which further generates ROS exacerbating the neuronal damage. For instance, copper ions, which are often dysregulated in AD, can bind to Aβ and catalyze the production of ROS, further contributing to oxidative stress and promoting Aβ aggregation. Similarly, there is growing evidence that iron, and zinc, also plays a significant role in Aβ formation and subsequent neurodegeneration. Elevated levels of ROS can activate various kinases that promote tau phosphorylation, leading to the formation of neurofibrillary tangles, another hallmark of AD pathology (Tamagno et al. 2021).

Oxidative stress also affects mitochondrial function, which is crucial for neuronal health. The electron transport chain in the mitochondria uses almost 98% of molecular oxygen. Oxidative stress impairs the mitochondrial function by altering the membrane permeability. This can result in a vicious cycle that encourages the formation of Aβ by increasing ROS generation and decreasing ATP production. Mitochondrial damage can also impair the clearance of Aβ, allowing for its accumulation and aggregation (Ionescu-Tucker and Cotman 2021). Furthermore, oxidative stress can induce apoptosis in neurons, leading to cell death and exacerbating cognitive decline associated with AD (Dhapola et al. 2024). In summary, oxidative stress plays a key role in the formation and accumulation of Aβ peptides in Alzheimer's disease. It causes mitochondrial dysfunction, increases the expression of enzymes implicated in Aβ synthesis, and encourages the aggregation of Aβ peptides (Fig. 10.2). This multifaceted relationship underscores the importance of targeting oxidative stress in therapeutic strategies aimed at mitigating the progression of Alzheimer's disease.

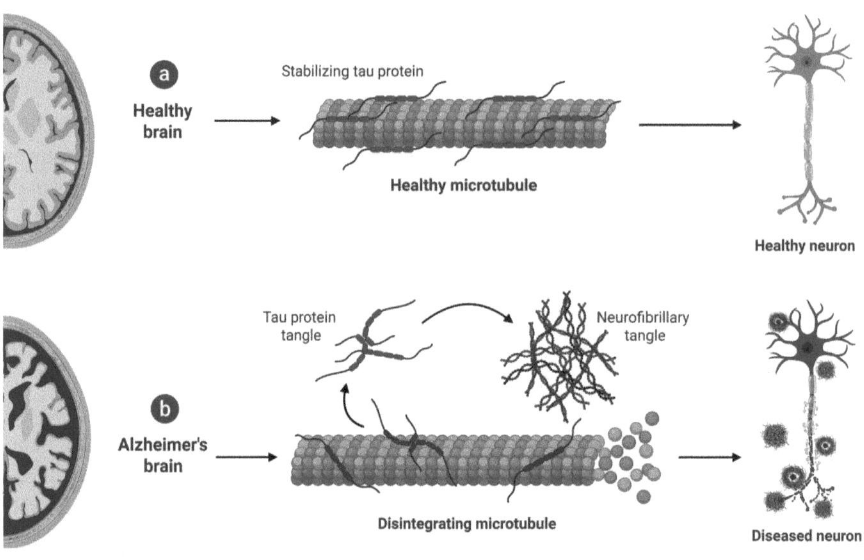

Fig. 10.2 Pathophysiology of Alzheimer's disease

10.5.2 Therapeutic Approaches Targeting Oxidative Stress in AD

As mentioned above, oxidative stress can precede Aβ plaque formation and tau phosphorylation. Furthermore, mitochondrial dysfunction, often a consequence of oxidative stress, leads to impaired energy metabolism and increased neuronal vulnerability, further promoting neurodegeneration (Persson et al. 2014). Thus, antioxidants can slow down the disease progression. Recent studies have explored the neuroprotective effects of natural antioxidants, in ameliorating oxidative damage and improving cognitive function in preclinical models (Zhu et al. 2025). For instance, compounds like curcumin and ginsenosides have been reported to activate autophagy and reduce oxidative stress in neuroblastoma cells overexpressing Aβ (Liu et al. 2024). Additionally, multi-target drugs that address several subpathologies, including oxidative stress, are being developed to enhance treatment efficacy (Angelova et al. 2024). Furthermore, it has been suggested that antioxidants be used in clinical settings to lessen oxidative stress and its harmful consequences on brain health. For example, the administration of Coenzyme Q10 has been suggested to improve mitochondrial function and reduce oxidative damage in AD patients (Fišar and Hroudová 2024). Several clinical trials also reported positive interventions with antioxidants in AD. Furthermore, dietary interventions incorporating antioxidants from natural sources, such as omega-3 fatty acids and phenolic compounds, have shown potential in countering oxidative stress and neuroinflammation associated with AD (Prodromou et al. 2024). Several other antioxidants with potential therapeutic applications in AD include Vitamin E, Vitamin C, Vitamin B12, β carotene, α-lipoic acid, caffeine, berberine, palmitin, silibinin, quercetin, melatonin, estrogen, selegiline, resveratrol, and polyphenols (Kamaljeet et al. 2024; Pritam et al. 2022).

10.6 Parkinson's Disease

The progressive loss of dopaminergic neurons in the substantia nigra pars compacta, a part of the brain responsible for controlling movement, is a hallmark of Parkinson's disease (PD), a chronic neurodegenerative condition (Kouli et al. 2018; "Parkinson's Disease | National Institute of Neurological Disorders and Stroke," n.d.; "Parkinson's disease explained," n.d.). The disease's characteristic motor symptoms, including bradykinesia (slowness of movement), resting tremor, muscle rigidity, and postural instability, are caused by this degradation. In addition to motor symptoms, patients may also experience cognitive decline, depression, and autonomic dysfunction (Kouli et al. 2018; "Parkinson's Disease | National Institute of Neurological Disorders and Stroke," n.d.; "Parkinson's disease explained," n.d.; Rodriguez-Oroz et al. 2009; Vasava et al. 2022).

While the exact cause of Parkinson's disease remains unknown, there is growing evidence to suggest that oxidative stress plays a pivotal role in the pathophysiology of the disorder (Blesa et al. 2015; Jenner 2003; Kouli et al. 2018; Rodriguez-Oroz et al. 2009). Oxidative stress refers to an imbalance between the production of reactive oxygen species (ROS) and the body's ability to neutralize them with antioxidant defenses. Excessive ROS can cause cellular damage to lipids, proteins, and DNA, which is thought to contribute to the neurodegeneration observed in PD (Brieger et al. 2012; Dias et al. 2013; Jenner 2003; Kieburtz and Wunderle 2013; Puspita et al. 2017). Figure 10.3 shows the schematic diagram of sources of oxidative stress in the brain and their role in Parkinson's disease and proposed therapeutic approach for Parkinson's disease.

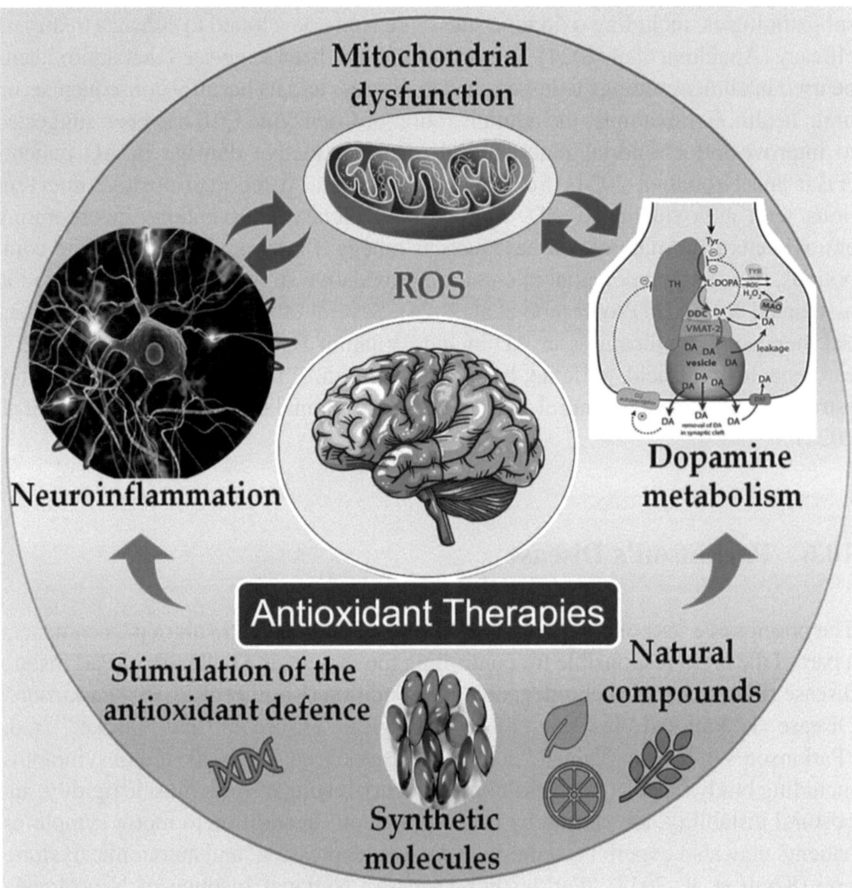

Fig. 10.3 An illustration of the brain's oxidative stress sources and how they contribute to Parkinson's disease

10.6.1 *Pathophysiology of Parkinson's Disease: The Role of Oxidative Stress*

10.6.1.1 Generation of ROS in the Brain

The brain is particularly vulnerable to oxidative stress due to its high oxygen consumption, high content of polyunsaturated fatty acids (which are prone to lipid peroxidation), and relatively low levels of antioxidant enzymes compared to other organs (Blesa et al. 2015; Brieger et al. 2012; Percário et al. 2020). In Parkinson's disease, ROS are generated through several mechanisms, including mitochondrial dysfunction, dopamine metabolism, and neuroinflammation (Barcia et al. 2004; Brieger et al. 2012; Segura-Aguilar et al. 2014). Table 10.3 shows the sources of oxidative stress in Parkinson's disease.

Table 10.3 Sources of oxidative stress in Parkinson's disease

Source of oxidative stress	Mechanism	ROS Produced	Impact on dopaminergic neurons	Consequences in Parkinson's disease
Mitochondrial dysfunction	Impairment of electron transport chain (ETC), particularly at complex I	Superoxide (O_2^-)	Elevated ROS generation causes oxidative damage to DNA, lipids, and proteins in the mitochondria	Neuronal death due to energy deficits and oxidative damage
Dopamine metabolism	Oxidation of dopamine via monoamine oxidase (MAO)	Hydrogen peroxide (H_2O_2), hydroxyl radicals (OH•)	H_2O_2 converted to OH• through Fenton reaction in the presence of iron; oxidative damage to proteins, lipids, and DNA	ROS damage, dopamine toxicity, formation of toxic quinones leading to neuron loss
Neuroinflammation	Activation of microglia due to neuronal injury or infection	ROS, pro-inflammatory cytokines	Microglia release ROS and cytokines, further exacerbating oxidative stress in neurons	Chronic inflammation increases neuronal damage and contributes to progression of Parkinson's disease
Iron accumulation	Excess iron in the substantia nigra reacts with H_2O_2	Hydroxyl radicals (OH•) via the Fenton reaction	Iron acts as a catalyst for ROS production, promoting oxidative stress and damaging neurons	Amplification of oxidative stress, leading to increased neuron degeneration

10.6.1.2 Mitochondrial Dysfunction

The cell's powerhouses, mitochondria, are in charge of oxidative phosphorylation, which produces ATP. However, they are also a significant source of ROS (Blesa et al. 2015; Dröse and Brandt 2008; Hall et al. 2012; Kieburtz and Wunderle 2013; Percário et al. 2020). A small number of electrons may escape the respiratory chain during electron transport, especially at Complex I, and combine with molecular oxygen to produce superoxide radicals (O_2^-). In Parkinson's disease, dysfunction in mitochondrial Complex I has been observed in post-mortem studies of the substantia nigra, leading to increased ROS production and subsequent neuronal damage (Blesa et al. 2015; Dröse and Brandt 2008; Kieburtz and Wunderle 2013; Trinh and Farrer 2013; Zuo and Motherwell 2013).

10.6.1.3 Dopamine Metabolism

Dopamine, the neurotransmitter primarily affected in Parkinson's disease, is highly susceptible to oxidation. Under normal conditions, dopamine is metabolized by monoamine oxidase (MAO) to produce hydrogen peroxide (H_2O_2), a type of ROS. One of the most destructive types of ROS, hydroxyl radicals (OH^-), can be produced when hydrogen peroxide undergoes the Fenton reaction with iron. Oxidized dopamine can also form toxic quinones, which can further contribute to neuronal damage (Blesa et al. 2015; Dröse and Brandt 2008; Segura-Aguilar et al. 2014).

10.6.1.4 Neuroinflammation

Oxidative stress has also been linked to inflammatory processes in the brain. Activated microglia (the brain's resident immune cells) release ROS and pro-inflammatory cytokines in response to injury or infection (Blesa et al. 2015; Brieger et al. 2012; Kieburtz and Wunderle 2013; Trinh and Farrer 2013). In Parkinson's disease, chronic microglial activation has been observed in the substantia nigra, suggesting that neuroinflammation may exacerbate oxidative damage to dopaminergic neurons (Barcia et al. 2004; Kieburtz and Wunderle 2013; Lotharius and Brundin 2002; Percário et al. 2020).

10.6.2 Vulnerability of Dopaminergic Neurons

For a number of reasons, dopaminergic neurons are especially vulnerable to oxidative stress.

10.6.2.1 High Energy Demand

Dopaminergic neurons in the substantia nigra have high metabolic rates and energy demands, making them reliant on efficient mitochondrial function. When mitochondria are impaired, these neurons are more vulnerable to ROS-induced damage.

10.6.2.2 Dopamine Autoxidation

As previously stated, dopamine is susceptible to oxidation, which results in the production of harmful quinones and ROS. The autoxidation of dopamine can further reduce the antioxidant capacity of neurons by depleting glutathione, a key cellular antioxidant (Blesa et al. 2015; Lotharius and Brundin 2002; Rodriguez-Oroz et al. 2009; Zuo and Motherwell 2013).

10.6.2.3 Iron Accumulation

The Fenton reaction, which produces extremely reactive hydroxyl radicals, can be catalyzed by the substantia nigra's high iron concentration. Post-mortem studies of Parkinson's patients have shown increased iron deposition in the substantia nigra, further contributing to oxidative stress (Blesa et al. 2015; Lotharius and Brundin 2002; Rodriguez-Oroz et al. 2009).

10.6.3 Lewy Bodies and Protein Aggregation

Lewy bodies, abnormal protein aggregates composed primarily of alpha-synuclein, are a hallmark of Parkinson's disease pathology (Billings et al. 2019; Deas et al. 2016; Ph.D, 2024). A protein called alpha-synuclein often contributes to dopamine release and synaptic function. However, under conditions of oxidative stress, alpha-synuclein can misfold and aggregate, forming Lewy bodies (Billings et al. 2019; Blesa et al. 2015). The accumulation of these aggregates is toxic to neurons and may further contribute to mitochondrial dysfunction and oxidative stress, creating a vicious cycle of neurodegeneration (Deas et al. 2016; Jenner 2003).

Oxidative modifications to alpha-synuclein, including nitration and oxidation of specific amino acid residues, have been detected in the brains of Parkinson's disease patients (Deas et al. 2016; Ph.D, 2024). These modifications are thought to promote the aggregation of alpha-synuclein and impair its normal function. The presence of Lewy bodies is also associated with mitochondrial dysfunction, further linking protein aggregation to oxidative stress and neuronal death in PD (Billings et al. 2019; Deas et al. 2016; Ph.D, 2024).

10.6.4 Dopaminergic Neuron Degeneration and Mitochondrial Dysfunction: Outcomes

The degeneration of dopaminergic neurons in Parkinson's disease is closely tied to mitochondrial dysfunction and oxidative stress (Halliday et al. 2005; Jenner 2003; Rodriguez-Oroz et al. 2009). The outcomes of these processes are significant and have far-reaching consequences for brain function and overall health.

10.6.4.1 Loss of Dopaminergic Neurons in the Substantia Nigra

The primary outcome of oxidative stress and mitochondrial dysfunction in Parkinson's disease is the progressive loss of dopaminergic neurons in the substantia nigra pars compacta (Blesa et al. 2015; Halliday et al. 2005; Segura-Aguilar et al. 2014). As these neurons die, the levels of dopamine in the brain decrease, leading to the hallmark motor symptoms of PD, such as bradykinesia, tremor, and rigidity (Kieburtz and Wunderle 2013; Lotharius and Brundin 2002; Zuo and Motherwell 2013). The extent of neuronal loss correlates with the severity of symptoms, and by the time patients present with motor symptoms, it is estimated that 60–80% of dopaminergic neurons have already been lost (Deas et al. 2016; Jenner 2003; Segura-Aguilar et al. 2014).

10.6.4.2 Impaired Mitochondrial Bioenergetics

Mitochondrial dysfunction not only increases ROS production but also impairs the ability of neurons to generate ATP, the energy currency of the cell (Dröse and Brandt 2008; Hall et al. 2012; Zuo and Motherwell 2013). Dopaminergic neurons have high energy requirements due to their extensive axonal networks and the constant need to release and recycle dopamine (Dröse and Brandt 2008; Segura-Aguilar et al. 2014). When ATP production is compromised, these neurons become more vulnerable to stress and less able to maintain essential cellular functions, such as ion homeostasis and synaptic transmission (Dröse and Brandt 2008; Hall et al. 2012; Zuo and Motherwell 2013).

10.6.4.3 Activation of Apoptotic Pathways

Prolonged oxidative stress can activate cell death pathways, particularly apoptosis (programmed cell death) (Chang and Chen 2020; Puspita et al. 2017; Zuo and Motherwell 2013). By releasing pro-apoptotic elements like cytochrome c from the mitochondria into the cytoplasm, mitochondrial damage is a key contributor in the activation of apoptotic pathways. The following sets off a series of processes that activate the caspases, a family of proteases responsible for breaking down cells.

Apoptosis of dopaminergic neurons is a key feature of Parkinson's disease and contributes to the progressive nature of the disorder (Chang and Chen 2020; Puspita et al. 2017; Segura-Aguilar et al. 2014).

10.6.4.4 Propagation of Neurodegeneration

The oxidative damage and mitochondrial dysfunction seen in Parkinson's disease are not confined to the substantia nigra but can spread to other parts of the brain, such as the locus coeruleus, cortex, and hippocampus. This propagation of neurodegeneration may explain the non-motor symptoms of PD, such as cognitive decline, mood disorders, and sleep disturbances (Barcia et al. 2004; Chang and Chen 2020; Kouli et al. 2018). The spread of Lewy body pathology to these regions may also contribute to the progression of the disease (Barcia et al. 2004; Halliday et al. 2005).

10.6.5 Therapeutic Approaches Targeting Oxidative Stress in Parkinson's Disease

Given the major part of oxidative stress in the pathogenesis of Parkinson's disease, therapeutic strategies aimed at reducing oxidative damage have been actively explored. These approaches include the use of antioxidants, mitochondrial-targeted therapies, and lifestyle interventions.

10.6.6 Antioxidant Therapies

One of the most straightforward approaches to combat oxidative stress is the use of antioxidants to neutralize ROS (Chang and Chen 2020; Chaudhary et al. 2023; Percário et al. 2020; Poljsak et al. 2013). The possible neuroprotective effects of a number of antioxidant substances in Parkinson's disease have been studied. Cell membranes are shielded from lipid peroxidation by vitamin E, a lipid-soluble antioxidant. High dosages of vitamin E may reduce the progression of Parkinson's disease, according to previous studies, but later clinical trials have not shown any appreciable advantages. However, vitamin E remains a potential adjunctive therapy for PD, particularly in combination with other treatments (Firuzi et al. 2011; Percário et al. 2020; Sharifi-Rad et al. 2020). Coenzyme Q10 is an important component of the mitochondrial electron transport chain and also functions as an antioxidant (Chaudhary et al. 2023; Lee et al. 2017; Shults et al. 2002). In preclinical models of Parkinson's disease, CoQ10 has been demonstrated to enhance mitochondrial activity and lessen oxidative damage. While early clinical trials suggested a potential benefit for CoQ10 in slowing disease progression, larger trials have produced mixed results. Further research is needed to determine the optimal dose and formulation of CoQ10 for PD patients (Lee et al. 2017; Percário et al. 2020; Sharifi-Rad et al. 2020). N-Acetylcysteine (NAC) is a precursor to glutathione, the brain's primary antioxidant. By increasing glutathione levels, NAC can reduce oxidative stress and protect neurons from ROS-induced damage (Shults et al. 2002; Spencer et al. 1998). Recent studies have shown that NAC can improve brain glutathione levels and reduce oxidative stress in PD patients, making it a promising candidate for further clinical trials (Percário et al. 2020; Shults et al. 2002; Spencer et al. 1998).

10.6.7 Mitochondrial-Targeted Therapies

Given the role of mitochondrial dysfunction in PD, therapies aimed at restoring mitochondrial function have also been explored (Jin et al. 2014; Percário et al. 2020; Snow and Murphy 2001; Thomas and Beal 2010). Creatine is a molecule that plays a key role in energy metabolism by facilitating the regeneration of ATP (Klivenyi et al. 2004; Snow and Murphy 2001). Preclinical research has demonstrated that creatine enhances mitochondrial function and guards against neuronal death in Parkinson's disease mice. However, clinical trials of creatine in PD patients have yielded disappointing results, with no significant improvement in disease progression (Jin et al. 2014; Klivenyi et al. 2004; Snow and Murphy 2001). MitoQ is a mitochondria-targeted antioxidant designed to accumulate within the mitochondria, where it can directly neutralize ROS. Preclinical studies have shown that MitoQ can protect dopaminergic neurons from oxidative stress and improve mitochondrial function (Feniouk and Skulachev, n.d.; Jin et al. 2014; Xi et al. 2018).

10.7 Pathophysiology of Huntington's Disease, Focusing on the Role of Oxidative Stress

A trinucleotide cytosine-adenine-guanine (CAG) repeat increase in the huntingtin (Htt) gene is the cause of Huntington's disease (HD), an autosomal dominant inherited progressive neurological condition. The CAG is located on the short arm of chromosome four, and inheritance is autosomal dominant (Stoker et al. 2022). Intermediate alleles have 27–35 CAG repeats, while the normal allele has fewer than 27. HD will develop with decreased penetrance in CAG repeats 36–39. Individuals with 40 or more CAG repeats will develop HD with full penetrance (Kumar et al. 2020). The mutant Htt protein known as huntingtin has an unusually long polyglutamine repeat as a result of the Htt gene mutation. It also causes disruptions in normal biological functions by interacting with other proteins (Teleanu et al. 2022). Massive neuronal degeneration in the striatum is caused by the aberrant protein (mutant huntingtin, mHTT), which then causes the cerebral cortex and thalamus to deteriorate. It is mainly characterized by impaired motor and cognitive traits, personality change, and psychiatric illness (Browne et al. 1999). HD often develops between the ages of 35 and 50, worsens over time, and becomes lethal following the onset of symptoms. Despite the fact that numerous factors, including protein misfolding, aberrant proteolysis, protein aggregation, transcriptional dysfunction, excitotoxic, and oxidative stress, have been linked to neuronal death in HD, the function of oxidative stress (OS) is the subject of the current book chapter.

Oxidative stress (OS) is the result of an imbalance brought on by an increase in oxidants relative to antioxidants. Growing concentrations of reactive species and short-lived molecules, including free radicals that cause oxidative damage, are indicative of this compromised redox equilibrium. Because the brain requires a significant amount of oxygen to function properly, it is vulnerable to the effects of free radicals and reactive oxygen species (ROS). The ROS involved in neurodegeneration includes hydrogen peroxide (H_2O_2), superoxide anion (O_2^-), and highly reactive hydroxyl radical ($HO^•$) (D'Egidio et al. 2023). These significantly contribute to neurodegeneration by deteriorating neuronal cells through the modulation of biomolecular function. The biomolecules (DNA, RNA, lipids, and proteins) and processes (nucleic acid oxidation, lipid peroxidation) in the cell play a substantial role as biomarkers for OS in HD patients. Lipid peroxidation, DNA damage, and specifically protein carbonylation were found to be more pronounced in HD (Túnez et al. 2011). Free radicals cause membrane lipid peroxidation, intracellular protein oxidation, and DNA oxidation at both the intracellular and extracellular levels. DNA repair processes are triggered by oxidative DNA damage, which results in the elimination of oxidized bases and the return of DNA's original structure and functionality. It has been demonstrated that mending damaged DNA may cause the mHTT gene's CAG trinucleotide repeats to expand and become unstable (Singh et al. 2019). Since it results in mitochondrial dysfunction, calcium homeostasis disruption, impaired antioxidant defenses, iron dysregulation, ferroptosis, and extensive cell damage, oxidative stress plays a crucial role in the pathophysiology of Huntington's disease.

10.7.1 Mutant Huntingtin Protein

Intranuclear and intracytoplasmic aggregates (beta-pleated sheet protein generated by hydrogen bonding), a pathogenic feature in the brains of HD patients, can result from a conformational shift in the mHTT protein caused by the enlarged glutamine repeats. The mHTT self-aggregates to produce insoluble fibrils and soluble oligomers that obstruct several important cellular processes. This leads to cell quiescence and, ultimately, cell death (Jarosińska and Rüdiger 2021). These aggregates may play a role in HD pathogenesis. These might disrupt normal synaptic transmission. Additionally, the aggregation process driven by polyglutamine might trigger different effects, such as transcriptional deregulation or proteasome impairment, affecting neuronal survival (Arrasate and Finkbeiner 2012). Additionally, transglutaminase may use polyglutamine as a substrate. Glyceraldehyde 3-phosphate dehydrogenase (GAPDH) and huntingtin-associated protein-1 (HAP-1) are the main potential protein interactors. HAP-1 does not have a preferential striatal distribution and is only present in the brain. Given that GAPDH is an essential glycolytic enzyme, its interaction with huntingtin may compromise metabolic activity (Browne et al. 1999).

10.7.2 Mitochondrial Abnormalities

The process by which cells create new mitochondria is known as mitochondrial biogenesis, and it frequently occurs in response to metabolic difficulties and transcription factors that support cellular stress resistance. PGC-1α, AMPK, and SIRT1 are key regulators of mitochondrial biogenesis, a process severely hindered in HD (Wells et al. 2024). In spite of their high oxygen consumption, mitochondria are the main source of ROS generation in mammalian cells. The fundamental shift in mitochondrial function generates more reactive oxygen species (ROS), which causes oxidative stress (Klemmensen et al. 2024). Toll-like receptor (TLR) receptor proteins and pro-inflammatory cytokines are stimulated by astrocytes and microglia in response to oxidative stress caused by elevated ROS. These proteins contribute to neuroinflammation, excitotoxicity, and neuronal degeneration by increasing ROS levels, calcium entry, and inflammatory mediators. The expression of important antioxidant enzymes such as SOD, catalase, and glutathione peroxidase is changed as a result of mHTT-induced transcriptional dysregulation and antioxidant defense. Deposition of transition metals like iron in the cytoplasm and mitochondria contributes to free radical generation (Paul and Snyder 2019). Increased NADPH oxidase (NOX) activity, particularly NOX2, has been observed in HD brains, directly contributing to elevated ROS levels. The mHTT gene disrupts cellular functions and may produce ROS, which increases oxidative stress. It also causes mHTT protein misfolding, aggregate formation in neuropils and neuronal nuclei, and neurotoxicity due to protein interaction. Bioenergy failure and damage to mitochondrial DNA (mtDNA) are caused by the interaction between the mHTT proteins and transporter II. By making neurons more susceptible to endogenous glutamate levels, impaired mitochondrial energy metabolism first lowers ATP production and may cause partial cell depolarization. Increased calcium influx follows, which causes the mitochondrial permeability transition pore to expand, disturbing ATP levels and releasing cytochrome c. This lowers ATP synthesis and triggers cell death, excitotoxicity, and apoptotic pathways. Oxidative stress is caused by the buildup of mHTT genes, neuroinflammation, elevated lipid levels, and mitochondrial malfunction. Further, it accelerates the disease's progression, causing additional neuronal damage (Joshi et al. 2025).

10.7.3 Oxidative Damage

Excessive ROS buildup is considered detrimental to cells and organisms because it induces oxidative stress. In physiological settings, this excess ROS causes redox-dependent signaling. However, ROS becomes harmful when it exceeds a certain level. Besides aberrant redox signaling, ROS can damage many cellular components, such as proteins, DNA, and lipids (Sun et al. 2023). Proteins undergo a variety of oxidative changes as a result of oxidative damage because different amino acid residues are susceptible to oxidation (depending on the ROS involved). By changing stability, activity, subcellular location, or the protein-protein interaction of oxidized proteins, these alterations may change how cells function. Nonetheless, upon persistent activation, these mechanisms may either aggravate the cellular damage or become impaired, eventually leading to a vicious cycle that promotes the progression of several pathologies (Calderaro et al. 2022).

10.8 Multiple Sclerosis (MS)

The central nervous system (CNS) is impacted by multiple sclerosis (MS), a chronic autoimmune disease. It is identified by neuroinflammation, demyelination, and axonal degeneration ("Atlas of MS 3 Edition," n.d.; Reich et al. 2018). Approximately 2.8 million individuals worldwide are afflicted with MS, which poses significant public health concerns. The complicated and multifaceted pathophysiology of multiple sclerosis (MS) results from the interplay of immune system dysregulation, environmental exposures, and genetic variables ("Atlas of MS 3 Edition," n.d.; Lane et al. 2022). Recent studies have shown the existence of close correlations between oxidative stress and inflammation, mitochondrial dysfunction, and tissue injury ("Atlas of MS 3 Edition," n.d.). In addition to highlighting the range of possible treatment approaches for oxidative damage, this chapter aims to contextualize the relationship between oxidative stress and multiple sclerosis.

10.8.1 Pathophysiology of Multiple Sclerosis

Lesions in the brain and spinal cord form as a result of immune-mediated myelin invasions in the central nervous system, which cause multiple sclerosis. This destruction of the myelin sheath leads to impaired neuronal signaling and neurological dysfunction. The following are important aspects of the pathophysiology of MS: Demyelination (loss of myelin sheath), Plaques, and Remyelination ("Atlas of MS 3 Edition," n.d.; Dighriri et al. 2023; Ghasemi et al. 2017; Lane et al. 2022; Tafti et al. 2025) (Fig. 10.4).

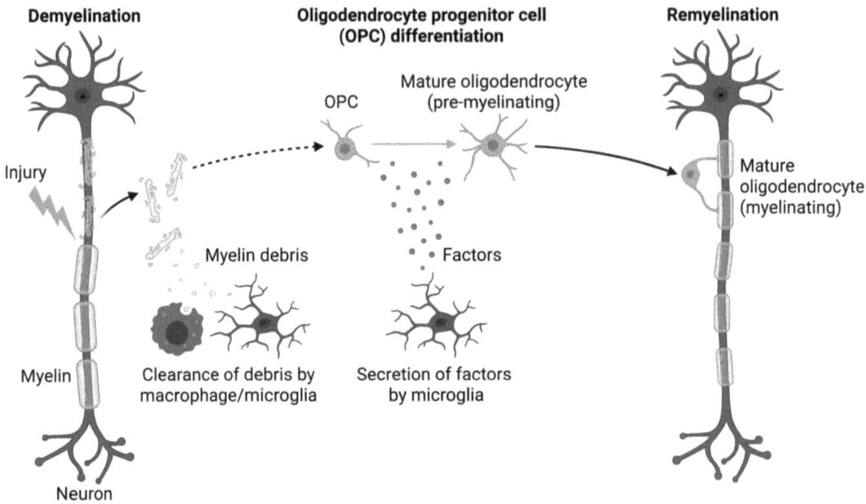

Fig. 10.4 Schematic representation of the demyelination and remyelination process (created using Bio Render)

10.8.1.1 Demyelination

The myelin sheath is a protective layer that surrounds the axons of neurons, enabling them to send electrical impulses quickly. This myelin is produced by oligodendrocytes, a type of cell that supports neurons. Demyelination refers to the loss or damage to the myelin sheath, which disrupts communication between neurons and other parts of the body, leading to various neurological disorders characterized by sensory, motor, and cognitive problems.

10.8.1.2 Plaques

In the demyelinating process, plaques are areas of scar tissue that form along the axons where myelin has been lost due to autoimmune attacks, inflammation, or other pathological processes. The resulting scar tissue is composed of a mix of inflammatory cells, glial cells, and extracellular matrix components. These plaques can vary in size, shape, and distribution and can be seen in imaging studies like MRI, which are hallmarks of MS conditions. Plaques are most commonly formed in white matter, particularly around the small veins, but they can affect anywhere within the central nervous system.

10.8.1.3 Remyelination

Remyelination is the restoration of myelin sheaths around damaged axons. This intrinsic process is regulated and controlled by the body and involves the activation, migration, and differentiation of oligodendrocyte precursor cells to form new myelin. While this process improves nerve function, the newly formed myelin sheath is often deficient and inefficient compared to the original.

10.8.2 Role of Oxidative Stress in Multiple Sclerosis Pathogenesis

Demyelination is marked by disruption of the BBB, an impediment that normally stops immune cells such as T cells, B cells, and macrophages from accessing the CNS. Immune cells release pro-inflammatory cytokines such as tumor necrosis factor-alpha (TNF-alpha), interleukin-6 (IL-6), interleukin-17 (IL-17), and interferon-gamma (IFN-gamma), which contribute to the apoptosis of oligodendrocytes and the breakdown of myelin. Additionally, microglia, the immune cells found in the CNS, become activated and worsen the condition by increasing oxidative stress through the release of reactive oxygen species (ROS) and nitric oxide (NO) (Adamczyk and Adamczyk-Sowa 2016; Ohl et al. 2016; Sanabria-Castro et al. 2024).

An imbalance between ROS generation and the antioxidant defense system results in oxidative stress. In MS, there is an overproduction of ROS from immune cells and dysfunctional mitochondria that results in lipid peroxidation, DNA lesions, and protein oxidation. Mitochondrial dysfunction is one of the most significant pathological hallmarks of MS. Here, damage to the mitochondria decreases the effectiveness of the electron transport chain resulting in an energy shortage and neuronal loss. Self-propagating oxidative damage drives neurodegeneration which leads to further damage of the system and advancement in the disease (Bhatti et al. 2017; Guo et al. 2013) (Fig. 10.5).

10.8.3 Therapeutic Approaches for Multiple Sclerosis

Different therapeutic strategies are considered to treat MS, including disease-modifying therapies (DMTs), antioxidant-based therapies, mitochondrial-targeted therapies, and lifestyle interventions (Table 10.4). Among them, DMTs are more effective and designed to suppress or modulate the inflammatory reactions associated with MS conditions.

10.9 Oxidative Stress in Neurodegenerative Diseases: Common Mechanisms and Distinct Pathways

Although oxidative stress is a commonality among these diseases, their molecular causes, impacted areas, and clinical results vary greatly (Tönnies and Trushina 2017). This chapter investigates disease-specific pathways, analyzes the common

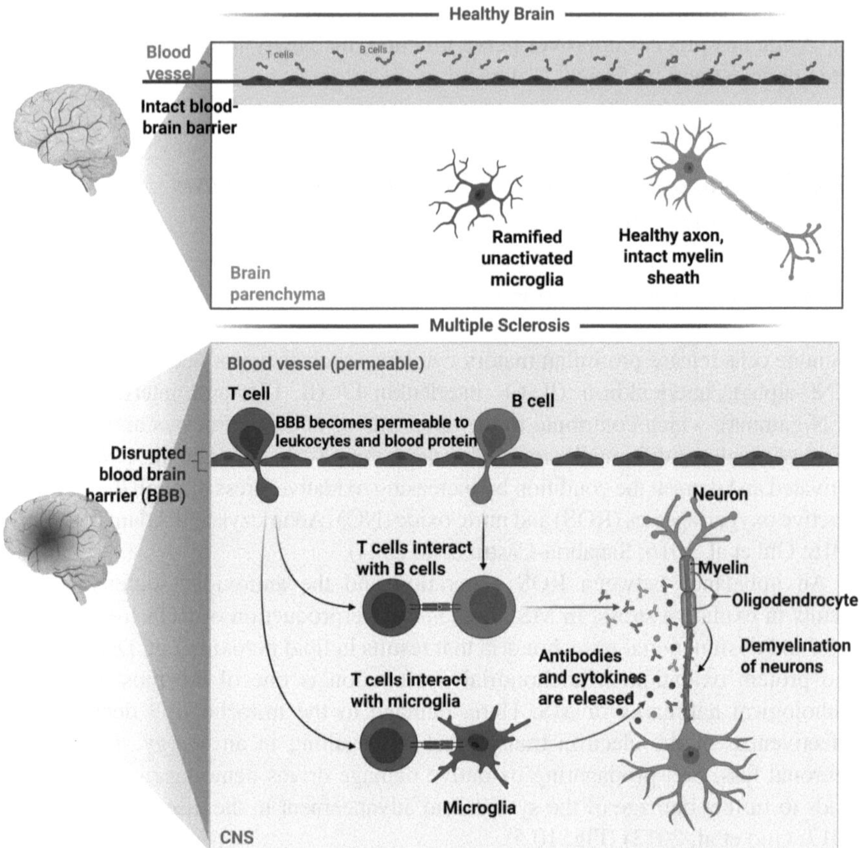

Fig. 10.5 Immune-mediated mechanisms in multiple sclerosis: Role of blood-brain barrier (BBB) disruption and neuroinflammation (created using BioRender)

oxidative stress mechanisms among several diseases, contrasts their effects, and assesses therapeutic approaches to offset this ubiquitous pathology.

10.9.1 Comparative Analysis of Oxidative Stress Impact

Reflecting their different pathologies, oxidative stress shows differently in AD, PD, HD, and MS (Butterfield and Boyd-Kimball 2018; Cheignon et al. 2018; Chen et al. 2012; Dong-Chen et al. 2023; Jurcau and Jurcau 2023; Misrani et al. 2021; Ohl et al. 2016; Pegoretti et al. 2020; Uttara et al. 2009) (Table 10.5).

Table 10.4 Current and emerging therapeutic strategies for multiple sclerosis

Drug	Mechanism of action	Common side effects	References
Disease-modifying therapies			
Glatiramer acetate	Acts as a ligand for MHC molecules, promotes regulatory cell induction, possible neuroprotective effects	Generally well-tolerated; ineffective for progressive MS	Finkelsztejn (2014)
Interferon-beta preparations (1b, 1a, Peginterferon beta-1a)	Modulates T- and B-cell function, alters cytokine expression, aids blood-brain barrier recovery	Flu-like symptoms, possible brief worsening of neurological symptoms	Filipi and Jack (2020)
Natalizumab	Blocks leukocyte adhesion to vascular endothelial cells, inhibiting migration into CNS	Mild headaches, flushing during IV administration	Ransohoff (2007); Wehner et al. (2009)
Mitoxantrone	Affects DNA repair and RNA synthesis, possibly impacts cellular and humoral immunity	Amenorrhea, alopecia, cardiotoxicity	Tafti et al. (2025)
Fingolimod	Inhibits T-cell migration, immunomodulatory effects	Lymphopenia, bradycardia, hepatotoxicity	Wehner et al. (2009)
Dimethyl fumarate (BG-12)	Activates the Nrf2 pathway, reducing oxidative stress and inflammation	Flushing, gastrointestinal symptoms, lymphopenia	Scannevin et al. (2012); Witherick et al. (2010)
Minocycline	Reduces microglial activation and oxidative damage	GI discomfort, dizziness	Chen et al. (2011)
Statins (atorvastatin, simvastatin)	Decrease inflammation and oxidative stress	Muscle pain, liver enzyme changes	Wang et al. (2011)
Antioxidant-based therapies			
Alpha-lipoic acid (ALA)	Potent antioxidant that crosses the blood-brain barrier	Skin rash, GI discomfort	Jiménez-Jiménez et al. (2024); Zuliani and Baroni (2015)
Vitamin D	Modulates immune response and reduces oxidative damage	Generally well-tolerated	Simon et al. (2012)
N-acetylcysteine (NAC)	Boosts GSH levels to neutralize ROS	Nausea, headache, GI discomfort	Khalatbari Mohseni et al. (2023)
Coenzyme Q10 (CoQ10)	Improves mitochondrial function and reduces oxidative damage	Mild GI symptoms	Salekzamani et al. (2025)
Mitochondria-targeted antioxidants			
MitoQ, SS31	Reduce mitochondrial ROS production and prevent neuronal apoptosis	Limited data	Mao et al. (2013)

(continued)

Table 10.4 (continued)

Drug	Mechanism of action	Common side effects	References
Dietary and metabolic strategies			
Caloric restriction, ketogenic diet	Improve mitochondrial bioenergetics and reduce oxidative damage	Requires adherence	Lin et al. (2022)
Mediterranean diet	Rich in polyphenols and omega-3 fatty acids that mitigate oxidative stress	Requires adherence	Abbasi et al. (2024)
Intermittent fasting	Enhances autophagy and reduces oxidative damage	Requires adherence	Lin et al. (2023)
Polyphenols (resveratrol, curcumin, quercetin)	Exhibit strong antioxidant and anti-inflammatory properties	Generally well-tolerated	La Rosa et al. (2023)
Stem cell and gene therapy			
Mesenchymal stem cells (MSCs)	Reduce oxidative stress and promote remyelination	Potential immune response risks	Gugliandolo et al. (2020); Islam et al. (2023)
Gene therapy targeting antioxidant pathways	Enhances expression of superoxide dismutase (SOD) and catalase	Limited data	Michaličková et al. (2020); Upadhayay and Mehan (2021)

10.9.2 Therapeutic Approaches Targeting Oxidative Stress

Given that oxidative stress is a major regulator of HD progression, several treatments may be modified to address OS. Antioxidants are used as therapeutics for managing and/or treating HD (Essa et al. 2019). These include enzymatic and non-enzymatic antioxidant treatments that scavenge ROS in neurons. Enzymatic antioxidants include superoxide mutase (SOD), glutathione peroxidase (Gpx), and catalase (CAT), whereas non-enzymatic antioxidants include Coenzyme Q10 ascorbic acid (vitamin C), α-tocopherol (vitamin E), glutathione (GSH), retinoic acid, carotenoids, and flavonoids (Gil-Mohapel et al. 2014). The other therapeutic approaches for HD involve targeting mHTT DNA and RNA, clearance of Htt protein, and DNA repair pathways. These approaches ultimately aim to reduce mHTT levels and, therefore, ameliorate all of its downstream pathogenic effects, which are multiple and varied (Cheng et al. 2024) (Table 10.6).

The new techniques, which include mRNA-targeting strategies, antisense oligonucleotides, and RNA interference systems, target mHTT and aggregates for degradation via the ubiquitin-proteasome or the autophagy-lysosomal systems. These methods include the proteolysis-targeting chimera, Trim-Away, autophagosome-tethering compound, autophagy-targeting chimera, lysosome-targeting chimera, and approach-targeting mHTT for chaperone-mediated autophagy (Jarosińska and Rüdiger 2021). Although these molecular strategies provide a knowledge-based approach to treat HD and other neurodegenerative diseases at the origin, an extensive study of neurogenerative disease and the role of OS are still unclear. More elaborate

Table 10.5 Impact of oxidative stress on different neurological disorders

Disease	Pathways	Impact	Region	Distinctive feature
Alzheimer's disease	Amyloid-beta (Aβ) hooks metal ions ($Cu2^{+}$, $Fe2^{+}$) in AD to produce ROS and activate microglial NADPH oxidase. Hyperphosphorylated tau disturbs mitochondrial activity, so increasing ROS in the hippocampal and cortex, important areas for memory and cognition	Aβ plaques and tau tangles driven by ROS cause neuronal loss and cognitive decline	Hippocampus and cortex, so undermining cognition and memory	Aβ-metal interactions enhance ROS specifically
Parkinsonism disease	While dopamine metabolism in the substantia nigra generates hydrogen peroxide via monoamine oxidase, PD features alpha-synuclein aggregates damaging mitochondrial Complex I. Loss of the antioxidant protein DJ-1 aggravates oxidative stress in dopaminergic neurons even more	ROS targets dopaminergic neurons, leading to motor symptoms (tremors, rigidity)	Substantia nigra; key driver is dopamine oxidation	Dopamine metabolism aggravates oxidative damage
Huntington's disease	Bindings of mutant huntingtin (mHTT) to mitochondria disturbs electron flow and raises ROS levels. With striatum vulnerable to excitotoxicity, impaired peroxisome proliferator-activated receptor gamma coactivator 1-alpha (PGC-1α) lowers antioxidant defenses	ROS impairs energy metabolism and antioxidant systems, leading to involuntary movements and cognitive impairment	Striatum, related to excitotoxicity	PGC-1α damage links oxidative stress to metabolic collapse
Multiple sclerosis	Oxidative stress develops in MS from immune-mediated demyelination. ROS from activated macrophages and microglia damages oligodendrocytes and myelin. Driven by inflammation, mitochondrial damage in axons aggravates ROS generation especially in white matter tracts	ROS destroys axons and myelin, producing motor and sensory deficits due to immunological attack	Driven by immunological attack, white matter tracts	Inflammation-mediated ROS targets oligodendrocytes, distinct from neuronal focus

Table 10.6 Therapeutic strategies, mechanism of action of Alzheimer's disease, Parkinsonism disease, Huntington disease, and multiple sclerosis

Disease	Therapeutic strategy	Mechanism of action	Targeted outcome
Alzheimer's disease	Metal chelators (e.g., clioquinol)	Reduce Aβ-metal interactions to lower ROS production	Protect hippocampal and cortical neurons from oxidative damage
	Antioxidants (e.g., vitamin E)	Neutralize ROS to prevent lipid peroxidation	Mitigate neuronal loss and cognitive decline
	Mitochondrial stabilizers (e.g., MitoQ)	Enhance mitochondrial function and reduce ROS leakage	Support energy production and cellular resilience
Parkinsonism disease	Coenzyme Q10	Supports mitochondrial Complex I activity, reducing ROS	Preserve dopaminergic neurons in the substantia nigra
	N-acetylcysteine (NAC)	Boosts glutathione (GSH) levels to neutralize ROS	Counter dopamine-related oxidative stress
Huntington disease	PGC-1α activators (e.g., resveratrol)	Upregulate antioxidant gene expression	Enhance defenses and reduce striatal ROS
	Creatine	Stabilizes mitochondrial energy production, lowering ROS	Mitigate metabolic collapse and neuronal damage
Multiple sclerosis	Dimethyl fumarate	Activates Nrf2 pathway to boost antioxidant defenses	Protect myelin and oligodendrocytes from ROS
	Anti-inflammatory agents (e.g., fingolimod)	Reduce microglial activation and ROS release	Reduce microglial activation and ROS release

studies in the future to understand the mechanism involved in HD will help provide more insights into drug development and new therapeutic targets. Therapeutics, catered to each disease, seek to interrupt oxidative stress by addressing its sources and strengthening defenses (Ashok et al. 2022; Forman and Zhang 2021).

10.10 Conclusion

In summary, oxidative stress plays a pivotal role in the pathogenesis of various neurodegenerative diseases, including AD, Parkinson's disease, Huntington's disease, and multiple sclerosis. These conditions are characterized by the accumulation of ROS, which lead to oxidative damage to lipids, proteins, and DNA, ultimately contributing to neuronal death and disease progression. One of the mechanisms underlying oxidative stress in these diseases involves mitochondrial dysfunction, protein misfolding, and inflammation, highlighting the complexity of their pathophysiology. Therapeutic strategies targeting oxidative stress have shown promise, including antioxidant therapies and lifestyle modifications such as diet and exercise.

However, the efficacy of these interventions remains inconsistent, underscoring the need for further research. Emerging treatments and future directions include targeting specific pathways involved in oxidative stress, such as the NF-kB pathway in AD, which may offer more effective therapeutic options.

The comparative analysis of oxidative stress across these diseases reveals both shared mechanisms, such as mitochondrial dysfunction, and disease-specific pathways, such as the role of amyloid-beta in AD and mutant huntingtin in Huntington's disease. Understanding these differences is crucial for developing targeted therapies. Future research should focus on elucidating the precise and common mechanisms by which oxidative stress contributes to neurodegeneration and exploring novel therapeutic strategies that can effectively mitigate oxidative damage. By advancing our knowledge in this area, we may uncover new avenues for preventing or treating these debilitating conditions, ultimately improving the quality of life for individuals affected by neurodegenerative diseases.

References

10 Early Signs and Symptoms of Alzheimer's & Dementia | alz.org [WWW Document], n.d. Alzheimer's Association. https://www.alz.org/alzheimers-dementia/10_signs. Accessed 7 Sept 2025

2024 Alzheimer's disease facts and figures (2024) Alzheimers Dement 20:3708–3821. https://doi.org/10.1002/alz.13809

Abbasi H, Shakouri F, Mosaddeghi-Heris R, Gholipour-Khalili E, Jahanshahlou F, Sanaie S, Naseri A, Talebi M (2024) Mediterranean-like diets in multiple sclerosis: a systematic review. Rev Neurol (Paris) 180:1021–1030. https://doi.org/10.1016/j.neurol.2023.07.017

Adamczyk B, Adamczyk-Sowa M (2016) New insights into the role of oxidative stress mechanisms in the pathophysiology and treatment of multiple sclerosis. Oxidative Med Cell Longev 2016: 1973834. https://doi.org/10.1155/2016/1973834

Al-Gubory KH, Garrel C, Faure P, Sugino N (2012) Roles of antioxidant enzymes in corpus luteum rescue from reactive oxygen species-induced oxidative stress. Reprod Biomed Online 25:551–560. https://doi.org/10.1016/j.rbmo.2012.08.004

Alzheimer's Disease Fact Sheet [WWW Document], 2023. . National Institute on Aging. https://www.nia.nih.gov/health/alzheimers-and-dementia/alzheimers-disease-fact-sheet. Accessed 7 Sept 2025

Alzheimer's disease—Symptoms and causes [WWW Document], n.d. . Mayo Clinic. https://www.mayoclinic.org/diseases-conditions/alzheimers-disease/symptoms-causes/syc-20350447. Accessed 7 Sept 2025

Anderson EJ, Kypson AP, Rodriguez E, Anderson CA, Lehr EJ, Neufer PD (2009) Substrate-specific derangements in mitochondrial metabolism and redox balance in the atrium of the type 2 diabetic human heart. J Am Coll Cardiol 54:1891–1898. https://doi.org/10.1016/j.jacc.2009.07.031

Angelova VT, Stoyanov BP, Simeonova R (2024) New insights into the development of donepezil-based hybrid and natural molecules as multi-target drug agents for Alzheimer's disease treatment. Molecules 29:5314. https://doi.org/10.3390/molecules29225314

Aparicio-Trejo OE, Tapia E, Sánchez-Lozada LG, Pedraza-Chaverri J (2018) Mitochondrial bioenergetics, redox state, dynamics and turnover alterations in renal mass reduction models of chronic kidney diseases and their possible implications in the progression of this illness. Pharmacol Res 135:1–11. https://doi.org/10.1016/j.phrs.2018.07.015

Arrasate M, Finkbeiner S (2012) Protein aggregates in Huntington's disease. Exp Neurol 238:1–11. https://doi.org/10.1016/j.expneurol.2011.12.013

Ashok A, Andrabi SS, Mansoor S, Kuang Y, Kwon BK, Labhasetwar V (2022) Antioxidant therapy in oxidative stress-induced neurodegenerative diseases: role of nanoparticle-based drug delivery systems in clinical translation. Antioxidants (Basel) 11:408. https://doi.org/10.3390/antiox11020408

Atlas of MS 3 Edition: Mapping Multiple Sclerosis Around The World | PDF | Multiple Sclerosis | Epidemiology [WWW Document], n.d. Scribd. https://www.scribd.com/document/500766430/Atlas-3rd-Edition-Epidemiology-report-EN-updated-30-9-20. Accessed 7 Sept 2025

Barcia C, Sánchez Bahillo A, Fernández-Villalba E, Bautista V, Poza Y Poza M, Fernández-Barreiro A, Hirsch EC, Herrero M-T (2004) Evidence of active microglia in substantia nigra pars compacta of parkinsonian monkeys 1 year after MPTP exposure. Glia 46:402–409. https://doi.org/10.1002/glia.20015

Barrera G (2012) Oxidative stress and lipid peroxidation products in cancer progression and therapy. ISRN Oncol 2012:137289. https://doi.org/10.5402/2012/137289

Bedard K, Krause K-H (2007) The NOX family of ROS-generating NADPH oxidases: physiology and pathophysiology. Physiol Rev 87:245–313. https://doi.org/10.1152/physrev.00044.2005

Bhattacharyya A, Chattopadhyay R, Mitra S, Crowe SE (2014) Oxidative stress: an essential factor in the pathogenesis of gastrointestinal mucosal diseases. Physiol Rev 94:329–354. https://doi.org/10.1152/physrev.00040.2012

Bhatti JS, Bhatti GK, Reddy PH (2017) Mitochondrial dysfunction and oxidative stress in metabolic disorders—a step towards mitochondria based therapeutic strategies. Biochim Biophys Acta Mol basis Dis 1863:1066–1077. https://doi.org/10.1016/j.bbadis.2016.11.010

Billings JL, Gordon SL, Rawling T, Doble PA, Bush AI, Adlard PA, Finkelstein DI, Hare DJ (2019) l-3,4-dihydroxyphenylalanine (l-DOPA) modulates brain iron, dopaminergic neurodegeneration and motor dysfunction in iron overload and mutant alpha-synuclein mouse models of Parkinson's disease. J Neurochem 150:88–106. https://doi.org/10.1111/jnc.14676

Blesa J, Trigo-Damas I, Quiroga-Varela A, Jackson-Lewis VR (2015) Oxidative stress and Parkinson's disease. Front Neuroanat 9:91. https://doi.org/10.3389/fnana.2015.00091

Bompart GJ, Prévot DS, Bascands JL (1990) Rapid automated analysis of glutathione reductase, peroxidase, and S-transferase activity: application to cisplatin-induced toxicity. Clin Biochem 23:501–504. https://doi.org/10.1016/0009-9120(90)80039-l

Brand MD (2010) The sites and topology of mitochondrial superoxide production. Exp Gerontol 45:466–472. https://doi.org/10.1016/j.exger.2010.01.003

Brieger K, Schiavone S, Miller FJ, Krause K-H (2012) Reactive oxygen species: from health to disease. Swiss Med Wkly 142:w13659. https://doi.org/10.4414/smw.2012.13659

Brigelius-Flohé R, Maiorino M (2013) Glutathione peroxidases. Biochim Biophys Acta 1830: 3289–3303. https://doi.org/10.1016/j.bbagen.2012.11.020

Browne SE, Ferrante RJ, Beal MF (1999) Oxidative stress in Huntington's disease. Brain Pathol 9: 147–163. https://doi.org/10.1111/j.1750-3639.1999.tb00216.x

Butterfield DA, Boyd-Kimball D (2018) Oxidative stress, amyloid-β peptide, and altered key molecular pathways in the pathogenesis and progression of Alzheimer's disease. J Alzheimer's Dis 62:1345–1367. https://doi.org/10.3233/JAD-170543

Calderaro A, Patanè GT, Tellone E, Barreca D, Ficarra S, Misiti F, Laganà G (2022) The neuroprotective potentiality of flavonoids on Alzheimer's disease. Int J Mol Sci 23:14835. https://doi.org/10.3390/ijms232314835

Chang K-H, Chen C-M (2020) The role of oxidative stress in Parkinson's disease. Antioxidants (Basel) 9:597. https://doi.org/10.3390/antiox9070597

Chaudhary P, Janmeda P, Docea AO, Yeskaliyeva B, Abdull Razis AF, Modu B, Calina D, Sharifi-Rad J (2023) Oxidative stress, free radicals and antioxidants: potential crosstalk in the pathophysiology of human diseases. Front Chem 11:1158198. https://doi.org/10.3389/fchem.2023.1158198

Cheignon C, Tomas M, Bonnefont-Rousselot D, Faller P, Hureau C, Collin F (2018) Oxidative stress and the amyloid beta peptide in Alzheimer's disease. Redox Biol 14:450–464. https://doi.org/10.1016/j.redox.2017.10.014

Chen X, Ma X, Jiang Y, Pi R, Liu Y, Ma L (2011) The prospects of minocycline in multiple sclerosis. J Neuroimmunol 235:1–8. https://doi.org/10.1016/j.jneuroim.2011.04.006

Chen X, Guo C, Kong J (2012) Oxidative stress in neurodegenerative diseases. Neural Regen Res 7: 376–385. https://doi.org/10.3969/j.issn.1673-5374.2012.05.009

Cheng Y, Zhang S, Shang H (2024) Latest advances on new promising molecular-based therapeutic approaches for Huntington's disease. J Transl Int Med 12:134–147. https://doi.org/10.2478/jtim-2023-0142

Conklin KA (2004) Chemotherapy-associated oxidative stress: impact on chemotherapeutic effectiveness. Integr Cancer Ther 3:294–300. https://doi.org/10.1177/1534735404270335

Couto N, Wood J, Barber J (2016) The role of glutathione reductase and related enzymes on cellular redox homoeostasis network. Free Radic Biol Med 95:27–42. https://doi.org/10.1016/j.freeradbiomed.2016.02.028

D'Egidio F, Castelli V, Cimini A, d'Angelo M (2023) Cell rearrangement and oxidant/antioxidant imbalance in Huntington's disease. Antioxidants (Basel) 12:571. https://doi.org/10.3390/antiox12030571

Dai D-F, Johnson SC, Villarin JJ, Chin MT, Nieves-Cintrón M, Chen T, Marcinek DJ, Dorn GW, Kang YJ, Prolla TA, Santana LF, Rabinovitch PS (2011) Mitochondrial oxidative stress mediates angiotensin II-induced cardiac hypertrophy and Galphaq overexpression-induced heart failure. Circ Res 108:837–846. https://doi.org/10.1161/CIRCRESAHA.110.232306

Dalle-Donne I, Rossi R, Giustarini D, Milzani A, Colombo R (2003) Protein carbonyl groups as biomarkers of oxidative stress. Clin Chim Acta 329:23–38. https://doi.org/10.1016/s0009-8981(03)00003-2

Deas E, Cremades N, Angelova PR, Ludtmann MHR, Yao Z, Chen S, Horrocks MH, Banushi B, Little D, Devine MJ, Gissen P, Klenerman D, Dobson CM, Wood NW, Gandhi S, Abramov AY (2016) Alpha-Synuclein oligomers interact with metal ions to induce oxidative stress and neuronal death in Parkinson's disease. Antioxid Redox Signal 24:376–391. https://doi.org/10.1089/ars.2015.6343

Dhapola R, Beura SK, Sharma P, Singh SK, Hari Krishna Reddy D (2024) Oxidative stress in Alzheimer's disease: current knowledge of signaling pathways and therapeutics. Mol Biol Rep 51:48. https://doi.org/10.1007/s11033-023-09021-z

Dias V, Junn E, Mouradian MM (2013) The role of oxidative stress in Parkinson's disease. J Parkinsons Dis 3:461–491. https://doi.org/10.3233/JPD-130230

Dighriri IM, Aldalbahi AA, Albeladi F, Tahiri AA, Kinani EM, Almohsen RA, Alamoudi NH, Alanazi AA, Alkhamshi SJ, Althomali NA, Alrubaiei SN, Altowairqi FK (2023) An overview of the history, pathophysiology, and pharmacological interventions of multiple sclerosis. Cureus 15:e33242. https://doi.org/10.7759/cureus.33242

Dong-Chen X, Yong C, Yang X, Chen-Yu S, Li-Hua P (2023) Signaling pathways in Parkinson's disease: molecular mechanisms and therapeutic interventions. Signal Transduct Target Ther 8: 73. https://doi.org/10.1038/s41392-023-01353-3

Dröse S, Brandt U (2008) The mechanism of mitochondrial superoxide production by the cytochrome bc1 complex. J Biol Chem 283:21649–21654. https://doi.org/10.1074/jbc.M803236200

Essa MM, Moghadas M, Ba-Omar T, Walid Qoronfleh M, Guillemin GJ, Manivasagam T, Justin-Thenmozhi A, Ray B, Bhat A, Chidambaram SB, Fernandes AJ, Song B-J, Akbar M (2019) Protective effects of antioxidants in Huntington's disease: an extensive review. Neurotox Res 35:739–774. https://doi.org/10.1007/s12640-018-9989-9

Feniouk BA, Skulachev VP (n.d.) Cellular and molecular mechanisms of action of mitochondria-targeted antioxidants. Curr Aging Sci 10:41–48. https://doi.org/10.2174/1874609809666160921113706

Filipi M, Jack S (2020) Interferons in the treatment of multiple sclerosis: a clinical efficacy, safety, and tolerability update. Int J MS Care 22:165–172. https://doi.org/10.7224/1537-2073.2018-063

Finkelsztejn A (2014) Multiple sclerosis: overview of disease-modifying agents. Perspect Medicin Chem 6:65–72. https://doi.org/10.4137/PMC.S13213

Firuzi O, Miri R, Tavakkoli M, Saso L (2011) Antioxidant therapy: current status and future prospects. Curr Med Chem 18:3871–3888. https://doi.org/10.2174/092986711803414368

Fišar Z, Hroudová J (2024) CoQ10 and mitochondrial dysfunction in Alzheimer's disease. Antioxidants (Basel) 13:191. https://doi.org/10.3390/antiox13020191

Forman HJ, Zhang H (2021) Targeting oxidative stress in disease: promise and limitations of antioxidant therapy. Nat Rev Drug Discov 20:689–709. https://doi.org/10.1038/s41573-021-00233-1

Fujii H, Nakai K, Fukagawa M (2011) Role of oxidative stress and indoxyl sulfate in progression of cardiovascular disease in chronic kidney disease. Ther Apher Dial 15:125–128. https://doi.org/10.1111/j.1744-9987.2010.00883.x

Gangwar RS, Bevan GH, Palanivel R, Das L, Rajagopalan S (2020) Oxidative stress pathways of air pollution mediated toxicity: recent insights. Redox Biol 34:101545. https://doi.org/10.1016/j.redox.2020.101545

Gargouri M, Soussi A, Akrouti A, Magné C, El Feki A (2019) Potential protective effects of the edible alga Arthrospira platensis against lead-induced oxidative stress, anemia, kidney injury, and histopathological changes in adult rats. Appl Physiol Nutr Metab 44:271–281. https://doi.org/10.1139/apnm-2018-0428

Ghasemi N, Razavi S, Nikzad E (2017) Multiple sclerosis: pathogenesis, symptoms, diagnoses and cell-based therapy. Cell J 19:1–10. https://doi.org/10.22074/cellj.2016.4867

Gil-Mohapel J, Brocardo PS, Christie BR (2014) The role of oxidative stress in Huntington's disease: are antioxidants good therapeutic candidates? Curr Drug Targets 15:454–468. https://doi.org/10.2174/1389450115666140115113734

Gonos ES, Kapetanou M, Sereikaite J, Bartosz G, Naparło K, Grzesik M, Sadowska-Bartosz I (2018) Origin and pathophysiology of protein carbonylation, nitration and chlorination in age-related brain diseases and aging. Aging (Albany NY) 10:868–901. https://doi.org/10.18632/aging.101450

Gonzalez-Hunt CP, Wadhwa M, Sanders LH (2018) DNA damage by oxidative stress: measurement strategies for two genomes. Curr Opin Toxicol 7:87–94. https://doi.org/10.1016/j.cotox.2017.11.001

Gryszczyńska B, Formanowicz D, Budzyń M, Wanic-Kossowska M, Pawliczak E, Formanowicz P, Majewski W, Strzyżewski KW, Kasprzak MP, Iskra M (2017) Advanced oxidation protein products and carbonylated proteins as biomarkers of oxidative stress in selected atherosclerosis-mediated diseases. Biomed Res Int 2017:4975264. https://doi.org/10.1155/2017/4975264

Gugliandolo A, Bramanti P, Mazzon E (2020) Mesenchymal stem cells in multiple sclerosis: recent evidence from pre-clinical to clinical studies. Int J Mol Sci 21:8662. https://doi.org/10.3390/ijms21228662

Guo C, Sun L, Chen X, Zhang D (2013) Oxidative stress, mitochondrial damage and neurodegenerative diseases. Neural Regen Res 8:2003–2014. https://doi.org/10.3969/j.issn.1673-5374.2013.21.009

Hall CN, Klein-Flügge MC, Howarth C, Attwell D (2012) Oxidative phosphorylation, not glycolysis, powers presynaptic and postsynaptic mechanisms underlying brain information processing. J Neurosci 32:8940–8951. https://doi.org/10.1523/JNEUROSCI.0026-12.2012

Halliday GM, Ophof A, Broe M, Jensen PH, Kettle E, Fedorow H, Cartwright MI, Griffiths FM, Shepherd CE, Double KL (2005) Alpha-synuclein redistributes to neuromelanin lipid in the substantia nigra early in Parkinson's disease. Brain 128:2654–2664. https://doi.org/10.1093/brain/awh584

Hazra TK, Izumi T, Maidt L, Floyd RA, Mitra S (1998) The presence of two distinct 8-oxoguanine repair enzymes in human cells: their potential complementary roles in preventing mutation. Nucleic Acids Res 26:5116–5122. https://doi.org/10.1093/nar/26.22.5116

Ionescu-Tucker A, Cotman CW (2021) Emerging roles of oxidative stress in brain aging and Alzheimer's disease. Neurobiol Aging 107:86–95. https://doi.org/10.1016/j.neurobiolaging.2021.07.014

Is Alzheimer's Disease Hereditary? [WWW Document], n.d. . Cleveland Clinic. https://my.clevelandclinic.org/health/diseases/9164-alzheimers-disease. Accessed 7 Sept 2025

Ishola IO, Olubodun-Obadun TG, Akinwande AS, Adeyemi OO (2023) Cajanus cajan (L) Millsp. Seed extract ameliorates scopolamine-induced amnesia through increase in antioxidant defense mechanisms and cholinergic neurotransmission. Niger J Physiol Sci 38:91–99. https://doi.org/10.54548/njps.v38i1.13

Islam MA, Alam SS, Kundu S, Ahmed S, Sultana S, Patar A, Hossan T (2023) Mesenchymal stem cell therapy in multiple sclerosis: a systematic review and meta-analysis. J Clin Med 12:6311. https://doi.org/10.3390/jcm12196311

Jarosińska OD, Rüdiger SGD (2021) Molecular strategies to target protein aggregation in Huntington's disease. Front Mol Biosci 8:769184. https://doi.org/10.3389/fmolb.2021.769184

Jenner P (2003) Oxidative stress in Parkinson's disease. Ann Neurol 53 Suppl 3:S26–S36; discussion S36-38. https://doi.org/10.1002/ana.10483

Jia D, Koonce NA, Griffin RJ, Jackson C, Corry PM (2010) Prevention and mitigation of acute death of mice after abdominal irradiation by the antioxidant N-acetyl-cysteine (NAC). Radiat Res 173:579–589. https://doi.org/10.1667/RR2030.1

Jiménez-Jiménez FJ, Alonso-Navarro H, Salgado-Cámara P, García-Martín E, Agúndez JAG (2024) Antioxidant therapies in the treatment of multiple sclerosis. Biomolecules 14:1266. https://doi.org/10.3390/biom14101266

Jin H, Kanthasamy A, Ghosh A, Anantharam V, Kalyanaraman B, Kanthasamy AG (2014) Mitochondria-targeted antioxidants for treatment of Parkinson's disease: preclinical and clinical outcomes. Biochim Biophys Acta 1842:1282–1294. https://doi.org/10.1016/j.bbadis.2013.09.007

Jomova K, Raptova R, Alomar SY, Alwasel SH, Nepovimova E, Kuca K, Valko M (2023) Reactive oxygen species, toxicity, oxidative stress, and antioxidants: chronic diseases and aging. Arch Toxicol 97:2499–2574. https://doi.org/10.1007/s00204-023-03562-9

Jones DP (2002) Redox potential of GSH/GSSG couple: assay and biological significance. Methods Enzymol 348:93–112. https://doi.org/10.1016/s0076-6879(02)48630-2

Joshi DC, Chavan MB, Gurow K, Gupta M, Dhaliwal JS, Ming LC (2025) The role of mitochondrial dysfunction in Huntington's disease: implications for therapeutic targeting. Biomed Pharmacother 183:117827. https://doi.org/10.1016/j.biopha.2025.117827

Jurcau A, Jurcau CM (2023) Mitochondria in Huntington's disease: implications in pathogenesis and mitochondrial-targeted therapeutic strategies. Neural Regen Res 18:1472–1477. https://doi.org/10.4103/1673-5374.360289

Kamaljeet, Singh S, Gupta GD, Aran KR (2024) Emerging role of antioxidants in Alzheimer's disease: insight into physiological, pathological mechanisms and management. Pharmaceutical Science Advances 2:100021. https://doi.org/10.1016/j.pscia.2023.100021

Kaźmierczak-Barańska J, Boguszewska K, Adamus-Grabicka A, Karwowski BT (2020) Two faces of vitamin C-antioxidative and pro-oxidative agent. Nutrients 12:1501. https://doi.org/10.3390/nu12051501

Khalatbari Mohseni G, Hosseini SA, Majdinasab N, Cheraghian B (2023) Effects of N-acetylcysteine on oxidative stress biomarkers, depression, and anxiety symptoms in patients with multiple sclerosis. Neuropsychopharmacol Rep 43:382–390. https://doi.org/10.1002/npr2.12360

Kieburtz K, Wunderle KB (2013) Parkinson's disease: evidence for environmental risk factors. Mov Disord 28:8–13. https://doi.org/10.1002/mds.25150

Klemmensen MM, Borrowman SH, Pearce C, Pyles B, Chandra B (2024) Mitochondrial dysfunction in neurodegenerative disorders. Neurotherapeutics 21:e00292. https://doi.org/10.1016/j.neurot.2023.10.002

Klivenyi P, Calingasan NY, Starkov A, Stavrovskaya IG, Kristal BS, Yang L, Wieringa B, Beal MF (2004) Neuroprotective mechanisms of creatine occur in the absence of mitochondrial creatine kinase. Neurobiol Dis 15:610–617. https://doi.org/10.1016/j.nbd.2003.12.014

Kondo N, Ishii Y, Kwon Y-W, Tanito M, Horita H, Nishinaka Y, Nakamura H, Yodoi J (2004) Redox-sensing release of human thioredoxin from T lymphocytes with negative feedback loops. J Immunol 172:442–448. https://doi.org/10.4049/jimmunol.172.1.442

Kouli A, Torsney KM, Kuan W-L (2018) Parkinson's disease: etiology, neuropathology, and pathogenesis. In: Stoker TB, Greenland JC (eds) Parkinson's disease: pathogenesis and clinical aspects. Codon Publications, Brisbane (AU)

Kumar R, Sinha VR, Dahiya L, Singh G, Sarwal A (2020) Impact of cyclodextrin derivatives on systemic release of duloxetine HCl via buccal route. Drug Dev Ind Pharm 46:931–945. https://doi.org/10.1080/03639045.2020.1764019

La Rosa G, Lonardo MS, Cacciapuoti N, Muscariello E, Guida B, Faraonio R, Santillo M, Damiano S (2023) Dietary polyphenols, microbiome, and multiple sclerosis: from molecular anti-inflammatory and neuroprotective mechanisms to clinical evidence. Int J Mol Sci 24:7247. https://doi.org/10.3390/ijms24087247

Lamptey RNL, Chaulagain B, Trivedi R, Gothwal A, Layek B, Singh J (2022) A review of the common neurodegenerative disorders: current therapeutic approaches and the potential role of nanotherapeutics. Int J Mol Sci 23:1851. https://doi.org/10.3390/ijms23031851

Lane J, Ng HS, Poyser C, Lucas RM, Tremlett H (2022) Multiple sclerosis incidence: a systematic review of change over time by geographical region. Mult Scler Relat Disord 63:103932. https://doi.org/10.1016/j.msard.2022.103932

Lee SQE, Tan TS, Kawamukai M, Chen ES (2017) Cellular factories for coenzyme Q10 production. Microb Cell Factories 16:39. https://doi.org/10.1186/s12934-017-0646-4

Li C, Xue Y, Ba X, Wang R (2022) The role of 8-oxoG repair systems in tumorigenesis and cancer therapy. Cells 11:3798. https://doi.org/10.3390/cells11233798

Lin W-S, Lin S-J, Liao P-Y, Suresh D, Hsu T-R, Wang P-Y (2022) Role of ketogenic diets in multiple sclerosis and related animal models: an updated review. Adv Nutr 13:2002–2014. https://doi.org/10.1093/advances/nmac065

Lin X, Wang S, Gao Y (2023) The effects of intermittent fasting for patients with multiple sclerosis (MS): a systematic review. Front Nutr 10:1328426. https://doi.org/10.3389/fnut.2023.1328426

Lismont C, Revenco I, Fransen M (2019) Peroxisomal hydrogen peroxide metabolism and signaling in health and disease. Int J Mol Sci 20:3673. https://doi.org/10.3390/ijms20153673

Liu Z, Cecarini V, Cuccioloni M, Bonfili L, Gong C, Angeletti M, Eleuteri AM (2024) Ginsenosides Rg1 and Rg2 activate autophagy and attenuate oxidative stress in Neuroblastoma cells overexpressing aβ(1-42). Antioxidants (Basel) 13:310. https://doi.org/10.3390/antiox13030310

Lotharius J, Brundin P (2002) Pathogenesis of Parkinson's disease: dopamine, vesicles and alpha-synuclein. Nat Rev Neurosci 3:932–942. https://doi.org/10.1038/nrn983

Lubos E, Loscalzo J, Handy DE (2011) Glutathione peroxidase-1 in health and disease: from molecular mechanisms to therapeutic opportunities. Antioxid Redox Signal 15:1957–1997. https://doi.org/10.1089/ars.2010.3586

Lushchak VI (2012) Glutathione homeostasis and functions: potential targets for medical interventions. J Amino Acids 2012:736837. https://doi.org/10.1155/2012/736837

Lustgarten M, Muller FL, Van Remmen H (2011) Chapter 8—an objective appraisal of the free radical theory of aging. In: Masoro EJ, Austad SN (eds) Handbook of the biology of aging (seventh edition), handbooks of aging. Academic Press, San Diego, pp 177–202. https://doi.org/10.1016/B978-0-12-378638-8.00008-7

Mao P, Manczak M, Shirendeb UP, Reddy PH (2013) MitoQ, a mitochondria-targeted antioxidant, delays disease progression and alleviates pathogenesis in an experimental autoimmune encephalomyelitis mouse model of multiple sclerosis. Biochim Biophys Acta 1832:2322–2331. https://doi.org/10.1016/j.bbadis.2013.09.005

Martins SG, Zilhão R, Thorsteinsdóttir S, Carlos AR (2021) Linking oxidative stress and DNA damage to changes in the expression of extracellular matrix components. Front Genet 12: 673002. https://doi.org/10.3389/fgene.2021.673002

Mazuryk O, Stochel G, Brindell M (2020) Variations in reactive oxygen species generation by urban airborne particulate matter in lung epithelial cells-impact of inorganic fraction. Front Chem 8:581752. https://doi.org/10.3389/fchem.2020.581752

Michaličková D, Hrnčíř T, Canová NK, Slanař O (2020) Targeting Keap1/Nrf2/ARE signaling pathway in multiple sclerosis. Eur J Pharmacol 873:172973. https://doi.org/10.1016/j.ejphar.2020.172973

Mills EE (1988) The modifying effect of beta-carotene on radiation and chemotherapy induced oral mucositis. Br J Cancer 57:416–417. https://doi.org/10.1038/bjc.1988.94

Misrani A, Tabassum S, Yang L (2021) Mitochondrial dysfunction and oxidative stress in Alzheimer's disease. Front Aging Neurosci 13:617588. https://doi.org/10.3389/fnagi.2021.617588

Mortensen A, Skibsted LH, Truscott TG (2001) The interaction of dietary carotenoids with radical species. Arch Biochem Biophys 385:13–19. https://doi.org/10.1006/abbi.2000.2172

Mortezaee K (2018) Nicotinamide adenine dinucleotide phosphate (NADPH) oxidase (NOX) and liver fibrosis: a review. Cell Biochem Funct 36:292–302. https://doi.org/10.1002/cbf.3351

Naito Y, Yoshikawa T (2002) Molecular and cellular mechanisms involved in *Helicobacter pylori*-induced inflammation and oxidative stress1, 2. Free Radic Biol Med 33:323–336. https://doi.org/10.1016/S0891-5849(02)00868-7

Nandi A, Yan L-J, Jana CK, Das N (2019) Role of catalase in oxidative stress- and age-associated degenerative diseases. Oxidative Med Cell Longev 2019:9613090. https://doi.org/10.1155/2019/9613090

Nozik-Grayck E, Suliman HB, Piantadosi CA (2005) Extracellular superoxide dismutase. Int J Biochem Cell Biol 37:2466–2471. https://doi.org/10.1016/j.biocel.2005.06.012

Nuszkiewicz J, Woźniak A, Szewczyk-Golec K (2020) Ionizing radiation as a source of oxidative stress-the protective role of melatonin and vitamin D. Int J Mol Sci 21:5804. https://doi.org/10.3390/ijms21165804

Ohl K, Tenbrock K, Kipp M (2016) Oxidative stress in multiple sclerosis: central and peripheral mode of action. Exp Neurol 277:58–67. https://doi.org/10.1016/j.expneurol.2015.11.010

Parkinson's Disease | National Institute of Neurological Disorders and Stroke [WWW Document], n.d.. https://www.ninds.nih.gov/health-information/disorders/parkinsons-disease. Accessed 7 Sept 2025

Parkinson's disease explained [WWW Document], n.d. NIH MedlinePlus Magazine. https://magazine.medlineplus.gov/article/parkinsons-disease-explained. Accessed 7 Sept 2025

Paul BD, Snyder SH (2019) Impaired redox signaling in Huntington's disease: therapeutic implications. Front Mol Neurosci 12:68. https://doi.org/10.3389/fnmol.2019.00068

Pegoretti V, Swanson KA, Bethea JR, Probert L, Eisel ULM, Fischer R (2020) Inflammation and oxidative stress in multiple sclerosis: consequences for therapy development. Oxidative Med Cell Longev 2020:7191080. https://doi.org/10.1155/2020/7191080

Percário S, da Silva Barbosa A, Varela ELP, Gomes ARQ, Ferreira MES, de Nazaré Araújo Moreira T, Dolabela MF (2020) Oxidative stress in Parkinson's disease: potential benefits of antioxidant supplementation. Oxidative Med Cell Longev 2020:2360872. https://doi.org/10.1155/2020/2360872

Persson T, Popescu BO, Cedazo-Minguez A (2014) Oxidative stress in Alzheimer's disease: why did antioxidant therapy fail? Oxidative Med Cell Longev 2014:427318. https://doi.org/10.1155/2014/427318

Ph.D, R.G., MD (2024) Lewy bodies, dementia, and Parkinson's—what does it all mean? American Parkinson Disease Association. https://www.apdaparkinson.org/article/understanding-parkinsons-disease-dementia-lewy-bodies/. Accessed 7 Sept 2025

Phaniendra A, Jestadi DB, Periyasamy L (2015) Free radicals: properties, sources, targets, and their implication in various diseases. Indian J Clin Biochem 30:11–26. https://doi.org/10.1007/s12291-014-0446-0

Pizzimenti S, Toaldo C, Pettazzoni P, Dianzani MU, Barrera G (2010) The "two-faced" effects of reactive oxygen species and the lipid peroxidation product 4-hydroxynonenal in the hallmarks of cancer. Cancers (Basel) 2:338–363. https://doi.org/10.3390/cancers2020338

Pizzino G, Irrera N, Cucinotta M, Pallio G, Mannino F, Arcoraci V, Squadrito F, Altavilla D, Bitto A (2017) Oxidative stress: harms and benefits for human health. Oxidative Med Cell Longev 2017:8416763. https://doi.org/10.1155/2017/8416763

Poljsak B, Šuput D, Milisav I (2013) Achieving the balance between ROS and antioxidants: when to use the synthetic antioxidants. Oxidative Med Cell Longev 2013:956792. https://doi.org/10.1155/2013/956792

Pritam P, Deka R, Bhardwaj A, Srivastava R, Kumar D, Jha AK, Jha NK, Villa C, Jha SK (2022) Antioxidants in Alzheimer's disease: current therapeutic significance and future prospects. Biology (Basel) 11:212. https://doi.org/10.3390/biology11020212

Prodromou SI, Chatzopoulou F, Saiti A, Giannopoulos-Dimitriou A, Koudoura LA, Pantazaki AA, Chatzidimitriou D, Vasiliou V, Vizirianakis IS (2024) Hepatotoxicity assessment of innovative nutritional supplements based on olive-oil formulations enriched with natural antioxidants. Front Nutr 11:1388492. https://doi.org/10.3389/fnut.2024.1388492

Puspita L, Chung SY, Shim J-W (2017) Oxidative stress and cellular pathologies in Parkinson's disease. Mol Brain 10:53. https://doi.org/10.1186/s13041-017-0340-9

Ramana KV, Srivastava S, Singhal SS (2014) Lipid peroxidation products in human health and disease 2014. Oxidative Med Cell Longev 2014:162414. https://doi.org/10.1155/2014/162414

Ransohoff RM (2007) Natalizumab for multiple sclerosis. N Engl J Med 356:2622–2629. https://doi.org/10.1056/NEJMct071462

Reich DS, Lucchinetti CF, Calabresi PA (2018) Multiple Sclerosis. N Engl J Med 378:169–180. https://doi.org/10.1056/NEJMra1401483

Riley PA (1994) Free radicals in biology: oxidative stress and the effects of ionizing radiation. Int J Radiat Biol 65:27–33. https://doi.org/10.1080/09553009414550041

Rodriguez-Oroz MC, Jahanshahi M, Krack P, Litvan I, Macias R, Bezard E, Obeso JA (2009) Initial clinical manifestations of Parkinson's disease: features and pathophysiological mechanisms. Lancet Neurol 8:1128–1139. https://doi.org/10.1016/S1474-4422(09)70293-5

Roy RG, Mandal PK, Maroon JC (2023) Oxidative stress occurs prior to amyloid aβ plaque formation and tau phosphorylation in Alzheimer's disease: role of glutathione and metal ions. ACS Chem Neurosci 14:2944–2954. https://doi.org/10.1021/acschemneuro.3c00486

Rukmangadachar LA, Bollu PC (2025) Amyloid beta peptide. In: StatPearls. StatPearls Publishing, Treasure Island (FL)

Salekzamani S, Pakkhesal S, Babaei M, Mirzaaghazadeh E, Mosaddeghi-Heris R, Talebi M, Sanaie S, Naseri A (2025) Coenzyme Q10 supplementation in multiple sclerosis; a systematic review. Mult Scler Relat Disord 93:106212. https://doi.org/10.1016/j.msard.2024.106212

Sanabria-Castro A, Alape-Girón A, Flores-Díaz M, Echeverri-McCandless A, Parajeles-Vindas A (2024) Oxidative stress involvement in the molecular pathogenesis and progression of multiple sclerosis: a literature review. Rev Neurosci 35:355–371. https://doi.org/10.1515/revneuro-2023-0091

Scannevin RH, Chollate S, Jung M, Shackett M, Patel H, Bista P, Zeng W, Ryan S, Yamamoto M, Lukashev M, Rhodes KJ (2012) Fumarates promote cytoprotection of central nervous system cells against oxidative stress via the nuclear factor (erythroid-derived 2)-like 2 pathway. J Pharmacol Exp Ther 341:274–284. https://doi.org/10.1124/jpet.111.190132

Schalk I, Zeng K, Wu SK, Stura EA, Matteson J, Huang M, Tandon A, Wilson IA, Balch WE (1996) Structure and mutational analysis of Rab GDP-dissociation inhibitor. Nature 381:42–48. https://doi.org/10.1038/381042a0

Schrader M, Fahimi HD (2006) Peroxisomes and oxidative stress. Biochim Biophys Acta 1763: 1755–1766. https://doi.org/10.1016/j.bbamcr.2006.09.006

Segura-Aguilar J, Paris I, Muñoz P, Ferrari E, Zecca L, Zucca FA (2014) Protective and toxic roles of dopamine in Parkinson's disease. J Neurochem 129:898–915. https://doi.org/10.1111/jnc.12686

Sharifi-Rad M, Anil Kumar NV, Zucca P, Varoni EM, Dini L, Panzarini E, Rajkovic J, Tsouh Fokou PV, Azzini E, Peluso I, Prakash Mishra A, Nigam M, El Rayess Y, Beyrouthy ME, Polito L, Iriti M, Martins N, Martorell M, Docea AO, Setzer WN, Calina D, Cho WC, Sharifi-Rad J (2020) Lifestyle, oxidative stress, and antioxidants: Back and forth in the pathophysiology of chronic diseases. Front Physiol 11:694. https://doi.org/10.3389/fphys.2020.00694

Shults CW, Oakes D, Kieburtz K, Beal MF, Haas R, Plumb S, Juncos JL, Nutt J, Shoulson I, Carter J, Kompoliti K, Perlmutter JS, Reich S, Stern M, Watts RL, Kurlan R, Molho E, Harrison M, Lew M, Parkinson Study Group (2002) Effects of coenzyme Q10 in early Parkinson disease: evidence of slowing of the functional decline. Arch Neurol 59:1541–1550. https://doi.org/10.1001/archneur.59.10.1541

Simon KC, Munger KL, Ascherio A (2012) Vitamin D and multiple sclerosis: epidemiology, immunology, and genetics. Curr Opin Neurol 25:246–251. https://doi.org/10.1097/WCO.0b013e3283533a7e

Simunkova M, Barbierikova Z, Jomova K, Hudecova L, Lauro P, Alwasel SH, Alhazza I, Rhodes CJ, Valko M (2021) Antioxidant vs. Prooxidant properties of the flavonoid, Kaempferol, in the presence of Cu(II) ions: a ROS-scavenging activity, Fenton reaction and DNA damage study. Int J Mol Sci 22:1619. https://doi.org/10.3390/ijms22041619

Singh A, Kukreti R, Saso L, Kukreti S (2019) Oxidative stress: a key modulator in neurodegenerative diseases. Molecules 24:1583. https://doi.org/10.3390/molecules24081583

Smith MT, Guyton KZ, Gibbons CF, Fritz JM, Portier CJ, Rusyn I, DeMarini DM, Caldwell JC, Kavlock RJ, Lambert PF, Hecht SS, Bucher JR, Stewart BW, Baan RA, Cogliano VJ, Straif K (2016) Key characteristics of carcinogens as a basis for organizing data on mechanisms of carcinogenesis. Environ Health Perspect 124:713–721. https://doi.org/10.1289/ehp.1509912

Snow RJ, Murphy RM (2001) Creatine and the creatine transporter: a review. Mol Cell Biochem 224:169–181. https://doi.org/10.1023/a:1011908606819

Spencer JP, Jenner P, Daniel SE, Lees AJ, Marsden DC, Halliwell B (1998) Conjugates of catecholamines with cysteine and GSH in Parkinson's disease: possible mechanisms of formation involving reactive oxygen species. J Neurochem 71:2112–2122. https://doi.org/10.1046/j.1471-4159.1998.71052112.x

Stohs SJ, Bagchi D (1995) Oxidative mechanisms in the toxicity of metal ions. Free Radic Biol Med 18:321–336. https://doi.org/10.1016/0891-5849(94)00159-h

Stoker TB, Mason SL, Greenland JC, Holden ST, Santini H, Barker RA (2022) Huntington's disease: diagnosis and management. Pract Neurol 22:32–41. https://doi.org/10.1136/practneurol-2021-003074

Sun Y, Zhang H, Zhang X, Wang W, Chen Y, Cai Z, Wang Q, Wang J, Shi Y (2023) Promotion of astrocyte-neuron glutamate-glutamine shuttle by SCFA contributes to the alleviation of Alzheimer's disease. Redox Biol 62:102690. https://doi.org/10.1016/j.redox.2023.102690

Suzuki YJ, Carini M, Butterfield DA (2010) Protein carbonylation. Antioxid Redox Signal 12:323–325. https://doi.org/10.1089/ars.2009.2887

Tafti D, Ehsan M, Xixis KL (2025) Multiple Sclerosis. In: StatPearls. StatPearls Publishing, Treasure Island (FL)

Tamagno E, Guglielmotto M, Vasciaveo V, Tabaton M (2021) Oxidative stress and Beta amyloid in Alzheimer's disease. Which comes first: the chicken or the egg? Antioxidants (Basel) 10:1479. https://doi.org/10.3390/antiox10091479

Taniyama Y, Griendling KK (2003) Reactive oxygen species in the vasculature: molecular and cellular mechanisms. Hypertension 42:1075–1081. https://doi.org/10.1161/01.HYP.0000100443.09293.4F

Teleanu DM, Niculescu A-G, Lungu II, Radu CI, Vladâcenco O, Roza E, Costăchescu B, Grumezescu AM, Teleanu RI (2022) An overview of oxidative stress, neuroinflammation, and neurodegenerative diseases. Int J Mol Sci 23:5938. https://doi.org/10.3390/ijms23115938

Thomas B, Beal MF (2010) Mitochondrial therapies for Parkinson's disease. Mov Disord 25(Suppl 1):S155–S160. https://doi.org/10.1002/mds.22781

Tönnies E, Trushina E (2017) Oxidative stress, synaptic dysfunction, and Alzheimer's disease. J Alzheimer's Dis 57:1105–1121. https://doi.org/10.3233/JAD-161088

Traber MG, Atkinson J (2007) Vitamin E, antioxidant and nothing more. Free Radic Biol Med 43: 4–15. https://doi.org/10.1016/j.freeradbiomed.2007.03.024

Trinh J, Farrer M (2013) Advances in the genetics of Parkinson disease. Nat Rev Neurol 9:445–454. https://doi.org/10.1038/nrneurol.2013.132

Túnez I, Sánchez-López F, Agüera E, Fernández-Bolaños R, Sánchez FM, Tasset-Cuevas I (2011) Important role of oxidative stress biomarkers in Huntington's disease. J Med Chem 54:5602–5606. https://doi.org/10.1021/jm200605a

Upadhayay S, Mehan S (2021) Targeting Nrf2/HO-1 anti-oxidant signaling pathway in the progression of multiple sclerosis and influences on neurological dysfunctions. Brain Disorders 3: 100019. https://doi.org/10.1016/j.dscb.2021.100019

Uttara B, Singh AV, Zamboni P, Mahajan RT (2009) Oxidative stress and neurodegenerative diseases: a review of upstream and downstream antioxidant therapeutic options. Curr Neuropharmacol 7:65–74. https://doi.org/10.2174/157015909787602823

Valko M, Leibfritz D, Moncol J, Cronin MTD, Mazur M, Telser J (2007) Free radicals and antioxidants in normal physiological functions and human disease. Int J Biochem Cell Biol 39:44–84. https://doi.org/10.1016/j.biocel.2006.07.001

Van Camp W, Inzé D, Van Montagu M (1997) The regulation and function of tobacco superoxide dismutases. Free Radic Biol Med 23:515–520. https://doi.org/10.1016/s0891-5849(97)00112-3

Vasava N, Gajera V, Lambole V, Desai T, Patel B, Akabari A (2022) Evaluation of anti-parkinsonian activity of Pueraria tuberosa (Roxb. ex Willd.) DC. on experimental animals. AP 11. https://doi.org/10.54085/ap.2022.11.2.37

Wang J, Xiao Y, Luo M, Luo H (2011) Statins for multiple sclerosis. Cochrane Database Syst Rev 2011:CD008386. https://doi.org/10.1002/14651858.CD008386.pub3

Wehner NG, Gasper C, Shopp G, Nelson J, Draper K, Parker S, Clarke J (2009) Immunotoxicity profile of natalizumab. J Immunotoxicol 6:115–129. https://doi.org/10.1080/15476910902977381

Wells RG, Neilson LE, McHill AW, Hiller AL (2024) Dietary fasting and time-restricted eating in Huntington's disease: therapeutic potential and underlying mechanisms. Transl Neurodegener 13:17. https://doi.org/10.1186/s40035-024-00406-z

Willcox JK, Ash SL, Catignani GL (2004) Antioxidants and prevention of chronic disease. Crit Rev Food Sci Nutr 44:275–295. https://doi.org/10.1080/10408690490468489

Willem M, Dewachter I, Smyth N, Van Dooren T, Borghgraef P, Haass C, Van Leuven F (2004) Beta-site amyloid precursor protein cleaving enzyme 1 increases amyloid deposition in brain parenchyma but reduces cerebrovascular amyloid angiopathy in aging BACE x APP[V717I] double-transgenic mice. Am J Pathol 165:1621–1631. https://doi.org/10.1016/s0002-9440(10)63419-0

Witherick J, Wilkins A, Scolding N, Kemp K (2010) Mechanisms of oxidative damage in multiple sclerosis and a cell therapy approach to treatment. Autoimmune Dis 2011:164608. https://doi.org/10.4061/2011/164608

Xi Y, Feng D, Tao K, Wang R, Shi Y, Qin H, Murphy MP, Yang Q, Zhao G (2018) MitoQ protects dopaminergic neurons in a 6-OHDA induced PD model by enhancing Mfn2-dependent mitochondrial fusion via activation of PGC-1α. Biochim Biophys Acta Mol basis Dis 1864:2859–2870. https://doi.org/10.1016/j.bbadis.2018.05.018

Yang S-R, Valvo S, Yao H, Kode A, Rajendrasozhan S, Edirisinghe I, Caito S, Adenuga D, Henry R, Fromm G, Maggirwar S, Li J-D, Bulger M, Rahman I (2008) IKK alpha causes chromatin modification on pro-inflammatory genes by cigarette smoke in mouse lung. Am J Respir Cell Mol Biol 38:689–698. https://doi.org/10.1165/rcmb.2007-0379OC

Zelko IN, Mariani TJ, Folz RJ (2002) Superoxide dismutase multigene family: a comparison of the CuZn-SOD (SOD1), Mn-SOD (SOD2), and EC-SOD (SOD3) gene structures, evolution, and expression. Free Radic Biol Med 33:337–349. https://doi.org/10.1016/s0891-5849(02)00905-x

Zhu Y, Tian M, Lu S, Qin Y, Zhao T, Shi H, Li Z, Qin D (2025) The antioxidant role of aromatic plant extracts in managing neurodegenerative diseases: a comprehensive review. Brain Res Bull 222:111253. https://doi.org/10.1016/j.brainresbull.2025.111253

Zorova LD, Popkov VA, Plotnikov EY, Silachev DN, Pevzner IB, Jankauskas SS, Babenko VA, Zorov SD, Balakireva AV, Juhaszova M, Sollott SJ, Zorov DB (2018) Mitochondrial membrane potential. Anal Biochem 552:50–59. https://doi.org/10.1016/j.ab.2017.07.009

Zuliani C, Baroni L (2015) Chapter 35—antioxidants for the prevention and treatment of multiple sclerosis: an overview. In: Watson RR, Preedy VR (eds) Bioactive nutraceuticals and dietary supplements in neurological and brain disease. Academic Press, San Diego, pp 341–353. https://doi.org/10.1016/B978-0-12-411462-3.00035-7

Zuo L, Motherwell MS (2013) The impact of reactive oxygen species and genetic mitochondrial mutations in Parkinson's disease. Gene 532:18–23. https://doi.org/10.1016/j.gene.2013.07.085

Chapter 11
Oxidative Stress During *Plasmodium falciparum* Pathogenesis

Sonia Devi, Mansi Thakkar, Anshika Billa, Nikunj Tandel, and Rajeev K. Tyagi

Abbreviations

GSH	Glutathione
H_2O_2/H2O2	Hydrogen peroxide
NAC	N-acetyl cysteine
NO	Nitric oxide
RNS	Reactive nitrogen species

Sonia Devi and Mansi Thakkar contributed equally with all other contributors.

S. Devi
Division of Cell Biology and Immunology, Biomedical Parasitology and Translational-Immunology Lab, CSIR-Institute of Microbial Technology (IMTECH), Chandigarh, India

Academy of Scientific and Innovative Research (AcSIR), Ghaziabad, India

M. Thakkar
Institute of Science, Nirma University, Ahmedabad, Gujarat, India

A. Billa
Departement of Medical Microbiology, Postgraduate Institute of Medical Education and Research (PGIMER), Chandigarh, India

N. Tandel (✉)
Department of Biotechnology, Veer Narmad South Gujarat University (VNSGU), Surat, Gujarat, India

R. K. Tyagi (✉)
University Center for Research and Development (UCRD), Biomedical Parasitology and Translational-Immunology Lab, University Institute of Biotechnology, Chandigarh University, Mohali, Punjab, India
e-mail: rajeev.e19474@cumail.in

N. Srivastava, S. A. Saraf (eds.), *Pharmacological Strategies for Prevention and Treatment of Oxidative Stress*, Oxidative Stress in Applied Basic Research and Clinical Practice, https://doi.org/10.1007/978-3-032-19837-2_11

ROS	Reactive oxygen species
SOD	Superoxide dismutase
TAS	Total antioxidant status

11.1 Introduction

Malaria, a disease with deep historical roots, has been a persistent threat to human health for centuries. Caused by the protozoan parasite, *P. falciparum*, it has significantly impacted human societies and evolution. This parasite, transmitted through the bites of infected *Anopheles* mosquitoes, has been a formidable obstacle to human progress (Sato 2021; Garrido-Cardenas et al. 2023). The history of *P. falciparum* is closely linked with human evolution and migration. Genetic evidence suggests that the parasite originated in Africa and has co-evolved with humans for hundreds of thousands of years (Loy et al. 2017). The development of agriculture and the growth of human settlements created ideal conditions for the proliferation of *Anopheles* mosquitoes, leading to the spread of malaria. Throughout history, malaria has been documented in various texts, with the term itself originating from the Italian word for "bad air," reflecting early misconceptions about its cause (Hempelmann and Krafts 2013). The discovery of the parasite's life cycle and its transmission by mosquitoes, by Alphonse Laveran and Sir Ronald Ross, respectively, marked significant milestones in the understanding and control of malaria (Cox 2010). Despite these advances, *P. falciparum* remains a resilient pathogen, capable of developing resistance to antimalarial drugs and adapting to changing environments (Marino et al. 2023; Jagannathan and Kakuru 2022; Kale et al. 2024). The ongoing battle against malaria underscores the importance of understanding the complex biology of *P. falciparum* and its interaction with the human host (Chahine and Le Roch 2022; Tyagi et al. 2018), including the role of oxidative stress in the pathogenesis of the disease (Vasquez et al. 2021). Oxidative stress is a critical factor in the pathogenesis of *P. falciparum* malaria, affecting both the parasite and the host. This stress is caused by both host and parasite-derived sources, primarily through the digestion of haemoglobin by the parasite, which releases free heme that catalyzes the formation of reactive oxygen species (ROS) (Wiser 2024). This chapter explores the relationship between oxidative stress and *P. falciparum*-mediated malaria infection, highlighting its implications for disease progression, severity, and potential therapeutic interventions.

11.2 Importance of Oxidative Stress in Human Health

Oxidative stress has emerged as a critical factor in the pathophysiology of numerous human diseases, making its study a pivotal area of research in biomedical sciences (Reddy 2023). The delicate balance between the production of ROS and the body's antioxidant defense mechanisms is fundamental to maintaining cellular homeostasis

(Reddy 2023). When this balance is disrupted, leading to an excessive accumulation of ROS, oxidative stress ensues, causing damage to cellular components, including lipids, proteins, and nucleic acids (Kıran et al. 2023). The importance of studying oxidative stress in human health cannot be overstated, as it is implicated in a wide range of diseases, including cancer, cardiovascular diseases, and neurodegenerative disorders (Reddy 2023). Oxidative stress occurs when there is an imbalance between the production of ROS and the ability of the biological system to detoxify these reactive species or repair the resulting damage. ROS, including free radicals such as superoxide (O_2^-) and hydroxyl radical (OH^-), are naturally produced during cellular metabolism, primarily in the mitochondria (Reddy 2023). While ROS play an essential role in cell signalling and homeostasis, their overproduction or inadequate neutralization can lead to oxidative damage.

11.2.1 Sources of Oxidants in the Body

The genesis of free radicals can be traced to both endogenous and exogenous sources. Endogenously, the human body synthesizes a variety of oxidants as metabolic by-products, which are broadly classified into ROS, reactive nitrogen species (RNS), and reactive sulfur species (RSS). These oxidants are generated primarily through normal physiological processes, such as the production of ROS during mitochondrial oxidative phosphorylation, where oxygen is reduced to produce adenosine triphosphate (ATP), resulting in the formation of superoxide anions ($O_2^{\bullet-}$) and hydrogen peroxide (H_2O_2) (Abdal Dayem et al. 2017). During inflammation, immune cells such as neutrophils and macrophages are activated to generate significant amounts of oxidants as a component of the host's defense mechanism against pathogens. This includes the generation of superoxide, hydrogen peroxide, and RNS such as nitric oxide (NO) and peroxynitrite ($ONOO^-$). Furthermore, RSS, which have been recognized more recently, also contribute to the oxidative stress milieu and are derived from metabolic pathways involving sulfur-containing molecules (Andrés et al. 2022b).

External sources of free radicals encompass a wide range of environmental contaminants, including heavy metals such as cadmium, mercury, lead, iron, and arsenic. These metals can induce oxidative stress by promoting the formation of ROS and other free radicals (Valko et al. 2005). Additionally, certain medications, including cyclosporine, tacrolimus, gentamicin, and bleomycin, have been shown to generate free radicals as a result of their metabolic breakdown or direct interaction with cellular components (Sadeg et al. 1993; Massicot et al. 1997; Pizzino et al. 2017). Chemical solvents and specific cooking techniques, such as smoking meats or using aged oils and fats, can also lead to the production of free radicals (Pizzino et al. 2017). Tobacco smoke, alcohol intake, and exposure to radiation are further exogenous factors that contribute to the generation of free radicals within the body. Upon entry, these external agents undergo degradation or metabolic processing, often resulting in the formation of free radicals as by-products and could lead to the formation of tumor (Pizzino et al. 2014; Willcox et al. 2004; Valko et al. 2004).

11.2.1.1 Types of ROS

ROS constitute a heterogeneous group of molecules that play critical roles in various cellular processes and are implicated in oxidative stress-induced damage. A notable subset of these species is the superoxide radical ($O_2^{\bullet-}$), which is predominantly generated during cellular respiration when electrons leak from the electron transport chain and react with molecular oxygen. This radical is primarily produced by the enzyme nicotinamide adenine dinucleotide phosphate (NADPH) oxidase during the respiratory burst in phagocytes, and it can initiate a cascade of oxidative reactions. As a free radical, superoxide can trigger a series of reactions that generate additional harmful species (Halliwell 2006). H_2O_2, a non-radical oxidant, is formed through the dismutation of superoxide radicals. Although it exhibits lower reactivity compared to superoxide, hydrogen peroxide can diffuse across cellular membranes, causing oxidative damage at distant sites within the cell. Under certain conditions, it can be converted into more reactive entities (Mittler and Zilinskas 1991; Das and Roychoudhury 2014).

Hydroxyl radicals ($OH^{\bullet}$) are among the most reactive forms of ROS, capable of interacting with nearly all types of biomolecules, including lipids, proteins, and DNA. Their extremely brief lifespan and restricted range render them particularly damaging in their immediate vicinity (Pinto et al. 2003). Hydroxyl radicals are primarily formed through Fenton reactions involving hydrogen peroxide and metal ions. Hypochlorous acid (HOCl) is another potent non-radical ROS generated by activated neutrophils during immune responses. It is produced from hydrogen peroxide in the presence of chloride ions and can cause significant oxidative damage to cellular components, including proteins and nucleic acids (Andrés et al. 2022a). Finally, peroxynitrite ($ONOO^-$) is formed through the interaction of NO and superoxide, serving as a strong oxidant that can modify proteins and lipids, thus contributing to cellular injury and the inflammatory response (Prolo et al. 2014).

Each of these species exhibits unique properties and reactivities that define their roles in both normal cellular activities and oxidative stress-related damage. A comprehensive understanding of these various types of ROS is essential for investigating their implications in health and disease.

11.2.2 Cellular Defense Mechanisms Against Oxidative Stress

Cells have evolved a sophisticated array of defense mechanisms to protect themselves against oxidative stress induced by ROS. Among these strategies, antioxidant enzymes play a pivotal role in neutralizing ROS. Superoxide dismutase (SOD) converts superoxide radicals into H_2O_2, thereby mitigating their harmful effects. Catalase (CAT) decomposes H_2O_2 into water and oxygen, preventing its accumulation. Glutathione peroxidase (GPx) reduces H_2O_2 and lipid peroxides using glutathione as a substrate (Ali et al. 2020). In addition to these enzymes, various small molecule antioxidants are crucial in scavenging free radicals. Glutathione (GSH), a tripeptide, protects cells by inhibiting the oxidation of thiol groups and maintaining redox equilibrium (Georgiou-Siafis and Tsiftsoglou 2023). Ascorbic acid (Vitamin C), a water-soluble antioxidant, directly scavenges free radicals and aids in the regeneration of other antioxidants (Gęgotek and Skrzydlewska 2022). Tocopherols (Vitamin E), fat-soluble antioxidants, shield cell membranes from oxidative damage (White et al. 1997). Certain antioxidants also act as metal chelators, binding to pro-oxidant metal ions such as iron and copper to inhibit their role in catalyzing free radical formation (Gulcin and Alwasel 2022).

Cells are further equipped with repair mechanisms to address oxidative damage. DNA repair enzymes rectify oxidative lesions in DNA, while chaperone proteins assist in refolding damaged proteins (Van Houten et al. 2018). Antioxidants can also directly scavenge reactive species, reducing their concentration and potential harm to cellular structures. Moreover, oxidative stress is often associated with inflammation. Cells can modulate inflammatory responses to reduce oxidative damage, exemplified by the relationship between oxidative stress and chronic inflammation (Hardbower et al. 2013). In summary, the cellular defense strategies employed to counteract oxidative stress involve a synergistic interplay of enzymatic and non-enzymatic antioxidants, metal chelation, repair mechanisms, and the modulation of inflammatory responses. These mechanisms collectively ensure the preservation of cellular integrity and functionality in the face of oxidative threats.

In the context of *P. falciparum* malaria, oxidative stress is particularly relevant due to the parasite's unique life cycle and its interaction with the human immune system. The parasite's metabolism and the host's immune response contribute significantly to the generation of oxidants.

11.3 Mechanisms of ROS Generation in Malaria

Oxidative stress in malaria is characterized by an excessive production of ROS/RNS, which can cause systemic and tissue damage, including in the lungs and brain. This imbalance can lead to severe clinical conditions such as cerebral malaria (Vasquez et al. 2021). The host's antioxidant defense system, comprising enzymes and small molecules, attempts to counteract this oxidative damage but may be insufficient. In malaria, ROS generation is linked to the parasite's metabolism, particularly the degradation of hemoglobin and the release of free heme (Wiser 2024). Additionally, the host's immune response contributes to ROS production, aiming to eliminate the parasite. The digestion of hemoglobin by *P. falciparum*, the parasite which is responsible for the most severe form of malaria, leads to ROS formation through several mechanisms (Wiser 2024; Vasquez et al. 2021). This process is a critical part of the parasite's life cycle within the human red blood cells (RBCs), where it obtains essential amino acids for its growth and development. However, this metabolic process also contributes to the pathogenesis of malaria, particularly through the generation of oxidative stress. It occurs in various sequential steps: *Hemoglobin Degradation and Heme Release*: *P. falciparum* digests hemoglobin within the acidic environment of its digestive vacuole. This process releases free heme, which is toxic if not properly managed. The parasite deals with the toxicity of heme by converting it into hemozoin, a less toxic, insoluble crystalline form. However, not all heme is converted into hemozoin, and some remains in its toxic form (Burda et al. 2020). *Fenton Reaction*: Free heme and other iron-containing compounds can participate in the Fenton reaction, a chemical reaction that, in the presence of H_2O_2, generates highly reactive hydroxyl radicals (OH•) (Burda et al. 2020). This reaction is catalyzed by the iron in heme and follows the equation: $Fe_2^{+} + H_2O_2 \rightarrow Fe_3^{+} + OH^{-} + OH\bullet$. The hydroxyl radical is one of the most ROS and can cause extensive damage to cellular components, including lipids, proteins, and DNA. *Haber–Weiss Reaction*: Another pathway through which heme can contribute to ROS formation is the Haber–Weiss reaction. This reaction involves the reduction of oxygen (O_2) by superoxide anion (O_2^{-}) to form H_2O_2, which can then participate in the Fenton reaction to produce hydroxyl radicals (Wunderlich et al. 2012). The presence of heme and other transition metals can facilitate this process. *Direct Oxidation*: The degradation of hemoglobin and the release of heme can also lead to the direct oxidation of various cellular components. The porphyrin ring of heme can interact with and oxidize lipids, proteins, and nucleic acids, contributing to the overall oxidative stress. *Immune Response Activation*: The presence of hemozoin and other byproducts of hemoglobin digestion can activate the host's immune response, leading to the production of ROS by immune cells such as macrophages and neutrophils. These cells produce ROS as part of their mechanism to kill pathogens, but in the process, they also contribute to the oxidative stress experienced by the host's cells (Olivier et al. 2014).

In summary, the digestion of hemoglobin by *P. falciparum* leads to ROS formation through direct chemical reactions involving free heme and other iron-containing compounds, as well as through the activation of the host's immune response (Fig. 11.1). These reactive species contribute to the oxidative stress associated with malaria, which can damage both the parasite and the host's cells, leading to the pathogenesis of the disease.

11.3.1 Role of Oxidative Stress in Parasite Development and Survival

Oxidative stress plays a dual role in malaria. While it can inhibit parasite growth and survival, *P. falciparum* has developed strategies to counteract oxidative stress, allowing it to thrive within the human host. It involves in initial invasion of RBCs to its eventual maturation and transmission to the mosquito vector. Oxidative stress contributes to the severity of malaria, impacting RBCs and leading to hemolysis. It

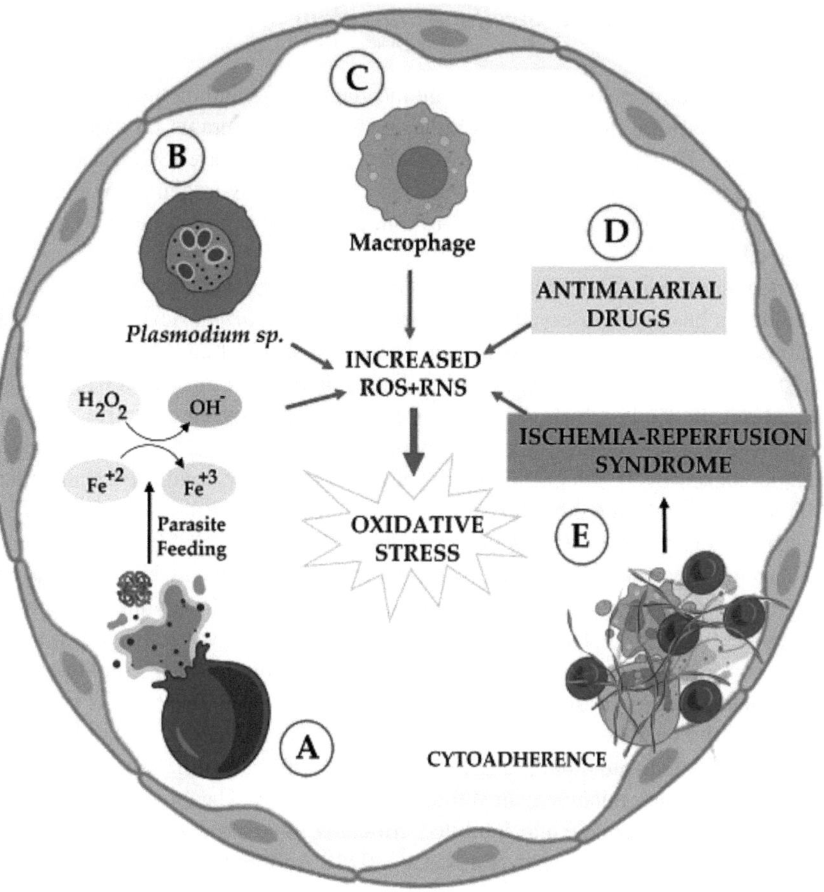

Fig. 11.1 Sources of oxidative stress in malaria. (**a**) Free-iron-induced oxidative stress (OS), as a consequence of hemoglobin digestion by the parasite and through the Fenton or Haber–Weiss reactions; (**c**) direct ROS/RNS production by the parasite; (**b**) host's inflammatory response-derived oxidative stress; (**d**) production of ROS/RNS by antimalarial drug metabolism; (**e**) cytoadherence- and/or anemia-derived oxidative stress, through ischemia-reperfusion syndrome [adapted with permission from (Gomes et al. 2022)]. Reactive oxygen species (ROS); reactive nitrogen species (RNS)

also plays a role in the development of complications such as cerebral malaria and anemia (Greve et al. 2000). Furthermore, moderate levels of oxidative stress have been shown to promote the formation of gametocytes, the sexual stage of the parasite that is responsible for transmission to the mosquito vector. This suggests that oxidative stress may play a role in the parasite's life cycle and its adaptation to the host environment.

11.3.2 Plasmodium-*Specific Antioxidant Mechanisms*

P. falciparum utilizes its own antioxidant mechanisms, including the glutathione and thioredoxin systems, to counteract oxidative stress and ensure its survival within the human host. The parasite also expresses superoxide dismutase (SOD) and catalase, which are enzymes that detoxify superoxide anions and hydrogen peroxide, respectively. These enzymes play a critical role in protecting the parasite from oxidative damage.

Artemisinin and its derivatives are crucial in treating malaria, but resistance to these drugs has been observed, particularly in *P. falciparum*. Several studies have established that oxidative stress can influence the survival and drug sensitivity of malaria parasites (Rocamora et al. 2018). In this direction, Pires and colleagues have carried out the study which explores the relationship between oxidative stress and the efficacy of artemisinin, a key antimalarial drug, particularly in the context of malaria caused by *P. falciparum* (Pires et al. 2024). They conducted standardized drug assays to evaluate the effects of oxidative stress and heat-shock conditions on the chemosensitivity of *P. falciparum* parasites (Pires et al. 2024). Specifically, synchronized parasite lines at the schizont stage were divided into three groups: one group was treated with oxidative stress (oxi-treated RBCs), while the other two groups were left untreated. One of the untreated cultures was subjected to a heat-shock condition by incubating at 41 °C for 4 h to mimic fever conditions. This experimental design allowed the researchers to assess how these stress conditions influenced the parasites' response to artemisinin treatment (Pires et al. 2024).

The results indicated that pre-induced oxidative stress significantly enhanced the effectiveness of artemisinin against the malaria parasites. The oxidative stress led to increased protein damage and inhibited critical cellular processes within the parasites, contributing to their death (Pires et al. 2024). Additionally, the study found that heat-shock responses also played a role in the survival of the parasites under stress, suggesting that these responses could be linked to the development of drug resistance. Moreover, the research highlights the mechanisms through which *P. falciparum* develops resistance to artemisinin, linking these mechanisms to oxidative stress responses (Pires et al. 2024). This connection is crucial, as it suggests that the parasites may adapt to oxidative damage, thereby complicating treatment strategies. The findings also align with earlier studies that indicated a relationship between oxidative stress and drug resistance, but they offer new insights into the specific pathways and cellular responses involved. In summary, the study emphasizes the dual role of oxidative stress in both enhancing the effectiveness of artemisinin and contributing to the development of resistance in malaria parasites (Pires et al. 2024). Overall, the research underscores the need for further studies to optimize the use of oxidative stress modulators in malaria therapy and to combat drug resistance effectively.

11.3.3 Host Antioxidant Defenses

The human host mounts a defense against oxidative stress through both enzymatic and non-enzymatic antioxidants. These defenses are critical for protecting the host from the damaging effects of ROS generated during malaria infection. Understanding the molecular pathways targeted by antioxidants and their interaction with oxidative pathways induced by malaria is crucial for developing effective therapeutic strategies. For examples, High levels of ROS can inhibit the growth and development of *P. falciparum* by damaging its DNA, proteins, and lipids. This oxidative damage can lead to the death of the parasite if it exceeds the parasite's antioxidant defense capabilities.

Recently, Kotepui et al. conducted a study to investigate the association between total antioxidant status (TAS) levels and *Plasmodium* infections, particularly in the context of malaria (Kotepui et al. 2024). The rationale for this research stems from the significant global health threat posed by malaria, which is known to induce oxidative stress in infected individuals. The authors sought to determine whether TAS could serve as a reliable biomarker for assessing the severity of malaria and the associated oxidative stress (Kotepui et al. 2024). By synthesizing data from various studies, the researchers aimed to clarify the relationship between TAS levels and malaria, considering factors such as *Plasmodium* species, age, geographical location, and methods for measuring TAS. This investigation is crucial for enhancing the understanding of oxidative stress in malaria and could contribute to improved diagnostic and therapeutic strategies (Kotepui et al. 2024). The review highlighted that high parasite density correlates with lower TAS levels, suggesting that increased parasitic load depletes the host's antioxidant resources (Kotepui et al. 2024). After treatment, patients exhibited higher TAS levels, indicating a recovery in antioxidant status as the parasitic load decreased. The study also emphasized the importance of both endogenous and exogenous antioxidants in maintaining oxidative balance, noting that malaria infection can overwhelm these defenses, leading to a measurable decrease in TAS (Kotepui et al. 2024). The authors concluded that TAS could serve as a potential biomarker for assessing malaria severity and called for further research to explore the mechanisms underlying the observed changes in TAS levels. This investigation is crucial for enhancing the understanding of oxidative stress in malaria and could contribute to improved diagnostic and therapeutic strategies.

11.3.3.1 Therapeutic Potential of Antioxidants in Malaria

Antioxidants represent a promising adjunct therapy in malaria treatment. Their combination with antimalarial drugs, the potential benefits of natural antioxidants, and the challenges and future directions in this area are discussed in this section. The oxidative stress caused by ROS formation can be targeted for therapeutic purposes in malaria treatment. This approach is based on the understanding that both the parasite *P. falciparum* and the human host are affected by oxidative stress during malaria infection (Gomes et al. 2022). Targeting oxidative stress pathways represents a promising approach for the development of new antimalarial therapies. However, the potential benefits of antioxidant supplementation in malaria patients remain controversial. Ongoing research is focused on identifying the specific targets within the parasite's antioxidant defense system that could lead to the development of novel antimalarial drugs.

Interestingly, recently Ofori et al., explores the immune mechanisms involved in *P. falciparum* malaria, with a focus on the role of neutrophils (Ofori et al. 2023). Neutrophils, which are white blood cells playing a major role in innate immunity, were already known for their ability to release ROS to combat pathogens. However, this study delves deeper into the specific role of neutrophils in malaria protection. The researchers hypothesized that neutrophils generate ROS through Fcγ receptor (FcγR) signaling, triggered by IgG antibodies, and that this process is essential for limiting the malaria parasite's growth, particularly in individuals living in malaria-endemic regions (Ofori et al. 2023). The engagement of these antibodies by the receptors on neutrophils triggers ROS production, which serves as a critical mechanism for limiting *P. falciparum* growth and contributing to naturally acquired immunity. The results confirmed that ROS generation by neutrophils via FcγR signaling is a vital mechanism for parasite killing and that individuals with higher levels of neutrophil ROS had better protection against febrile malaria. This mechanism likely enhances immunity in individuals repeatedly exposed to malaria in endemic areas, where IgG antibodies against *P. falciparum* are developed over time. This finding suggests a potential target for therapeutic strategies aimed at enhancing immune protection in populations vulnerable to malaria (Ofori et al. 2023).

By targeting the mechanisms of ROS generation or enhancing antioxidant defenses, it is possible to develop new strategies for malaria treatment and management. Here are several ways in which oxidative stress can be targeted for therapeutic purposes:

1. *Antioxidant Therapy*: For the improvement of host's antioxidant and immunological response in case of malaria, different studies have evinced the probable future benefits of supplementation with antioxidant compounds from vitamins, mushrooms, plants and drugs which can aid in modulating the host's response by strengthening the endogenous antioxidant defense against oxidative stress or by acting indirectly against parasitic destruction. Administering antioxidants can help protect the host's cells from oxidative damage. Antioxidants such as vitamin C, vitamin E, and glutathione can neutralize free radicals and reduce oxidative stress (Gomes et al. 2022). It has been observed that plants are an important source of new molecules, and in one of study for the evaluation of antimalarial effects, where the survival rate of mice increased having cerebral malaria, plant extract was used of *Terminalia abiba, which* showed antioxidant and anti-inflammatory properties by reducing ROS/RNS (Camara et al. 2019). *Terminalia avicennioides*, a member of the same botanical family, exhibits notable phytochemical properties. Its leaves, pods, and bark are rich in gallic

acid, which has been confirmed to possess antioxidant capabilities through standardized assays (Sadiq et al. 2017; Ouattara et al. 2014). The presence of phenolic and flavonoid compounds in five plant species—*Allamanda cathartica*, *Bixa orellana*, *Cymbopogon citratus*, *Momordica charantia*, and *Ficus exasperate*—suggests potential therapeutic applications for severe malaria (Omonhinmin et al. 2015). Flavonoids from *Ficus exasperate* and *Dacryodes edulis* (Conrad and Uche 2013), *Trichilia heudelotii*, *Polyalthia longifolia*, & *Markhamia tomentosa* (Bankole et al. 2016), and *Dissotis rotundifolia* (Djehoue et al. 2020) also exhibit promising bioactivities. Essential oils derived from various plants have demonstrated antioxidant and parasiticidal properties (Bankole et al. 2016; Kpadonou et al. 2019). Ethanolic extracts and alkaloid fractions from the powdered bark of *Aspidosperma nitidum*, a medicinal plant from the Amazon, have shown antiplasmodial activity against chloroquine-resistant *P. falciparum* clones. In vivo studies have indicated a reduction in parasitemia by 80% on the fifth day of *P. berghei* infection in mice, although this effect was not maintained until the eighth day (Do Nascimento Brandão et al. 2020). In vitro investigations have explored the resistance of *P. falciparum* clones to chloroquine and the antiplasmodial activity of *A. nitidum*'s ethanolic extract (Gaur et al. 2015). Additionally, the in vitro antimalarial activity of eleutherin and isoeleutherin naphthoquinones, isolated from *Eleutherine plicata*, has been linked to the cytochrome bc1 complex. Previous research has indicated that quinones from medicinal plants can exhibit similar in vitro antimalarial effects, highlighting their potential as novel antimalarial agents (Gomes et al. 2021a, b).

Further, the use of commercial medications containing antioxidants has been explored as a potential adjunct in antimalarial treatment. These include deferoxamine, fenozyme, folate, L-glutathione, melatonin, and N-acetyl cysteine (NAC). Among these, NAC has shown promising results as a mucolytic free-radical scavenger for adjuvant therapy in malaria. NAC is also recognized as an exogenous precursor of glutathione (GSH), a crucial component of the antioxidant defense system (Isah and Ibrahim 2014; Nagy 2013; Aldini et al. 2018). In a study by Gomes et al., the supplementation of NAC and *Agaricus sylvaticus* in *P. berghei*-infected mice demonstrated a reduction in parasitemia and an increase in the total capacity of the mice. However, *A. sylvaticus* was more effective in reducing oxidative damage. Conversely, NAC was unable to protect pregnant mice infected with *P. chabaudi* (Sarr et al. 2017). An intriguing study by Firth et al., aimed to assess whether combined artemisinin and NAC therapy was more effective than artemisinin alone. The combined therapy showed reduced malondialdehyde (MDA) levels in lung and brain tissue in experimental malaria caused by *P. berghei* (Fitri et al. 2013). Similarly, another study identified a synergistic effect of combined chloroquine and NAC therapy, resulting in a reduction of parasitemia and MDA levels in *P. berghei*-infected mice. However, the effectiveness of antioxidant therapy in malaria treatment is still under investigation, and more research is needed to determine the optimal dosages and delivery methods.

2. *Enhancing Host Antioxidant Defenses*: Strategies that aim to boost the body's natural antioxidant defenses could offer protection against oxidative stress. This could involve activating enzymes such as superoxide dismutase, catalase, and glutathione peroxidase, which play key roles in detoxifying ROS (Ighodaro and Akinloye 2018). The importance of redox reactions in the mechanisms of drug action and resistance in *P. falciparum* is play a pivotal role. There are several main challenges in studying the intracellular glutathione redox potential in *P. falciparum*. One significant challenge is the difficulty of accurately measuring redox potential in living cells using traditional methods, which often disrupt cellular integrity and fail to provide relevant sub-cellular data (Kasozi et al. 2013). This limitation hampers the understanding of redox metabolism and its implications for drug action and resistance in malaria parasites. Therefore, a need for a reliable and non-invasive method to monitor redox changes in real-time is emphasized, as it is crucial for advancing research in malaria biology and treatment strategies (Kasozi et al. 2013). In this direction, an interesting study carried out which mainly focus on exploring the redox potential of glutathione in the malaria parasite, which is crucial for understanding its biology and response to antimalarial drugs. The researchers developed a novel biosensor, hGrx1-roGFP2, to measure the cytosolic glutathione redox potential (EGSH) in real-time without disrupting the cells, addressing the limitations of traditional methods that often compromise cellular integrity (Kasozi et al. 2013). The hGrx1-roGFP2 probe, which consists of human glutaredoxin-1 linked to a redox-sensitive green fluorescent protein, was specifically designed to be targeted to the cytosol of the parasites, ensuring its stability and responsiveness to the redox environment (Kasozi et al. 2013). This innovative approach enables researchers to study the effects of various redox-active compounds and antimalarial drugs on the cytosolic glutathione redox potential in real time, thereby providing valuable insights into the redox metabolism of malaria parasites and their response to treatment. These findings suggest that the hGrx1-roGFP2 biosensor is a valuable tool for studying redox metabolism in living parasites, which could lead to better understanding and treatment strategies for malaria (Kasozi et al. 2013). Overall, this research highlights the intricate relationship between redox potential and the efficacy of antimalarial therapies, paving the way for future studies in this area.
3. *Targeting Parasite Antioxidant Systems*: As mentioned earlier, *P. falciparum* has its own antioxidant systems, including glutathione and thioredoxin pathways, which are crucial for its survival in the oxidative environment of the human host. Drugs that specifically inhibit these parasite antioxidant systems could increase the parasite's susceptibility to oxidative stress, making it more vulnerable to the host's immune response or to other antimalarial drugs (Pal 2023). Targeting the parasite's antioxidant systems is a strategic approach in the development of new antimalarial drugs. Here are some specific examples of drugs and compounds that have been investigated for their ability to target the parasite's antioxidant systems: *Glutathione System Inhibitors*: *Buthionine Sulfoximine (BSO)* is a drug that inhibits γ-glutamylcysteine synthetase, an enzyme involved in the synthesis of glutathione. By depleting glutathione levels, BSO may increase the susceptibility

of *P. falciparum* to oxidative stress (Chandra et al. 2015). *Glutathione Peroxidase Inhibitors*: Compounds that inhibit glutathione peroxidase, an enzyme that reduces hydrogen peroxide, can increase oxidative stress within the parasite. *Thioredoxin System Inhibitors*: *Auranofin*: Originally developed as an anti-rheumatic drug, auranofin has been shown to inhibit the thioredoxin reductase of *P. falciparum*, leading to increased oxidative stress and parasite death (Sannella et al. 2008). *Triapine*: This compound inhibits ribonucleotide reductase, an enzyme involved in DNA synthesis, and also has effects on the thioredoxin system, contributing to oxidative stress in the parasite (Joardar et al. 2020). *Superoxide Dismutase Inhibitors*: *Dioxygenases*: Certain dioxygenases can act as superoxide dismutase mimics, but when inhibited, they can increase superoxide levels and oxidative stress in the parasite. *Catalase Inhibitors*: *Aminotriazole*: This compound can inhibit catalase, an enzyme that breaks down hydrogen peroxide, thus increasing the levels of this reactive oxygen species within the parasite (Ravindar et al. 2023; Chu et al. 2019). *Combinations with Traditional Antimalarials: Artemisinin Combination Therapy (ACT)*: Artemisinin and its derivatives are known to generate reactive oxygen species within the parasite. Combining artemisinin with other compounds that either increase ROS production in the parasite or inhibit its antioxidant defenses could potentially increase the drug's effectiveness. Combining these drugs with agents that inhibit the parasite's antioxidant systems could enhance their efficacy (Pal 2023).

4. *Dietary and Nutritional Interventions*: Ensuring adequate intake of nutrients that have antioxidant properties, such as certain vitamins and minerals, could support the host's defense against oxidative stress during malaria infection. This approach could be particularly relevant in malaria-endemic regions where malnutrition is common (Vasquez et al. 2021; Shankar 2000).
5. *Nanomedicine and Targeted Drug Delivery*: Advanced drug delivery systems, including nanoparticles, can be designed to specifically target and deliver antioxidants or oxidant-generating compounds to the site of infection, enhancing their therapeutic effects while minimizing systemic side effects (Kannan et al. 2019; Manconi et al. 2019; Tiwari et al. 2024). It investigates a novel approach to enhancing curcumin's antimalarial properties through nanoformulation. For instance, curcumin, a bioactive compound isolated from *Curcuma longa* (turmeric), is renowned for its potent antioxidant and anti-inflammatory attributes. However, its clinical application is hampered by its low bioavailability. Thus, Manconi and colleagues conducted a study to address the escalating resistance to conventional antimalarial drugs, which underscores the urgent need for innovative therapeutic approaches (Manconi et al. 2019). This investigation sought to enhance the bioavailability and stability of curcumin by encapsulating it within Eudragit-coated nutriosomes. The rationale behind this strategy was to facilitate improved intestinal absorption, protect curcumin from gastrointestinal degradation, and augment its systemic efficacy (Manconi et al. 2019). Furthermore, it was postulated that the nanoformulation could amplify curcumin's antioxidant properties, shielding tissues from the oxidative damage associated with malaria

infection. The Eudragit coating conferred stability and protection to curcumin within the gastrointestinal tract, leading to increased bioavailability. In malaria-infected mice, the nanoformulated curcumin demonstrated superior survival rates and a more pronounced reduction in parasitemia compared to unencapsulated curcumin. (Manconi et al. 2019). These results indicate that curcumin-loaded Eudragit-nutriosomes represent a promising strategy for enhancing curcumin's therapeutic potential, particularly in the context of malaria. By improving bioavailability and antioxidant activity, this nanoformulation offers a dual-action approach to protect the gut from oxidative stress and combat systemic malaria infection more effectively. This innovative formulation could be a significant step forward in the development of alternative treatments, especially for drug-resistant malaria strains.

It is important to note that while these compounds show promise in targeting the parasite's antioxidant systems, their use in clinical settings requires further research to ensure efficacy, safety, and to minimize the risk of developing drug resistance. Additionally, the development of new antimalarial drugs must consider the complex interplay between the parasite's antioxidant systems, the host's immune response, and the potential for off-target effects. In all, targeting oxidative stress represents a promising approach for the development of new malaria treatments. However, this strategy requires a careful balance, as excessive reduction of oxidative stress could potentially benefit the parasite, while too much oxidative stress could harm the host. Therefore, future research should focus on understanding the complex interplay between oxidative stress, the parasite, and the host's immune response to develop effective and safe therapeutic interventions.

11.4 Conclusion

The role of oxidative stress in *P. falciparum* malaria is complex and multifaceted, affecting both the parasite and the host. Understanding the mechanisms of ROS generation and the antioxidant defenses of both the parasite and the host opens up new avenues for therapeutic interventions. In conclusion, while oxidative stress serves as an important defense mechanism against *Plasmodium*, it can also be detrimental if excessive. Restoring the balance of oxidative stress through antioxidant supplementation could serve as a beneficial adjunct therapy in preventing severe complications of malaria. Potential therapies could include the administration of antioxidant-rich diets, supplementation with vitamins C and E, or drugs that enhance the endogenous antioxidant capacity. Future research prospects and potential therapeutic interventions targeting oxidative stress in malaria are explored, highlighting the potential for antioxidants to play a significant role in the fight against this ancient disease.

Acknowledgments Rajeev Tyagi would like to express his gratitude to DBT, New Delhi, Ramalingaswami Re-entry Fellowship Project (No. BT/RLF/Re-entry/27/2018*)* and Indian Council of Medical Research (ICMR), New Delhi extramural grant (35/1/2020-Nano/BMS) for generously supporting this study. Rajeev K. Tyagi would like to express his thanks to the central MIL facility of CSIR-IMTECH, Chandigarh. Sonia Thakur gratefully acknowledges the support of the Department of Science and Technology (DST) for providing the Inspire Fellowship (DST/INSPIRE Fellowship/ [IF210486]) for her research. Nikunj Tandel would like to thank the Department of Biotechnology (DBT), VNSGU, Surat, Gujarat for providing facility to carry out his research work.

References

Abdal Dayem A, Hossain MK, Lee SB, Kim K, Saha SK, Yang GM, Choi HY, Cho SG (2017) The role of Reactive Oxygen Species (ROS) in the biological activities of metallic nanoparticles. Int J Mol Sci 18:120

Aldini G, Altomare A, Baron G, Vistoli G, Carini M, Borsani L, Sergio F (2018) N-Acetylcysteine as an antioxidant and disulphide breaking agent: the reasons why. Free Radic Res 52:751–762

Ali SS, Ahsan H, Zia MK, Siddiqui T, Khan FH (2020) Understanding oxidants and antioxidants: classical team with new players. J Food Biochem 44:e13145

Andrés CMC, Pérez De La Lastra JM, Juan CA, Plou FJ, Pérez-Lebeña E (2022a) Hypochlorous acid chemistry in mammalian cells-influence on infection and role in various pathologies. Int J Mol Sci 23

Andrés CMC, Pérez De La Lastra JM, Juan CA, Plou FJ, Pérez-Lebeña E (2022b) The role of reactive species on innate immunity. Vaccines (Basel) 10:1735

Bankole A, Adekunle A, Sowemimo A, Umebese C, Abiodun O, Gbotosho G (2016) Phytochemical screening and in vivo antimalarial activity of extracts from three medicinal plants used in malaria treatment in Nigeria. Parasitol Res 115:299–305

Burda PC, Crosskey T, Lauk K, Zurborg A, Söhnchen C, Liffner B, Wilcke L, Pietsch E, Strauss J, Jeffries CM, Svergun DI, Wilson DW, Wilmanns M, Gilberger TW (2020) Structure-based identification and functional characterization of a Lipocalin in the malaria parasite Plasmodium falciparum. Cell Rep 31:107817

Camara A, Haddad M, Reybier K, Traoré MS, Baldé MA, Royo J, Baldé AO, Batigne P, Haidara M, Baldé ES, Coste A, Baldé AM, Aubouy A (2019) Terminalia albida treatment improves survival in experimental cerebral malaria through reactive oxygen species scavenging and anti-inflammatory properties. Malar J 18:431

Chahine Z, Le Roch KG (2022) Decrypting the complexity of the human malaria parasite biology through systems biology approaches. Front Syst Biol 2:940321

Chandra R, Kumar S, Puri SK (2015) Buthionine sulfoximine increases the efficacy of arteether antimalarial activity in arteether-resistant Plasmodium vinckei by glutathione depletion. Malar J 6:4

Chu XM, Wang C, Wang WL, Liang LL, Liu W, Gong KK, Sun KL (2019) Triazole derivatives and their antiplasmodial and antimalarial activities. Eur J Med Chem 166:206–223

Conrad OA, Uche AI (2013) Assessment of in vivo antioxidant properties of Dacryodes edulis and Ficus exasperata as anti-malaria plants. Asian Pacific J Trop Dis 3:294–300

Cox FE (2010) History of the discovery of the malaria parasites and their vectors. Parasit Vectors 3: 5

Das K, Roychoudhury A (2014) Reactive oxygen species (ROS) and response of antioxidants as ROS-scavengers during environmental stress in plants. Front Environ Sci 2

Djehoue R, Amoussa AMO, Sanni A, Lagnika L (2020) Phytochemical composition and antioxidant property of Dissotis rotundifolia used for malaria management in South Benin. J Med Plants 8:23–29

Do Nascimento Brandão DL, Martins MT, Silva AO, Almeida AD, De Paula RC, De Oliveira AB, Coelho-Ferreira MR, Gomes ATA, De Vasconcelos F, Pereira WLA (2020) Anti-malarial activity and toxicity of Aspidosperma nitidum Benth: a plant used in traditional medicine in the Brazilian Amazon. Res Soc Dev 9:e5059108817

Fitri LE, Iskandar A, Permatasari N, Gunawan JA, Indrawan K (2013) EFEK KOMBINASI ARTEMISININ DAN N-acetylcysteineTERHADAP KADAR Malondialdehyde (MDA) OTAK DAN PARU MENCIT GALUR Balb/c YANG DIINFEKSI Plasmodium berghei. Jurnal Kedokteran Brawijaya 24

Garrido-Cardenas JA, González-Cerón L, García-Maroto F, Cebrián-Carmona J, Manzano-Agugliaro F, Mesa-Valle CM (2023) Analysis of fifty years of severe malaria worldwide research. Pathogens 12:373

Gaur R, Cheema HS, Kumar Y, Singh SP, Yadav DK, Darokar MP, Khan F, Bhakuni RS (2015) In vitro antimalarial activity and molecular modeling studies of novel artemisinin derivatives. RSC Adv 5:47959–47974

Gęgotek A, Skrzydlewska E (2022) Antioxidative and anti-inflammatory activity of ascorbic acid. Antioxidants (Basel) 11:1993

Georgiou-Siafis SK, Tsiftsoglou AS (2023) The key role of GSH in keeping the redox balance in mammalian cells: mechanisms and significance of GSH in detoxification via formation of conjugates. Antioxidants [Online] 12:1953

Gomes ARQ, Brígido HPC, Vale VV, Correa-Barbosa J, Percário S, Dolabela MF (2021a) Antimalarial potential of quinones isolated from plants: an integrative review. Res Soc Dev 10:e38210212507

Gomes ARQ, Da Rocha Galucio NC, De Albuquerque KCO, Brígido HPC, Varela ELP, Castro ALG, Vale VV, Bahia MO, Burbano RMR, De Molfeta FA (2021b) Toxicity evaluation of Eleutherine plicata Herb. extracts and possible cell death mechanism. Toxicol Rep 8:1480–1487

Gomes ARQ, Cunha N, Varela ELP, Brígido HPC, Vale VV, Dolabela MF, De Carvalho EP, Percário S (2022) Oxidative stress in malaria: potential benefits of antioxidant therapy. Int J Mol Sci 23

Greve B, Kremsner PG, Lell B, Luckner D, Schmid D (2000) Malarial anaemia in African children associated with high oxygen-radical production. Lancet 355:40–41

Gulcin İ, Alwasel SH (2022) Metal ions, metal chelators and metal chelating assay as antioxidant method. Processes [Online] 10:132

Halliwell B (2006) Reactive species and antioxidants. Redox biology is a fundamental theme of aerobic life. Plant Physiol 141:312–322

Hardbower DM, De Sablet T, Chaturvedi R, Wilson KT (2013) Chronic inflammation and oxidative stress: the smoking gun for helicobacter pylori-induced gastric cancer? Gut Microbes 4:475–481

Hempelmann E, Krafts K (2013) Bad air, amulets and mosquitoes: 2,000 years of changing perspectives on malaria. Malar J 12:232

Ighodaro OM, Akinloye OA (2018) First line defence antioxidants-superoxide dismutase (SOD), catalase (CAT) and glutathione peroxidase (GPX): their fundamental role in the entire antioxidant defence grid. Alex J Med 54:287–293

Isah MB, Ibrahim MA (2014) The role of antioxidants treatment on the pathogenesis of malarial infections: a review. Parasitol Res 113:801–809

Jagannathan P, Kakuru A (2022) Malaria in 2022: increasing challenges, cautious optimism. Nat Commun 13:2678

Joardar N, Guevara-Flores A, Martínez-González JJ, Sinha Babu SP (2020) Thiol antioxidant thioredoxin reductase: a prospective biochemical crossroads between anticancer and antiparasitic treatments of the modern era. Int J Biol Macromol 165:249–267

Kale S, Uplekar SM, Bandyopadhyay N, Rao PN, Ali SZ, Sharma SK, Tandel N, Patel A, Singh R, Dank A, Ravishankaran S, Priya GSL, Asokan A, Eapen A, Singh OP, Carlton JM, Mallick PK (2024) Antimalarial drug resistance profiling of Plasmodium falciparum infections in India using ion torrent deep sequencing. Front Malaria 2

Kannan D, Yadav N, Ahmad S, Namdev P, Bhattacharjee S, Lochab B, Singh S (2019) Pre-clinical study of iron oxide nanoparticles fortified artesunate for efficient targeting of malarial parasite. EBioMedicine 45:261–277

Kasozi D, Mohring F, Rahlfs S, Meyer AJ, Becker K (2013) Real-time imaging of the intracellular glutathione redox potential in the malaria parasite Plasmodium falciparum. PLoS Pathog 9: e1003782

Kıran TR, Otlu O, Karabulut AB (2023) Oxidative stress and antioxidants in health and disease. J Lab Med 47:1–11

Kotepui KU, Mahittikorn A, Mala W, Lasom S, Masangkay FR, Majima HJ, Kotepui M (2024) Total antioxidant status levels in malaria: a systematic review and meta-analysis. Malar J 23:198

Kpadonou D, Kpoviessi S, Bero J, Agbani P, Gbaguidi F, Kpadonou-Kpoviessi B, Sinsin B, Frédérich M, Quetin-Leclercq J (2019) Chemical composition, in vitro antioxidant and antiparasitic properties of the essential oils of three plants used in traditional medicine in Benin. J Med Plant 13:384–395

Loy DE, Liu W, Li Y, Learn GH, Plenderleith LJ, Sundararaman SA, Sharp PM, Hahn BH (2017) Out of Africa: origins and evolution of the human malaria parasites Plasmodium falciparum and Plasmodium vivax. Int J Parasitol 47:87–97

Manconi M, Manca ML, Escribano-Ferrer E, Coma-Cros EM, Biosca A, Lantero E, Fernàndez-Busquets X, Fadda AM, Caddeo C (2019) Nanoformulation of curcumin-loaded eudragit-nutriosomes to counteract malaria infection by a dual strategy: improving antioxidant intestinal activity and systemic efficacy. Int J Pharm 556:82–88

Marino A, Calvo M, Migliorisi G, Campanella E, Franco A, Cappello E, Trovato L, Patamia I (2023) Evidence of Plasmodium falciparum 12 years following the initial infection: a case report and mini-review of the literature. World Acad Sci J 5:35

Massicot F, Martin C, Dutertre-Catella H, Ellouk-Achard S, Pham-Huy C, Thevenin M, Rucay P, Warnet JM, Claude JR (1997) Modulation of energy status and cytotoxicity induced by FK506 and cyclosporin A in a renal epithelial cell line. Arch Toxicol 71:529–531

Mittler R, Zilinskas BA (1991) Purification and characterization of pea cytosolic ascorbate peroxidase. Plant Physiol 97:962–968

Nagy P (2013) Kinetics and mechanisms of thiol–disulfide exchange covering direct substitution and thiol oxidation-mediated pathways. Antioxid Redox Signal 18:1623–1641

Ofori EA, Garcia-Senosiain A, Naghizadeh M, Kana IH, Dziegiel MH, Adu B, Singh S, Theisen M (2023) Human blood neutrophils generate ROS through FcγR-signaling to mediate protection against febrile P. falciparum malaria. Commun Biol 6:743

Olivier M, Van Den Ham K, Shio MT, Kassa FA, Fougeray S (2014) Malarial pigment hemozoin and the innate inflammatory response. Front Immunol 5:25

Omonhinmin A, Dike I, Rotimi S (2015) Phytochemical, cytotoxicity and antioxidant activities of five anti-malaria plants. Res J Med Plant 9:81–89

Ouattara LP, Sanon S, Mahiou-Leddet V, Gansané A, Baghdikian B, Traoré A, Nébié I, Traoré AS, Azas N, Ollivier E, Sirima SB (2014) In vitro antiplasmodial activity of some medicinal plants of Burkina Faso. Parasitol Res 113:405–416

Pal C (2023) Redox modulating small molecules having antimalarial efficacy. Biochem Pharmacol 218:115927

Pinto E, Sigaud-Kutner TC, Leitao MA, Okamoto OK, Morse D, Colepicolo P (2003) Heavy metal–induced oxidative stress in algae 1. J Phycol 39:1008–1018

Pires CV, Cassandra D, Xu S, Laleu B, Burrows JN, Adams JH (2024) Oxidative stress changes the effectiveness of artemisinin in plasmodium falciparum. mBio 15:e0316923

Pizzino G, Bitto A, Interdonato M, Galfo F, Irrera N, Mecchio A, Pallio G, Ramistella V, De Luca F, Minutoli L, Squadrito F, Altavilla D (2014) Oxidative stress and DNA repair and detoxification gene expression in adolescents exposed to heavy metals living in the Milazzo-Valle del Mela area (Sicily, Italy). Redox Biol 2:686–693

Pizzino G, Irrera N, Cucinotta M, Pallio G, Mannino F, Arcoraci V, Squadrito F, Altavilla D, Bitto A (2017) Oxidative stress: harms and benefits for human health. Oxidative Med Cell Longev 2017:8416763

Prolo C, Alvarez MN, Radi R (2014) Peroxynitrite, a potent macrophage-derived oxidizing cytotoxin to combat invading pathogens. Biofactors 40:215–225

Ravindar L, Hasbullah SA, Rakesh KP, Hassan NI (2023) Triazole hybrid compounds: a new frontier in malaria treatment. Eur J Med Chem 259:115694

Reddy VP (2023) Oxidative stress in health and disease. Biomedicine 11

Rocamora F, Zhu L, Liong KY, Dondorp A, Miotto O, Mok S, Bozdech Z (2018) Oxidative stress and protein damage responses mediate artemisinin resistance in malaria parasites. PLoS Pathog 14:e1006930

Sadeg N, Pham Huy C, Martin C, Warnet JM, Claude JR (1993) Effect of cyclosporin A and its metabolites and analogs on lipid peroxidation in rabbit renal microsomes. Drug Chem Toxicol 16:165–174

Sadiq MB, Tharaphan P, Chotivanich K, Tarning J, Anal AK (2017) In vitro antioxidant and antimalarial activities of leaves, pods and bark extracts of Acacia nilotica (L.) Del. BMC Complement Altern Med 17:372

Sannella AR, Casini A, Gabbiani C, Messori L, Bilia AR, Vincieri FF, Majori G, Severini C (2008) New uses for old drugs. Auranofin, a clinically established antiarthritic metallodrug, exhibits potent antimalarial effects in vitro: mechanistic and pharmacological implications. FEBS Lett 582:844–847

Sarr D, Cooper CA, Bracken TC, Martinez-Uribe O, Nagy T, Moore JM (2017) Oxidative stress: a potential therapeutic target in placental malaria. Immunohorizons 1:29–41

Sato S (2021) Plasmodium-a brief introduction to the parasites causing human malaria and their basic biology. J Physiol Anthropol 40:1

Shankar AH (2000) Nutritional modulation of malaria morbidity and mortality. J Infect Dis 182: S37–S53

Tiwari S, Kumar R, Devi S, Sharma P, Chaudhary NR, Negi S, Tandel N, Marepally S, Pied S, Tyagi RK (2024) Biogenically synthesized green silver nanoparticles exhibit antimalarial activity. Discov Nano 19:136

Tyagi RK, Tandel N, Deshpande R, Engelman RW, Patel SD, Tyagi P (2018) Humanized mice are instrumental to the study of plasmodium falciparum infection. Front Immunol 9:2550

Valko M, Izakovic M, Mazur M, Rhodes CJ, Telser J (2004) Role of oxygen radicals in DNA damage and cancer incidence. Mol Cell Biochem 266:37–56

Valko M, Morris H, Cronin MT (2005) Metals, toxicity and oxidative stress. Curr Med Chem 12: 1161–1208

Van Houten B, Santa-Gonzalez GA, Camargo M (2018) DNA repair after oxidative stress: current challenges. Curr Opin Toxicol 7:9–16

Vasquez M, Zuniga M, Rodriguez A (2021) Oxidative stress and pathogenesis in malaria. Front Cell Infect Microbiol 11:768182

White E, Shannon JS, Patterson RE (1997) Relationship between vitamin and calcium supplement use and colon cancer. Cancer Epidemiol Biomarkers Prev 6:769–774

Willcox JK, Ash SL, Catignani GL (2004) Antioxidants and prevention of chronic disease. Crit Rev Food Sci Nutr 44:275–295

Wiser MF (2024) The digestive vacuole of the malaria parasite: a specialized lysosome. Pathogens 13

Wunderlich J, Rohrbach P, Dalton JP (2012) The malaria digestive vacuole. Front Biosci Schol Ed 4:1424–1448

Chapter 12
Challenges and Limitations in Targeting Oxidative Stress

Manisha Pandey, Pooja, Sunita Nirban, Meghaditya Roy, Rasna Kumari, Bapi Gorain, Neha Jain, and Rohit Kumar Verma

Abbreviations

2-OH-E2	2-hydroxy estradiol
8-OHdG	8-hydroxy-2′-deoxyguanosine
8-oxo-dG	8-oxo-7,8-dihydro-2′-deoxyguanosine
AD	Alzheimer's disease
ADME	Absorption, distribution, metabolism, and excretion
AMPK	AMP-activated protein kinase
BAX	Bcl-2-associated X protein
BBB	Blood–brain barrier
BHT	Butylated hydroxytoluene
BP	Blood pressure
Ca^{2+}	Calcium ion
CAT	Catalase
CKD	Chronic kidney disease
Cl^{-}	Chloride ion
CLD	Chronic liver disease

M. Pandey (✉) · Pooja · S. Nirban · M. Roy · R. Kumari
Department of Pharmaceutical Sciences, Central University of Haryana, Mahendergarh, India
e-mail: drmanisha@cuh.ac.in

B. Gorain
Department of Pharmaceutical Sciences and Technology, Birla Institute of Technology, Mesra, Ranchi, India

N. Jain
Department of Pharmaceutics, Amity Institute of Pharmacy, Amity University, Noida, India

R. K. Verma
School of Life-long Learning, Central University of Haryana, Mahendergarh, India

N. Srivastava, S. A. Saraf (eds.), *Pharmacological Strategies for Prevention and Treatment of Oxidative Stress*, Oxidative Stress in Applied Basic Research and Clinical Practice, https://doi.org/10.1007/978-3-032-19837-2_12

COPD	Chronic obstructive pulmonary disease
CVD	Cardiovascular disease
DAG	Diacylglycerol
DM	Diabetes mellitus
DNA	Deoxyribonucleic acid
DR	Diabetic retina
ELISA	Enzyme-linked immunosorbent assay
ETC	Electron transfer chain
GC–MS	Gas Chromatography–Mass Spectrometry
GPx	Glutathione peroxidase
GSH	Glutathione
H_2O_2	Hydrogen peroxide
HOCl	Hypochlorous acid
HOGG1	Oxoguanine glycosylase 1
HPLC	High-Performance Liquid Chromatography
IL-1β	Interleukin-1β
JAK/STAT	Janus kinase–signal transducer and activator of transcription pathway
LC–MS/MS	Liquid Chromatography–Tandem Mass Spectrometry
LPO	Lipoperoxidation
MAPK	Mitogen-activated protein kinase
MDA	Malondialdehyde
mV	Millivolt
NADPH	Nicotinamide adenine dinucleotide phosphate
NF-κB	Nuclear factor kappa B
NO•	Nitric oxide
Nrf2	Nuclear factor erythroid 2-related factor 2
$O_2{\bullet}^-$	Superoxide radical
OH•	Hydroxyl radical
$ONOO^-$	Peroxynitrite
OS	Oxidative stress
PD	Parkinson's disease
PI3K/Akt/mTOR	Phosphoinositide 3-kinase/Protein kinase B/Mammalian target of rapamycin
RA	Rheumatoid arthritis
RNS	Reactive nitrogen species
RONS	Reactive oxygen and nitrogen species
ROS	Reactive oxygen species
SASP	Senescence-associated secretory phenotype
SE	Status epilepticus
SOD	Superoxide dismutase
T1DM	Type 1 diabetes mellitus
T2DM	Type 2 diabetes mellitus
TLE	Temporal lobe epilepsy
TNF-α	Tumor necrosis factor-alpha
TPP	Triphenylphosphonium

12.1 Introduction

The phrase "oxidative stress" leads to oxidant generation and antioxidant defense imbalance of both. The term used by Helmut Sies1 majorly results in biological system damage. The characteristic indication of oxidative stress (OS) is the occurrence of an imbalance between the development and elimination of reactive oxygen species (ROS) or reactive nitrogen species (RNS) (Phaniendra et al. 2015). ROS are more reactive components due to their chemical composition and can originate from nitrogen or oxygen metabolism. Nitric oxide (NO•), superoxide radical (O2•-), and hydroxyl radical (OH•) are examples of free radicals that can be classified as ROS and RNS. On the other hand, other non-free radicals are also present, including peroxynitrite (ONOO−) and hydrogen peroxide (H2O2). ROS trigger metabolic processes within the mitochondria that reduce oxygen via electron transfer chain (ETC). The endoplasmic reticulum along with peroxisomes is the other source of ROS (García-Sánchez et al. 2020). Several metabolic, and chronic diseases or malignancies have been associated with OS. Although the concept of oxygen-free reactive species has been around for more than 50 years, it was not until the last 20 years that their connection to the onset of illness was established. As a result, research on the protective properties of antioxidants has been extensively conducted (Sharifi-Rad et al. 2020). In many biological processes, free reactive species are crucial. Many of these are essential to survival, like the phagocytes' intracellular eradication of bacteria, particularly by granulocytes and macrophages. Reactive species may potentially play a significant role in redox signalling, a subset of biological communication mechanisms. When present in small-to-moderate quantities, ROS helps regulate a multitude of cellular functions and processes related to homeostasis maintenance (Dröge 2002).

Overproduction of ROS causes cellular proteins to change structurally and functionally, which causes cellular dysfunction and interferes with essential physiological functions. DNA, proteins, and lipids are damaged by high ROS levels. Specifically, ROS can increase the permeability and fluidity of the membrane by breaking the lipid membrane. Site-specific amino acid alterations, peptide strand breakage, deposition of the products obtained from the crosslinked reaction, change in electric charge, enzyme deactivation, and susceptibility to proteolysis are all examples of protein damage. Finally, ROS can degrade DNA by crosslinking DNA-protein, oxidizing deoxyribose, rupturing strands, deleting nucleotides, and changing bases (Bano et al. 2021). (O2•-) and hydroxyl (OH•) are the two major types of oxygen free radicals. They are produced under the conditions of chemical decrease from molecular oxygen. Overexposure to these free radicals can result in apoptosis and cell damage, which can contribute to a variety of disorders, including diabetes, cancer, stroke, myocardial infarction, and other serious ailments (Fujii et al. 2022). Free radicals and DNA are considered to interact to cause mutations that alter the cell cycle, resulting in neoplasia, which is the cause of many malignancies. The body possesses multiple enzymatic processes to reduce the harm caused by radicals and to guard against the excessive synthesis of free radicals because these molecules are essential for life. Antioxidants are essential to these defence systems. In healthy individuals, oxidants and antioxidants are delicately balanced to defend against the damaging effect of ROS. Therefore, the constant generation of reactive species in aerobic organisms needs to be balanced by an equivalent rate of expenditure of antioxidants. Antioxidants, whether enzymatic or not, are compounds that both hunt for and neutralize free radicals and repair the harm they produce. Numerous natural and exogenous antioxidants provide defence against oxidative harm and chronic illnesses (Phaniendra et al. 2015). Naturally occurring ROS generation through the involvement of the mitochondrial respiratory chain because reactive species can be advantageous for metabolism while also being detrimental to cells under certain circumstances. Secondly, ROS overpower antioxidant systems during pathological or stressful situations, creating an imbalance that results in OS and irreversible alterations in cell components such as proteins, carbohydrates, and lipids. Additionally, ROS can interfere with regular cellular signalling pathways (Sharifi-Rad et al. 2020).

The amount of ROS in the cell is required for several biological activities, including immunity, death, protein phosphorylation, and transcription factor activation. Superoxide dismutase (SOD), catalase (Cat), and glutathione peroxidase (GPx) are the primary endogenous antioxidant enzymes that neutralize reactive oxygen species (ROS). SOD is a member of the metalloenzyme family that converts O2•- to oxygen and H2O2, with three forms of SOD found in mammals: cytoplasmic SOD (SOD1), mitochondrial SOD (SOD2), and extracellular SOD (SOD3). Other nonenzymatic compounds that scavenge free radicals, such as vitamins, melatonin, and glutathione (GSH), can also neutralize ROS. ROS persist in the human system longer and oxidize vulnerable biomolecules when antioxidant defences are unable to effectively neutralize them. The overabundance of ROS can harm genetic material, lipids in the membrane, and cellular proteins and impair the normal functioning of cells. A mediator of vascular vasorelaxation that is dependent on the endothelium is the NO• radical. The nitric oxide synthase (NOS) enzyme generally produces NO•. Under OS circumstances, NO• and the radical O2•-react to produce ONOO−, which damages endothelium (Andrés Juan et al. 2021).

One of the most important parameters for the damage of lipids is lipoperoxidation (LPO) is one way that OS damages lipids. Carbon–carbon double bonds, specifically in fatty acids that are polyunsaturated, are a defining characteristic of LPO. Hydroperoxides, in addition to propanal, hexanal, malondialdehyde (MDA), and 4-hydroxynonenal, are the principal LPO products. In addition to it, there is structural damage to DNA when ROS binds with guanine nucleotides. 8-hydroxy-2′--deoxyguanosine (8-OHdG) or 8-oxo-7,8-dihydro-2′-deoxyguanosine (8-oxodG) are frequently formed via guanine oxidation. The oxo guanine glycosylase (hOGG1) enzyme is involved in the restoration of several metabolites that are regarded as OS biomarkers in ordinary conditions (Cherian et al. 2019).

The existence of OS in several chronic illnesses may hasten its development (Fig. 12.1). Rheumatoid arthritis regarded as an autoimmune disease, causes tissue damage due to the intimate relationship between OS and the inflammatory process. Hyperglycaemia and the advancement of type 2 diabetes mellitus (DM) are

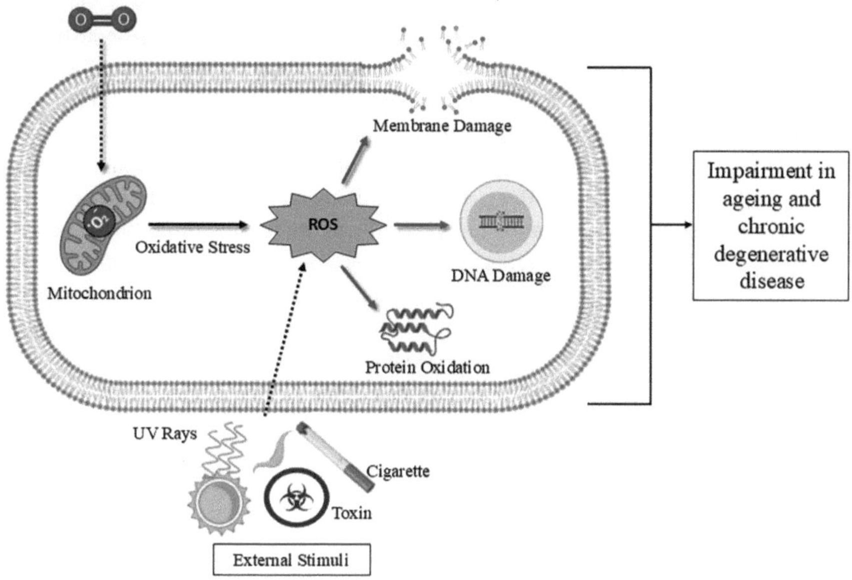

Fig. 12.1 Schematic representation of role of oxidative stress on aging and chronic disease. (Leyane et al. (2022))

associated with OS. The primary process by which OS is involved in cardiovascular disease (CVD) is through its impact on hypertension and the development of atheroma leaflets. Increased ROS generation is linked to the pathological generation of several severe diseases, involved cancer, aging, HIV infection, as well as neurological diseases. Conversely, exogenous variables can change ROS production, such as prescribed pharmaceutical therapies for specific chronic diseases. This chapter aims to explain the function of OS in several clinical processes, including cancer, HIV, DM, rheumatoid arthritis, high blood pressure, and some neurodegenerative illnesses (García-Sánchez et al. 2020). The role of oxidative stress on aging and chronic disease have been explained in Fig. 12.1.

12.2 Role of OS in Disease Generation

12.2.1 Cardiovascular Disease (CVD) and OS

Approximately all of the critical steps in the pathophysiology of intense thrombotic events and atherosclerosis, like dyslipidemia, lead to the formation of atheroma, LDL oxidation, dysfunctioning of endothelial, rupturing of plaque, injury of myocardial ischemia, and thrombosis that recurs (i.e., the secondary clot that frequently formed after starting thrombolysis), have been linked to OS. The relationship between different coronary disorders risk factors, like high BP, diabetes, and cigarette smoking, and the disease's clinical sequelae, which include vasoconstriction, thrombosis, plaque rupture, and vascular remodelling, has been associated with OS (Dubois-deruy et al. 2020). In the scenario, of CVD, OS plays an important purpose as a modulator of aberrant platelet activity and defective endothelium-dependent vasodilation. Superoxide anion reduces the biological activity of NO and is a main cause of OS. It also has direct effects. Overproducing superoxide in the blood vessels promotes further platelet activation and recruitment, increasing thrombus generation (Scioli et al. 1995). By scavenging ROS, numerous of which alter platelet action, antioxidants may subsequently block platelets. These antioxidant mechanisms in platelets are intimately related, even though lipid- and water-soluble antioxidants are found in separate subcellular sites. Antioxidants can also indirectly suppress platelets by modifying ROS metabolism, which affects platelet function in numerous ways (Chauhan et al. 2015).

12.2.2 OS in Cancer

Cancer is regarded as a complicated process that results in changes in both the cells and the molecules, which can be caused by stimuli that are exogeneous and exogeneous. It is most widely recognized that one of the components that are responsible for the development of cancer is oxidative damage of DNA. Abnormalities in chromosomes and OS-induced oncogene activation can both contribute to cancer. The generation of hydrolyzed DNA bases is the most important process in chemical carcinogenesis, which often results in DNA oxidation. These adducts alter the physiological transcriptome profile and lead to changes in genes, which inhibit normal cell growth (Azmanova and Pitto-Barry 2022). Numerous alterations to DNA structure, such as base and sugar lesions, crosslinking of DNA–protein, breakdown of strands, and base-free sites, can also be brought on by oxidative stress. Various levels of carcinogenesis, like the conversion of normal cells into tumour cells, as well as the development, proliferation, invasion, angiogenesis, and metastasis of tumour cells, are stimulated by oxidants. oxidative stress is linked to cancer in virtually every stage. Because cancer cells develop greater oxidants than healthy cells do. The primary sources found in cancer cells are 5-lipoxygenase, NOX4, and mitochondria which are elevated oxidants. Oxidants in the locus can also originate from normal cells, like endothelial cells and inflammatory immune cells, that are present in or around the tumour mass. Increasing oxidative indicators have been seen in several cancer forms (Jiang et al. 2023). For example, it has been exhibited that greater H2O2 is exhaled by non-small-cell lung cancer patients in comparison with control subjects (Chan et al. 2009). Additionally, increased stages of 8-OHdG (Matsui et al. 2000). were detected in breast cancer tissues and 8-OHdG was importantly reduced in prostate cancers (Ohtake et al. 2018).

12.2.3 OS in Chronic Kidney Disease (CKD)

CKD is regarded as a progressive reduction in functioning in renal or GFR of less than 60 mL/min over less than three months. According to published research, elevated intracellular ROS levels are essential for the onset of CKD. In the kidney, Cellular mitochondria along with NADPH oxidases are the two main components of ROS generation. Normal conditions, however, cause any ROS-facilitated damage to be corrected by the renal antioxidant system, which includes CAT, GPX, and SOD. The overabundance of ROS generated by mitochondria is indicative of renal damage (Tirichen et al. 2021). Several biometabolites get oxidized resulting in anatomical alteration and structural makeup because of an imbalance in the redox system. Research on aldosterone-infused mice models revealed that proteinuria and podocyte destruction are preceded by mitochondrial dysfunction. Renal damage and atherosclerosis pathogenesis in CKD are facilitated by ROS, which are produced as a byproduct of this process. There is a connection between the build-up of O2•− and OS in renal failure along with it SOD downregulation and NOX overexpression in CKD (Leyane et al. 2022).

Furthermore, ONOO− is produced when O2•− and NO• interact, resulting in nitrosative stress, which is the nitrosylation of redox-sensitive thiols by NO. ONOO− oxidizes proteins, lipids, and DNA to cause oxidative damage that leads to cell injury, necrosis, and apoptosis. Also, it modifies cellular signalling cascades. Moreover, hypochlorous acid (HOCl) is produced through the metabolism of Cl− and H2O2 by myeloperoxidase (MPO), which contributes to chlorinated stress (Daenen et al. 2018). Studies by demonstrated increased HOCl-derived protein oxidation in CKD patients' kidney tissues. The increased production of AGEs in renal failure causes carbonyl stress, which in turn fuels inflammation in CDK. Due to the generation of RONS, the activation of resident cells and cells called immune cells, i.e., neutrophils and macrophages prolong the oxidative state. In patients with uraemia, the overproduction of mitochondrial ROS stimulates the NOD-, LRR-, and pyrin domain-containing protein 3 (NLRP3) inflammasome, hence promoting CKD-mediated chronic micro-inflammation (Granata et al. 2015). Angiotensin II was one of the factors that reduced NO• generation, and hypertension is another factor that encourages the rise in ROS generation in CKD.

12.2.4 ROS and Aging

Greater than 60 years ago a theory was proposed named "free radical theory of ageing,", which identified ROS as the root source of harm build-up in connective tissues and cell components. Ageing and ageing-related degenerative disorders would result from this accumulation of cell losses (Tsoukalas et al. 2019). Even if there are many theories of ageing nowadays, oxidative stress remains the fundamental idea behind the majority of them (Payne and Chinnery 2015). In the cells, decreased state of OS is mostly associated with the mitochondria and NOX systems. Numerous studies have demonstrated that age-associated disorders have increased NOX activity. NOX is a family of numerous membrane-linked enzymes found in the mitochondria of various types of cells (Liguori et al. 2018). High-molecular protein aggregates have been observed to develop in cells during the ageing process. These aggregates are mostly composed of different lipids and proteins (Davalli et al. 2016). A variety of reactive metabolites have oxidized or altered the majority of the aggregated proteins, and these modifications may attach to cellular proteins. Therefore, the breakdown of these aggregates is essential for maintaining protein homeostasis. The proteasome, which solely targets unfolded proteins for degradation, is the primary site for the breakdown of damaged cell proteins (Saez and Vilchez 2014). Protein aggregation production in cells is increased by proteasome inhibition, which also stops newly generated oxidized proteins from being further broken down (Chondrogianni et al. 2014). In addition, ageing causes proteasome malfunction. Protease activation slows down the ageing process and lengthens life, whereas proteasomal malfunction is linked to age progression and accumulation of proteins. It has been demonstrated in numerous invertebrate models and cell lines that overexpressing certain proteasomal regulatory or catalytic subunits or treating them with particular substances increases proteasome activity (Saez and Vilchez 2014). This has led to the trend of taking antioxidant supplements in conjunction with the popularisation of the "free radical theory of ageing" (Yukuyama et al. 2016). The majority of current studies suggest that supplements of antioxidants do not decrease the prevalence of disorders associated with age.

12.2.5 Diabetes and Oxidative Stress

Hyperglycaemia, or abnormally high blood glucose, is the precursor of diabetes. Several pro-inflammatory substances like cytokines, chronic hyperglycaemia, alteration in signalling pathways, OS, advanced glycation end products (AGEs) as well as cellular cell death are associated with advanced problems in diabetes. Moreover, hyperglycaemia promotes the synthesis of diacylglycerol (DAG) and raises the generation of ROS and OS via activating the protein kinase c (PKC) pathway and NOX. Chronic hyperglycaemia in type II diabetes is brought on by a reduction in the absorption of glucose into muscle and adipose tissue (Pérez-Matute et al. 2009). It is hypothesized that in type II diabetes, oxidative stress and ROS exposure cause faulty β cells to form in the pancreatic β cells, preventing them from producing and/or releasing enough insulin. Insulin signalling is bidirectionally modulated in RONS. The fact that RONS block the insulin signalling pathway is becoming increasingly clear, suggesting that they may be mediators in the development of resistance to insulin. But RONS mediates this pathway, making sure that insulin has the desired physiological and cellular consequences. Glucose oscillations play a significant role in modifying the main cause of diabetes and can directly cause elevated OS (Potenza et al. 2008) (Leyane et al. 2022).

12.2.6 OS in Neurological Disease

Alzheimer's disease (AD), memory loss, amyotrophic lateral sclerosis (ALS), and Parkinson's disease (PD) are examples of several neurological disorders that are directly associated with OS (Olufunmilayo et al. 2023).

12.2.7 Alzheimer's Disease (AD)

Numerous clinical studies in AD have illustrated the significant role in the generation of dementia as OS is responsible for the damaging of neurons. The generation of β-amyloid, a damaging peptide constantly detected in AD patient's brains is attributed to free radical activity and is known to have a role in the neurodegeneration that is seen throughout the beginning and generation of health aliments (Ionescu-Tucker and Cotman 2021). AD is said to be caused by an excess of ROS, which leads to the aggregation of tau in neurofibrillary tangles and Aβ in amyloid plaques. Moreover, this may result in impaired mitochondrial activity and/or increased metabolic demand. In one study, tau-elevated case SOD1 expression proved to be beneficial against Neurodegeneration in animal trials. It is commonly known that a mutation in SOD1 (Cu/Zn) causes familial amyotrophic lateral sclerosis (Khola et al. 2024). There are several shared characteristics among the severe neurodegenerative diseases of the CNS, including inflammation, OS, dysfunctional synapses, misfolded proteins, and impaired autophagia (Dhapola et al. 2024).

12.2.8 Parkinson's Disease (PD)

In PD, the destruction of dopaminergic neurons is involved through oxidative stress. Cell death can result when biological factors impair the normal function of the neuron redox potential. The substantia nigra (SN) of PD has been illustrate to have both oxidative and nitrative damage. Parkinson's disease results in a decrease of dopaminergic neurons in the substantia nigra. Dopamine is an unstable chemical that, in this area of the brain, produces dopamine quinones and ROS by autooxidation, which results in OS. The OS and mitochondrial malfunction play key roles in Parkinson's disease neurodenergization. Through decreased activity and regulation of antioxidant genes, including SOD2, glutathione peroxidase, and catalase, ageing lowers the capacity to sustain antioxidant mechanisms against OS (Puspita et al. 2017; Dias et al. 2013).

12.2.9 Sexual Maturation and OS

Many writers have noted that OS may be the cause of a delayed beginning of puberty and sexual maturation. It appears that this is the case for both pregnant women and children exposed to a similar metallic element throughout the prepubertal stage. Cadmium is well known to generate OS and free radicals (Interdonato et al. 2015; Pizzino et al. 2017).

12.2.10 Respiratory Disease (RD) and OS

Numerous results of studies demonstrated the link between OS and illnesses of the lungs, including asthma and COPD, that are identified through systemic and local severe inflammation. It is a recognized fact that inflammation is raised by oxidants through the stimulation of a variety of kinases which adds pathways and AP-1 and NF-kappa B that are transcription factors (Marzocco et al. 2023).

12.2.11 Rheumatoid Arthritis (RA) and OS

RA is defined as a long-term inflammatory disease condition that affects nearby tissues and joints. It has distinctive characteristics through the invasion of stimulated T cells and macrophages. The elevated stages of prostaglandin and isoprostane in the synovial fluid of affected organisms indicate that reactive species at the region of inflammation act as a significant purpose in the starting and advancement of this condition. Reactive species can alter the production of self-antigen-type proteins in autoimmune disorders, either modifying their antigenic profile or boosting immune response. In sensitive people, exogenous antioxidants like allergens can affect immunity (Fonseca et al. 2019).

12.3 Mediators and Targets of OS

The OS phenomenon entails the overproduction of ROS and RNS such as NO, $ONOO^-$, $O_2^{\cdot -}$, H_2O_2, and $HO^{\cdot}$(Yu and Xiao 2021). This reactive species enhances necrosis and apoptosis, inflammation of cells, destruction of DNA base, DNA and chromosomal aberration, single and double-strand breaks in DNA, cross-links between DNA and proteins, lipid membranes, mitochondrial function and collagen structure. Oxidative stress causes several human disorders like cancer, neurological disorders, COPD, chronic renal disease and cardiovascular diseases (Klran et al. 2023).

Hormonal imbalance occurs due to unopposed oestrogen, which affects gene expression in apoptosis and cell proliferation, leading to uncontrolled cell growth and carcinogenesis. Endogenous oestradiol metabolites such as 2-hydroxy oestradiol (2-OH-E2), as well as 4-hydroxy oestrogens (4-OH-E2), produce ROS and RNS, leading to DNA oxidative damage that may result in cell death or endometrial cancer. Various studies revealed how 2-MeO-E2 activates mitochondria-dependent caspases, Bak, BAX and apoptosis of ovarian cancer cells (Bukato et al. 2024). An essential inflammatory mediator, NF- κB controls proinflammatory chemokines and cytokines, contributing to the maintenance of chronic inflammation in RA. Enhanced OS in RA leads to reduced antioxidant defence and triggers the release of mediators and proinflammatory cytokines like NF-κB, IL-1β and TNF- α, thereby worsening the development of RA (Zhou et al. 2021).

Nanoplastics, a newly recognised global environmental hazard, are now prevalent in daily lives and cause a serious health risk, once inhaled and trapped in the respiratory system. Specifically, 800 nm diameter polystyrene nanoplastics (PS-NPs) have been found to enter the cytoplasm of A549, activate the transcription of genes associated with an antioxidant network, and significantly affect cell viability. Hence, Polystyrene nanoplastics regulate apoptosis, oxidative stress, and ageing in the epithelial cell line of human alveoli (Milillo et al. 2024). Red blood cells (RBC_S) mediate the development of diabetic complications. Due to high blood sugar levels in diabetes, ROS can damage red blood cells, resulting in altered membrane integrity and RBC function like glycation of haemoglobin (HbA1c). RBCs produce cytokines and inflammatory mediators when subjected to OS, leading to endothelial dysfunction and chronic inflammation that can raise the risk of diabetic complications (Obeagu 2024). Amyloid-β (Aβ) proteins and hyperphosphorylated tau proteins accumulate in AD, causing apoptosis and synapse loss. Tau and Aβ accumulation initiate apoptotic pathways, leading to the death of neurons. This process initiates different proteins, involving caspases like effector, initiator and inflammatory caspases and Bcl-2 family proteins like Bad, Bax, Bid, Bcl-XL, and Bcl-XS. These proteins activate a series of events that leads to cell death (Kumari et al. 2023).

Ferroptosis produced by OS is a crucial factor in neurological diseases. A protein Nrf2 enters the nucleus in response to OS and binds with ARE and MAF to trigger antioxidant metabolic genes, therefore mitigating oxidative stress. Additionally, Nrf2 activates GPX4 enzyme and can be inhibited by Keap1. Fenton reaction containing Fe2+ and TCA cycle in mitochondria both can cause OS which produces reactive oxygen species. NAC, a precursor of L-cysteine and GPX4 influences oxidative stress via GPX4 and system XC-. Under the influence of different regulatory parameters, NOX can produce ROS, leading to OS (Li et al. 2022). Autoimmune disorders like rheumatic arthritis, systemic lupus erythematosus and psoriasis are produced through the combination of genetic and environmental factors causing overstimulation of immune cells and chronic inflammation. In the vicious cycle of autoimmune disorder, oxidative stress enhances the formation of eicosanoid, lipid mediators, acrolein, 4-hydroxynonenal-protein adducts and endocannabinoids, which can increase inflammation and also act as extracellular signalling molecules and autoantigens (Wójcik et al. 2021).

Specialised pre-resolving mediators (SPMs) are a new class of endogenous lipids that comprise arachidonic acid (AA), eicosapenoic acid (EPA) and docosahexaenoic acid (DHA). Lipoxins (LXs) comprise two mediators such as LXA_4 and LXB_4, both synthesized from AA. E-series resolvins (RvE) are biosynthesised by adding molecular oxygen to EPA. In experimental models of skin inflammation, liver injury, vascular dysfunction and renal ischemia/reperfusion injury, MaR1 suppressed O_2^- release and resulting tissue damage. Protectins (PD) include two mediators such as PD1 and PDX (Leuti et al. 2019). CKD is a major public health concern issue and highlights the recognition of pathogenic mechanisms and potential treatment targets. Nevertheless, oxidative stress raises the ROS levels and low-molecular-weight antioxidants cannot successfully eradicate ROS levels, leading to CKD development and progression (Kishi et al. 2024). The neurodegenerative disorder, glaucoma, is complex and characterized by visual field loss and progressive optic nerve damage. Deformations in lipid mediators endocannabinoids, lysophospholipids, neuroprotectin D1, leukotrienes, sphingolipids, and lipoxins, leading to oxidative stress, retinal ganglion cell loss, neuroinflammation and vascular dysfunction. Prostaglandins and analogues primarily cause a reduction in intra-ocular pressure (IOP) to manage glaucoma (Mathew and Sivak 2024). Regular exposure to chemicals, destructive contaminants, and environmental pathogenic entities can cause pathological modification of the epithelial barrier, elevating the chances of allergies. ILC2s and eosinophils regulate allergic inflammation via lipid mediators, specifically prostaglandins (PGs) and cysteinyl leukotrienes (CysLTs) (Ling-Yu Kan et al. 2024).

12.4 Therapeutic Strategies Against Oxidative Stress

The formation of wounds occurs when elements that cause injuries compromise the integrity of normal skin tissue. If there is an appropriate rescue mechanism in place, wounds have the potential to compromise the skin's regular functioning and create a gap in the self-defence system. Vasoconstriction, inflammation, coagulation, wound remodelling and cellular proliferation are all parts of the biological mechanism that self-regulates during wound healing. OS is widely known as a key component in wound healing regulation. Remarkably, research conducted in the last ten years has demonstrated that the beginning of OS has two distinct functions in treating the wound, depending upon the OS setting. OS state in mammals is caused by either a build-up of ROS or a decrease in antioxidant capability. Changes in the state of ROS and OS are closely linked to the production of granulation tissue, angiogenesis, and inflammation throughout the wound healing process. The control of OS has been found to have therapeutic promise recently (Wang et al. 2023).

One of the characteristics of senescence is that somatic cells start to lose their capacity to multiply when they reach a particular number of mitotic levels. Senescent cells exhibit a variety of remarkable morphological changes in addition to losing the ability to divide, including flattened and enlarged cell morphology, elevated ROS production, the build-up of ROS-mediated damage derivatives (granules and lipofuscin) and the SASP (senescence-associated secretory phenotype). As we age, senescent cells proliferate, impairing our ability to regenerate tissue and fostering an inflammatory milieu that promotes ageing and age-related illnesses. The discovery that the deliberate elimination of senescent cells in vivo lowers inflammation and boosts immune system performance, delaying the onset of ageing and hence lengthening lifespan, is evidence in favour of this causal link. Moreover, age-related fragility and bone loss are prevented by senolytic therapies that eradicate senescent cells, hence corroborating the causal relationship between senescence and ageing (Lee et al. 2023).

Since the late 1980s, ROS, have been linked to the aetiology and dissemination of HIV-1. ROS can modify the structure of proteins, lipids, and glycans, which may result in genetic changes and impair the operation of essential systems. Numerous critical biological processes are impacted by ROS levels, such as cytokine signalling, neuromodulation, hormone action, apoptosis, ion transport, protein phosphorylation, and general cell signalling. ROS are generated spontaneously in the mitochondrial oxidative metabolism in response to a variety of physiological stimuli, in addition to their function as a secondary messenger in the MAP kinase cascade. They also directly modify redox-sensitive transcription factors and regulate gene expression. Electrons that break free from the mitochondrial ETC and interact with molecular oxygen are the primary source of ROS, which include H2O2 and O2−. Moreover, NADPH oxidases (NOX) can also be activated, which releases superoxide when their enzymes convert NADPH to molecular oxygen by transferring electrons. Cells have developed a wide range of defence mechanisms to fend off the harmful effects of ROS and maintain stable control over their production. The main components of these defence mechanisms are antioxidants and enzymes that scavenge ROS. The scavenging enzymes include GPx, which uses glutathione's reducing power to eliminate H2O2, SODs, which change (•O2−) into less reactive H2O2, and CAT, which converts H2O2 to water and molecular oxygen. Additional defence mechanisms include the glutathione/glutaredoxin system, the thioredoxin (TRX) system, and peroxiredoxins. To preserve cellular homeostasis and a balanced ROS level, these complex mechanisms cooperate (Harshithkumar et al. 2024).

ROS-Modulated Regulatory Pathways and Their Effect on Cellular Activities Numerous important cellular pathways, including AMPK, JAK/STAT, NF-κB, PI3K/Akt/mTOR, and MAPK, are modulated by ROS. Numerous physiological processes, including energy balance, immunological responses, inflammation, antioxidant defences, growth of cells, survival, regulation of cell cycle, and DNA repair, are regulated by these pathways. These pathways' biological functions are significantly impacted by variations in ROS levels (Harshithkumar et al. 2024).

In millions of people worldwide, unprovoked, and recurrent seizures are the hallmark of epilepsy, a chronic neurological illness. Serious cognitive, social, psychological, and financial repercussions result from epilepsy. When uncontrolled, epileptic seizures can significantly reduce quality of life. The raise in the frequency and synchronization of firing of the neurons as well as an excess of excitatory over inhibitory neurotransmitters are the causes of epilepsy. The most prevalent type of epilepsy, known as temporal lobe epilepsy (TLE), is characterised by focal-onset seizures that primarily affect the temporal lobe and are associated with memory and learning impairments. TLE accounts for 60% of all epilepsy cases and is characterised by recurrent, unexpected partial complex seizures. The epileptic seizure type known as status epilepticus (SE), which is characterised by convulsive activity lasting more than five minutes, has a significant morbidity and fatality rate. Neuronal death resulting from epileptic seizures can stimulate epileptogenesis and subsequent seizures. Among the postulated processes of epileptogenesis include modifications to synapses, neurotransmitters, receptors, OS mitochondrial dysfunction, cytokine signalling, and apoptosis. A growing body of evidence suggests that the onset of epilepsy is linked with OS and excessive formation of ROS. OS causes brain alterations, such as inflammation, neurodegeneration, and a decreased seizure threshold, which leads to epileptogenesis before seizures manifest. OS speeds up neurodegeneration, neuronal excitability, and the onset of seizures via modifying Ca2+ homeostasis. Empirical data suggests that there is a connection between epilepsy and brain inflammation. Animal models and epileptic patients have both displayed indications of neuroinflammation. To release several inflammatory components different cells like astrocytes, microglia, and endothelial cells, cause peripheral immune cells, and BBB are responsible for chronic inflammation (Prakash et al. 2023).

In the twenty-first century, diabetes is increasingly regarded as a rapidly growing global health issue. Its adult incidence was 8.8% worldwide in 2017 and is expected to rise to 9.9% by 2045. The WHO research states that diabetic retinopathy is the main cause of visual impairment in adults, accounting for at least 5% of the 37 million cases of blindness worldwide. Moreover, by 2030, diabetes is expected to rank seventh among all causes of death by the World Health Organisation. Type 1 diabetes mellitus (T1DM) and type 2 diabetes mellitus (T2DM) are the two forms of the disease. The fact that elevated OS contributes significantly to the aetiology of diabetes and its consequences is becoming more widely recognised. Overproduction of ROS results in OS, a biological process that contributes to the generation many diseases and issues linked with them. ROS destroy cells by interacting with multiple components of the cell. Thankfully, organisms have an antioxidant mechanism that protects them against ROS. The retina has an effective antioxidant defence system made up of antioxidant enzymes and free radical scavengers, however, prolonged hyperglycaemia weakens this system and raises oxidative stress (Gao et al. 2023).

12.4.1 Epilepsy and OS

The onset of conditions like epilepsy can be accelerated by OS, which is characterised as an imbalance in the production and removal of RNS. Aerobically active organs are specifically susceptible to the generation of ROS and free radicals due to the early electron leakage from the ETC. $O2^{\bullet-}$ is one effect of the electrons being transferred to O2. SOD counteracts this by converting $O2^{\bullet-}$ to H2O2, which is then transformed into water and oxygen by GPX and CAT. Normally, a cell uses 1–5% of its oxygen consumption to produce ROS, but in certain situations, such as when there is a Ca2+ excess, this rate can be increased by changes in mitochondrial homeostasis. Oxidative damage may occur in complexes I and III of the electron transport chain, namely in the iron-sulphur clusters. The brain is especially vulnerable to OS because of its high metabolic needs, which cause it to actively engage in aerobic metabolism. Furthermore, the peripheral blood of TLE patients had increased levels of oxidative damage indicators. SE patients have reduced plasma activity of GSH, SOD, and CAT, as well as a decreased total antioxidant capacity in their blood. Epileptic convulsions can cause OS, which can worsen pre-existing brain damage and trigger a series of further seizures. Increased NOX activity and mitochondrial dysfunction during acute seizures result in the formation of excess ROS (Prakash et al. 2023).

12.4.2 The Generation of ROS and its Relevance to OS

OS denotes a reductive balance skew, which suggests either an excess of ROS production or a malfunctioning antioxidant defence system. The main causes of OS include aging, drug use, and exposure to danger. A consequence of mitochondrial respiration is ROS. As a result, rising ROS generation and increased mitochondrial activity are related. Alternatively, when the equilibrium of the mitochondria is disturbed, there is also an increased production of ROS. Antioxidants like GPx, CAT, Prx, SOD, and CAT as well as non-enzymatic substances regulate the amount of ROS in a normal cell. The physiological range of ROS is a necessary part of the receptor-mediated signal transduction cascade. Overloading ROS, however, can interfere with antioxidants'capacity to fulfil their rheostatic role, which can result in DNA damage, lipid and protein peroxidation, and even cell death. Drawing from previous studies, we arrived at the following conclusions: (1) ROS is physiologically necessary for the maintenance of cell function (e.g., proliferation, migration, differentiation, and adhesion); (2) low levels of ROS can cause cell cycle arrest; (3) an increase in ROS initiates cell-mediated defence responses; and (4) an excess of ROS promotes cell death and, in severe cases, contiguous tissue necrosis. ROS have dose-dependent effects on cellular function (Wang et al. 2023).

12.4.3 OS and ROS in Diabetic Retinopathy

The imbalance between intrinsic antioxidant capacity and free radicals, which are highly reactive molecules produced by our bodies through a variety of processes, is referred to as "oxidative stress." Over the past 40 years, it has been the focus of substantial research in many medical specialities, including ophthalmology. OS is a highly complex phenomenon that impacts several cell types and is linked with other forms of stress. Retinal damage may be lessened by antioxidant therapy because OS plays a significant role in the pathogenesis of diabetic retina (DR). It is now widely acknowledged that ROS involvement contributes significantly to the development of DR. When there is prolonged hyperglycaemia, the retina's microvessels are more susceptible to OS, which increases the generation of ROS. Free radicals, or ROS for short, are oxidant molecules with an additional electron that makes them incredibly unstable and reactive. They take electrons from surrounding molecules to bring stability back, which causes an oxidative chain to form. Notably, intracellular antioxidants balance the creation and neutralisation of free radicals (ROS), which are created continuously during normal metabolic activities (Gao et al. 2023).

12.4.4 Targeting Mitochondrial OS for Aging and Age-Related Diseases

Premature destruction of tissue and organ function due to mitochondrial OS results from defective mitochondria are a major contributor to mitochondrial OS. Considering the causative relationship, managing mitochondrial OS effectively could be a helpful strategy for treating age-related diseases and ageing itself. One effective and potentially effective tactic is the stimulation of mitophagy to control mitochondrial OS. One of the most important processes for preserving mitochondrial homeostasis is mitophagy, which is the selective autophagic destruction of damaged mitochondria. In particular, the initial phase of PINK1/PARKIN-dependent mitophagy is triggered by the recruitment of PTEN-induced putative kinase protein 1 (PINK1) and ubiquitin ligase PARKIN by defective mitochondria caused by elevated levels of mitochondrial ROS. In injured mitochondria, PINK1 triggers PARKIN recruitment to the OMM, where OMM proteins are ubiquitinated. OMM proteins that form polyubiquitin chains attract autophagy receptor proteins, which then bind with the phagophore's LC3 (microtubule-associated protein 1A/1B light chain 3). Subsequently, phagophores, autophagosomes, and autolysosomes form to destroy damaged mitochondria, which reduces the amount of ROS produced by the mitochondria. These results are supported by the observation that mice treated with the mitophagy inducer trehalose showed better age-related atherosclerosis and decreased production of superoxide (Lee et al. 2023).

12.4.5 OS in Liver Cell Populations During Chronic Liver Disease (CLD)

Increased and persistent OS is one of CLD's traits. Even though it might not be enough to directly cause cell death in some situations, it can cause a shift in redox homeostasis to a continuously dysregulated state, which in turn upregulates several target genes involved in the progression of CLD, promoting inflammation, fibrogenesis, and angiogenesis. OS is usually brought on by the etiological agent's direct effects on parenchymal cells. This increased intracellular generation of ROS and other redox-related intermediates is associated with oxidative stress. This phenomenon sustains hepatocyte damage and cell death overtime in addition to promoting the progression of fibrogenesis and inducing an inflammatory response. Changes in hepatocyte metabolism and/or apoptosis are caused by many intracellular signalling pathways that are set off by alterations in the redox balance. For example, OS has been associated with iron overload-mediated ferroptosis and endoplasmic reticulum stress, as well as with pyroptosis and NLRP3 inflammasome activation. Profibrogenic reactions and phenotypic changes in Hepatic Stellate Cells (HSC) might be directly caused by the long-term generation of ROS and redox-related reactive mediators by damaged hepatocytes and activated non-parenchymal cells. Furthermore, higher ROS generation is a characteristic of activated HSC and is linked with raised mitochondrial activity and NOX overexpression. There is a rise in ROS in mitochondria due to fibroblast stimulation in the number of tissues. Additionally, modifications to the mitochondria have been observed in the creation of matrix and HSC proliferation (Blas-García et al. 2023). The role of ROS in exacerbating and prolonging liver damage is represented in Fig. 12.2.

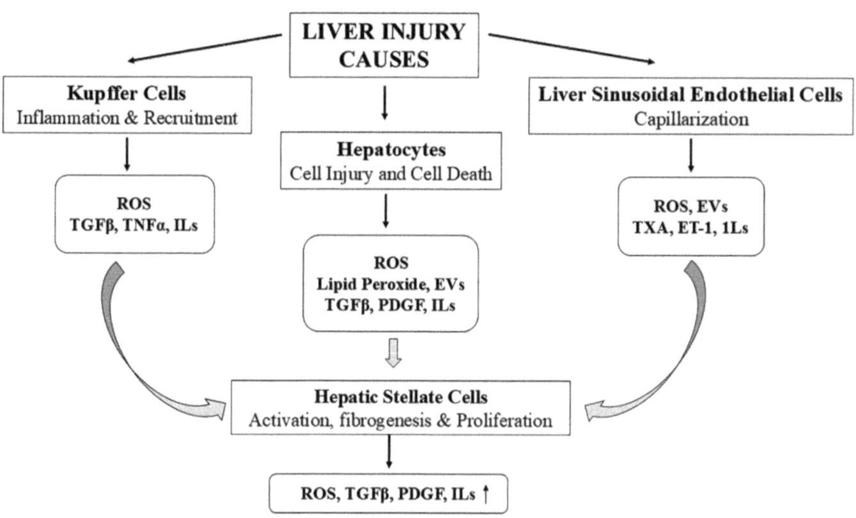

Fig. 12.2 Illustration of intercellular communication in hepatic damage and role of ROS in exacerbating and prolonging liver damage. (Blas-García and Apostolova (2023))

12.5 Rationale of Targeting OS

Several well-recognized CNS disorders such as stroke and AD are the results due to the common factor known as OS. OS is still a prominent treatment target for neurological disorders because the brain is specifically vulnerable to oxidative damage and because it is known that elevated levels of ROS promote neuropathology. Furthermore, OS has been connected to many disorders' early phases. An increasing amount of research suggests that the basic or secondary pathophysiological mechanisms behind acute brain injury are significantly influenced by oxidative stress. In addition, there is a reduction in the antioxidant system of the body results due to several components including environmental and genetic factors that lead to promotion in OS advancement (Briyal et al. 2023).

OS can be the cause of a numerous type of cardiovascular diseases. For normal functioning, cardiac muscles rely on oxidative phosphorylation as they have a higher density of mitochondria. ROS are created by cells when they undergo oxidative phosphorylation, a process that damages the cell's proteins, lipids, and DNA. Inflammation, impaired vasomotor performance, and endothelial dysfunction follow from this. An imbalance between the body's defence mechanisms against the creation of ROS exacerbates oxidative stress and is commonly seen in conditions like hypertension and atherosclerosis. Moreover, the progression of cardiovascular disease is intensified because of OS which is responsible for triggering signalling pathways associated with myocardial remodelling, hypertrophy, and death. Furthermore, data points to the involvement of oxidative stress in some cardiovascular conditions, including cardiac arrhythmia, atherosclerosis, ischemia-reperfusion injury, and heart failure. For example, OS-induced DNA damage is necessary for vascular remodelling following a reperfusion injury. Understanding the connection between OS and cardiovascular health is crucial to reducing the harmful effects of OS and preventing or treating cardiac diseases (Jin and Kang 2024).

In OS therapies, there are several strategies including exogenous as well as antioxidant have been explored. The major aim is to avoid undesirable side effects along with healthy cell signalling that was maintained by redox balance through antioxidant-based treatment. The relationship between ROS and antioxidants is complex and multidimensional, both in terms of location and function. For regular cell signalling functions cell wants a normal range of ROS, yet too much ROS can be just as detrimental to joint health as not enough. Antioxidants and ROS have intra- and extracellular locations that need to be considered. The efficacy of ROS targeting techniques hinges on their capacity to access the relevant compartments where the ROS are generated. One intriguing treatment approach for OA and other diseases resulting from redox signalling abnormalities is to target oxidative stress. These treatments' poor absorption and poor retention in the joint area have hindered their application in OA. Current treatment constraints may be solved by developing biomaterial systems that imitate antioxidants. These difficulties highlight the fact that, as Halliwell explains, determining which ROS, in what quantity, and precisely which molecular target to target is crucial for developing treatments to address the imbalance in any of these species. Furthermore, efforts are currently underway to monitor ROS in vivo and in vitro, which will require more developments in redox biology. The incapacity to sensitively and precisely quantify ROS generated within the system is a major drawback in this field. Most ROS have short half-lives, which makes it challenging to assess them in vivo as well as in vitro (Petrova Zustiak et al. 2023).

To transfer antioxidant chemicals in mitochondria the only well-researched carrier is triphenylphosphonium (TPP). This is specifically true for hydrophobic molecules, which means that they have a difficult time passing through cell membranes and into the cytoplasm, which contains the mitochondria. As TPP has a + ve charge that was attracted to mitochondrial transmembrane with high potential, along with it same as other lipophilic cations TPP is also used in targeting the delivery to mitochondria. The electrochemical gradient is responsible for the deposition of lipophilic cations in the mitochondria in between the inner mitochondrial membrane (-150 to -180 mV) and the cellular membrane (-30 to -60 mV). As a result, after entering the cell, the TPP moiety makes use of the plasma membrane potential, which is in charge of the broad uptake of bioactive molecules, to enter the mitochondrial matrix. Towards the matrix, there will be polar head faces of TPP and the hydrophobic functional group of the TPP–TPP-antioxidant conjugate is placed between the lipidic tails.

12.6 Limitations and Challenges in OS Valuation

It is very important to evaluate OS in different disease conditions, but the main advantage is also considered as the disadvantage of evaluating the large number of biomarkers in several samples through various methodologies. These different methods have several merits and demerits associated with sensibility, costs, and practicability. However, various issues linked with preanalytical, analytical as well as post-analytical issues result in complications in the interpretation of results and it also requires standardization (Melania Gaggini et al. 2020).

12.6.1 Pre-analytical Factors

These are the variables that influence the sample in the laboratory before analysis. These factors are categorized into several categories shown in Table 12.1.

12.6.1.1 Collection of Sample

The collection of samples plays an important role in pre-analytical factors that affect the final product. However, participants' posture such as sitting, standing or supine, fasting status, etc. considered preanalytical determinants. Particularly for saliva and

Table 12.1 List of pre-analytical factors

S. No	Factors	Examples	Ref.
1.	Collection of samples.	Type of biological sample, anticoagulant collection tube, withdrawal of daytime.	Ialongo (2017)
2.	Handling of sample and long-term and short-term storage and stability.	Temperature conditions, collection time from laboratory to delivery, delay before assaying, and time and temperature conditions.	Ene et al. (2017), Vassalle et al. (2020)
3.	Processing of sample.	Haemolyzed and centrifugation samples.	Vassalle et al. (2020)

sample type, different parameters must be considered such as an increase in secretion through food ingestion, carrier presence, chewing gum mastication, and blood contamination possibilities (Ialongo 2017). For right sample conservation in the case of urine 24-hour collection, there are certain difficulties in ensuring the appropriate stability for analysis. Despite this, blood is the most preferable diffused specimen in urine and saliva samples to evaluate oxidative stress. EDTA (ethylenediaminetetraacetic acid) is an anticoagulant found to be less referred to as reducing the capacity of ferrous ion is stabilized by iron chelation. For some analytes, circadian rhythm also diet interference should be considered as well, and in the morning withdrawal protocol considered based on fasting status is preferred (Vassalle et al. 2020).

12.6.1.2 Handling of Sample and Long-Term and Short-Term Stability

In the shortest period, appropriate samples reach to laboratory also after collection cold chain interruption is not allowed. To ensure stability and avoid auto-oxidation of the sample with time, refrigerated centrifugation along with temperatures of −20 and −80 °C should be considered. By employing these precautions, several oxidative stress biomarkers show better stability at stored temperatures of −20 and −80 °C for or up to a year (Ene et al. 2017).

12.6.1.3 Processing of Sample

The important factors that play an important role in pre-analytical factors are haemolysis and lipid content, samples that show highlighted signs of these interpretations must not be tested. For evaluation improvement, sample pre-treatment can be done; for instance, in the case of oxidative stress biomarkers such as MDA, BHT (butylated hydroxytoluene) used as stabilizer. Freeze-thaw cycle repetition should be avoided (Vassalle et al. 2020).

12.6.2 Analytical Factors

These are defined as those factors involved in using methods or processes. Various methods are related to biomarkers for oxidative stress evaluation with different specificities and sensitivity containing both merits and demerits (Ghezzi 2020).

The most commonly used approach is to determine oxidative stress through lipid peroxidation along with MDA, 4-hydroxynoneanal (HNE), and 8-iso-Prostaglandin F2-alpha (Melania Gaggini et al. 2020). Based on this, results have evidenced the importance of biomarkers in several degenerative or severe disease conditions in the assessment of OS (Bigagli et al. 2019). Specifically, MDA measurement by spectrophotometric or fluorometric analysis is linked with cheapness and easy-to-use which will favour its availability in several samples. Preferably, this method contains lower specificities along with other techniques, for example, GC-MS LC-MS. The lower specificity is maintained by several methods and assays named adduct separation phases, for example, utilizing HPLC (Tsikas 2017). Whereas maximum methods used to determine MDA measurement involve the use of derivatization of reagents, that could react with carbonyl groups in MDA (Sobsey et al. 2016).

Additionally, for the determination of isoprostanes GC-MS and LC-MS /MS are the best methods for evaluation of oxidative stress, however, these are expensive and need special operators as well as instrumentations. For the assessment of F2-isoprostanes ELISA kits are very diffuse even if it is not more accurate and specific. ELISA is based on the principle of antigen-antibody reaction in a competitive binding test. The particular antigen is recognized by antibodies and specific metabolites are determined by assay through GC-MS several differences are seen in the results. Sometimes, the concentration observed by ELISA is overestimated, and antibodies observed by ELISA is cross-react along with metabolites, however, GC/MS shown as more selective (M Gaggini et al. 2020).

12.6.3 Post-analytical Factors

In the case of post-analytical factors, the major limitation is in lack of referred agreement in reliable reference values. As per this, we have two possibilities first is single value availability and the second is monitoring that covers serial measurements in the same patients. When we determine a single value, the main limitation is to have a threshold value that could distinguish biomarkers as positive or negative with the provided cut-off or even better than it (Cristina 2018). Secondly, dynamic variation must be interpreted for biomarkers with time.

Significantly, results interpretation, the resulted value may be improved by several factors like genetics including familiarity, lifestyle habits such as smoking habits, stress, alcohol abuse, drugs, physical activity, and anxiety, physiological includes environmental factors, diet, age, and gender, circadian rhythm as well as intra- and inter-unit variability. Different measurement methods have certain limitations or challenges in the elucidation of results which would cause confusion and sometimes it leads to misclassification. However, several determinants are affected by biomarker circulating value example distribution volume, metabolism, and clearance must be considered (Vassalle et al. 2020).

Several used measurements may limit the result interpretation leading to confusion and misclassification. Moreover, various determinants are affected by biomarker circulating value, for example, metabolism or as its distribution volume that should be considered.

12.6.4 Factors Affecting Result Interpretation

OS plays a great act in the pathogenesis of various severe diseases and ageing processes for which various antioxidant therapies act effectively. For the maintenance of health and well-being, preventing the worsening of any disease conditions and complications exogenous supplementation is considered. Despite this many epidemiology studies were unable to prove the benefits of supplementation of antioxidants specifically in cardiovascular disorders (Perillo et al. 2020).Some other reasons or components affect the interpretation of results which include antioxidant type, dose choice, multiple and single approach, supplementation approach, etc. in case of oxidative stress. Many other findings also show that supplementation is important in primary and secondary prevention of any degenerative disease, whereas several other trials found contradictory results. In the case of cardiovascular disorders, several molecules were found effective ascorbic acid, folic acid, vitamin A, vitamin E, selenium, zinc, and beta-carotene alone or in combination (Goszcz et al. 2015). Additionally, in neurodegenerative disease antioxidant efficacy is uncertain, based on this vitamins C, E, and B12 are effectively used as potential molecules in AD (Ferreira et al. 2015).

For instance, the vitamin E experimental model can examine the supply of vitamin E alone or along with other components such as antioxidants or anti-inflammatory, effective in enhancing memory and cognition by improving oxidative stress (Gugliandolo et al. 2017).

The most important factor that is required to be carefully evaluated is dosage. For instance, higher doses of vitamin E show effective results that's why vitamin E is not enough to reach blood concentration and can't have beneficial effects. It is also estimated that threshold value in supplementation in which benefits can be expressed such as 800 IU/d estimated in the case of vitamin E. Oppositely, a high dose of vitamin E is linked with raised deaths or side effects as well as adverse effects, worsening oxidative stress status as acting like pro-oxidants in several conditions. To avoid such conditions, the concentration of plasma vitamin E or other antioxidants at the time of administration and baseline should be monitored. Study time initiation is another problem that is addressed (Browne et al. 2019).

In the field of cancer, based on several studies showed the benefits of the supplementation of antioxidants with many clinical trials (Perillo et al. 2020). It is well recognized that oxidative stress may play a great role in the progression of several stages of cancer including the promotion of the disease. However, the production of ROS cannot be regarded as more dangerous as a regular low level of generation of ROS is compulsory for the stimulation of endogenous antioxidants in upregulation. Additionally, there is redox interconnection which means a particular balance between the generation and scavenging of ROS. Therefore, the determination of recovery and maintenance of redox homeostasis is difficult and should be taken as a new belief in the field of oncology (Chatzianagnostou et al. 2015).

12.7 Limitations and Challenges in OS Therapeutic Management

In clinical situations, controlling OS comes with several restrictions and difficulties. Many diseases, such as cardiovascular, neurodegenerative, cancer, metabolic disorders, and ageing, are associated with OS, and the major reason behind it is an imbalance in the generation of ROS and damage to the body's antioxidant defences (Chaudhary et al. 2023). ROS comprises reactive species like $NO^{\cdot}$, $OH^{\cdot}$ and $O_2^{\cdot-}$ radicals and non-free radicals like $ONOO^{-}$, and H_2O_2 (Phaniendra et al. 2015). Inside the mitochondria, ROS trigger metabolic processes that result in a decreased amount of oxygen via the electron transport channel (Brand 2010), SOD, and GPx all mitigate ROS as they are antioxidant endogenous enzymes (Small et al. 2012).

There are several antioxidant treatment approaches under investigation, among which are in phases of clinical trials. One of the strategies is to remove free radicals before they react with other molecules or free radicals like the removal of superoxide radicals before it reacts with nitric oxide radicals and the removal of hydrogen peroxide before it forms hydroxyl radical. Another strategy includes raised generation in the production of antioxidant enzymes and supplementing antioxidants via diet (Forman and Zhang 2021). There are several dietary antioxidants like vitamins like retinoids, carotenoids, tocopherol, and ascorbic acid; minerals like manganese, copper, zinc and selenium; endogenous like glutathione, ubiquinone, and alpha-lipoic acid; plant-derived such as flavonoids, polyphenols, curcumin, resveratrol, and lycopene; and omega-3-fatty acids which can be supplemented via food and drinks (Pham-Huy et al. 2008).

However, these antioxidant supplements face barriers such as being unable to achieve effective therapeutic concentration in vivo and managing its dose, scavenging by small molecules, disease level, and patient-related factors like age, and gender.

It can be difficult to achieve an effective antioxidant concentration under oxidative stress for several reasons in addition with the intricacy of the oxidative stress process, the physicochemical properties of antioxidants, and the biochemical milieu of the body. The plant-based antioxidants such as flavonoids and polyphenols demonstrated poor absorption via oral route owing to their limited solubility and extensive first-pass metabolism (Zhao et al. 2019). Certain antioxidants may not be as efficient in structures like the brain because they need help overcoming biological barriers like the blood-brain barrier (BBB). Although showing high concentrations, vitamin E, a fat-soluble antioxidant, may not reach sufficient levels in water-rich regions like the brain (Feng et al. 2023). Nanoparticle-based targeting could be of great help in such cases.

Another important constraint of the antioxidant system is the limited effectiveness of quenching by extremely small molecules. The assertion that an antioxidant scavengers hydroxyl radical is essentially pointless because nearly every molecule reacts with hydroxyl radical at nearly the same pace. Therefore, the sole defense that works against hydroxyl radicals is to stop them from developing, and the best method to do that is the elimination of hydrogen peroxide (Forman and Zhang 2021).

The strength to which oxidative stress adds up to this disease restricts the efficacy of antioxidant defences. It is considerably more common for oxidative stress to be a secondary cause of ailment than the fundamental cause, so mitigating oxidative stress might not have an important effect on the course of the illness. If antioxidant defences are taken into consideration in therapeutic studies, this limitation is arguably the most important one that is frequently disregarded (Mirmiran et al. 2022). Antioxidants might connect differently with the body as the condition advances due to changes in stages of OS and the surroundings of cells. The issue is to identify the extent to which antioxidant therapies must be improved to lessen the disease manifestations in which OS is not the major reason for disease (Sharifi-Rad et al. 2020).

Antioxidants are widely recognized for their capacity to counteract reactive species and mitigate the impact of oxidative stress. However, in specific circumstances, certain antioxidants may demonstrate prooxidant properties. Therefore, rather than shielding cells from oxidative damage, they might be a part of the process that produces RNS or ROS, which in turn causes oxidative stress (Sotler et al. 2019). In high doses, alpha-tocopherol is also recognized as a deadly prooxidant and a strong antioxidant. It turns into a radical when it reacts with ROS, and if there is insufficient vitamin C for it to regenerate, it stays in the reactive condition (J and G 1994).

Although vitamin C is a strong antioxidant, depending on the dosage, it may also act as a prooxidant. When taken at decreased doses (30–100 mg/kg body weight), it can have an antioxidant effect; when taken at large doses (1000 mg/kg body weight), it can have a prooxidant impact (Asplund et al. 2002). Additionally, vitamin C has a prooxidant impact when combined with iron, reducing ferric to ferrous, or with copper, reducing cupric to cuprous ions. By the Fenton reaction, the reduced transition metals subsequently transform hydrogen peroxide to OH^- (Duarte and Lunec 2005). There is reduce in the body's natural defences against reactive species making the environment more susceptible to oxidation through Trolox which is a Vitamin E and Vitamin C water-soluble supplement. Because of their prooxidative characteristics, some antioxidants may cause mild oxidative stress (Poljšak and Raspor 2008). Beta-carotene's prooxidant activity is dependent on how it interacts with biological membranes and whether co-antioxidants like vitamin C are present.

The antioxidant properties of beta-carotene diminish with increasing oxygen concentration. Increased death rates were seen following continuous usage of beta-carotene, retinoids, and tocopherol supplements, reported in a meta-analysis and a systematic review (Bjelakovic et al. 2007).

The gradual loss of tissues and organ functionality occurs with ageing. The ageing theory of free radicals also regarded as the OS model of ageing is predicated on the hypothesis of structure damage that holds the build-up of several bio metabolites, for example, DNA, lipids and proteins through ROS and RNS subsequently generates age-related functional deficits (Beckman and Ames 1998). Although the precise mechanism underlying oxidative stress-induced ageing remains unclear, elevated ROS and RNS levels likely trigger cellular senescence, a physiological response to replication-related impairments that halts the growth of cells. Due to their reduced physiological ability, older patients may need stronger or larger dosages of antioxidant treatment, but they are also more vulnerable to side effects (McHugh and Gil 2018). The ADME of drugs vary with age, and this can impact the pharmacokinetics of antioxidant treatments. An extended half-life of antioxidants may result from reduced breakdown and elimination rates in older people. To prevent toxicity, precise dosage modifications are required (Shi and Klotz 2011). Drug-drug interactions are more likely to occur in older individuals due to their need for many drugs to treat various long-term medical conditions. Vitamin C supplements may interfere with chemotherapy and radiation therapy in cancer patients ("Antioxidant Supplements: What You Need To Know | NCCIH," 2024).

References

Andrés Juan C, de la Lastra JMP, Plou FJ, Pérez-Lebeña E, Reinbothe S (2021) The chemistry of reactive oxygen species (ROS) revisited: outlining their role in biological macromolecules (DNA, lipids and proteins) and induced pathologies. Int J Mol Sci 22(9):4642. https://doi.org/10.3390/ijms22094642

Antioxidant Supplements: What You Need To Know | NCCIH [WWW Document], 2024 https://www.nccih.nih.gov/health/antioxidant-supplements-what-you-need-to-know. Accessed 10 Nov 2024

Asplund KUM, Jansson PJ, Lindqvist C, Nordstöm T (2002) Measurement of ascorbic acid (vitamin C) induced hydroxyl radical generation in household drinking water. Free Radic Res 36:1271–1276. https://doi.org/10.1080/1071576021000036425

Azmanova M, Pitto-Barry A (2022) Oxidative stress in cancer therapy: friend or enemy? Chembiochem 23. https://doi.org/10.1002/CBIC.202100641

Bano A, Gupta A, Rai S, Fatima T, Sharma S, Pathak N, Bano A, Gupta A, Rai S, Fatima T, Sharma S, Pathak N (2021) Mechanistic role of reactive oxygen species and its regulation <em>via</em> the antioxidant system under environmental stress. https://doi.org/10.5772/INTECHOPEN.101045

Beckman KB, Ames BN (1998) The free radical theory of aging matures. Physiol Rev 78:547–581. https://doi.org/10.1152/PHYSREV.1998.78.2.547

Bigagli E, Lodovici M (2019) Circulating oxidative stress biomarkers in clinical studies on type 2 diabetes and its complications. Oxidative Med Cell Longev 2019. https://doi.org/10.1155/2019/5953685

Bjelakovic G, Nikolova D, Gluud LL, Simonetti RG, Gluud C (2007) Mortality in randomized trials of antioxidant supplements for primary and secondary prevention: systematic review and meta-analysis. JAMA 297:842–857. https://doi.org/10.1001/JAMA.297.8.842

Blas-García A, Apostolova N (2023) Novel therapeutic approaches to liver fibrosis based on targeting oxidative stress. Antioxidants (Basel) 12(8):1567. https://doi.org/10.3390/antiox12081567

Brand MD (2010) The sites and topology of mitochondrial superoxide production. Exp Gerontol 45:466–472. https://doi.org/10.1016/J.EXGER.2010.01.003

Briyal S, Ranjan AK, Gulati A (2023) Oxidative stress: a target to treat Alzheimer's disease and stroke. Neurochem Int 165. https://doi.org/10.1016/j.neuint.2023.105509

Browne D, Mcguinness B, Woodside JV, Gareth, Mckay J, Mckay GJ (2019) Vitamin E and Alzheimer's disease: what do we know so far? Taylor Fr 14:1303–1317. https://doi.org/10.2147/CIA.S186760

Bukato K, Kostrzewa T, Gammazza AM, et al (2024) Endogenous estrogen metabolites as oxidative stress mediators and endometrial cancer biomarkers. Cell Commun Signal 22. https://doi.org/10.1186/s12964-024-01583-0

Chan HP, Tran V, Lewis C, Thomas PS (2009) Elevated levels of oxidative stress markers in exhaled breath condensate. J Thorac Oncol 4:172–178. https://doi.org/10.1097/JTO.0b013e3181949eb9

Chatzianagnostou K, Del Turco S, Pingitore A, Sabatino L, Vassalle C (2015) The Mediterranean lifestyle as a non-pharmacological and natural antioxidant for healthy aging. Antioxidants 4: 719–736. https://doi.org/10.3390/ANTIOX4040719

Chaudhary P, Janmeda P, Docea AO, Yeskaliyeva B, Abdull Razis AF, Modu B, Calina D, Sharifi-Rad J (2023) Oxidative stress, free radicals and antioxidants: potential crosstalk in the pathophysiology of human diseases. Front Chem 11:1158198. https://doi.org/10.3389/FCHEM.2023.1158198/BIBTEX

Chauhan A, Semwal D, Mishra S, Semwal R (2015) Ayurvedic research and methodology: present status and future strategies. Ayu 36:364. https://doi.org/10.4103/0974-8520.190699

Cherian D, Peter T, Narayanan A, Madhavan S, Achammada S, Vynat G (2019) Malondialdehyde as a marker of oxidative stress in periodontitis patients. J Pharm Bioallied Sci 11:S297–S300. https://doi.org/10.4103/JPBS.JPBS_17_19

Chondrogianni N, Sakellari M, Lefaki M, Papaevgeniou N, Gonos ES (2014) Proteasome activation delays aging in vitro and in vivo. Free Radic Biol Med 71:303–320. https://doi.org/10.1016/J.FREERADBIOMED.2014.03.031

Cristina V (2018) Oxidative stress and cardiovascular risk prediction: the long way towards a "radical" perspective. Int J Cardiol 273:252–253. https://doi.org/10.1016/J.IJCARD.2018.09.033

Da Fonseca LJS, Nunes-Souza V, Goulart MOF, Rabelo LA (2019) Oxidative stress in rheumatoid arthritis: what the future might hold regarding novel biomarkers and add-on therapies. Oxidative Med Cell Longev 2019. https://doi.org/10.1155/2019/7536805

Daenen K, Andries A, Mekahli D, Van Schepdael A, Jouret F, Bammens B (2018) Oxidative stress in chronic kidney disease. Pediatr Nephrol 346(34):975–991. https://doi.org/10.1007/S00467-018-4005-4

Davalli P, Mitic T, Caporali A, Lauriola A, D'arca D (2016) ROS, cell senescence, and novel molecular mechanisms in aging and age-related diseases. Wiley Online Libr 2016. https://doi.org/10.1155/2016/3565127

Dhapola R, Beura SK, Sharma P, Singh SK, HariKrishnaReddy D (2024) Oxidative stress in Alzheimer's disease: current knowledge of signaling pathways and therapeutics. Mol Biol Rep 51. https://doi.org/10.1007/S11033-023-09021-Z

Dias V, Junn E, Mouradian MM (2013) The role of oxidative stress in Parkinson's disease. J Parkinsons Dis 3:461–491. https://doi.org/10.3233/JPD-130230
Dröge W (2002) Free radicals in the physiological control of cell function. Physiol Rev 82:47–95. https://doi.org/10.1152/PHYSREV.00018.2001
Duarte TL, Lunec J (2005) Review: when is an antioxidant not an antioxidant? A review of novel actions and reactions of vitamin C. Free Radic Res 39:671–686. https://doi.org/10.1080/10715760500104025
Dubois-deruy E, Peugnet V, Turkieh A, Pinet F (2020) Oxidative stress in cardiovascular diseases. Antioxidants 9:864–874. https://doi.org/10.3390/ANTIOX9090864
Ene E, Jansen HJM, Beekhof PK, Viezeliene D, Muzakova V, Skalicky J (2017) Long-term stability of oxidative stress biomarkers in human serum. Res. 2017 Taylor Fr 51:970–977. https://doi.org/10.1080/10715762.2017.1398403
Feng J, Zheng Y, Guo M, Ares I, Martínez M, Lopez-Torres B, Martínez-Larrañaga MR, Wang X, Anadón A, Martínez MA (2023) Oxidative stress, the blood–brain barrier and neurodegenerative diseases: the critical beneficial role of dietary antioxidants. Acta Pharm Sin B 13:3988–4024. https://doi.org/10.1016/J.APSB.2023.07.010
Ferreira MES, de Vasconcelos AS, da Costa Vilhena T, da Silva TL, da Silva Barbosa A, Gomes ARQ, Dolabela MF, Percário S (2015) Oxidative stress in Alzheimer's disease: should we keep trying antioxidant therapies? Cell Mol Neurobiol 35:595–614. https://doi.org/10.1007/S10571-015-0157-Y
Forman HJ, Zhang H (2021) Targeting oxidative stress in disease: promise and limitations of antioxidant therapy. Nat Rev Drug Discov 20:689–709. https://doi.org/10.1038/S41573-021-00233-1
Fujii J, Homma T, Osaki T (2022) Superoxide radicals in the execution of cell death. Antioxidants 11:501–511. https://doi.org/10.3390/ANTIOX11030501
Gaggini M, Sabatino L, Vassalle C (2020) Conventional and innovative methods to assess oxidativestressbiomarkersintheclinicalcardiovascular setting. BioTechniques 68:223–231. https://doi.org/10.2144/BTN-2019-0138
Gao J, Tao L, Jiang Z (2023) Alleviate oxidative stress in diabetic retinopathy: antioxidant therapeutic strategies. Redox Rep 28. https://doi.org/10.1080/13510002.2023.2272386
García-Sánchez A, Guillermina Miranda-Díaz A, Germán Cardona-Muñoz E, Santibañez JF (2020) The role of oxidative stress in physiopathology and pharmacological treatment with pro-and antioxidant properties in chronic diseases. Oxid Med Cell Longev. https://doi.org/10.1155/2020/2082145
Ghezzi P (2020) Environmental risk factors and their footprints in vivo—a proposal for the classification of oxidative stress biomarkers. Redox Biol 34:101442. https://doi.org/10.1016/J.REDOX.2020.101442
Goszcz K, Deakin SJ, Duthie GG, Stewart D, Leslie SJ, Megson IL (2015) Antioxidants in cardiovascular therapy: panacea or false hope? Front Cardiovasc Med 2. https://doi.org/10.3389/FCVM.2015.00029/FULL
Granata S, Masola V, Zoratti E, Scupoli MT, Baruzzi A, Messa M, Sallustio F, Gesualdo L, Lupo A, Zaza G (2015) NLRP3 inflammasome activation in dialyzed chronic kidney disease patients. PloS One 10. https://doi.org/10.1371/journal.pone.0122272
Gugliandolo A, Bramanti P, Mazzon E (2017) Role of vitamin e in the treatment of alzheimer's disease: evidence from animal models. Int J Mol Sci 18. https://doi.org/10.3390/IJMS18122504
Harshithkumar R, Shah P, Jadaun P, Mukherjee A (2024) ROS chronicles in HIV infection: genesis of oxidative stress, associated pathologies, and therapeutic strategies. Curr Issues Mol Biol 46:8852–8873. https://doi.org/10.3390/CIMB46080523
Ialongo C (2017) Preanalytic of total antioxidant capacity assays performed in serum, plasma, urine and saliva. Clin Biochem 50:356–363. https://doi.org/10.1016/J.CLINBIOCHEM.2016.11.037
Interdonato M, Pizzino G, Bitto A, Galfo F, Irrera N, Mecchio A, Pallio G, Ramistella V, De Luca F, Santamaria A, Minutoli L, Marini H, Squadrito F, Altavilla D (2015) Cadmium delays puberty onset and testis growth in adolescents. Clin Endocrinol 83:357–362. https://doi.org/10.1111/CEN.12704

Ionescu-Tucker A, Cotman CW (2021) Emerging roles of oxidative stress in brain aging and Alzheimer's disease. Neurobiol Aging 107:86–95. https://doi.org/10.1016/j.neurobiolaging.2021.07.014

J, B., G, B. (1994) The alpha-tocopherol, beta-carotene cancer prevention study in Finland. Nutr Rev 52:242–245. https://doi.org/10.1111/J.1753-4887.1994.TB01430.X

Jiang H, Zuo J, Li B, Chen R, Luo K, Xiang X, Lu S, Huang C, Liu L, Tang J, Gao F (2023) Drug-induced oxidative stress in cancer treatments: angel or devil? Redox Biol 63. https://doi.org/10.1016/J.REDOX.2023.102754

Jin S, Kang PM (2024) A systematic review on advances in management of oxidative stress-associated cardiovascular diseases. Antioxidants 13. https://doi.org/10.3390/ANTIOX13080923

Khola N, Moar K, Kumar Maurya P (2024) Oxidative stress and neurological disorders: therapeutic strategies and pharmacological intervention. Brain Hear 2:2704. https://doi.org/10.36922/BH.2704

Kishi S, Nagasu H, Kidokoro K, Kashihara N (2024) Oxidative stress and the role of redox signalling in chronic kidney disease. Nat Rev Nephrol 20:101–119. https://doi.org/10.1038/S41581-023-00775-0

Klran TR, Otlu O, Karabulut AB (2023) Oxidative stress and antioxidants in health and disease. J Lab Med 47:1–11. https://doi.org/10.1515/LABMED-2022-0108/HTML

Kumari S, Dhapola R, Reddy DHK (2023) Apoptosis in Alzheimer's disease: insight into the signaling pathways and therapeutic avenues. Apoptosis 28:943–957. https://doi.org/10.1007/S10495-023-01848-Y

Lee YH, Kuk MU, So MK, Song ES, Lee H, Ahn SK, Kwon HW, Park JT, Park SC (2023) Targeting mitochondrial oxidative stress as a strategy to treat aging and age-related diseases. mdpi.com. https://doi.org/10.3390/antiox12040934

Leuti A, Maccarrone M, Chiurchiù V (2019) Proresolving lipid mediators: endogenous modulators of oxidative stress. Wiley Online Libr. https://doi.org/10.1155/2019/8107265

Leyane T, Jere S, Houreld NN (2022) Oxidative stress in ageing and chronic degenerative pathologies: molecular mechanisms involved in counteracting oxidative stress and chronic inflammation. Int J Mol Sci 23(13):7273. https://doi.org/10.3390/ijms23137273

Li J, Jia B, Cheng Y, Song Y, Li Q, Luo C (2022) Targeting molecular mediators of ferroptosis and oxidative stress for neurological disorders. Wiley Online Libr. https://doi.org/10.1155/2022/3999083

Liguori I, Russo G, Curcio F, Bulli G, Aran L, Della-Morte D, Gargiulo G, Testa G, Cacciatore F, Bonaduce D, Abete P (2018) Oxidative stress, aging, and diseases. Clin Interv Aging:757–772. https://doi.org/10.2147/CIA.S158513

Ling-Yu Kan L, Li P, Sze-Man Hon S, Yin-Tung Lai A, Li A, Ching-Yau Wong K, Huang D, Wong C-K (2024) Deciphering the interplay between the epithelial barrier, immune cells, and metabolic mediators in allergic disease. mdpi.com. https://doi.org/10.3390/ijms25136913

Marzocco S, Lorenzetti S, Sierra-Vargas MP, Montero-Vargas M, Debray-García Y, Carlos Vizuet-De-Rueda J, Loaeza-Román A, Terán LM (2023) Oxidative stress and air pollution: its impact on chronic respiratory diseases. Int J Mol Sci. https://doi.org/10.3390/ijms24010853

Mathew DJ, Sivak JM (2024) Lipid mediators in glaucoma: unraveling their diverse roles and untapped therapeutic potential. Prostaglandins Other Lipid Mediat 171. https://doi.org/10.1016/J.PROSTAGLANDINS.2024.106815

Matsui A, Ikeda T, Enomoto K, Hosoda K, Nakashima H, Omae K, Watanabe M, Hibi T, Kitajima M (2000) Increased formation of oxidative DNA damage, 8-hydroxy-2′- deoxyguanosine, in human breast cancer tissue and its relationship to GSTP1 and COMT genotypes. Cancer Lett 151:87–95. https://doi.org/10.1016/S0304-3835(99)00424-3

McHugh D, Gil J (2018) Senescence and aging: causes, consequences, and therapeutic avenues. J Cell Biol 217:65–77. https://doi.org/10.1083/JCB.201708092

Milillo C, Aruffo E, Di Carlo P, et al (2024) Polystyrene nanoplastics mediate oxidative stress, senescence, and apoptosis in a human alveolar epithelial cell line. Front public Heal 12. https://doi.org/10.3389/fpubh.2024.1385387

Mirmiran P, Hosseini-Esfahani F, Esfandiar Z, Hosseinpour-Niazi S, Azizi F (2022) Associations between dietary antioxidant intakes and cardiovascular disease. Sci Rep 12. https://doi.org/10.1038/S41598-022-05632-X

Obeagu EI (2024) Red blood cells as biomarkers and mediators in complications of diabetes mellitus: a review. Med (United States) 103:E37265. https://doi.org/10.1097/MD.0000000000037265

Ohtake S, Kawahara T, Ishiguro Y, Takeshima T, Kuroda S, Izumi K, Miyamoto H, Uemura H (2018) Oxidative stress marker 8-hydroxyguanosine is more highly expressed in prostate cancer than in benign prostatic hyperplasia. Mol Clin Oncol 9. https://doi.org/10.3892/MCO.2018.1665

Olufunmilayo EO, Gerke-Duncan MB, Holsinger RMD (2023) Oxidative stress and Antioxidants in neurodegenerative disorders. Antioxidants 12. https://doi.org/10.3390/ANTIOX12020517

Payne BAI, Chinnery PF (2015) Mitochondrial dysfunction in aging: much progress but many unresolved questions. Biochim Biophys Acta 1847:1347. https://doi.org/10.1016/J.BBABIO.2015.05.022

Pérez-Matute P, Zulet MA, Martínez JA (2009) Reactive species and diabetes: counteracting oxidative stress to improve health. Curr Opin Pharmacol 9:771–779. https://doi.org/10.1016/J.COPH.2009.08.005

Perillo B, Di Donato M, Pezone A, Di Zazzo E, Giovannelli P, Galasso G, Castoria G, Migliaccio A (2020) ROS in cancer therapy: the bright side of the moon. Exp Mol Med 52:192–203. https://doi.org/10.1038/S12276-020-0384-2

Petrova Zustiak S, Jain E, Lee Aldrich J, Panicker A, Ovalle R, Sharma B, Crayton Pruitt J (2023) Drug delivery strategies and nanozyme technologies to overcome limitations for targeting oxidative stress in osteoarthritis. mdpi.com. https://doi.org/10.3390/ph16071044

Pham-Huy LA, He H, Pham-Huy C (2008) Free radicals, antioxidants in disease and health. Int J Biomed Sci 4:89. https://doi.org/10.59566/ijbs.2008.4089

Phaniendra A, Jestadi DB, Periyasamy L (2015) Free radicals: properties, sources, targets, and their implication in various diseases. Indian J Clin Biochem 30:11–26. https://doi.org/10.1007/S12291-014-0446-0

Pizzino G, Irrera N, Cucinotta M, Pallio G, Mannino F, Arcoraci V, Squadrito F, Altavilla D, Bitto A (2017) Oxidative stress: harms and benefits for human health. Wiley Online Libr. https://doi.org/10.1155/2017/8416763

Poljšak B, Raspor P (2008) The antioxidant and pro-oxidant activity of vitamin C and trolox in vitro: a comparative study. J Appl Toxicol 28:183–188. https://doi.org/10.1002/JAT.1264

Potenza M, Gagliardi S, Nacci C, Carratu M, Montagnani M (2008) Endothelial dysfunction in diabetes: from mechanisms to therapeutic targets. Curr Med Chem 16:94–112. https://doi.org/10.2174/092986709787002853

Prakash C, Sharma D, Kumar P, Madireddy S, Madireddy S (2023) Therapeutic strategies to ameliorate neuronal damage in epilepsy by regulating oxidative stress, mitochondrial dysfunction, and neuroinflammation. mdpi.com. https://doi.org/10.3390/brainsci13050784

Puspita L, Chung S, brain JS-M (2017) Oxidative stress and cellular pathologies in Parkinson's disease, vol 10. Springer. https://doi.org/10.1186/s13041-017-0340-9

Saez I, Vilchez D (2014) The mechanistic links between proteasome activity, aging and agerelated diseases. Curr Genomics 15:38–51. https://doi.org/10.2174/138920291501140306113344

Scioli MG, Storti G, D'amico F, Guzmán RR, Centofanti F, Doldo E, María E, Miranda C, Orlandi A (1995) Oxidative stress and new pathogenetic mechanisms in endothelial dysfunction: potential diagnostic biomarkers and therapeutic targets. mdpi.com. https://doi.org/10.3390/jcm9061995

Sharifi-Rad M, Anil Kumar NV, Zucca P, Varoni EM, Dini L, Panzarini E, Rajkovic J, Tsouh Fokou PV, Azzini E, Peluso I, Prakash Mishra A, Nigam M, El Rayess Y, El Beyrouthy M, Polito L, Iriti M, Martins N, Martorell M, Docea AO, Setzer WN, Calina D, Cho WC, Sharifi-Rad J (2020) Lifestyle, oxidative stress, and antioxidants: back and forth in the pathophysiology of chronic diseases. Front Physiol 11. https://doi.org/10.3389/FPHYS.2020.00694/FULL

Shi S, Klotz U (2011) Age-related changes in pharmacokinetics. Curr Drug Metab 12:601–610. https://doi.org/10.2174/138920011796504527

Small DM, Coombes JS, Bennett N, Johnson DW, Gobe GC (2012) Oxidative stress, anti-oxidant therapies and chronic kidney disease. Nephrology (Carlton) 17:311–321. https://doi.org/10.1111/J.1440-1797.2012.01572.X

Sobsey CA, Han J, Lin K, Swardfager W, Levitt A, Borchers CH (2016) Development and evaluation of a liquid chromatography–mass spectrometry method for rapid, accurate quantitation of malondialdehyde in human plasma. J Chromatogr B Anal Technol Biomed Life Sci 1029–1030:205–212. https://doi.org/10.1016/J.JCHROMB.2016.07.013

Sotler R, Poljšak B, Dahmane R, Jukić T, Pavan Jukić D, Rotim C, Trebše P, Starc A (2019) Prooxidant activities of antioxidants and their impact on health. Acta Clin Croat 58:726. https://doi.org/10.20471/ACC.2019.58.04.20

Tirichen H, Yaigoub H, Xu W, Wu C, Li R, Li Y (2021) Mitochondrial reactive oxygen species and their contribution in chronic kidney disease progression through oxidative stress. Front Physiol 12. https://doi.org/10.3389/FPHYS.2021.627837/FULL

Tsikas D (2017) Assessment of lipid peroxidation by measuring malondialdehyde (MDA) and relatives in biological samples: analytical and biological challenges. Anal Biochem 524:13–30. https://doi.org/10.1016/J.AB.2016.10.021

Tsoukalas D, Fragkiadaki P, Docea AO, Alegakis AK, Sarandi E, Thanasoula M (2019) Discovery of potent telomerase activators: unfolding new therapeutic and anti-aging perspectives. spandidos-publications.com 20:3701–3708. https://doi.org/10.3892/mmr.2019.10614

Vassalle C, Maltinti M, Sabatino L (2020) Targeting oxidative stress for disease prevention and therapy: where do we stand, and where do we go from here. Molecules 25. https://doi.org/10.3390/MOLECULES25112653

Wang G, Yang F, Zhou W, Xiao N, Luo M, Tang Z (2023) The initiation of oxidative stress and therapeutic strategies in wound healing. Biomed Pharmacother 157. https://doi.org/10.1016/j.biopha.2022.114004

Wójcik P, Gegotek A, Žarkovi'c NŽ, Zbieta Skrzydlewska E (2021) Oxidative stress and lipid mediators modulate immune cell functions in autoimmune diseases. mdpi.com. https://doi.org/10.3390/ijms22020723

Yu C, Xiao JH (2021) The Keap1-Nrf2 system: a mediator between oxidative stress and aging. Oxid Med Cell Longev 2021:6635460. https://doi.org/10.1155/2021/6635460

Yukuyama MN, Ghisleni DDM, Pinto TJA, Bou-Chacra NA (2016) Nanoemulsion process selection and application in cosmetics—a review. Int J Cosmet Sci 38:13–24. https://doi.org/10.1111/ICS.12260

Zhao J, Yang J, Xie Y (2019) Improvement strategies for the oral bioavailability of poorly water-soluble flavonoids: an overview. Int J Pharm 570. https://doi.org/10.1016/J.IJPHARM.2019.118642

Zhou H, Luo D, Chen D, Tan X, Bai X, Liu Z, Yang X, Liu W (2021) Current advances of nanocarrier technology-based active cosmetic ingredients for beauty applications. Clin Cosmet Investig Dermatol 14:867–887. https://doi.org/10.2147/CCID.S313429

Chapter 13
Antioxidant Therapeutic Strategies: Current Scenario and Future Perspective

Aparna Batra, Ashna Pradeep, Namrata Gautam, Debopriya Dutta, and Sushama Talegaonkar

Abbreviations

8-ISO	8-Isoprostanes
AKT	Protein Kinase B
ALD	Acute Lung Injury
AnxV-Rb1-LPs	Ginsenoside Rb1-anchored Nano-sponge-like Liposomes
ARDS	Acute Respiratory Distress Syndrome
ARE	Antioxidant Response Elements
BACH1	BTB and CNC Homology 1
BBB	Blood-Brain Barrier
BE	Baicalein
BE-MC	Baicalein-encapsulated Micelles
BHT	Butylated Hydroxytoluene
BPNs	Biopolymeric Nanoparticles
BSA NPs	Bovine Serum Albumin Nanoparticles
CAD	Coronary Artery Disease
CAT	Catalase
CEP	Cepharanthine
CEP-DPI	Cepharanthine Dry Powder Inhaler
CHD	Congenital Heart Disease
COPD	Chronic Obstructive Pulmonary Disease

A. Batra · A. Pradeep · D. Dutta · S. Talegaonkar (✉)
Department of Pharmaceutics, School of Pharmaceutical Sciences, Delhi Pharmaceutical Sciences & Research University, New Delhi, India
e-mail: stalegaonkar@dpsru.edu.in

N. Gautam
School of Pharmacy, Noida International University, Noida, Uttar Pradesh, India

N. Srivastava, S. A. Saraf (eds.), *Pharmacological Strategies for Prevention and Treatment of Oxidative Stress*, Oxidative Stress in Applied Basic Research and Clinical Practice, https://doi.org/10.1007/978-3-032-19837-2_13

CoQ10	Coenzyme Q10
COX-2	Cyclooxygenase-2
CVD	Cardiovascular Diseases
CYBA	Cytochrome B-245 Alpha Chain
CYBB	Cytochrome B-245 Beta Chain
DM	Diabetes Mellitus
EA	Ellagic Acid
EGCG	Epigallocatechin Gallate
EUK-134	Synthetic Antioxidant Mimic
Fe-GA NCPs	Iron-Gallic Acid Coordination Nanoparticles
G6PD	Glucose-6-Phosphate Dehydrogenase
GPx	Glutathione Peroxidase
GSH	Glutathione
GSSG	Oxidized Glutathione
hBECs	Human Bronchial Epithelial Cells
HF	Heart Failure
HP-β-CD	Hydroxypropyl-β-Cyclodextrin
hSOD1	Human Cu-Zn-Superoxide Dismutase
ICAM1	Intercellular Adhesion Molecule 1
IL-10	Interleukin-10
IL6	Interleukin-6
iNOS	Inducible Nitric Oxide Synthase
IVC	Intravenous Vitamin C
KA	Kainic Acid
KEAP1	Kelch-like ECH-associated Protein 1
Lipo DOX	Liposomal Doxorubicin
Lipo DOX-FA	Liposomal Doxorubicin with Ferulic Acid
LOX	Lipoxygenase
mAb	Monoclonal Antibody
MAPKs	Mitogen-Activated Protein Kinases
MDA	Malondialdehyde
MI	Myocardial Infarction
MitoApocynin	Mitochondria-targeted Apocynin
MitoQ	Mitochondria-targeted Quinone
MMP-2	Matrix Metalloproteinase-2
MMP-9	Matrix Metalloproteinase-9
NAC	N-Acetylcysteine
NADPH	Nicotinamide Adenine Dinucleotide Phosphate
NanoCAPs	Nano-co-encapsulated Antioxidant Particles
NF-κB	Nuclear Factor kappa B
NIA	Niacinamide
NOS	Nitric Oxide Synthase
NOX	NADPH Oxidase
NRF2	Nuclear Factor Erythroid 2–Related Factor 2

PCG-1α	Peroxisome Proliferator-activated Receptor Gamma Coactivator 1-alpha
PECAM1	Platelet Endothelial Cell Adhesion Molecule 1
PGE2	Prostaglandin E2
PI3K	Phosphoinositide-3-Kinase
PRX	Peroxiredoxin
PUFA	Polyunsaturated Fatty Acids
PVAX	Nanoparticles for Vanillyl Alcohol Delivery
RA	Rheumatoid Arthritis
RNS	Reactive Nitrogen Species
ROS	Reactive Oxygen Species
scFv	Single Chain Variable Fragment
SE	Status Epilepticus
SFN	Sulforaphane
SOD	Superoxide Dismutase
SODS	Superoxide Dismutases
STAT3/5	Signal Transducer and Activator of Transcription 3/5
TDDS	Transdermal Drug Delivery System
TH-EVNs	Taraxacum officinale Extracellular Vesicle-like Nanoparticles
TRAP	Total Redox Antioxidant Potential
Trx	Thioredoxin
TXN RD1	Thioredoxin Reductase 1
VED	Vascular Endothelial Dysfunction
XOR	Xanthine Oxidoreductase
α-TOH	α-Tocopherol
βTrCP	Beta-transducin Repeat-containing Protein

13.1 Introduction

13.1.1 Antioxidants

Antioxidants are substances that play a vital role in protecting the body against oxidative stress and damage caused by free radicals originated from external or internal sources (Chandimali et al. 2025). These molecules have been extensively studied for their potential health benefits, including their ability to prevent chronic diseases such as cancer, heart disease, and diabetes (Ayoka et al. 2022; Long et al. 2025; Morsy et al. 2025).

Antioxidants are broadly classified into two major categories namely enzymatic and non-enzymatic antioxidants. Enzymatic antioxidants are located in the cytoplasm and include glutathione peroxidase (GPx), peroxide-scavenging enzyme-catalase (CAT), and superoxide dismutase (SOD). The non-enzymatic antioxidants contain mainly melatonin, selegiline, etc. are also responsible for neutralizing

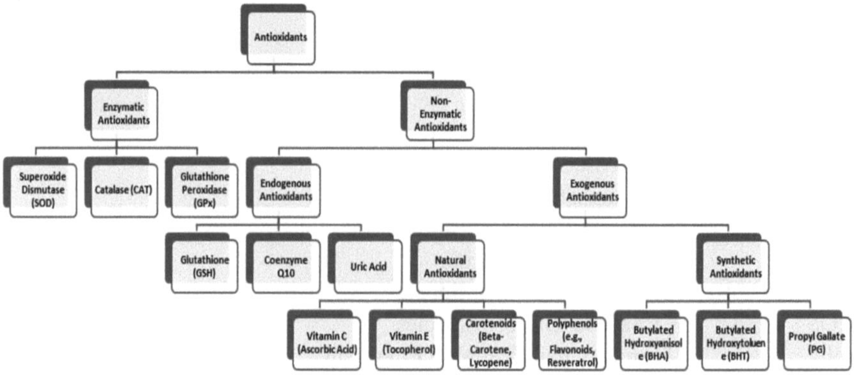

Fig. 13.1 Classification of antioxidants

reactive species (Pruteanu et al. 2023). Natural antioxidants are typically obtained from plant sources but are also found in microorganisms, some animal tissues, and most plants. Blueberries, plums, broccoli, and other citrus fruits are known to be rich in antioxidants. Some of the most important natural antioxidant groups include tocopherols and tocotrienols, ascorbic acid, flavonoids, carotenoids, and phenolic acids (Arias et al. 2022; Cory et al. 2018; Gautam et al. 2025a). The classification of antioxidants into enzymatic and non-enzymatic categories is illustrated in Fig. 13.1.

13.1.2 Antioxidant Therapeutic Strategies

Antioxidant therapeutic strategies have attracted considerable interest in modern medicine due to their potential to alleviate oxidative stress. This stress arises when the generation of reactive oxygen species (ROS) surpasses the body's capacity to neutralize them through endogenous and exogenous antioxidants. Over time, this imbalance can result in cellular damage, inflammation, and tissue dysfunction. Therapeutic approaches aimed at boosting the body's antioxidant defense systems involve both direct supplementation with antioxidants and enhancing the body's natural antioxidant production. Additionally, innovative approach like gene therapy is developed to enhance the delivery and efficacy of antioxidants to specific tissues. As a result, a major approach in therapeutic interventions involves the prevention of harmful oxidant generation, the inhibition of signalling pathways triggered by oxidants that lead to inflammation or cell demise, and the augmentation of both antioxidant enzymes and their corresponding substrates (Jena et al. 2023). Clinical trials are conducted for various treatments, including GC4419 (a SOD mimic) for squamous cell cancers (NCT01921426), and sulphoraphane (an activator of the NRF2 transcription factor) for COPD (NCT01335971). Various antioxidant therapeutic strategies are under investigation, with some currently in clinical trials. These strategies include:

SOD and SOD-catalase mimetics

Various antioxidant enzyme mimics, including SOD and SOD-catalase mimetics, are undergoing clinical trials due to their potential in alleviating oxidative stress. Although these mimics, like metalloporphyrins and Mn-salen complexes, are less effective than natural enzymes and sometimes act as pro-oxidants. They are valuable in extracellular environments and may impact mitochondrial function (Policar et al. 2022). Orgotein, the first SOD-based drug, demonstrated anti-inflammatory effects in trials but is now primarily used in veterinary medicine due to inconsistent results in human studies (Afonso et al. 2007).

Glutathione Peroxidase Mimics

Several mimics of glutathione peroxidase (GPX) have been developed, with ebselen being the most notable. Ebselen, a selenoorganic compound, has broad substrate specificity and can reduce oxidative damage, prevent hearing loss, and decrease inflammation in animal studies. It also induces phase II detoxification enzymes (Forman and Zhang 2021). Clinical trials for conditions like Meniere's disease, bipolar disorder, and acute ischemic stroke have shown that oral ebselen is well tolerated and effective. ALT-2074, a newer and more potent analogue of ebselen, has demonstrated efficacy in reducing inflammation, oxidative damage, and neuronal death in vitro and in animal models (Santi et al. 2021).

Iron Chelation

The metal ion dissociation from proteins, can contribute to hydroxyl radical (•OH) production. While certain chelating agents may enhance this activity others act to suppress it. Utilizing inhibitory chelators could theoretically prevent –OH production; however, because iron is vital for many biological functions, chelation therapy is typically reserved for preventing iron overload in conditions like sickle cell and thalassemia, which demands regular hemotransfusion. Currently, three iron chelating drugs, deferrioxamine, deferiprone, and deferasirox, are often used in clinical settings (Yang et al. 2018).

Increasing GSH

Cells generally maintain millimolar concentrations of glutathione (GSH), a crucial antioxidant that protects against oxidative stress by neutralizing free radicals and ROS (Kozlov et al. 2024). However, during periods of intense oxidative stress, the demand for GSH increases significantly, which can lead to its depletion. This reduction in GSH levels impairs the cell's ability to combat oxidative damage, contributing to the progression of various diseases (Jomova et al. 2023). To counteract this depletion, several strategies have been developed to maintain or restore GSH levels. One effective approach involves the use of GSH esters, which are stable forms of GSH that can be absorbed into cells and converted back into active GSH, thereby replenishing intracellular levels. Another approach focuses on providing agents that supply cysteine, a key precursor for GSH synthesis. Cysteine is vital for the production of GSH, and increasing its availability can enhance the cell's capacity to regenerate GSH and maintain its antioxidant defences (Giustarini et al. 2023). These strategies have demonstrated potential in managing and mitigating the effects of oxidative stress across various diseases, including neurodegenerative disorders, cardiovascular conditions, and certain cancers, by improving the overall redox balance within cells and tissues.

N-acetylcysteine (NAC)

NAC is a well-studied antioxidant that is water-soluble and quickly absorbed. Inside cells, NAC is deacetylated to produce cysteine, replenishing GSH and reducing cysteine conjugates in plasma (Tenório et al. 2021). NAC is used to treat conditions such as acetaminophen toxicity, cystic fibrosis (administered via airways), and nephropathy. It binds transition and heavy metal ions, producing complexes that are easily eliminated from the body. NAC has been demonstrated to be a strong anti-inflammatory agent that works by inhibiting NF-kB activation and cytokine production (Tieu et al. 2023).

NRF2 Activators

NRF2 (Nuclear Factor Erythroid 2–Related Factor 2) activators are increasingly recognized as potential therapeutic agents for enhancing the body's antioxidant defenses and mitigating various oxidative stress-related conditions (Kang et al. 2020). NRF2 is a transcription factor that plays a crucial role in regulating the expression of antioxidant and detoxification enzymes. By activating NRF2, these agents stimulate the production of several key antioxidant enzymes, thereby improving the body's ability to counteract oxidative damage and maintain cellular health (Kansanen et al. 2013; Wu et al. 2024). When NRF2 is activated, it translocates to the nucleus of cells, where it binds to antioxidant response elements (ARE) in the DNA. This binding promotes the transcription of genes that encode for antioxidant enzymes and other protective proteins. These enzymes include glutathione peroxidase, superoxide dismutase (SOD), and catalase, which work together to neutralize harmful free radicals and reactive oxygen species (ROS) (Medoro et al. 2024).

NRF2 activators can be grouped into five distinct categories based on their mechanisms of action in enhancing NRF2 activity. The first category involves the modification of KEAP1 (Kelch-like ECH-associated protein 1), which is responsible for NRF2 degradation under normal conditions; by altering KEAP1, these activators prevent the breakdown of NRF2, allowing it to accumulate and activate its target genes. The second category disrupts the interaction between βTrCP (Beta-transducin repeat-containing protein) and NRF2, which is crucial for NRF2 degradation through the ubiquitin-proteasome pathway. The third mechanism involves the sequestration of KEAP1 by p62, a protein that competes with NRF2 for KEAP1 binding, thereby stabilizing NRF2 and promoting its activity. The fourth category includes activators that induce the new synthesis of NRF2 in a manner that avoids its degradation by KEAP1, ensuring a continuous supply of NRF2. Lastly, some NRF2 activators inhibit BACH1 (BTB and CNC homology 1), a transcription factor that normally suppresses NRF2 activity by competing for the ARE (antioxidant response elements) in the DNA, thus allowing NRF2 to exert its protective effects. Each of these mechanisms offers a unique approach to enhancing NRF2 signalling and boosting cellular antioxidant defenses (Lee and Hu 2020). Natural sources such as tea, cocoa, broccoli, grape seeds, and turmeric can activate NRF2 signalling and induce antioxidant enzymes. Several of these natural activators are undergoing clinical trials for disease treatment and prevention, including several trials specifically for turmeric extract.

13.2 Antioxidant Therapies in Chronic Disease

13.2.1 Antioxidant Therapy in Cardiovascular Diseases

In today's world, cardiovascular diseases are one of the leading causes of illness and mortality (Kang et al. 2019). Reactive oxygen species (ROS), which are generated during excessive oxidative stress, are responsible for the pathophysiology of a number of cardiovascular disorders, including atherosclerosis, cardiac hypertrophy, cardiomyopathy, heart failure, ventricular remodelling, ischemia/reperfusion injury, and myocardial infarction (Jain et al. 2015; Peoples et al. 2019; Wang and Kang 2020).

Role of antioxidants in CVD

In healthy individuals, many exogenous and endogenous antioxidants have been found to reduce the chances of occurrence of cardiovascular diseases by serving as a natural defense against free radical induced damage. By giving electrons to the free radicals, antioxidants counteract their effects and create far fewer reactive radicals. Studies have shown that people who consumed large amounts of vitamin C, vitamin E, and β-carotene from their diet were at a reduced risk of developing coronary artery disease (CAD) and may live longer. The antioxidant enzymes in the cell play a crucial role in controlling redox equilibrium. Thioredoxin (Trx), glutathione peroxidase (GPX), peroxiredoxin (PRX), catalase, and SOD are examples of antioxidant enzymes. They play a crucial part in protecting against oxidative stress by breaking down the ROS (Wang and Kang 2020). In addition to maintaining a high glutathione (GSH) to oxidized glutathione (GSSG) ratio (GSH:GSSG ratio), antioxidants also work to quench superoxide and secondary oxidants. This redox modulation may be helpful at the cellular level in preventing vascular endothelial dysfunction (VED), which is a factor in the onset of CVD (Varadharaj et al. 2017).

Antioxidant therapies in oxidative stress-associated CVDs have been found to involve few distinct approaches in clinical settings: (a) blocking the production of oxidative stress by using inhibitors of xanthine oxidase and nitric oxide synthase; (b) increasing endogenous antioxidant capacity through the administration of N-acetyl cysteine; (c) increasing antioxidant capacity by administering exogenous antioxidant supplements (D'Oria et al. 2020; Pavlidis 2022; Wang and Kang 2020).

1. Xanthine oxidase and nitric oxide synthase inhibitors:

Xanthine Oxidase Inhibition

Hypoxanthine and xanthine are oxidized to uric acid under the catalytic action of the enzyme xanthine oxidoreductase (XOR), which is composed of xanthine oxidase and xanthine dehydrogenase. The intermediates of this process, superoxide and hydrogen peroxide, raise the degree of oxidative stress (Berry and Hare 2004; Bredemeier et al. 2018).

Medication like allopurinol that inhibits xanthine oxidase lowers the risk of cardiovascular illnesses caused by oxidative stress, including endothelial dysfunction, hypertension, and heart failure (Saito et al. 2021). According to research that employed a systematic review and meta-analysis, xanthine oxidase inhibitor medication significantly affected a number of cardiovascular-related parameters, suggesting that it may be helpful in the prevention of cardiovascular illnesses (Tanaka and Node 2022).

Nitric Oxide Synthase Inhibition
When the nitric oxide synthase (NOS) is uncoupled and/or when the supply of substrates or cofactors is limited, NOS enzymes can also generate superoxide and other oxidative products and radicals. This can cause the aforementioned balance to change and contribute to oxidative stress which is caused by the increase in the production of ROS and the decrease in nitric oxide. Hence, inhibiting NOS activity can help restore the balance between NO and ROS, thereby reducing oxidative stress in the cardiovascular system (Berry and Hare 2004).

2. Boosting Antioxidant capacity with N-Acetyl Cysteine
   The amino acid cysteine, a rate-limiting substrate for the production of glutathione, a significant endogenous antioxidant, is formed from N-acetylcysteine (NAC). NAC boosts glutathione synthesis by making cysteine more available, which strengthens the body's built-in antioxidant defenses against oxidative stress (Talasaz et al. 2011; Zhu et al. 2022). Free radicals and reactive oxygen species can be neutralized by glutathione, preventing them from damaging cells (El Sharkwy and Abd El Aziz 2019). The administration of NAC has demonstrated the ability to mitigate oxidative damage to the heart and blood vessels, thereby lowering the likelihood of cardiovascular ailments such as heart failure and atherosclerosis (Talasaz et al. 2011).
3. Exogenous antioxidants used in CVD:
   Exogenous antioxidants have been shown to reduce oxidative stress and inflammation, improve endothelial function (antithrombotic action), and may protect against cardiovascular events, making them promising agents for the prevention of CVD.

   Lycopene is one of the antioxidants used in cardiovascular diseases as it can help prevent oxidative damage to cells and tissues in the cardiovascular system by neutralizing oxygen species (Arballo et al. 2021; Heber and Lu 2002; Przybylska and Tokarczyk 2022). The mechanism of lycopene's antioxidant action in cardiovascular diseases is illustrated in Fig. 13.2. More research has now been conducted that demonstrates the link between omega-3 polyunsaturated fatty acids (PUFA) consumption and a lower risk of cardiovascular illnesses (CVDs), such as heart failure (HF), Myocardial Infarction (MI), Congenital Heart Disease (CHD), hypertension, and arrhythmias (Mistretta et al.

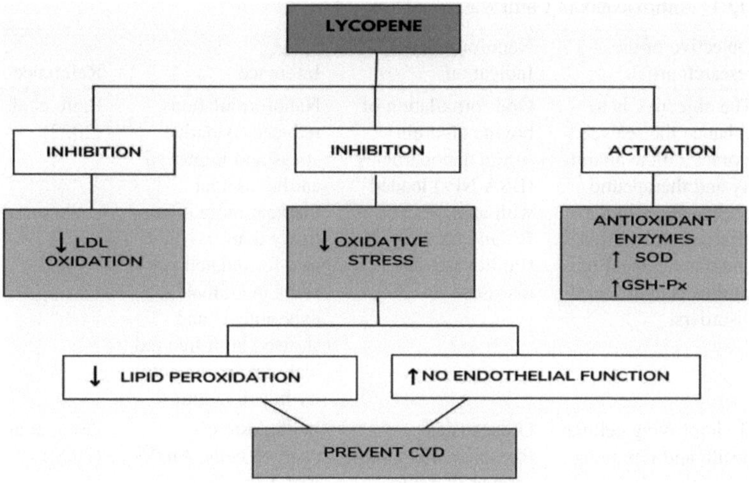

Fig. 13.2 Mechanism of lycopene as an antioxidant in CVD (Przybylska and Tokarczyk 2022)

2023a). Vitamins A, E, and C, as well as zinc, are antioxidant-rich substances that may decrease the onset and course of cardiovascular disease.

The evolution of antioxidant therapy for cardiovascular protection has advanced significantly with nanotechnology. Traditional antioxidants like CoQ10, beta carotene, lycopene, quercetin, resveratrol, and vitamins C and E have shown preventive and therapeutic cardiovascular benefits. Nano-antioxidants now enhance these effects with tunable catalytic and redox properties, enabling better antioxidant delivery. This innovation particularly benefits natural antioxidants like flavonoids, which exhibit significant cardioprotective potential through epidemiological and experimental studies. Nanoformulations offer a promising pathway to increase the efficacy of these compounds, opening new avenues for CVD prevention and therapy (Table 13.1).

The response of lycopene in reducing oxidative stress by inhibiting LDL oxidation and activating antioxidant enzymes thus contributing to the prevention of cardiovascular diseases has been depicted in the Fig. 13.2.

13.2.2 *Antioxidant Therapy in Neurodegenerative Disorders*

The term "neurodegenerative diseases" refers to a class of clinical ailments that present with varying symptoms but are all ultimately defined by a gradual loss of neuronal function and eventual generalized impairment of the body (Panaiotov et al. 2024).

Table 13.1 Antioxidants in Cardiovascular Diseases

S. no	Objective of the research article	Nanoformulation & Indication	Inference	Reference
1	The objective is to enhance the seaweed extract's bioavailability and therapeutic effectiveness in the effective prevention and management of cardiovascular disorders.	Oral formulation of bovine albumin serum nanoparticles (BSA NPs) loaded with an extract of *Eisenia bicyclis* Cardiovascular diseases	Nanoformulations reduced oxidative stress and improved cardiovascular markers more effectively than non-formulated versions in animal experiments and showed high free radical scavenging activity in cell cultures.	Pinto et al. (2022)
2	By improving cellular health and removing harmful cholesterol crystals more effectively, this review highlights the potential of nano-sponge-like liposomes as a novel treatment for cardiovascular disorders like atherosclerosis that are linked to high cholesterol.	Ginsenoside Rb1-anchored nano-sponge-like liposomes (AnxV-Rb1-LPs). Atherosclerosis	In cholesterol-charged cells, AnxV-Rb1-LPs effectively dissolves and removes cholesterol crystals from atherosclerotic plaques, inflammation, and apoptosis.	Gong et al. (2022)
3	This review assesses Fe-GA CPNs' potential to overcome oxidative stress linked to ischemia injury by scavenging ROS and re-establishing redox balance in neuronal cells.	Iron-gallic acid coordination nanoparticles (Fe-GA NCPs). Ischemic stroke	Reactive oxygen species (ROS) were reduced while neurological outcomes were improved in an ischemic rat stroke model, indicating that Fe-GA NCPs had strong antioxidative qualities.	Kang et al. (2019)
4	In this study, the proresolving peptide Ac2-26-carrying targeted nanoparticles' capacity to prevent advanced atherosclerosis in hypercholesterolemic mice models is investigated, with a focus on how they improve plaque stability and reduce inflammation.	Ac2-26 peptide encapsulated in collagen IV-targeted nanoparticles. Advanced atherosclerosis	Mice with high cholesterol levels that were loaded with Ac2-26 nanoparticles showed a significant improvement in plaque stability and a suppression in oxidative stress.	Fredman et al. (2015)

(continued)

Table 13.1 (continued)

S. no	Objective of the research article	Nanoformulation & Indication	Inference	Reference
5	This study suggests designing and synthesizing the nanoparticles using a microfluidic chip, in order to determine whether targeted nanoparticles loaded with the anti-inflammatory cytokine interleukin-10 (IL-10) can effectively reduce inflammation in lesions by altering macrophage activity and improving inflammation resolution.	Interleukin-10 (IL-10) encapsulated in targeted nanoparticles. Advanced atherosclerosis	Microfluidic chip-derived IL-10 nanoparticles significantly decreased plaque load and enhanced inflammation resolution in advanced atherosclerosis models.	Kamaly et al. (2016)
6	In contrast to the gold-standard monoclonal antibody systems, which are used to target atherosclerotic lesions, this article examines the potential applications of creating liposomes conjugated with bio-orthogonally bio ligands that target PECAM1 and ICAM 1, with a focus towards assessing the specificity and potency of scFv.	Radiolabeled liposomes with scFv and mAb ligands. Vascular targeting in atherosclerosis and imaging	In the pulmonary vasculature, conjugates of liposomes and single chain variable fragments (scFv) have ten times the specificity of conventional antibodies for the PECAM-1 and ICAM-1 sites.	Hood et al. (2018)
7	The objective of this paper is to create targeted PEG-liposomes that contain the antioxidant EUK-134. This will improve delivery to the vascular endothelium, lowering inflammation and oxidative stress in acute pulmonary injury and enhancing therapeutic results.	Liposomes encapsulating EUK-134. Acute pulmonary inflammation	In models of acute pulmonary damage, the endothelial-targeted liposomes effectively distribute EUK-134, considerably lowering oxidative stress and inflammatory indicators.	Howard et al. (2014)

(continued)

Table 13.1 (continued)

S. no	Objective of the research article	Nanoformulation & Indication	Inference	Reference
8	The article is focused on the design and synthesis of co-assembled peptide amphiphile nanofibers specifically targeting atherosclerotic plaques and their evaluation for the enhancement of cholesterol efflux from macrophages to induce plaque regression.	Coassembled peptide amphiphile nanofibers. Atherosclerosis	They can target atherosclerotic plaques effectively by enhancing localized drug delivery with a minimal number of systemic side effects.	So et al. (2018)
9	The study looked at the mechanisms of action in relation to inflammation and lipid metabolism and assessed how well anionic nanoliposomes worked to slow the progression of atherosclerosis in LDLR-deficient mice provided a high-fat diet.	Anionic nanoliposomes like hydrogenated soy phosphatidylcholine. Atherosclerosis	Fat rich diet provided to LDLR-deficient animals, anionic nanoliposomes increase lipid clearance and decrease inflammation, which inhibits the development and progression of atherosclerotic lesions.	Krishna et al. (2018)
10	The objective of the study is to avoid neointimal hyperplasia following arterial injury by designing a tailored delivery system for nitric oxide (NO) using supramolecular nanofibers.	S-nitrosylated supramolecular nanofibers. Prevention of restenosis post-angioplasty	The study shows that S-nitrosylated nanofibers significantly decrease neointimal hyperplasia without causing systemic adverse effects by efficiently delivering nitric oxide to the site of arterial damage.	BahnsonEdward et al. (2016)
11	The aim is to create PVAX nanoparticles as a targeted treatment for ischemia/reperfusion (I/R) injury, where the release of vanillyl alcohol (VA) from a reaction with hydrogen peroxide (H2O2) lowers reactive oxygen species (ROS) and has an anti-inflammatory impact.	H2O2-responsive nanoparticles. Myocardial ischemia-reperfusion injury (I/R)	In the CVD models, these antioxidant nanoparticles further reduce inflammation and organ damage by releasing active molecules at the sites of oxidative stress.	Lee et al. (2013)

The pathogenesis of neurodegenerative illnesses is characterized by selective neuronal susceptibility to several stimuli, including oxidative stress (Grabska-Kobyłecka et al. 2023; Guo et al. 2013). Epidemiological research has clearly shown the role of inflammation and oxidative stress in the onset and advancement of neurodegenerative diseases (Berr 2000; Head 2009). Brain tissue is extremely sensitive to oxidative damage because of its high amount of polyunsaturated fatty acids, rapid oxygen consumption, and relatively low concentration of antioxidant molecules (Uttara et al. 2009). Furthermore, an increase in reactive oxygen species (ROS) build-up in the brain causes the blood-brain barrier (BBB) to become more permeable, which causes neuroinflammation and neuronal degeneration (Małkiewicz et al. 2019). The development of neurodegenerative diseases is significantly accelerated by variables such as impaired mitochondrial activity, cellular damage, and dysfunctional DNA repair mechanisms brought on by elevated oxidative stress (Kim et al. 2015).

Antioxidants have emerged as a promising therapeutic approach for neurodegenerative disorders, working primarily by counteracting oxidative stress-induced damage. Their mechanism involves interacting with and stabilizing free radicals, preventing potential cellular damage. Sulforaphane can activate genes and molecules with anti-inflammatory, anti-apoptotic, and antioxidant characteristics through its molecular target Nrf2. For the treatment of neurodegenerative diseases like multiple sclerosis, Parkinson's disease, and Alzheimer's disease, SFN is able to complement already available therapies. Curcumin's neuroprotective effects are due to its interaction and modulation with molecular targets, including growth factors, transcription factors, inflammatory cytokines, kinases, and the antioxidant system. Genistein is found to cross the blood–brain barrier and may have neuroprotective properties that may either postpone the onset of neurodegenerative diseases or lessen their symptoms (Chico et al. 2026).

In formulation development, several key considerations must be addressed. Antioxidants are effective at low concentrations, requiring careful dosing in formulations. Despite promising results, there are hurdles in developing effective antioxidant formulations for neurological application, however researchers are continually working for novel antioxidant therapies for neurodegerative disorders (Table 13.2).

13.2.3 Diabetes Mellitus

A variety of common endocrine diseases that are characterised by consistently increased blood sugar levels are referred to as diabetes mellitus.

Diabetes is caused by either inadequate or insufficient insulin production, or by both. Diabetes mellitus (DM), an autoimmune disorder with multiple aetiologies associated with irregularities in protein, lipid, and carbohydrate metabolism, has long-term effects (Mooradian 2006).

Under stress, RON production overpowers antioxidant defenses, resulting in a redox imbalance that triggers oxidative stress and permanent alterations in cell

Table 13.2 Antioxidants in Neurodegenerative Disorders

S. No	Objective of the research article	Nanoformulation & Indication	Inference	Reference
1	The review explores the neuroprotective effects of Moringa oleifera through NF-kB/Nrf2/HO-1 signalling pathway, emphasizing its bioactive compounds, mechanisms, safety, and possible directions for future research and clinical use.	*Moringa oleifera* Alzheimer's and Parkinson's diseases	These components exhibit neuroprotective effects in several types of neurodegenerative diseases by supporting mitochondrial activity, neurogenesis, and antioxidant enzyme levels in the brain.	Mundkar et al. (2022)
2	The objective of this review is to assess the effectiveness of herbal extract-based nanoformulations in the treatment of neurodegenerative diseases (NDs)	Various herbal extracts like Curcumin, Quercetin, Resveratrol Alzheimer's and Parkinson's, Multiple sclerosis	Antioxidants are supplied to the specific areas of the affected brain by nanoparticles, boosting their therapeutic efficiency while lowering side effects.	Moradi et al. (2020)
3	This study examined the effectiveness of novel nanotechnology approaches, including lipid nanoparticles, nanobodies, and nanoantibodies, in the treatment of neurodegenerative diseases.	Nanobodies, nanoantibodies, and lipid nanoparticles Alzheimer's disease, Parkinson's disease	In addition to lowering inflammation and improving drugs distribution across the blood-brain barrier, nanotechnology can reverse nerve cell death.	Asefy et al. (2021)
4	The review analyses the ability of nanomedicine in targeting the root causes of neurodegenerative diseases and explore its synergistic effects with gene therapy for effectively addressing mutated genes in Alzheimer's and Parkinson's diseases.	Nanoscale drug delivery systems Alzheimer's and Parkinson's Diseases	Nanomedicine targets mutations in genes and proteins of neurodegenerative diseases to potentially stop further progression of these diseases.	Jagaran and Singh (2021)
5	It evaluates the role of multifaceted nanoparticles in targeting different pathways which lead to neurodegenerative diseases for improved therapeutic effect.	Engineered nanoparticles Neurodegenerative diseases	Targeting molecular pathways involved in both localized and systemic neurodegenerative disorders, nanosystems enhance drug absorption, overcome BBB challenges, and promote treatment approaches.	Mistretta et al. (2023b)

(continued)

Table 13.2 (continued)

S. No	Objective of the research article	Nanoformulation & Indication	Inference	Reference
6	The article provides an overview about the recent developments in nanotherapeutics which aim to address the limitations and adverse effects of current treatments for neurological conditions.	Magnetic nanoparticles Neurodegenerative disorders	Due of their superparamagnetic abilities, the nanoparticles' magnetic qualities enable them to be considered as potential stimuli-responsive tailored treatments.	Vashist et al. (2023)
7	The paper examines the neuroprotective effects of liposomal formulations containing polyphenol-rich grape pomace extracts against neurodegeneration in Parkinson's disease.	Polyphenol-rich extracts from grape seeds & skins Parkinson's disease	By improving brain targeting and BBB crossing, the functionalized liposomes effectively lower ROS levels and prevent neurodegeneration.	Marino et al. (2021)
8	The review evaluates the efficiency of designer exosomes in delivering catalase mRNA to the brain, which reduces neurotoxicity and neuroinflammation and can be used to treat Parkinson's disease.	Engineered exosomes delivering CAT mRNA Parkinson's disease	Both in vitro and in vivo models of Parkinson's disease can be treated using exosomes that efficiently carry CAT mRNA into target cells, reducing neurotoxicity and neuroinflammation.	Kojima et al. (2018), Li et al. (2020)
9	This review takes into account techniques that improve brain targeting and the potential of engineered nanomaterials as therapeutic agents in Neurodegenerative disorders.	Liposomes, nanogels, polymer based NPs, Dendrimers, nanocomposite, metal based NPs Alzheimer's, Parkinson's, Huntington's, and amyotrophic lateral sclerosis	Targeting specific brain areas allows for the change of specific neural pathways, resulting in desired therapeutic outcomes. It also improves the effectiveness of existing neural stem cell (NSC) differentiation-based treatments.	Nayab et al. (2023)
10	The objective of this paper is to tackle the difficulties presented by the blood-brain barrier and the blood-cerebrospinal fluid barrier in treating Alzheimer's disease by investigating the possibility of nanocarrier systems to improve drug delivery to the brain.	Nanocarriers like hydrogel, carbon nanotube, gold NP, liposome nanocage Alzheimer's disease	These nanocarriers help the medicament to cross the BBB and delivers it to the target site.	Taliyan et al. (2022)

(continued)

Table 13.2 (continued)

S. No	Objective of the research article	Nanoformulation & Indication	Inference	Reference
11	The review highlights the potential of a nanoparticle-based intranasal drug delivery system provides over the conventional therapies alone to overcome Alzheimer's.	Nanoparticles Alzheimer's disease	High therapeutic effectiveness, improved bioavailability, and minimal systemic adverse effects are the results of this pathway, which delivers the medication directly into the brain through the olfactory system.	Pandey et al. (2020)
12	The paper aims to present a comprehensive overview of the most recent drug delivery methods based on nanotechnology for the treatment of different kinds of dementia.	Polymeric nanoparticles, solid lipid nanoparticles, nanostructured lipid carriers, microemulsions, nanoemulsions, and liquid crystals Alzheimer's disease, dementia	Nanoscale drug delivery systems remains in systemic circulation for longer durations and can penetrate the blood-brain barrier.	Siafaka et al. (2020)

components (Forman and Zhang 2021). Oxidative stress is the primary source of deterioration to various cell structures such as, DNA, membranes, proteins and lipids. Such damage may have an effect on cell health, which can impede cell activity and trigger various biological processes through the formation of secondary reactive species. It is an important mechanism linked to a number of severe chronic conditions, including neurological issues, cancer, heart disease, and liver disease.

Role of Antioxidants in Diabetes Mellitus

Antioxidants are molecules that can donate an electron to neutralize ROS, thus preventing cellular damage. They play a vital role in managing oxidative stress and its related complications in diabetes. Antioxidants show their action by following ways- It can be Free radical scavenging in which antioxidants neutralize free radicals by donating an electron, thereby preventing cellular damage; Antioxidants perform metal chelation by chelating the transition metals and reducing their availability to catalyse ROS formation. Antioxidants can also influence the expression of genes involved in oxidative stress responses (Arabshomali et al. 2023).

It is apparent there is substantial protection against diabetes from dietary antioxidants such as tocopherols, β-carotene and ascorbic acid and flavonoids.

Studies on vitamin E supplementation demonstrate protection of cells by reducing cytotoxicity, lipid peroxidation, protein oxidation, and protein glycation. Coenzyme Q, commonly known as ubiquinone, has the potential to reduce oxidative stress by "recoupling" mitochondrial oxidative phosphorylation and by suppressing reactive oxidant radicals, therefore lowering the generation of superoxide (Shafras et al. 2024). A broad and diverse class of phytochemicals i.e. Polyphenols is found in plant-based diets, including flavonoids, phenolic acid, and stilbens. In addition to enhancing acute insulin secretion and insulin sensitivity, polyphenols reduce hyperglycaemia. Activation of insulin receptors and glucose uptake in insulin-sensitive tissues, inhibition of the breakdown of carbohydrates, stimulation of insulin secretion, modulation of glucose release from the liver, reduction in intestinal glucose absorption, modification of intracellular signalling pathways, and gene expression are some of the potential mechanisms (Aryaeian et al. 2017).

When it comes to diabetes mellitus, vitamin E enhances glucose metabolism, protects the cell from oxidative damage, helps to reduce the vascular problems brought on by hyperglycemia, enhances beta cell activity, and enhances insulin action (Asbaghi et al. 2023). Since increased glucose metabolism increases antioxidant levels, particularly vitamin E, have been favorably connected with lower plasma glucose levels.

Coenzyme Q- Lipoic acid, which is essential for the activity of mitochondrial dehydrogenase reactions, is an additional substance that scavenges free radicals activity. It was shown that lipoic acid enhanced glucose transfer in cultured muscle cells by promoting the translocation of GLUT4 travels to the plasma membrane from internal pools. In cultured adipocytes, lipoic acid therapy safeguarded protecting the insulin receptor from oxidative damage and preserving its integrity of function (Shafras et al. 2024).

Antioxidant nanoformulations offer innovative solutions for managing diabetes mellitus by enhancing bioavailability, stability, and efficacy. As shown in Table 13.3, these formulations, such as liposomal EGCG and quercetin, resveratrol nano suspensions, and chitosan-stabilized selenium nanoparticles, effectively combat oxidative stress, improve glucose metabolism, and reduce diabetes-related complications. By leveraging nanotechnology, these therapies provide targeted and sustained benefits, highlighting their potential in improving diabetes management and patient outcomes.

13.2.4 Cancer

Being one of the major causes of death, cancer has drawn a lot of scientific interest and research. The fight against cancer remains difficult despite the thousands of papers on the subject since its genesis and progression are still not completely clear, making it a concerning and often fatal disease in humans.

Cancer is usually associated with increased oxidative stress, which is defined as the excess production of reactive oxygen species (ROS) due to changes in metabolism or inappropriate signaling. Endogenous antioxidants (AOs) are unable to neutralize such ROS, which results in changes to the redox homeostasis equilibrium necessary for biological processes and, ultimately, disease (Luo et al. 2022).

Table 13.3 Antioxidants in Diabetes Mellitus

S. No	Title/Objective of Research/Review Literature	Nano formulation & Indication	Inference	Reference
1.	To investigate the impact of a green tea compound, Epigallocatechin- Gallate nanoformulation on Matrix Metalloproteinases and oxidative stress in Experimental Diabetes Mellitus	Epigallocatechin Gallate (Liposomes) Diabetes Mellitus	Due of its distinct modulatory effect on MMP-2 and -9 as well as oxidative stress/antioxidant indicators, the liposomal nanoformulation of EGCG may be used as an adjuvant therapy in DM.	Bulboaca et al. (2020)
2	Quercetin-enveloped Soluplus ®/P407 micelle design synthesis, optimisation, characterisation, and in-vivo evaluation for diabetes treatment.	Quercetin (miscelles) Diabetes mellitus	For the management of diabetes, quercetin-loaded Soloplus miscelles increase their bioavailability and offer extended release.	Singh et al. (2018)
3	Synthesis of a Nanoformulation of Resveratrol to Treat Diabetic Retinopathy	Resveratrol Nano-suspension Diabetic retinopathy	Resveratrol Nanosuspension showed decreased cell proliferation and migration indicating its potential for future medical implementation.	Gonzalez-Perez et al. (2024)
4	Sulforaphane-assisted improvement in glycaemic management and decreased hepatic glucose production in people with Type-2 diabetes	Liposomes, nanoparticles, Sustained-Release Sulforaphane Tablets/Capsules through oral route Type 2 Diabetic mellitus	Sulphoraphane nanoparticles demonstrated suppressed hepatic gluconeogenesis by activating Nrf2, reducing glucose production by up to 41%, and improving glucose tolerance.	Axelsson et al. (2017)
5	Liposomal quercetin: A potentially effective treatment for type 2 diabetes mellitus that reduces inflammation and hepatic insulin resistance	Quercetin liposomes delivered through oral route Hepatic insulin resistance and inflammation in type-2 diabetes mellitus	By controlling Akt expression and reducing inflammation, particularly through the NF-κB pathway, quercetin liposomes demonstrate improved effectiveness in modifying insulin resistance. Additionally, it also influences the expression of PGE2 and COX-2.	Rocha et al. (2024)

(continued)

Table 13.3 (continued)

S. No	Title/Objective of Research/Review Literature	Nano formulation & Indication	Inference	Reference
6	The objective is to study the reduction of cardio-hepatic damage by the Chitosan-stabilized selenium nanoparticles through the upregulation of caspase, Bax/Bcl-2, and Fas/FasL pathways in a type-2 diabetes mellitus model.	Nano-co-encapsulated antioxidant particles (NanoCAPs) comprising coenzyme Q10 (CoQ10) and ellagic acid (EA) chitosan nanoparticles. Diabetes mellitus	Compared to a mono-therapeutic approach, combined therapy using metformin and chitosan stabilized nanoparticles produced a more impressive anti-diabetic impact and restored insulin resistance to a greater degree, making it a potentially effective therapeutic option.	Mohamed et al. (2021)
7	Co-encapsulated antioxidant nanoparticles (NanoCAPs) containing Ellagic Acid and Coenzyme Q10: Preparation and Assessment in Streptozotocin-Induced Diabetic Rats as a Potential Therapeutic Approach for Diabetes and Its Complications.	Nano-co-encapsulated antioxidant particles (NanoCAPs) comprising coenzyme Q10 (CoQ10) and ellagic acid (EA) chitosan nanoparticles.	In streptozotocin-induced diabetic rats, the combination of EA and CoQ10 demonstrated positive effects on lowering lipid peroxidation, dyslipidaemia, and delaying organ damage.	Ratnam et al. (2008)
8	Designing a Crystalline Nano-Suspension of Lycopene for the Possible Treatment of Diabetes Associated with Oxidative Stress in Experimental Animals	Lycopene Nano suspension Diabetes mellitus	Studies conducted in vivo have demonstrated an apparent decrease in elevated blood sugar levels.	Mishra and Kumari (2021)

Superoxide dismutases (SODS), glutathione, thioredoxins, NADPH, and other metabolic and genetic reprogramming processes are some of the ways that cells revert to an eventual overproduction of ROS. Generally, ROS levels are tightly regulated by antioxidant enzymatic and non-enzymatic systems (Sezgin-Bayindir et al. 2021).

Researchers have looked into using antioxidants to cure cancer as oxidative stress is linked to both the progression of cancer and carcinogenic processes. Numerous treatment approaches have been investigated in pre-clinical and clinical research investigations, and some of them have the potential to be anticancer medications. To effectively use them, though, suitable delivery mechanisms must be created so that the active antioxidants may be easily transferred to the target areas both during and after delivery (Sezgin-Bayindir et al. 2021). The modulation in ROS levels and its impact is shown in Fig. 13.3.

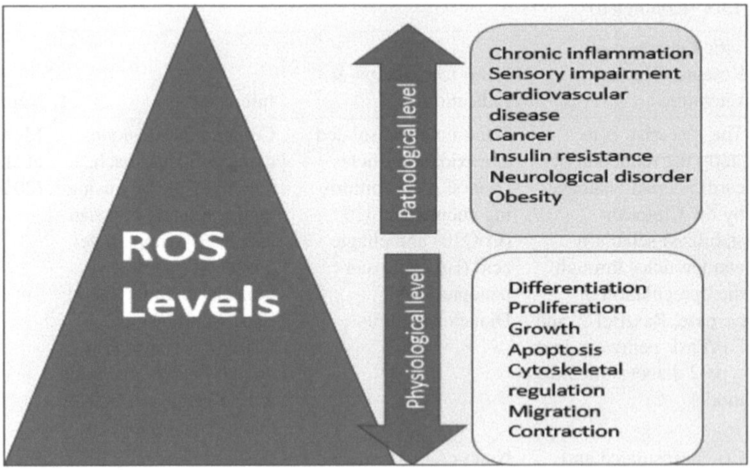

Fig. 13.3 Physiological and pathological ROS levels

There are various ways through which antioxidants play an important role in cancer modulation by various ways—It can perform its function by Neutralizing free radicals due to which damage to DNA and other cellular components can be prevented. Certain antioxidants can affect cellular pathways that control cell growth and programmed cell death (apoptosis), potentially inhibiting the proliferation of cancer cells. By bolstering the immune system, antioxidants may improve the body's natural ability to defend against cancerous cells.

A group of researchers have hypothesized that sulphoraphane co-loaded with teriflunomide in lignin nanoparticles will synergistically potentiate the cytotoxic effects in triple negative breast cancer. Sulphoraphane, a chemopreventive medication, decreases ROS and prevents cell dormancy, boosting apoptosis and lowering cancer recurrence following chemotherapy (Dutta et al. 2024). With similar aim, Siddique and her team hypothesised that lignin's antioxidant properties could induce dormancy in tumour cells, allowing them to be exposed to chemotherapeutic agents and promote cell apoptosis, preventing cancer relapse (Siddiqui et al. 2018). Lignin, a multifunctional biopolymer with antioxidant and anti-cancer properties, is being increasingly used as a drug carrier (Barapatre et al. 2016). Numerous nano-formulations have demonstrated to significantly improve the selectivity and cellular uptake of medications by cancer cells, lowering the risk of unexpected systemic side effects in cancer patients. While phytoconstituents have a wide range of beneficial active ingredients that can be used to treat cancer, polyphenols, isoquinoline alkaloids, and iso-thiocyanates have shown the best results (Nair et al. 2024).

There are certain antioxidants which are crucial as antioxidant in cancer such as curcumin, quercetin and resveratrol. The yellow-coloured polyphenol curcumin is isolated from *Curcuma longa* L. that is used to treat a variety of illnesses due to its antibacterial, anti-inflammatory, antioxidant, anticancer, and neuroprotective properties. Quercetin has been shown to play a significant role in cancer treatment. The literature has described the anti-inflammatory, anti-tumour, anti-aging, antioxidant, and anticancer activities of resveratrol (3,5,4′-trihydroxystilbene) as a polyphenol (Cecerska-Heryć et al. 2024).

Curcumin- Due to its extreme hydrophobicity, curcumin is not very soluble in aqueous solutions. Nevertheless, several investigations have assessed the bioavailability and solubility of curcumin-CD complexes, revealing a notable increase in curcumin solubility (up to 10,000-fold) when compared to the uncomplexed (free) Curcumin's anticancer properties are ascribed to its capacity to induce apoptosis and inhibit tumour growth via modulating many biochemical and signalling pathway constituents (Tønnesen et al. 2002).

Quercetin- Using HP-β-CD, quercetin and rutin inclusion complexes were created, and the outcomes demonstrated dose-dependent cytotoxicity even on human lung cancer (A549 cells) and breast cancer (MDA-MB-231) cell lines (Zhang et al. 2024).

Resveratrol- Numerous molecular pathways have been linked to resveratrol, such as the tyrosine kinase, nuclear factor kappa N (NF-kB), and phosphoinositide-3-kinase (PI3K)/AKT.STAT3/5. The three stages of carcinogenesis that resveratrol directly affects are initiation, promotion, and progression. Resveratrol's anti-proliferative and pro-apoptotic properties were suggested to stem from its capacity to regulate the expression of pro-oncogenic and tumor suppressor microRNAs, nuclear respiratory factors (NRF-) 1 and 2, nuclear factor-kappa B (NF-kB), and gamma activator 1 alpha (PCG-1α), among other things, in vitro experiments (Rauf et al. 2018). Therefore, curcumin and resveratrol, extensively researched nutraceuticals, are effective ROS scavengers that protect cells from oxidative stress. Curcumin and resveratrol have been studied in over 100 clinical trials, demonstrating their safety, tolerability, and efficacy against a variety of chronic conditions in humans (Saxena et al. 2023).

Antioxidants play a crucial role in counterbalancing oxidative stress through both enzymatic (endogenous) and non-enzymatic (exogenous) mechanisms. Both natural and synthetic antioxidants work by reducing oxygen pathways, helping to manage autoimmune responses. Recent advancements in nanotechnology have significantly improved antioxidant therapy by enhancing bioavailability and targeted delivery, for instance nanoparticles coupled with specific antigens have been reported to regulate T-cell responses in Multiple Sclerosis. Table 13.4, literature data that reveals numerous lipid-based nano-systems comprising antioxidant mixtures for several tutors.

13.2.5 Auto Immune Disorders

The body is protected from illness and infection by a functioning immune system. However, when the immune system stops working properly, it starts attacking healthy tissues, cells, and organs. These incidents, which are referred to as autoimmune diseases, can impact any area of the body, impairing physiological function and perhaps becoming fatal. Autoimmune diseases are influenced by a combination

Table 13.4 Antioxidants in Cancer

	Title/Objective of Research/Review Literature	Nano formulation & Indication	Inference	Reference
1.	Targeted and pH-responsive Rutin administration for breast cancer treatment employing carbon dots functionalized with folic acid	Carbon dots of Rutin Breast cancer	Carbon dots (CDs) of Rutin, were prepared and evaluated for anti-cancer activity in breast tissue cell lines. The results confirmed the anti-cancerous activity of Rutin via inducing oxidative stress and mitochondrial damage (Antioxidant Mechanism).	Tejwan et al. (2022)
2	Evaluation of cytotoxic impact on MCF-7 breast cancer cell lines using a synergistic composition of antioxidants, silver nanoparticles, and chitosan in a nanoparticle-based formulation	Chitosan-based nano-formulations encapsulating antioxidants (ascorbic acid, tocopherol, catechol) Breast cancer	The nano-formulations demonstrated up to 76% encapsulation efficiency, targeted drug delivery to breast cancer cell and enhanced antioxidant activity as compared to the base materials with excellent hemocompatibility.	Nayak et al. (2016)
3	Nano-formulations based on resveratrol as a new therapeutic approach for ovarian cancer: SIRT-1/Beclin/MMP-9/P53/AKT signaling and autophagy induction	Resveratrol-based nano-formulations (Liposomal) in Ovarian cancer	Liposomal resveratrol significantly improves therapeutic effects over conventional resveratrol by better targeting and inhibiting cancer cell growth, oxidative damage, and inflammatory responses.	Kadry (2024)
4	To investigate the potential protection against rat oral cancer with systemic and topical (slow-release) curcumin nano-niosome antioxidants	Curcumin based nano-niosomes (Mouthwash) Oral cancer	Curcumin-loaded niosomes were effective in suppressing proliferation of cancer cells and delaying the development of severe dysplasia.	Fazli et al. (2022)
5	Curcumin-loaded mesoporous silica: a potent medication delivery method for the treatment of cancer	Mesoporous silica material loaded with curcumin Invitro studies	In vitro tests on curcumin-loaded mesoporous silica material demonstrated that it suppressed cancer cell migration and viability as compared to free curcumin, suggesting the effectiveness of the drug delivery method.	Kotcherlakota et al. (2016)

(continued)

Table 13.4 (continued)

	Title/Objective of Research/Review Literature	Nano formulation & Indication	Inference	Reference
6	Nano-Formulations of Natural Antioxidants for the Treatment of Liver Cancer	Resveratrol Liver Cancer	Resvratrol Nanoformulations, such as those using human serum albumin (HSA) or glycyrrhizic acid (GA) conjugates, enhances delivery, bioavailability, and tumor-targeting efficiency. These nanoformulations promote apoptosis and inhibit tumor growth in liver cancer models, offering a promising therapeutic approach.	Cristani et al. (2024)
7	Promote Resveratrol's Anti-Hepatoma Effect in Vitro and In Vivo with Nano-Gold Loaded Resveratrol	Resveratrol gold Nano particles Liver Cancer	In HepG2 cells, resveratrol gold nanoparticles had a more potent effect in preventing cell division and death.	Zhang et al. (2019)
8	Techniques for managing gastrointestinal cancer with resveratrol	Resveratrol based liposomes Gastrointestinal cancer	Resveratrol formulated as nanoparticles, liposomes is formulated to improve bioavailability and target cancer cells in GI tract effectively.	Roshani et al. (2022)
9	Nanotechnology's Potential and Challenges as a Tocotrienol Delivery System for Cancer Treatment	Tocotrienols (Vitamin E) based nano lipid carriers Cancer	Transferrin targeted niosomes have shown marked tumor regression.	Maniam et al. (2018)
10	Role of curcumin in cancer treatment	Curcumin loaded nanoemulsions, Liposomes, Nanoparticles though oral, intravenous & topic route. Cancer prevention	Curcumin shows strong anticancer properties by Inhibiting NF-κB and STAT3 pathways	Zoi et al. (2021)
11	Therapeutic utilisation of quercetin-loaded nanoarchaeosomes for breast cancer: a method of ROS-mediated cell death	Nanoarchaeosomes of Quercetin Breast Cancer	Since NAQ-induced death by ROS generation and by effectively inhibiting the cells at G0/G1 phases, nanoarchaeosomes can be a strong anticancer agent.	Ariraman et al. (2024)

(continued)

Table 13.4 (continued)

	Title/Objective of Research/Review Literature	Nano formulation & Indication	Inference	Reference
12	To explore nanoformulation-based transdermal drug delivery systems with the objective of developing nontoxic and effective materials for antitumor therapy.	Large Unilamellar Vesicles (LUVs) of Quercetin, Ascorbic acid and Tannic acid	Among the various formulations analyzed, quercetin-loaded large unilamellar vesicles (QUE-LUVs) showed the most promising results by markedly decreasing the viability of cancer cells in comparison to untreated controls. Thus, the selective toxicity highlights their potential as a targeted approach in cancer treatment.	Mendes et al. (2024)

of genetic (e.g., CD25, STAT4), epigenetic (e.g., DNA methylation, histone changes), and environmental factors (e.g., drugs, hormones, toxins). While current treatments like glucocorticoids, NSAIDs, immune suppressants, and biologics help manage symptoms, a permanent cure is still lacking. As a possible treatment for autoimmune disorders, antioxidant therapy has attracted attention. Recently, natural compounds—especially polyphenols—have gained attention for their potential as alternative therapies. Known for their antioxidant, anti-inflammatory, and immune-regulating properties, polyphenols show promise in managing conditions such as vitiligo, ulcerative colitis, and multiple sclerosis (Khan et al. 2019).

Autoimmune diseases, such as rheumatoid arthritis, lupus, multiple sclerosis, and others, involve the immune system mistakenly attacking the body's own tissues, leading to inflammation and tissue damage. Oxidative stress plays a significant role in the pathogenesis of these disorders, where an imbalance between free radicals and antioxidants can exacerbate inflammation and tissue damage (Gautam et al. 2025b).

Oxidative Stress and Its Role in Autoimmune Disorder: Reactive oxygen species (ROS) and reactive nitrogen species (RNS) are two ways that oxidative stress functions in autoimmune diseases: Although they are byproducts of regular cellular metabolism, excessive production of them can be hazardous. Antioxidants and ROS/RNS are in equilibrium in a healthy body. But this equilibrium is upset in autoimmune diseases, which results in oxidative stress.

Impact on Cells and Tissues: Excessive ROS and RNS can damage lipids, proteins, and DNA, leading to cell dysfunction or death. This damage contributes to the inflammatory processes characteristic of autoimmune diseases. For example, in RA, ROS can induce synovial inflammation and joint destruction (Srivastava et al. 2011).

Antioxidant Therapy Approaches

Polyphenols are the substances found in foods like green tea, berries, and turmeric, polyphenols like curcumin and resveratrol have shown promise in modulating oxidative stress and inflammation in autoimmune conditions. Numerous pharmacological and medicinal qualities, such as antioxidant and anti-inflammatory effects, are possessed by polyphenols. Polyphenols, being immunomodulatory agents, are becoming more and more useful as medicinal tools in the treatment of autoimmune diseases such as multiple sclerosis (MS), ulcerative colitis, and vitiligo. The host's immune response is regulated by intracellular pathways that polyphenols activate. These pathways include the arachidonic acid dependent pathway, the mitogen-activated protein kinases (MAPKs) pathway, the phosphatidylinositol 3-kinase/protein kinase B (PI3K/Akt) signalling pathway, and epigenetic modulation (Khan et al. 2019).

Coenzyme Q10 (CoQ10) is found in membranes and plasma lipoproteins, it is a potent lipophilic antioxidant and a carrier of electrons in the mitochondria. Due to the fact that CoQ10 is naturally produced and derived from food, there is increased interest in its potential as a treatment for diseases linked to increased oxidative stress and mitochondrial dysfunction (López-Pedrera et al. 2021).

Various antioxidants showing impact in auto immune disorders are given below in the Table 13.5.

Table 13.5 Antioxidants in Autoimmune Disorders

Sr. No.	Title/Objective of Research/ Review Literature	Nano-formulation &Indication	Inference	Reference
1	A thorough, evidence-based review of the application of curcumin-based nanomedicines in the treatment of inflammatory and immuno-modulated diseases	Curcumin nanoparticles Rheumatoid arthritis	Curcumin nanoparticles demonstrated enhanced bioavailability and efficacy in reducing inflammatory cytokines like TNF-α and IL-6.	Laurindo et al. (2023)
2	To study the Nanocomplex of ultra-small iron-quercetin metal natural substance that regulates macrophages and antioxidants in rheumatoid arthritis	Iron-quercetin natural coordination nanoparticles Rheumatoid arthritis	By preventing the NF-κB pathway from being activated in cells, iron quercetin nanoparticles reduce the development of pro-inflammatory macrophages and further lower the amount of pro-inflammatory substances.	Han et al. (2023)

13.3 Innovative Delivery Systems for Antioxidants

13.3.1 Nanotechnology-Based Parenteral Delivery Systems

Antioxidants are essential for protecting the body from diseases linked to oxidative stress. Their quick metabolism, instability, and restricted bioavailability, however, can make them less useful as medicines. For an antioxidant to have the desired effects on metabolism and excretion, it must have adequate absorption, distribution, and clearance through various biofilm levels inside the body. The water solubility, permeability, lipophilicity, and physicochemical stability of the active chemicals all have a major impact on these properties.

Innovative delivery techniques have been developed to get over these challenges, enhancing the antioxidants' potential for therapeutic application. This section looks at a number of innovative antioxidant delivery strategies, outlining their characteristics, benefits and disadvantages.

13.3.1.1 Liposomes

Antioxidant liposomes are particularly promising for treating a wide range of disorders where oxidative stress is a prominent issue. They have been largely utilized to discover the molecular mechanism of action for different pro- or antioxidants (Stone and Smith 2004).

Antioxidants like α-tocopherol (α-TOH) and butylated hydroxytoluene (BHT) have also been employed to prevent unsaturated fatty acid moieties in liposome phospholipids from oxidising while being stored or sonicated. This chapter, however, focusses on the potential medical uses of antioxidant liposomes and its use as delivery system in the treatment of pathological condition related to oxidative stress (Suntres 2011).

Certain antioxidants can be included into liposome formulations. These include:

Antioxidants Soluble in Lipids which includes Vitamin E (tocopherols and tocotrienols), ubiquinones, retinoids, carotenoids (lutein and β-carotene), lipid-soluble flavonoids, soy isoflavones, and synthetic antioxidants like probucol and butylated hydroxytoluene (BHT) are among the lipid-soluble antioxidants that can be added to liposomes. These antioxidants can be efficiently incorporated by liposomes, enabling them to shield cellular membranes from oxidative damage.

Antioxidants Soluble in Water

Water-soluble flavonoids, lipoic acid, pro-cysteine, urate, glutathione, ascorbate (vitamin C), and N-acetylcysteine (NAC) are a few of them. These antioxidants assist antioxidant enzymes and help squelch free radicals.

Zazuli et al. (2025) reported liposomal encapsulation of vitamin C has been shown to notably enhance its absorption through human skin. This is largely due to the liposome's hydrophobic nature, which promotes better interaction with skin cells like corneocytes. Additionally, vitamin C delivered via liposomes demonstrates stronger antioxidant and anti-inflammatory properties, particularly in protecting the skin from damage caused by UVA and UVB radiation.

Entrapped Enzymes of Antioxidants
To improve the transport and effectiveness of antioxidant enzymes, liposomes can be used to encapsulate enzymes like catalase and superoxide dismutase (SOD). These enzymes assist in protecting against reactive oxygen species (ROS) and protect cells from oxidative damage.

The liposomes helps antioxidants perform their action by blocking free radicals and reactive nitrogen species, antioxidants in liposomes can limit lipid peroxidation, shield DNA and proteins from oxidative damage, and control inflammatory responses.

Its overall protective impact is enhanced by the combination of water- and lipid-soluble antioxidants.

Antioxidant liposomes can be administered topically, intra-tracheally, intravenously, by inhalation in an aerosol form, subcutaneously, or by intramuscular injection. The liposomes have been a promising drug delivery system while keeping the drug release at the suitable rate in order to maintain therapeutic efficiency. Miatmoko et al. (2025) reported the combined therapy response of doxorubicin with antioxidants like ferulic acid has been shown to lessen its adverse effects while improving its overall therapeutic impact. Due to the poor bioavailability of both agents, the use of efficient drug delivery systems is essential for reaching the target tissues effectively. Compared to standard doxorubicin liposomes (Lipo DOX), the formulation co-loaded with ferulic acid (Lipo DOX-FA) released the drug more quickly, indicating a faster onset of therapeutic action while still maintaining stable encapsulation in circulation. Additionally, incorporating ferulic acid into Lipo DOX lowered the IC50 value in cancer cells, reflecting increased cytotoxic activity. Importantly, this enhancement did not increase toxicity in normal HEK cells, suggesting selective action toward malignant cells.

13.3.1.2 Biopolymeric Nanoparticles

Biopolymeric nanoparticles (BPNs) have been produced primarily for their ability to offer a variety of advantages, such as biocompatibility, biodegradability, and low toxicity, in addition to the usual advantages of nanoparticles. Therefore, biopolymers have attracted a lot of attention for their many applications in biomedicine, including drug delivery and antibacterial activity, because they are natural, biodegradable, and ecologically beneficial (Sreena and Nathanael 2023).

Biopolymeric nanoparticles (BPNs) are the most common type of nanoparticles; they are made of natural polymers found in biological species such as proteins (like collagen, gelatin, β-casein, zein, and albumin), polysaccharides (like hyaluronic acid, chitosan, alginate, starch, and heparin), and protein-mimicking polypeptides (like cationic polypeptides like polylysine and polyornithine (polymers). This is because BPNs can prevent degradation of the active components and increase their bioavailability, they are very effective at delivering antioxidants.

There are few antioxidants which are delivered by BPNs such as Chitosan/ Alginate Nanoparticles: Applied to the oral administration of curcumin, an anti-inflammatory and anti-tumour agent strong antioxidant and Lecithin/Chitosan Nanoparticles: Designed to release therapeutic medicines over a longer period of time, they may exhibit experienced antioxidant action.

Biopolymeric nanoparticles enhance the delivery of antioxidants through several ways by Protection of Active Compounds since encapsulation protects antioxidants from degradation, such as photodegradation and oxidation. Making it a controlled release formulation which provide controlled and sustained release of antioxidants, improving their bioavailability and efficacy. Surface modification of nanoparticles allows for targeted delivery to specific tissues or cells, enhancing the therapeutic effect while minimizing side effects. Nanoparticles improve the permeability of antioxidants across biological barriers, such as the gastrointestinal tract or cellular membranes (Sridhar et al. n.d.)

Zazuli et al. (2025) elaborated that incorporating curcumin into polymeric nanoparticles, particularly those formed from cellulose-based materials like ethyl cellulose and methylcellulose, greatly enhances its protection from environmental degradation such as light and oxidation. These nanoparticles are created via a self-assembly process and significantly improve curcumin's stability and antioxidant potential. Notably, they offer more effective free radical neutralization than free curcumin, largely due to improved skin absorption. This encapsulation strategy proves beneficial in maintaining curcumin's activity and enhancing its therapeutic effects when applied topically.

13.3.1.3 Ethosomes

A peculiar drug delivery method called ethosomes is intended to improve the skin-transmitted delivery of bioactive substances like antioxidants. The ethosomes can vary from nanometres to micrometers.

The ethosomes are peculiar due to their high ethanol concentration because ethanol is known to disrupt the organisation of the skin's lipid bilayer. As a result, it allows a vesicle to pass through the stratum corneum when incorporated into its membrane (Verma and Pathak 2010).

A unique drug delivery method called ethosomes is intended to improve the skin-transmitted delivery of bioactive substances like antioxidants. Ethosomes is a drug delivery method which allows the synergistic actions of ethanol, lipid vesicles and are responsible for the improved delivery of antioxidants. The ethosomes are able to enter the stratum corneum due to the interaction between ethanol and the lipid molecules in the skin, which increases fluidity and decreases density of the lipid multilayer. The antioxidants that were encapsulated are subsequently released into the skin's deeper layers when the ethosomes fuse with the lipids in the skin (Touitou et al. 2000).

Different antioxidants have been efficiently delivered using ethosomes. Because of their improved stability and deeper skin penetration, the antioxidants included in ethosomes are more effective at scavenging free radicals and shielding the skin from oxidative damage. It helps in the better penetration into the skin a deep and effective manner. There are certain antioxidants which can be given as antioxidants that includes—Vitamin C which is known for its skin-brightening and anti-aging properties, Vitamin E, a potent antioxidant is delivered by this way to improve its ability to combat oxidative stress. Coenzyme Q10 shows its anti-aging effects in improved manner when given through ethosomal delivery method (Akhtar et al. 2022).

13.3.1.4 Cubosomes

A particular kind of nanostructured particle called a cubosome is created when amphiphilic lipids self-assemble to create a cubic phase in water. These particles' huge interior surface area and unique internal cubic form make them perfect for a range of medical applications, including the delivery of antioxidants. The cubic liquid crystals formed by isotropic and transparent cubosomes are physically stable in the presence of excess water. They are non-toxic and biodegradable, and they have the ability to dissolve hydrophobic, hydrophilic, and amphiphilic compounds. Because of their thermodynamic stability, cubosomes have an infinite shelf life (Umar et al. 2022). Cubosome colloidal dispersions can be stabilised by adding polymers. Polymers like poloxamer 407 stabilise amphiphilic lipids like glyceryl monooleate, which make up cubosomes. These components allow the formation of a bicontinuous cubic phase structure, providing a high surface area for drug encapsulation (Singh et al. 2023; Sivadasan et al. 2023).

Cubosomes helps antioxidant delivery based on the solubility of antioxidants. The antioxidants that are encapsulated are shielded from deterioration and have a controlled release mechanism. This regulated release, which guarantees the antioxidants' focused and continuous delivery, is frequently brought about by environmental variables like pH, temperature, or particular enzymes. There are multiple advantages which cubosomes offer such as it provides the increased stability to antioxidants by shielding them from external environment, also functions in increasing bioavailability of antioxidants by encouraging its absorption and retaining it at the target side, this nature is especially helpful for such antioxidants that have low oral bioavailability and poor water solubility (Sercombe et al. 2015).

13.3.1.5 Niosomes

An expanding class of unique vesicular systems is represented by niosomes. They are made up of biodegradable and generally harmless nonionic surfactants. They were created as stable and affordable liposome substitutes.

Over the past three decades, vesicular nanocarriers called niosomes have gained considerable attention as potential drug delivery system owing to their unique advantages. Their fluid compartment encloses amphiphilic molecules that form lamellar (bilayer) structures. These amphiphilic molecules, sometimes referred to as surfactants, have self-assembling qualities and can aggregate into a range of configurations, such as micelles or a planar lamellar bilayer. They include both hydrophobic groups (tails) and hydrophilic groups (heads). Solubitan esters and analogues, sugar-, polyoxyethylene-, polyglycerol-, and crown ether-based surfactants are among the surfactants that may be employed as possible drug delivery systems (Azeem et al. 2009).

Occasionally, these surfactants may be combined with membrane additives like cholesterol or its derivatives. As cationic > anionic > nonionic surfactants decrease in order of potential irritation, nonionic surfactants are favoured.

The vesicular structure of niosomes makes it capable of encapsulating both lipophilic and hydrophilic molecules. When lipophilic compounds partition into the lipophilic domain of bilayers, they become confined, whereas hydrophilic medicines are often contained in the inner watery core or adsorbed on the bilayer surfaces.

For the development of novel cosmeceutical products, niosomes based on Tween 60 containing curcumin, resveratrol, and alpha-tocopherol as single agents and in combination were designed.

When compared to formulations containing a single antioxidant, the results demonstrated that coencapsulation of alpha-tocopherol/curcumin and resveratrol/curcumin altered the entrapment efficiency values and the physicochemical attributes of niosomes. Furthermore, a synergistic antioxidant effect amongst the antioxidants led to an enhanced capacity to lower free radicals. The resveratrol/curcumin formulation demonstrated a 40% inhibition of free radicals; the alpha-tocopherol/curcumin sample had the highest level of performance, with almost 100% inhibition (Tavano et al. 2014).

Niosomes are superior to other nanocarriers in the following ways: The surfactants used in the production of niosomes are biodegradable, immunogenicity-free, and biocompatible. No improper solvents are used during the large-scale, production of niosomes. Liposomes and niosomes can encapsulate a large quantity of material in a small vesicular volume and have same physicochemical properties, depending on the composition of the bilayer and the methods employed during production. They are both amphiphilic vesicles that can be used for long-term and targeted drug delivery (Name and Jyoti Mali 2022).

Uses for delivery of antioxidants- Skin Care: The delivery of antioxidants in niosome-based skin care treatments has been the subject of substantial research. Niosomes loaded with quercetin, for example, have demonstrated increased antioxidant activity and skin penetration, making them useful for photoprotective and anti-aging therapies (Chaudhary et al. 2022).

Oral Delivery: Niosomes increase the oral bioavailability of antioxidants which are poorly soluble. Research has indicated that curcumin's solubility, stability, and bioavailability are increased when it is encapsulated in niosomes, which improves therapeutic effects (Ghumman et al. 2023).

Transdermal administration: By improving antioxidants' penetration through the stratum corneum, niosomes enable the transdermal administration of antioxidants. This may be used to treat skin conditions and provide antioxidants to the body through the skin (Muzzalupo and Tavano 2015).

Cancer Therapy: Antioxidant delivery via niosomes has been investigated for cancer therapy. Antioxidants such as resveratrol that are encapsulated in niosomes are more stable and bioavailable, which may increase their effectiveness in lowering oxidative stress and preventing the growth of cancer cells (Muzzalupo and Tavano 2015).

13.3.1.6 Nanostructured Lipid Carriers

Nanomedicine has grown in popularity because it is effective at delivering medications and other bioactive substances to target tissues more precisely and under control by encasing or attaching them to nanostructures. These drug delivery methods use colloidal drug carrier systems called nanocarriers, which have submicron particle sizes, usually less than 1000 nm (Alexis et al. 2008). Antioxidant delivery can be enhanced with the use of Nanostructured Lipid Carriers (NLCs), a sophisticated lipid-based medication delivery technology. Because NLCs are made up of a combination of liquid and solid lipids, they have the ability to encapsulate a variety of bioactive substances and improve their stability, bioavailability, and controlled release characteristics (Nguyen et al. 2022).

13.3.2 Transdermal Dosage Forms/Delivery Systems

A transdermal drug delivery system (TDDS) is a device that offers an alternate method of administering medication, which can have an active or passive design. The skin barrier can be crossed with the help of this gadget to administer medication.

With the use of TDDS, patients may get a controlled release of their medications through the skin, potentially reducing the impacts of first-pass metabolism, minimizing systemic adverse effects, increasing patient compliance, and improving dose effectiveness by permitting stable blood drug profiles over the course of therapy. With regard to TDDS, nanoparticles have the potential to greatly increase the bioavailability and decrease immunogenicity of macromolecular medications by facilitating their passage through the stratum corneum (Wang et al. 2016).

13.3.2.1 Creams and Ointments

In dermatology and cosmetic science, the application of antioxidants through creams and ointments is becoming a significant area of research, with the goal of strengthening the skin's defenses against reactive oxygen species (ROS) and associated skin conditions. The topical administration of antioxidants for cutaneous activity presents an enormous challenge due to the molecules' inherent susceptibility to breakdown. A new era in drug delivery design was brought about by the quest for a novel delivery technology that will both maintain the stability of antioxidants and improve their skin deposition (Allemann and Baumann 2008).

Antioxidants are introduced into creams and ointments for their innumerable benefits like prevention of skin aging, protection against UV rays, reduction of inflammation, and improved skin texture (Addor 2017). The majority of skin care products with antiaging claims are based on exogenous antioxidants, which our bodies are unable to produce, such as vitamins, polyphenols, and flavonoids (Montenegro 2014). The skin is continuously exposed to environmental elements, mostly UV radiation and air contaminants, as it is the body's outermost layer of defense. The development of wrinkles, hyperpigmentation, dryness, and loss of skin tone are all brought on by oxidative stress, which also quickens the aging process. By neutralizing the production of free radicals, the use of cosmetics containing antioxidant actives helps delay the appearance of indications of aging.

Ferulic Acid

Plant cells are the source of the chemical molecule ferulic acid. It is a member of the hydroxycinnamic acid group of compounds. The primary function of phenolic acids, particularly those derived from cinnamic acid, is to exhibit antioxidant activity. This is mostly determined by the number of hydroxyl and methoxy groups that are bonded to the phenyl ring (Razzaghi-Asl et al. 2013). The body can absorb ferulic acid more readily than any other phenolic acid, and it remains in the bloodstream for a longer period of time. It is often used as a brightening, photoprotective, and photoaging-delaying ingredient in skin care products (sunscreens). The complex antioxidant action mechanism of ferulic acid is primarily focused on preventing the production of nitrogen or reactive oxygen species (ROS), but it also involves neutralizing or "sweeping" away free radicals. Additionally, protonated metal ions like Cu(II) or Fe(II) are chelated by this acid. Not only does ferrulic acid scavenge free radicals, but it also inhibits the enzymes that cause free radical formation and increases the activity of scavenger enzymes. Its chemical structure has a direct bearing on it (Zduńska et al. 2018). This acid is used in many day creams and night creams along with other ingredients like retinol, vitamin E and resveratrol. It is also used in brightening creams as they reduce hyperpigmentation.

Niacinamide A common vitamin used to treat skin conditions is niacinamide (NIA), which is soluble in water. Moreover, NIA has antioxidant properties and aids in DNA damage repair (Zhen et al. 2019). Niacinamide helps to regulate our skin's epidermal cell turnover. It helps the skin regenerate and produce new cells by lowering oxidative stress on the skin brought on by the sun, pollution, aging, etc. Moreover, niacinamide has anti-inflammatory effects. It helps control acne, rosacea, eczema, and red, inflamed skin. In the long term, niacinamide at a 5% concentration may help reduce hyperpigmentation and dark patches, according to certain research findings.

A number of antioxidants are involved in the skin defense and renewal by diverse mechanisms of action. Vitamin C (ascorbic acid) induces collagen synthesis and has anti-inflammatory activity, leading to repair and tightness of the skin (Zduńska et al. 2018). Green tea catechin extract, and especially EGCG, protects the skin from UV-induced damage and has collagenase inhibitory action. In the same manner, Vitamin E exhibits age-reducing effects both through stimulating collagen formation and inhibiting its breakdown, so promoting general skin health and resilience (Ngoc et al. 2023).

13.3.2.2 Hydrogels and Lotions

A hydrogel is a three-dimensional (3D) network of hydrophilic polymers that can expand in water and retain a huge quantity of water while keeping the structure because individual polymer chains are cross-linked chemically or physically (Bahram et al. 2016). Hydrogels are a good medication delivery mechanism for treating skin wounds because of this property. The pathophysiology of chronic wounds has been linked to oxidative stress, which is known to hinder wound healing by damaging cellular components, postponing the inflammatory phase of healing, and preventing the development of new blood vessels. Thus, managing chronic wounds necessitates a multidisciplinary approach that provides optimal treatment, considers the lesion's underlying causes, and promotes wound healing.

Antioxidant release kinetics benefit from the use of hydrogels. One way to solve the problem of long-term therapeutic level maintenance at the wound site is by the regulated and sustained release of antioxidants from biopolymers. Prolonged anti-oxidant protection is guaranteed by this regulated release, which also reduces the possibility of adverse effects from high-dose, quick delivery (Liu and Chen 2024).

Pranantyo et al. (2024) developed a hydrogel dressing for topical application with dual functionality, including antibacterial cationic polyimidazolium and antioxidative N-acetylcysteine into a crosslinked network. This dressing has built-in antibiofilm and antioxidative properties. In diabetic mouse wounds infected with biofilms of Pseudomonas aeruginosa or methicillin-resistant Staphylococcus aureus, this treatment helped the wounds heal. Furthermore, the dressing induced keratinocyte differentiation and re-epithelialization in a human skin equivalent model, providing a flexible and non-contaminated wound care option (Comotto et al. 2019).

In clinical practice, hydrogel-based wet wound dressings, like alginate, are frequently utilized to treat skin conditions. These alginate dressings have anti-inflammatory and antibacterial properties because they include natural antioxidants including t-resveratrol and curcumin. Over time, the hydrogel keeps its superior mechanical qualities and oxygen permeability (Tan et al. 2024).

Extracted dandelions' extracellular vesicle-like nanoparticles (TH-EVNs) exhibit antivirulence properties by selectively attaching to the exotoxins of Staphylococcus aureus (S. aureus), shielding the host cell from harm. Significant binding power and stability are shown in the neutralization of TH-EVNs against exotoxins, demonstrating a full detoxifying effect in vivo. Following this, a TH-EVNs-loaded dressing for exotoxin-invasive S. aureus wounds is created using gelatin and methacryloyl hydrogel. In addition to encouraging cell migration and proliferation, hydrogel dressings provide strong mechanical and physical qualities that allow for wound retention and controlled release of TH-EVNs (Liu and Chen 2024; Morena et al. 2022; Tan et al. 2024).

13.3.3 Inhalation Dosage Form

As a way to administer medication, inhalation evolved into aerosol therapy in the twentieth century, which has long been a part of respiratory care. Using inhalation device, the therapeutic component is applied topically to the respiratory tract. Tiny drug particles may be suspended in a solution, powder, or other form when the medication is administered. The way an inhaler is designed, how well it delivers medication to the lower respiratory tract depends on a number of factors, including the ease of use, internal resistance of dry powder inhalers (DPIs), medication formulation, particle size, and the velocity of the aerosol plume produced by pressurized metered-dose inhalers (pMDIs) (Nelson 2016).

13.3.3.1 Aerosols and Nebulizers

Antioxidant administration by inhalation, or via the nose or lungs, offers a more viable strategy to reduce the detrimental impacts of air pollution on human health since it may directly raise antioxidant levels in the airway surface liquid (ASL), shielding the lung tissue from oxidative damage caused by air pollution. Some of these antioxidants have anti-inflammatory and antibacterial qualities in addition to directly scavenging reactive oxygen species (ROS). This suggests that they may mitigate oxidative stress and reduce the risk of bacterial and viral respiratory infections brought on by exposure to air pollution.

Vitamin D: Numerous investigations have looked into the use of vitamin D pre- or co-treatment in response to different types of pollutants. Particulate matter exposure is the condition in which vitamin D's protective benefits are most noticeable. When vitamin D was administered, transcriptome analysis of primary Human Bronchial Epithelial Cells (hBECs) exposed to urban PM revealed decreased expression of the pro-inflammatory cytokine gene IL6 and increased expression of the gene for the antioxidant pathway G6PD (Pfeffer et al. 2018). Reduced lipid oxidation, an enhanced antioxidant response, and decreased IL-6 levels were also seen in the same research (Schichlein et al. 2023).

Baicalein: A flavonoid called baicalein (BE) was isolated from the roots of Scutellaria radix. Prior research has previously shown that BE's neuroprotective qualities are a result of a variety of pharmacological actions it exerts on the brain, including as anti-inflammatory and antioxidant activities. Patients with central nervous system (CNS) neurodegenerative illnesses exhibit widespread symptoms of neuro-inflammation and oxidative stress in their brains. Because of its anti-inflammatory and antioxidant properties, BE has thus been proposed as a possible therapy for CNS neurodegenerative illnesses. Despite its neuroprotective properties, its low oral bioavailability and poor solubility hinder its utilization in neurodegenerative disorders. Therefore, in order to improve its distribution throughout the brain, BE was encapsulated into poly (ethylene glycol)-block-poly (D, L-lactide) micelles (BE-MC) and administered via nasal inhalation. The findings suggested that nasal delivery of BE-MC might be a helpful therapy for CNS neurodegenerative illnesses as it demonstrated similar in-vitro antioxidant activity to BE solution without killing healthy cells (Zhang et al. 2020).

Superoxide dismutase: In the acute phase of acute respiratory distress syndrome (ARDS), which is a potentially fatal respiratory failure syndrome, neutrophil infiltration into the alveolar space leads to the uncontrollable release of reactive oxygen species (ROS) and proteases, which overwhelm antioxidant defenses and injure the alveolar epithelium and lung endothelium. To prevent ROS damage and to cure Escherichia coli-induced ARDS, a research was conducted utilizing a new recombinant human Cu-Zn-superoxide dismutase (SOD) fusion protein by aerosolized SOD administration. The results demonstrated that hSOD1 decreased oxidative damage and mitigated E. coli-induced lung injury in rats, offering a solid foundation for future research into the therapeutic potential of hSOD1 in the treatment of ARD (McCarthy et al. 2023).

Glutathione: An antioxidant called glutathione is given to the lungs using a nebulizer as part of a therapy called nebulized glutathione. Individuals with asthma, emphysema, chronic respiratory allergies, chronic bronchitis, and any type of lung scarring from prior injuries have also shown positive results with it (*Nebulized Glutathione Treatment for Lung Health.*, n.d.). Nebulization of glutathione and N-acetylcysteine is also done to treat the early respiratory stress caused by COVID-19 (Lana et al. 2021).

13.3.3.2 Dry Powder Inhalers

The dry powder inhaler disperses medication into the lungs in the form of dry powder, requiring little coordination between the patient's breathing and the device's actuation. The dry powder inhaler's formulation, which is subjected to greater dispersion forces to break down into individual particles, delivers the aerosolized medication powder (Velaga et al. 2018).

The alkaloid cepharanthine (CEP) derived from Stephania cepharantha Hayata is a bisbenzylisoquinoline. Its pharmacological effects are multifaceted, encompassing antiviral, immunomodulatory, antioxidant, and anti-inflammatory properties (Bailly 2019). It could lessen the degree of lung damage and cytokine storm brought on by a SARS-CoV-2 infection. Freeze-drying method was used by the researchers to prepare the cepharanthine dry powder inhaler (CEP-DPI). Based on current research, the CEP-DPI in particular has been suggested as a therapeutic strategy for treating acute lung damage (ALD) linked to serious respiratory infections such as COVID-19. CEP is relevant for more research and clinical trials in the treatment of ALI because of its targeted administration and antioxidant properties (Liang et al. 2023).

13.4 Current Scenarios in Antioxidant Therapeutic Strategies

13.4.1 Clinical Applications and Efficacy

There are various clinical applications of antioxidants, some are listed below:

13.4.1.1 Clinical Trials and Research

In order to improve the effects of chemotherapy and radiation therapy, high-dose intravenous vitamin C is being investigated in ongoing clinical trials as a supplemental therapy for cancer treatment. High-dose IVC has been generally well-tolerated in clinical trials, with minimal and manageable side effects such as increased urination, dry mouth, and temporary increases in blood pressure (Böttger et al. 2021).

Studies are investigating curcumin, a compound in turmeric, for its anti- inflammatory and antioxidant properties, and its role in preventing and treating various cancers. Curcumin's capacity to inhibit cyclooxygenase-2 (COX-2), lipoxygenase (LOX), and inducible nitric oxide synthase (iNOS) is most likely the mechanism by which it exerts an anti-inflammatory impact. Important enzymes that mediate inflammatory processes include LOX, iNOS, and COX-2. A number of human cancers and inflammatory diseases have been linked to improper overexpression of COX-2 and/or iNOS in their pathogenesis (Menon and Sudheer 2007).

Found in grapes and berries, resveratrol is being researched for its potential to inhibit cancer cell growth and induce apoptosis. Resveratrol's anti-proliferative and pro-apoptotic properties were suggested to stem from its capacity to regulate the expression of pro-oncogenic and tumour suppressor microRNAs, nuclear respiratory factors (NRF) 1 and 2, nuclear factor-kappa B (NF-kB), and gamma activator 1 alpha (PCG-1α), among other things, in vitro experiments (Rauf et al. 2018).

13.4.1.2 Combination Therapies of Antioxidants

Antioxidants with Chemotherapy: Research is focused on combining antioxidants with traditional chemotherapy to enhance efficacy and reduce side effects. For instance, using antioxidants like selenium and vitamin E to mitigate the toxicity of chemotherapeutic agents. In the event of gastric tumours, the carcinogen-activating enzymes can be inhibited by the antioxidant activity of flavonoids such as quercetin and naringin, which are found in apples and grapefruit, respectively. Additionally, CYP1A1 (cytochrome P450 enzyme) expression inhibits the conversion of toxins into carcinogens, preventing DNA damage (Zahra et al. 2021).

Synergistic Effects: Combining multiple antioxidants (e.g., vitamin C, vitamin E, and selenium) to achieve synergistic effects in reducing oxidative stress and improving therapeutic outcomes.

13.4.1.3 Targeted Delivery Systems

Nanotechnology: Utilizing biopolymeric nanoparticles to improve the delivery and bioavailability of antioxidants directly to tumour sites, enhancing their therapeutic potential and minimizing side effects.

Bioactive compounds derived from various sources and natural resources, such as gallic acid (GA) derived from dried rose flower extracts (Rosa rugosa), have also demonstrated potential antioxidant capacity, scavenging ability of active oxygen species and electrophiles, inhibition of nitrosation reactions, and reduction of lipid peroxidation level (Khalil et al. 2020).

Liposomes and Micelles: Developing liposomal and micellar formulations to encapsulate antioxidants, ensuring targeted delivery and sustained release, thereby increasing their effectiveness in cancer treatment.

13.5 Market Trends and Commercial Products

Growing health consciousness, the desire for natural and organic products, technical developments, and the growth of the food and beverage, pharmaceutical, and cosmetics industries are some of the major trends and drivers influencing the market for antioxidants.

The following are some noteworthy trends:

Growing Health Awareness: People are becoming more conscious of the advantages antioxidants have for their health, including how they can lower oxidative stress and fend off chronic illnesses. The demand for goods strong in antioxidants has increased as a result.

Natural and Organic Products: Antioxidants obtained from fruits, vegetables, and other natural sources are becoming increasingly popular. Because they have fewer negative effects and are thought to have health advantages, consumers prefer them to synthetic antioxidants.

Functional Foods and Beverages: The market for antioxidants is significantly influenced by the functional food and beverage sector. Antioxidant-fortified goods including teas, juices, and nutritional supplements are becoming more and more well-liked among consumers who are concerned about their health.

Cosmetics & Personal Care: Because of its anti-aging and skin-protective qualities, antioxidants are frequently utilized in cosmetics and personal care products. This market is expanding quickly as more people look for items that support youthful, healthy skin.

Antioxidants are used in pharmaceutical applications because of their ability to provide therapeutic benefits across various disease and health conditions. This covers their application in formulas for neuroprotection, cancer prevention, and cardiovascular health.

13.5.1 Commercial Products

The antioxidant market is categorized into natural and synthetic types. Natural antioxidants consist of compounds such as Vitamin C, Vitamin E, and polyphenols, whereas synthetic variants include substances like butylated hydroxytoluene (BHT) and butylated hydroxy anisole (BHA).

13.6 Future Prospects of Antioxidant Therapeutic Strategies

Diverse innovative methods and mechanisms underlying the actions of antioxidants are establishing the groundwork for future developments in antioxidant therapy. With several of them presently undergoing clinical trials, the possibilities for antioxidant therapy techniques are being investigated increasingly.

13.6.1 Personalized Antioxidant Therapy

A new area of medicine called "personalized medicine" makes assessments about disease prevention, diagnosis, and therapy based on a person's genetic profile (*Personalized Medicine*, n.d.).

Biomarker based approaches: Male reproductive potential may be predicted by Malondialdehyde (MDA), glutathione (GSH), and total redox antioxidant potential (TRAP) indicators in addition to the seminogram used in regular semen analysis, particularly when these markers are combined with increasing sensitivity and specificity (Krzyściak et al. 2020). Increased plasma *8-Isoprostanes*

(8-ISO) levels may suggest that lipid peroxidation plays a major role in the emergence of potentially malignant oral disorders, including leucoplakia (Senghore et al. 2018).

Genomic-Based Personalization: In research by (Baig et al. 2019) it was shown that, in comparison to lean patients, obese people had higher postprandial levels of pro-oxidant (CYBB and CYBA) and anti-oxidant (TXN RD1) gene expression.

Disease-Specific Approaches: Antioxidants Targeted to Mitochondria for Parkinson's Disease Treatment: In Parkinson's disease models, small antioxidant compounds that target mitochondria, such as MitoQ and MitoApocynin, have been shown to have neuroprotective benefits (Jin et al. 2014).

13.6.2 Potential Synergistic Combination Therapy

A therapeutic intervention known as a combination treatment involves giving the patient more than one therapy. Several distinct tablets, each having a different medicine, or single pills containing multiple pharmaceuticals are examples of combination therapy regimens. Polyphenols have shown to be viable options for cancer treatment, when properly synergized with chemotherapy medications. They possess excellent proapoptotic, anti-inflammatory, antiproliferative, antiangiogenic, antimetastatic, and antioxidant qualities. When used with traditional antitumorals like cisplatin, doxorubicin, paclitaxel, 5-fluorouracil, and others, they can have a synergistic effect. In order to provide a regulated release and a combination regenerative and therapeutic effect, they can also be placed into different biomaterials (Vladu et al. 2022). Inhibiting NADPH oxidase (NOX) and activating nuclear factor erythroid 2-related factor 2 (Nrf-2) simultaneously is the most effective way to improve antioxidant defense and lower the risk of epilepsy. Combination treatment, especially after kainic acid (KA)-induced status epilepticus (SE), can reduce oxidative stress, stop epileptogenesis, and alter chronic epilepsy (Shekh-Ahmad et al. 2019). Curcumin-CD complexes have been discovered to improve the therapeutic efficacy of gemcitabine in an experimentally produced model of lung cancer when combined with other medicines for the treatment of cancer (Rocks et al. 2012). When curcumin and paclitaxel, a chemotherapeutic taxane medication, were encapsulated with poly (β-cyclodextrin triazine), synergistic effects were shown in several cancer models of the ovary, prostate, lung, and breast (Boztas et al. 2013; Sezgin-Bayindir et al. 2021).

13.6.3 Stimuli- Responsive Antioxidant Delivery

Stimuli-responsive antioxidant delivery systems are emerging strategies developed for the controlled liberation of antioxidants induced by specific environmental stimulants, such as pH-temperature dependent or enzyme-sensitive. These are

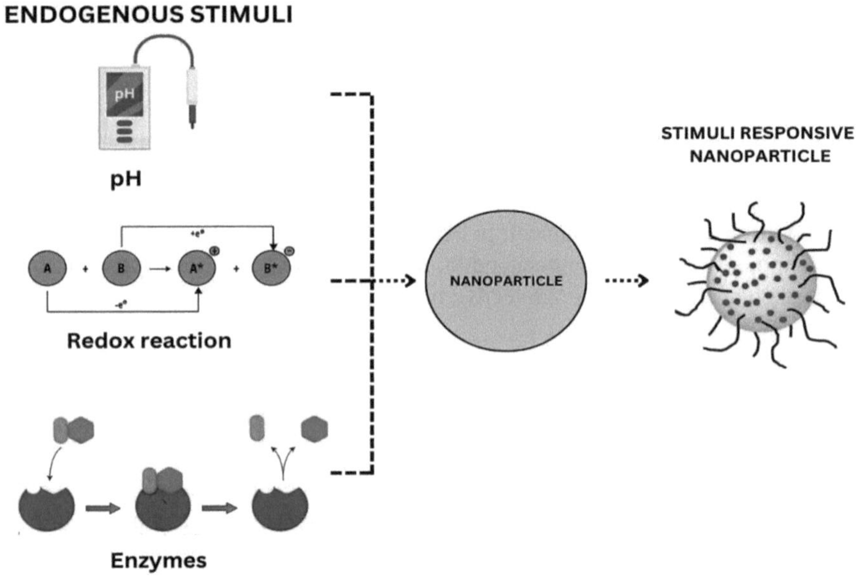

Fig. 13.4 Endogenous stimuli responsive nanoparticle systems (Pham et al. 2020)

experimental approaches to increase the efficacy and specificity of antioxidants, mostly in conditions associated with oxidative stress. The design and mechanisms of these advanced systems are illustrated in Fig. 13.4.

Antitumor Therapy and Drug Delivery: Utilizing Stimuli-Responsive Nanomaterials

1. pH-Responsive Nanoparticles: Because tumour tissues typically have an acidic environment, these nanoparticles can release their payload in reaction. By limiting negative effects on healthy tissues, this focused release increases the medicinal drugs' efficacy.
2. Redox-Responsive Systems: In tumour microenvironment, there is presence of high reactive oxygen species content wherein these nanoparticles are released to show their action.
3. Enzyme-Responsive Nanoparticles: These systems contain the appropriate enzymes that are continually activated in several malignant tissues, which then would enable a release of therapeutic drugs (Pham et al. 2020).

13.7 Conclusion

Antioxidant therapeutic strategies show significant promise in addressing many oxidative stress-related disorders. While the current applications demonstrate efficacy in various clinical settings, the field is rapidly evolving. Future developments in personalized antioxidant therapy, novel delivery systems, and combination treatments offer the potential for more targeted and effective interventions. The integration of advanced delivery systems and combination therapies, along with a deeper understanding of individual patient needs, is likely to enhance the efficacy of antioxidant interventions. However, ongoing research is needed to optimize dosages, understand long-term effects, and develop more efficient delivery mechanisms to fully harness the therapeutic potential of antioxidants across a range of medical conditions. The growing market demand and focus on natural sources further indicate the potential for antioxidants in both preventive and therapeutic applications. The chapter provides a comprehensive overview of antioxidant therapeutic strategies, covering their mechanisms, applications, and future prospects. It discusses various types of antioxidants, and their roles in combating oxidative stress associated with chronic diseases like cardiovascular disorders, neurodegenerative conditions, diabetes, cancer, and autoimmune diseases. The chapter also explores over innovative delivery systems for antioxidants, including nanotechnology-based formulations. The market trends for antioxidant products are outlined, noting the growing demand for natural and organic antioxidants in various industries. The chapter concludes by examining future prospects, including emerging antioxidants and personalized antioxidant therapy approaches.

References

Addor FAS (2017) Antioxidants in dermatology. An Bras Dermatol. https://doi.org/10.1590/abd1806-4841.20175697

Afonso V, Champy R, Mitrovic D, Collin P, Lomri A (2007) Reactive oxygen species and superoxide dismutases: role in joint diseases. Joint Bone Spine. https://doi.org/10.1016/j.jbspin.2007.02.002

Akhtar N, Akhtar N, Menaa F, Alharbi W, Alaryani FSS, Alqahtani AM, Ahmad F (2022) Fabrication of ethosomes containing tocopherol acetate to enhance transdermal permeation: in vitro and ex vivo characterizations. Gels 8. https://doi.org/10.3390/gels8060335

Alexis F, Rhee JW, Richie JP, Radovic-Moreno AF, Langer R, Farokhzad OC (2008) New frontiers in nanotechnology for cancer treatment. Urol Oncol Semin Orig Investig. https://doi.org/10.1016/j.urolonc.2007.03.017

Allemann IB, Baumann L (2008) Antioxidants used in skin care formulations. Skin Therapy Lett 13:5–9

Arabshomali A, Bazzazzadehgan S, Mahdi F, Shariat-Madar Z (2023) Potential benefits of antioxidant phytochemicals in type 2 diabetes. Molecules. https://doi.org/10.3390/molecules28207209

Arballo J, Amengual J, Erdman JW (2021) Lycopene: a critical review of digestion, absorption, metabolism, and excretion. Antioxidants. https://doi.org/10.3390/antiox10030342

Arias A, Feijoo G, Moreira MT (2022) Exploring the potential of antioxidants from fruits and vegetables and strategies for their recovery. Innov Food Sci Emerg Technol. https://doi.org/10.1016/j.ifset.2022.102974

Ariraman S, Seetharaman A, Babunagappan KV, Sudhakar S (2024) Quercetin-loaded nanoarchaeosomes for breast cancer therapy: a ROS mediated cell death mechanism. Mater Adv 5:6944–6956

Aryaeian N, Sedehi SK, Arablou T (2017) Polyphenols and their effects on diabetes management: a review. Med J Islam Repub Iran 31:886–892. https://doi.org/10.14196/mjiri.31.134

Asbaghi O, Nazarian B, Yousefi M, Anjom-Shoae J, Rasekhi H, Sadeghi O (2023) Effect of vitamin E intake on glycemic control and insulin resistance in diabetic patients: an updated systematic review and meta-analysis of randomized controlled trials. Nutr J. https://doi.org/10.1186/s12937-023-00840-1

Asefy Z, Hoseinnejhad S, Ceferov Z (2021) Nanoparticles approaches in neurodegenerative diseases diagnosis and treatment. Neurol Sci 42:2653–2660

Axelsson AS, Tubbs E, Mecham B, Chacko S, Nenonen HA, Tang Y, Fahey JW, Derry JMJ, Wollheim CB, Wierup N, Haymond MW, Friend SH, Mulder H, Rosengren AH (2017) Sulforaphane reduces hepatic glucose production and improves glucose control in patients with type 2 diabetes. Sci Transl Med 9:eaah4477. https://doi.org/10.1126/scitranslmed.aah4477

Ayoka TO, Ezema BO, Eze CN, Nnadi CO (2022) Antioxidants for the prevention and treatment of non-communicable diseases. J Explor Res Pharmacol. https://doi.org/10.14218/jerp.2022.00028

Azeem A, Anwer MK, Talegaonkar S (2009) Niosomes in sustained and targeted drug delivery: some recent advances. J Drug Target. https://doi.org/10.3109/10611860903079454

BahnsonEdward SM, KassamHussein A, MoyerTyson J, MorganCourtney E, VercammenJanet M, FlynnMegan E, StuppSamuel I, KibbeMelina R (2016) Targeted nitric oxide delivery by supramolecular nanofibers for the prevention of restenosis after arterial injury. Antioxid Redox Signal 24(8):401–418

Bahram M, Mohseni N, Moghtader M (2016) An introduction to hydrogels and some recent applications. In: Emerging concepts in analysis and applications of hydrogels. https://doi.org/10.5772/64301

Baig S, Rizi EP, Chia C, Shabeer M, Aung N, Loh TP, Magkos F, Vidal-Puig A, Seet RCS, Meng Khoo C, Toh SA (2019) Genes involved in oxidative stress pathways are differentially expressed in circulating mononuclear cells derived from obese insulin-resistant and lean insulin-sensitive individuals following a single mixed-meal challenge. Front Endocrinol (Lausanne) 10. https://doi.org/10.3389/fendo.2019.00256

Bailly C (2019) Cepharanthine: an update of its mode of action, pharmacological properties and medical applications. Phytomedicine. https://doi.org/10.1016/j.phymed.2019.152956

Barapatre A, Meena AS, Mekala S, Das A, Jha H (2016) In vitro evaluation of antioxidant and cytotoxic activities of lignin fractions extracted from Acacia nilotica. Int J Biol Macromol 86: 443–453. https://doi.org/10.1016/j.ijbiomac.2016.01.109

Berr C (2000) Cognitive impairment and oxidative stress in the elderly: results of epidemiological studies. Biofactors. https://doi.org/10.1002/biof.5520130132

Berry CE, Hare JM (2004) Xanthine oxidoreductase and cardiovascular disease: molecular mechanisms and pathophysiological implications. J Physiol. https://doi.org/10.1113/jphysiol.2003.055913

Böttger F, Vallés-Martí A, Cahn L, Jimenez CR (2021) High-dose intravenous vitamin C, a promising multi-targeting agent in the treatment of cancer. J Exp Clin Cancer Res. https://doi.org/10.1186/s13046-021-02134-y

Boztas AO, Karakuzu O, Galante G, Ugur Z, Kocabas F, Altuntas CZ, Yazaydin AO (2013) Synergistic interaction of paclitaxel and curcumin with cyclodextrin polymer complexation in human cancer cells. Mol Pharm 10. https://doi.org/10.1021/mp400101k

Bredemeier M, Lopes LM, Eisenreich MA, Hickmann S, Bongiorno GK, d'Avila R, Morsch ALB, da Silva Stein F, Campos GGD (2018) Xanthine oxidase inhibitors for prevention of cardiovascular events: a systematic review and meta-analysis of randomized controlled trials. BMC Cardiovasc Disord 18. https://doi.org/10.1186/s12872-018-0757-9

Bulboaca AE, Boarescu P-M, Porfire AS, Dogaru G, Barbalata C, Valeanu M, Munteanu C, Râjnoveanu RM, Nicula CA, Stanescu IC (2020) The effect of nano-epigallocatechin-gallate on oxidative stress and matrix metalloproteinases in experimental diabetes mellitus. Antioxidants 9:172

Cecerska-Heryć E, Wiśniewska Z, Serwin N, Polikowska A, Goszka M, Engwert W, Michałów J, Pękała M, Budkowska M, Michalczyk A, Dołęgowska B (2024) Can compounds of natural origin be important in chemoprevention? Anticancer properties of quercetin, resveratrol, and curcumin-a comprehensive review. Int J Mol Sci. https://doi.org/10.3390/ijms25084505

Chandimali N, Bak SG, Park EH, Lim HJ, Won YS, Kim EK, Park SI, Lee SJ (2025) Free radicals and their impact on health and antioxidant defenses: a review. Cell Death Discov. https://doi.org/10.1038/s41420-024-02278-8

Chaudhary A, Gupta M, Kumari K, Sachan AK (2022) Preparation and evaluation of niosomes for the delivery of antioxidant to the skin for antiaging purpose. Asian J Pharm Res Dev 10. https://doi.org/10.22270/ajprd.v10i3.1127

Chico L, Schirinzi E, Balestrini L, Polzella M, Siciliano G (2026) Nrf2-activating natural compounds in neurodegenerative diseases: targeting oxidative stress and protein aggregation. Int J Mol Sci 27(3):1592. https://doi.org/10.3390/ijms27031592

Comotto M, Saghazadeh S, Bagherifard S, Aliakbarian B, Kazemzadeh-Narbat M, Sharifi F, Mousavi Shaegh SA, Arab-Tehrany E, Annabi N, Perego P, Khademhosseini A, Tamayol A (2019) Breathable hydrogel dressings containing natural antioxidants for management of skin disorders. J Biomater Appl 33. https://doi.org/10.1177/0885328218816526

Cory H, Passarelli S, Szeto J, Tamez M, Mattei J (2018) The role of polyphenols in human health and food systems: a mini-review. Front Nutr. https://doi.org/10.3389/fnut.2018.00087

Cristani M, Citarella A, Carnamucio F, Micale N (2024) Nano-formulations of natural antioxidants for the treatment of liver cancer. Biomolecules 14:1031

D'Oria R, Schipani R, Leonardini A, Natalicchio A, Perrini S, Cignarelli A, Laviola L, Giorgino F (2020) The role of oxidative stress in cardiac disease: from physiological response to injury factor. Oxid Med Cell Longev. https://doi.org/10.1155/2020/5732956

Dutta D, Siddiqui L, Shah S, Talegaonkar S (2024) Synergistic anticancer activity of Sulphoraphane and Teriflunomide co loaded lignin nanoparticles against triple negative breast cancer: targeted nanoparticle delivery and drug repurposing. Med Hypotheses 189:111404. https://doi.org/10.1016/j.mehy.2024.111404

El Sharkwy IA, Abd El Aziz WM (2019) Randomized controlled trial of N-acetylcysteine versus l-carnitine among women with clomiphene-citrate-resistant polycystic ovary syndrome. Int J Gynecol Obstet. https://doi.org/10.1002/ijgo.12902

Fazli B, Irani S, Bardania H, Moosavi M-S, Rohani B (2022) Prophylactic effect of topical (slow-release) and systemic curcumin nano-niosome antioxidant on oral cancer in rat. BMC Complement Med Ther 22:109

Forman HJ, Zhang H (2021) Targeting oxidative stress in disease: promise and limitations of antioxidant therapy. Nat Rev Drug Discov. https://doi.org/10.1038/s41573-021-00233-1

Fredman G, Kamaly N, Spolitu S, Milton J, Ghorpade D, Chiasson R, Kuriakose G, Perretti M, Farokhzad O, Tabas I (2015) Targeted nanoparticles containing the proresolving peptide Ac2-26 protect against advanced atherosclerosis in hypercholesterolemic mice. Sci Transl Med 7:275ra20

Gautam N, Dutta D, Mittal S, Alam P, Emad NA, Al-Sabri MH, Verma SP, Talegaonkar S (2025a) QbD-Driven preparation, characterization, and pharmacokinetic investigation of daidzein-l oaded nano-cargos of hydroxyapatite. Sci Rep 15:2967. https://doi.org/10.1038/s41598-025-85463-8

Gautam N, Prashant S, Antra C, Surajita S, Divya V, Monalisa M, Debopriya D, Manish S, Talegaonkar S (2025b) Investigating the osteogenic potential of bone-targeted daidzein loaded hydroxyapatite nanoparticles for postmenopausal osteoporosis: pharmacodynamic, biochemical, and genotoxicity evaluations. J Drug Target:1–16. https://doi.org/10.1080/1061186X.2025.2503499

Ghumman SA, Ijaz A, Noreen S, Aslam A, Kausar R, Irfan A, Latif S, Shazly GA, Shah PA, Rana M, Aslam A, Altaf M, Kotwica-Mojzych K, Bin Jardan YA (2023) Formulation and characterization of curcumin niosomes: antioxidant and cytotoxicity studies. Pharmaceuticals 16. https://doi.org/10.3390/ph16101406

Giustarini D, Milzani A, Dalle-Donne I, Rossi R (2023) How to increase cellular glutathione. Antioxidants. https://doi.org/10.3390/antiox12051094

Gong F, Wang Z, Mo R, Wang Y, Su J, Li X, Omonova CTQ, Khamis AM, Zhang Q, Dong M, Su Z (2022) Nano-sponge-like liposomes remove cholesterol crystals for antiatherosclerosis. J Control Release 349:940–953. https://doi.org/10.1016/J.JCONREL.2022.07.021

Gonzalez-Perez J, Lopera-Echavarría AM, Arevalo-Alquichire S, Araque-Marín P, Londoño ME (2024) Development of a resveratrol nanoformulation for the treatment of diabetic retinopathy. Materials 17:1420

Grabska-Kobyłecka I, Szpakowski P, Król A, Książek-Winiarek D, Kobyłecki A, Głąbiński A, Nowak D (2023) Polyphenols and their impact on the prevention of neurodegenerative diseases and development. Nutrients. https://doi.org/10.3390/nu15153454

Guo CY, Sun L, Chen XP, Zhang DS (2013) Oxidative stress, mitochondrial damage and neurodegenerative diseases. Neural Regen Res 8. https://doi.org/10.3969/j.issn.1673-5374.2013.21.009

Han Z, Gao X, Wang Y, Cheng S, Zhong X, Xu Y, Zhou X, Zhang Z, Liu Z, Cheng L (2023) Ultrasmall iron-quercetin metal natural product nanocomplex with antioxidant and macrophage regulation in rheumatoid arthritis. Acta Pharm Sin B 13:1726–1739. https://doi.org/10.1016/j.apsb.2022.11.020

Head E (2009) Oxidative damage and cognitive dysfunction: antioxidant treatments to promote healthy brain aging. Neurochem Res. https://doi.org/10.1007/s11064-008-9808-4

Heber D, Lu QY (2002) Overview of mechanisms of action of lycopene. Exp Biol Med. https://doi.org/10.1177/153537020222701013

Hood ED, Greineder CF, Shuvaeva T, Walsh L, Villa CH, Muzykantov VR (2018) Vascular targeting of radiolabeled liposomes with bio-orthogonally conjugated ligands: single chain fragments provide higher specificity than antibodies. Bioconjug Chem 29:3626–3637

Howard MD, Greineder CF, Hood ED, Muzykantov VR (2014) Endothelial targeting of liposomes encapsulating SOD/catalase mimetic EUK-134 alleviates acute pulmonary inflammation. J Control Release 177:34–41

Jagaran K, Singh M (2021) Nanomedicine for neurodegenerative disorders: focus on Alzheimer's and Parkinson's diseases. Int J Mol Sci 22:9082

Jain A, Mehra N, Swarnakar N (2015) Role of antioxidants for the treatment of cardiovascular diseases: challenges and opportunities. Curr Pharm Des 21. https://doi.org/10.2174/1381612821666150803151758

Jena AB, Samal RR, Bhol NK, Duttaroy AK (2023) Cellular Red-Ox system in health and disease: the latest update. Biomed Pharmacother. https://doi.org/10.1016/j.biopha.2023.114606

Jin H, Kanthasamy A, Ghosh A, Anantharam V, Kalyanaraman B, Kanthasamy AG (2014) Mitochondria-targeted antioxidants for treatment of Parkinson's disease: preclinical and clinical outcomes. Biochim Biophys Acta Mol basis Dis. https://doi.org/10.1016/j.bbadis.2013.09.007

Jomova K, Raptova R, Alomar SY, Alwasel SH, Nepovimova E, Kuca K, Valko M (2023) Reactive oxygen species, toxicity, oxidative stress, and antioxidants: chronic diseases and aging. Arch Toxicol. https://doi.org/10.1007/s00204-023-03562-9

Kadry MO (2024) Resveratrol-based nano-formulations as an emerging therapeutic strategy for ovarian carcinoma: autophagy stimulation and SIRT-1/Beclin/MMP-9/P53/AKT signaling. Cancer Nanotechnol 15:36

Kamaly N, Fredman G, Fojas JJR, Subramanian M, Choi WII, Zepeda K, Vilos C, Yu M, Gadde S, Wu J (2016) Targeted interleukin-10 nanotherapeutics developed with a microfluidic chip enhance resolution of inflammation in advanced atherosclerosis. ACS Nano 10:5280–5292

Kang LEI, Jiang D, Ni D, Ehlerding E, Wang R, Cai W (2019) Antioxidative neuroprotective role of iron-gallic acid coordination nanoparticles and PET/MR usage in an ischemia stroke model. J Nucl Med 60:113

Kang JS, Nam LB, Yoo OK, Keum YS (2020) Molecular mechanisms and systemic targeting of NRF2 dysregulation in cancer. Biochem Pharmacol. https://doi.org/10.1016/j.bcp.2020.114002

Kansanen E, Kuosmanen SM, Leinonen H, Levonenn AL (2013) The Keap1-Nrf2 pathway: mechanisms of activation and dysregulation in cancer. Redox Biol. https://doi.org/10.1016/j.redox.2012.10.001

Khalil I, Yehye WA, Etxeberria AE, Alhadi AA, Dezfooli SM, Julkapli NBM, Basirun WJ, Seyfoddin A (2020) Nanoantioxidants: recent trends in antioxidant delivery applications. Antioxidants. https://doi.org/10.3390/antiox9010024

Khan H, Sureda A, Belwal T, Çetinkaya S, Süntar İ, Tejada S, Devkota HP, Ullah H, Aschner M (2019) Polyphenols in the treatment of autoimmune diseases. Autoimmun Rev 18:647–657. https://doi.org/10.1016/j.autrev.2019.05.001

Kim GH, Kim JE, Rhie SJ, Yoon S (2015) The role of oxidative stress in neurodegenerative diseases. Exp Neurobiol. https://doi.org/10.5607/en.2015.24.4.325

Kojima R, Bojar D, Rizzi G, Hamri GC-E, El-Baba MD, Saxena P, Ausländer S, Tan KR, Fussenegger M (2018) Designer exosomes produced by implanted cells intracerebrally deliver therapeutic cargo for Parkinson's disease treatment. Nat Commun 9:1305

Kotcherlakota R, Barui AK, Prashar S, Fajardo M, Briones D, Rodríguez-Diéguez A, Patra CR, Gómez-Ruiz S (2016) Curcumin loaded mesoporous silica: an effective drug delivery system for cancer treatment. Biomater Sci 4:448–459

Kozlov AV, Javadov S, Sommer N (2024) Cellular ROS and antioxidants: physiological and pathological role. Antioxidants. https://doi.org/10.3390/antiox13050602

Krishna SM, Moxon JV, Jose RJ, Li J, Sahebkar A, Jaafari MR, Hatamipour M, Liu D, Golledge J (2018) Anionic nanoliposomes reduced atherosclerosis progression in Low Density Lipoprotein Receptor (LDLR) deficient mice fed a high fat diet. J Cell Physiol 233:6951–6964

Krzyściak W, Papież M, Bąk E, Morava E, Krzyściak P, Ligęzka A, Gniadek A, Vyhouskaya P, Janeczko J (2020) Sperm antioxidant biomarkers and their correlation with clinical condition and lifestyle with regard to male reproductive potential. J Clin Med 9:1–19. https://doi.org/10.3390/jcm9061785

Lana JFSD, Lana AVSD, Rodrigues QS, Santos GS, Navani R, Navani A, da Fonseca LF, Azzini GOM, Setti T, Mosaner T, Simplicio CL, Setti TM (2021) Nebulization of glutathione and N-Acetylcysteine as an adjuvant therapy for COVID-19 onset. Adv Redox Res. https://doi.org/10.1016/j.arres.2021.100015

Laurindo LF, de Carvalho GM, de Oliveira Zanuso B, Figueira ME, Direito R, de Alvares Goulart R, Buglio DS, Barbalho SM (2023) Curcumin-based nanomedicines in the treatment of inflammatory and immunomodulated diseases: an evidence-based comprehensive review. Pharmaceutics 15:229

Lee S, Hu L (2020) Nrf2 activation through the inhibition of Keap1–Nrf2 protein–protein interaction. Med Chem Res. https://doi.org/10.1007/s00044-020-02539-y

Lee D, Bae S, Hong D, Lim H, Yoon JH, Hwang O, Park S, Ke Q, Khang G, Kang PM (2013) H2O2-responsive molecularly engineered polymer nanoparticles as ischemia/reperfusion-targeted nanotherapeutic agents. Sci Rep 3:2233

Li C-W, Li L-L, Chen S, Zhang J-X, Lu W-L (2020) Antioxidant nanotherapies for the treatment of inflammatory diseases. Front Bioeng Biotechnol 8:200

Liang D, Wang W, Chen G, Li J, Dou G, Gan H, Han P, Du L, Gu R (2023) Cepharanthine dry powder inhaler for the treatment of acute lung injury. Molecules 28. https://doi.org/10.3390/molecules28114441

Liu B, Chen K (2024) Advances in hydrogel-based drug delivery systems. Gels. https://doi.org/10.3390/gels10040262

Long Y, Shi H, Ye J, Qi X (2025) Exploring strategies to prevent and treat ovarian cancer in terms of oxidative stress and antioxidants. Antioxidants. https://doi.org/10.3390/antiox14010114

López-Pedrera C, Villalba JM, Patiño-Trives AM, Luque-Tévar M, Barbarroja N, Aguirre MÁ, Escudero-Contreras A, Pérez-Sánchez C (2021) Therapeutic potential and immunomodulatory role of coenzyme q10 and its analogues in systemic autoimmune diseases. Antioxidants. https://doi.org/10.3390/antiox10040600

Luo M, Zhou L, Huang Z, Li B, Nice EC, Xu J, Huang C (2022) Antioxidant therapy in cancer: rationale and progress. Antioxidants. https://doi.org/10.3390/antiox11061128

Małkiewicz MA, Szarmach A, Sabisz A, Cubała WJ, Szurowska E, Winklewski PJ (2019) Blood-brain barrier permeability and physical exercise. J Neuroinflammation. https://doi.org/10.1186/s12974-019-1403-x

Maniam G, Mai C-W, Zulkefeli M, Dufès C, Tan DM-Y, Fu J-Y (2018) Challenges and opportunities of nanotechnology as delivery platform for tocotrienols in cancer therapy. Front Pharmacol 9:1358

Marino A, Battaglini M, Desii A, Lavarello C, Genchi G, Petretto A, Ciofani G (2021) Liposomes loaded with polyphenol-rich grape pomace extracts protect from neurodegeneration in a rotenone-based in vitro model of Parkinson's disease. Biomater Sci 9. https://doi.org/10.1039/d1bm01202a

McCarthy SD, Tilbury MA, Masterson CH, MacLoughlin R, González HE, Laffey JG, Wall JG, O'Toole D (2023) Aerosol delivery of a novel recombinant modified superoxide dismutase protein reduces oxidant injury and attenuates Escherichia coli induced lung injury in rats. J Aerosol Med Pulm Drug Deliv 36. https://doi.org/10.1089/jamp.2022.0069

Medoro A, Saso L, Scapagnini G, Davinelli S (2024) NRF2 signaling pathway and telomere length in aging and age-related diseases. Mol Cell Biochem 479:2597–2613. https://doi.org/10.1007/s11010-023-04878-x

Mendes LA, Farnesi-de-Assunção TS, Oliveira PA, Rotta IS, Moreto JA, Devienne KF, Paiva AD, Slade NBL (2024) Antitumor potential of lipid nanoformulations with natural antioxidants. Nano Trends 7:100040. https://doi.org/10.1016/j.nwnano.2024.100040

Menon VP, Sudheer AR (2007) Antioxidant and anti-inflammatory properties of curcumin. Adv Exp Med Biol. https://doi.org/10.1007/978-0-387-46401-5_3

Miatmoko A, Christy PK, Isnaini A, Hariawan BS, Cahyani DM, Ahmad M, Diyah NW, Adrianto MF, Deevi RK, Hamid IS, Ekowati J (2025) Characterization and in vitro anticancer study of PEGylated liposome dually loaded with ferulic acid and doxorubicin. Sci Rep 15. https://doi.org/10.1038/s41598-024-82228-7

Mishra SB, Kumari N (2021) Engineering of crystalline nano-suspension of lycopene for potential management of oxidative stress–linked diabetes in experimental animals. BioNanoScience 11: 345–354. https://doi.org/10.1007/s12668-021-00843-4

Mistretta M, Farini A, Torrente Y, Villa C (2023a) Multifaceted nanoparticles: emerging mechanisms and therapies in neurodegenerative diseases. Brain. https://doi.org/10.1093/brain/awad014

Mistretta M, Farini A, Torrente Y, Villa C (2023b) Multifaceted nanoparticles: emerging mechanisms and therapies in neurodegenerative diseases. Brain 146:2227–2240

Mohamed AA-R, Khater SI, Arisha AH, Metwally MMM, Mostafa-Hedeab G, El-Shetry ES (2021) Chitosan-stabilized selenium nanoparticles alleviate cardio-hepatic damage in type 2 diabetes mellitus model via regulation of caspase, Bax/Bcl-2, and Fas/FasL-pathway. Gene 768:145288

Montenegro L (2014) Nanocarriers for skin delivery of cosmetic antioxidants. J Pharm Pharmacogn Res. https://doi.org/10.56499/jppres14.033_2.4.73

Mooradian AD (2006) Antioxidants and diabetes. Nestle Nutr Workshop Ser Clin Perform Programme. https://doi.org/10.4103/2230-8210.104057

Moradi SZ, Momtaz S, Bayrami Z, Farzaei MH, Abdollahi M (2020) Nanoformulations of herbal extracts in treatment of neurodegenerative disorders. Front Bioeng Biotechnol 8:238

Morena AG, Pérez-Rafael S, Tzanov T (2022) Lignin-based nanoparticles as both structural and active elements in self-assembling and self-healing multifunctional hydrogels for chronic wound management. Pharmaceutics 14. https://doi.org/10.3390/pharmaceutics14122658

Morsy HM, Zaky MY, Yassin NYS, Khalifa AYZ (2025) Nanoparticle-based flavonoid therapeutics: pioneering biomedical applications in antioxidants, cancer treatment, cardiovascular health, neuroprotection, and cosmeceuticals. Int J Pharm 670:125135. https://doi.org/10.1016/j.ijpharm.2024.125135

Mundkar M, Bijalwan A, Soni D, Kumar P (2022) Neuroprotective potential of Moringa oleifera mediated by NF-kB/Nrf2/HO-1 signaling pathway: a review. J Food Biochem. https://doi.org/10.1111/jfbc.14451

Muzzalupo R, Tavano L (2015) Niosomal drug delivery for transdermal targeting: recent advances. Res Rep Transderm Drug Deliv. https://doi.org/10.2147/rrtd.s64773

Nair A, Singh R, Gautam N, Saxena S, Mittal S, Shah S, Talegaonkar S (2024) Multifaceted role of phytoconstituents based nano drug delivery systems in combating TNBC: a paradigm shift from chemical to natural. Naunyn Schmiedeberg's Arch Pharmacol 397:9207–9226. https://doi.org/10.1007/s00210-024-03234-0

Name A, Jyoti Mali MD (2022) A novel approach for niosomes as a transdermal drug delivery: the future scenario. 10:2320–2882

Nayab DE, Din F u, Ali H, Kausar WA, Urooj S, Zafar M, Khan I, Shabbir K, Khan GM (2023) Nano biomaterials based strategies for enhanced brain targeting in the treatment of neurodegenerative diseases: an up-to-date perspective. J Nanobiotechnol 21:477

Nayak D, Minz AP, Ashe S, Rauta PR, Kumari M, Chopra P, Nayak B (2016) Synergistic combination of antioxidants, silver nanoparticles and chitosan in a nanoparticle based formulation: characterization and cytotoxic effect on MCF-7 breast cancer cell lines. J Colloid Interface Sci 470:142–152. https://doi.org/10.1016/j.jcis.2016.02.043

Nelson HS (2016) Inhalation devices, delivery systems, and patient technique. Ann Allergy Asthma Immunol 117. https://doi.org/10.1016/j.anai.2016.05.006

Ngoc LTN, Moon JY, Lee YC (2023) Antioxidants for improved skin appearance: intracellular mechanism, challenges and future strategies. Int J Cosmet Sci 45. https://doi.org/10.1111/ics.12848

Nguyen VH, Thuy VN, Van TV, Dao AH, Lee BJ (2022) Nanostructured lipid carriers and their potential applications for versatile drug delivery via oral administration. Open Nano. https://doi.org/10.1016/j.onano.2022.100064

Panaiotov S, Tancheva L, Kalfin R, Petkova-Kirova P (2024) Zeolite and neurodegenerative diseases. Molecules. https://doi.org/10.3390/molecules29112614

Pandey M, Choudhury H, Verma RK, Chawla V, Bhattamisra SK, Gorain B, Raja MAG, Amjad MW (2020) Nanoparticles based intranasal delivery of drug to treat Alzheimer's disease: a recent update. CNS Neurol Disord-Drug Targets 19:648–662

Pavlidis G (2022) Oxidative stress and antioxidant therapy in cardiovascular diseases—clinical challenge. J Clin Med. https://doi.org/10.3390/jcm11133784

Peoples JN, Saraf A, Ghazal N, Pham TT, Kwong JQ (2019) Mitochondrial dysfunction and oxidative stress in heart disease. Exp Mol Med. https://doi.org/10.1038/s12276-019-0355-7

Pfeffer PE, Lu H, Mann EH, Chen YH, Ho TR, Cousins DJ, Corrigan C, Kelly FJ, Mudway IS, Hawrylowicz CM (2018) Effects of vitamin D on inflammatory and oxidative stress responses of human bronchial epithelial cells exposed to particulate matter. PLoS One 13. https://doi.org/10.1371/journal.pone.0200040

Pham SH, Choi Y, Choi J (2020) Stimuli-responsive nanomaterials for application in antitumor therapy and drug delivery. Pharmaceutics. https://doi.org/10.3390/pharmaceutics12070630

Pinto S, Gaspar MM, Ascensão L, Faísca P, Reis CP, Pacheco R (2022) Nanoformulation of seaweed Eisenia bicyclis in albumin nanoparticles targeting cardiovascular diseases: in vitro and in vivo evaluation. Mar Drugs 20:608. https://doi.org/10.3390/MD20100608/S1

Policar C, Bouvet J, Bertrand HC, Delsuc N (2022) SOD mimics: from the tool box of the chemists to cellular studies. Curr Opin Chem Biol. https://doi.org/10.1016/j.cbpa.2021.102109

Pranantyo D, Yeo CK, Wu Y, Fan C, Xu X, Yip YS, Vos MIG, Mahadevegowda SH, Lim PLK, Yang L, Hammond PT, Leavesley DI, Tan NS, Chan-Park MB (2024) Hydrogel dressings with intrinsic antibiofilm and antioxidative dual functionalities accelerate infected diabetic wound healing. Nat Commun 15. https://doi.org/10.1038/s41467-024-44968-y

Pruteanu LL, Bailey DS, Grădinaru AC, Jäntschi L (2023) The biochemistry and effectiveness of antioxidants in food, fruits, and marine algae. Antioxidants. https://doi.org/10.3390/antiox12040860

Przybylska S, Tokarczyk G (2022) Lycopene in the prevention of cardiovascular diseases. Int J Mol Sci. https://doi.org/10.3390/ijms23041957

Ratnam DV, Chandraiah G, Sonaje K, Viswanad B, Bhardwaj V, Ramarao P, Ravikumar MNV (2008) A potential therapeutic strategy for diabetes and its complications in the form of co-encapsulated antioxidant nanoparticles (NanoCAPs) of ellagic acid and coenzyme Q10: preparation and evaluation in streptozotocin induced diabetic rats. J Biomed Nanotechnol 4: 33–43

Rauf A, Muhammad I, Masood Sadiq B, Muhammad N, Dennis GP, Mubarak MS (2018) Resveratrol as an anti-cancer agent: a review. Crit Rev Food Sci Nutr 58:1428–1447. https://doi.org/10.1080/10408398.2016.1263597

Razzaghi-Asl N, Garrido J, Khazraei H, Borges F, Firuzi O (2013) Antioxidant properties of Hydroxycinnamic acids: a review of structure- activity relationships. Curr Med Chem 20. https://doi.org/10.2174/09298673113209990141

Rocha S, Luísa Corvo M, Freitas M, Fernandes E (2024) Liposomal quercetin: a promising strategy to combat hepatic insulin resistance and inflammation in type 2 diabetes mellitus. Int J Pharm 661:124441. https://doi.org/10.1016/j.ijpharm.2024.124441

Rocks N, Bekaert S, Coia I, Paulissen G, Gueders M, Evrard B, Van Heugen JC, Chiap P, Foidart JM, Noel A, Cataldo D (2012) Curcumin-cyclodextrin complexes potentiate gemcitabine effects in an orthotopic mouse model of lung cancer. Br J Cancer 107. https://doi.org/10.1038/bjc.2012.379

Roshani M, Jafari A, Loghman A, Sheida AH, Taghavi T, Tamehri Zadeh SS, Hamblin MR, Homayounfal M, Mirzaei H (2022) Applications of resveratrol in the treatment of gastrointestinal cancer. Biomed Pharmacother 153:113274. https://doi.org/10.1016/j.biopha.2022.113274

Saito H, Tanaka K, Iwasaki T, Oda A, Watanabe S, Kanno M, Kimura H, Shimabukuro M, Asahi K, Watanabe T, Kazama JJ (2021) Xanthine oxidase inhibitors are associated with reduced risk of cardiovascular disease. Sci Rep 11. https://doi.org/10.1038/s41598-020-80835-8

Santi C, Scimmi C, Sancineto L (2021) Ebselen and analogues: pharmacological properties and synthetic strategies for their preparation. Molecules. https://doi.org/10.3390/molecules26144230

Saxena S, Singh R, Dutta D, Gautam N, Setya S, Talegaonkar S (2023) Nutraceuticals and their applications: recent trends and challenges. In: Anxiety, gut microbiome, and nutraceuticals: recent trends and clinical evidence. https://doi.org/10.1201/9781003333821-1

Schichlein KD, Smith GJ, Jaspers I (2023) Protective effects of inhaled antioxidants against air pollution-induced pathological responses. Respir Res. https://doi.org/10.1186/s12931-023-02490-7

Senghore T, Li YF, Sung FC, Tsai MH, Hua CH, Liu CS, Huang RJ, Yeh CC (2018) Biomarkers of oxidative stress associated with the risk of potentially malignant oral disorders. Anticancer Res 38. https://doi.org/10.21873/anticanres.12771

Sercombe L, Veerati T, Moheimani F, Wu SY, Sood AK, Hua S (2015) Advances and challenges of liposome assisted drug delivery. Front Pharmacol. https://doi.org/10.3389/fphar.2015.00286

Sezgin-Bayindir Z, Losada-Barreiro S, Bravo-Díaz C, Sova M, Kristl J, Saso L (2021) Nanotechnology-based drug delivery to improve the therapeutic benefits of NRF2 modulators in cancer therapy. Antioxidants. https://doi.org/10.3390/antiox10050685

Shafras M, Sabaragamuwa R, Suwair M (2024) Role of dietary antioxidants in diabetes: an overview. Food Chem Adv 4:100666. https://doi.org/10.1016/j.focha.2024.100666

Shekh-Ahmad T, Lieb A, Kovac S, Gola L, Wigley WC, Abramov AY, Walker MC (2019) Combination antioxidant therapy prevents epileptogenesis and modifies chronic epilepsy. Redox Biol 26. https://doi.org/10.1016/j.redox.2019.101278

Siafaka PI, Mutlu G, Okur NÜ (2020) Alzheimer's disease and its related dementia types: a review on their management via nanotechnology based therapeutic strategies. Curr Alzheimer Res 17: 1239–1261

Siddiqui L, Mishra H, Mishra PK, Iqbal Z, Talegaonkar S (2018) Novel 4-in-1 strategy to combat colon cancer, drug resistance and cancer relapse utilizing functionalized bioinspiring lignin nanoparticle. Med Hypotheses 121:10–14. https://doi.org/10.1016/j.mehy.2018.09.003

Singh J, Mittal P, Vasant Bonde G, Ajmal G, Mishra B (2018) Design, optimization, characterization and in-vivo evaluation of quercetin enveloped Soluplus®/P407 micelles in diabetes treatment. Artif Cells Nanomed Biotechnol 46:S546–S555

Singh S, Sachan K, Verma S, Singh N, Singh PK (2023) Cubosomes: an emerging and promising drug delivery system for enhancing cancer therapy. Curr Pharm Biotechnol 25. https://doi.org/10.2174/0113892010257937231025065352

Sivadasan D, Sultan MH, Alqahtani SS, Javed S (2023) Cubosomes in drug delivery—a comprehensive review on its structural components, preparation techniques and therapeutic applications. Biomedicines. https://doi.org/10.3390/biomedicines11041114

So MM, Mansukhani NA, Peters EB, Albaghdadi MS, Wang Z, Rubert Pérez CM, Kibbe MR, Stupp SI (2018) Peptide amphiphile nanostructures for targeting of atherosclerotic plaque and drug delivery. Adv Biosyst 2:1700123

Sreena R, Nathanael AJ (2023) Biodegradable biopolymeric nanoparticles for biomedical applications-challenges and future outlook. Materials. https://doi.org/10.3390/ma16062364

Sridhar K, Stephen Inbaraj B, Kumar N, Tripathi M (n.d.) Biopolymeric nanoparticles based effective delivery of bioactive compounds toward the sustainable development of anticancerous therapeutics

Srivastava RM, Singh S, Dubey SK, Misra K, Khar A (2011) Immunomodulatory and therapeutic activity of curcumin. Int Immunopharmacol 11:331–341. https://doi.org/10.1016/j.intimp.2010.08.014

Stone W, Smith M (2004) Therapeutic uses of antioxidant liposomes. Mol Biotechnol 27:217–230. https://doi.org/10.1385/MB:27:3:217

Suntres ZE (2011) Liposomal antioxidants for protection against oxidant-induced damage. J Toxicol. https://doi.org/10.1155/2011/152474

Talasaz AH, Khalili H, Fahimi F, Mojtaba S (2011) Potential role of N-acetylcysteine in cardiovascular disorders. Therapy. https://doi.org/10.2217/thy.11.12

Taliyan R, Kakoty V, Sarathlal KC, Kharavtekar SS, Karennanavar CR, Choudhary YK, Singhvi G, Riadi Y, Dubey SK, Kesharwani P (2022) Nanocarrier mediated drug delivery as an impeccable therapeutic approach against Alzheimer's disease. J Control Release 343:528–550

Tan S, Liu Z, Cong M, Zhong X, Mao Y, Fan M, Jiao F, Qiao H (2024) Dandelion-derived vesicles-laden hydrogel dressings capable of neutralizing Staphylococcus aureus exotoxins for the care of invasive wounds. J Control Release 368. https://doi.org/10.1016/j.jconrel.2024.02.045

Tanaka A, Node K (2022) Xanthine oxidase inhibition for cardiovascular disease prevention. Lancet. https://doi.org/10.1016/S0140-6736(22)01778-0

Tavano L, Muzzalupo R, Picci N, De Cindio B (2014) Co-encapsulation of lipophilic antioxidants into niosomal carriers: percutaneous permeation studies for cosmeceutical applications. Colloids Surf B Biointerfaces 114. https://doi.org/10.1016/j.colsurfb.2013.09.055

Tenório MCDS, Graciliano NG, Moura FA, de Oliveira ACM, Goulart MOF (2021) N-acetylcysteine (Nac): impacts on human health. Antioxidants. https://doi.org/10.3390/antiox10060967

Tejwan N, Sadhukhan P, Sharma A, Singh Th A, Hatimuria M, Pabbathi A, Das J, Sil PC (2022) pH-responsive and targeted delivery of rutin for breast cancer therapy via folic acid-functionalized carbon dots. Diam Relat Mater 129:109346. https://doi.org/10.1016/j.diamond.2022.109346

Tieu S, Charchoglyan A, Paulsen L, Wagter-Lesperance LC, Shandilya UK, Bridle BW, Mallard BA, Karrow NA (2023) N-Acetylcysteine and its immunomodulatory properties in humans and domesticated animals. Antioxidants. https://doi.org/10.3390/antiox12101867

Tønnesen HH, Másson M, Loftsson T (2002) Studies of curcumin and curcuminoids. XXVII. Cyclodextrin complexation: solubility, chemical and photochemical stability. Int J Pharm 244:127–135. https://doi.org/10.1016/S0378-5173(02)00323-X

Touitou E, Dayan N, Bergelson L, Godin B, Eliaz M (2000) Ethosomes — novel vesicular carriers for enhanced delivery: characterization and skin penetration properties. J Control Release 65: 403–418. https://doi.org/10.1016/S0168-3659(99)00222-9

Umar H, Wahab HA, Gazzali AM, Tahir H, Ahmad W (2022) Cubosomes: design, development, and tumor-targeted drug delivery applications. Polymers (Basel). https://doi.org/10.3390/polym14153118

Uttara B, Singh A, Zamboni P, Mahajan R (2009) Oxidative stress and neurodegenerative diseases: a review of upstream and downstream antioxidant therapeutic options. Curr Neuropharmacol 7. https://doi.org/10.2174/157015909787602823

Varadharaj S, Kelly OJ, Khayat RN, Kumar PS, Ahmed N, Zweier JL (2017) Role of dietary antioxidants in the preservation of vascular function and the modulation of health and disease. Front Cardiovasc Med. https://doi.org/10.3389/fcvm.2017.00064

Vashist A, Manickam P, Raymond AD, Arias AY, Kolishetti N, Vashist A, Arias E, Nair M (2023) Recent advances in nanotherapeutics for neurological disorders. ACS Appl Bio Mater 6:2614–2621. https://doi.org/10.1021/acsabm.3c00254

Velaga SP, Djuris J, Cvijic S, Rozou S, Russo P, Colombo G, Rossi A (2018) Dry powder inhalers: an overview of the in vitro dissolution methodologies and their correlation with the biopharmaceutical aspects of the drug products. Eur J Pharm Sci 113. https://doi.org/10.1016/j.ejps.2017.09.002

Verma P, Pathak K (2010) Therapeutic and cosmeceutical potential of ethosomes: an overview. J Adv Pharm Technol Res. https://doi.org/10.4103/0110-5558.72415

Vladu AF, Ficai D, Ene AG, Ficai A (2022) Combination therapy using polyphenols: an efficient way to improve antitumoral activity and reduce resistance. Int J Mol Sci. https://doi.org/10.3390/ijms231810244

Wang W, Kang PM (2020) Oxidative stress and antioxidant treatments in cardiovascular diseases. Antioxidants. https://doi.org/10.3390/antiox9121292

Wang M, Marepally SK, Vemula PK, Xu C (2016) Inorganic nanoparticles for transdermal drug delivery and topical application. In: Nanoscience in dermatology. https://doi.org/10.1016/B978-0-12-802926-8.00005-7

Wu L, Hu Z, Song X, Liao YJ, Xiahou JH, Li Y, Zhang ZH (2024) Targeting Nrf2 signaling pathways in the role of bladder cancer: from signal network to targeted therapy. Biomed Pharmacother. https://doi.org/10.1016/j.biopha.2024.116829

Yang J, Zhang G, Dong D, Shang P (2018) Effects of iron overload and oxidative damage on the musculoskeletal system in the space environment: data from spaceflights and ground-based simulation models. Int J Mol Sci. https://doi.org/10.3390/ijms19092608

Zahra KF, Lefter R, Ali A, Abdellah EC, Trus C, Ciobica A, Timofte D (2021) The involvement of the oxidative stress status in cancer pathology: a double view on the role of the antioxidants. Oxidative Med Cell Longev. https://doi.org/10.1155/2021/9965916

Zazuli Z, Hartati R, Rowa CR, Asyarie S, Satrialdi (2025) The potential application of nanocarriers in delivering topical antioxidants. Pharmaceuticals. https://doi.org/10.3390/ph18010056

Zduńska K, Dana A, Kolodziejczak A, Rotsztejn H (2018) Antioxidant properties of ferulic acid and its possible application. Skin Pharmacol Physiol. https://doi.org/10.1159/000491755

Zhang D, Zhang J, Zeng J, Li Z, Zuo H, Huang C, Zhao X (2019) Nano-gold loaded with resveratrol enhance the anti-hepatoma effect of resveratrol in vitro and in vivo. J Biomed Nanotechnol 15: 288–300

Zhang L, Yang S, Huang L, Ho PCL (2020) Poly (ethylene glycol)-block-poly (D, L-lactide) (PEG-PLA) micelles for brain delivery of baicalein through nasal route for potential treatment of neurodegenerative diseases due to oxidative stress and inflammation: an in vitro and in vivo study. Int J Pharm 591. https://doi.org/10.1016/j.ijpharm.2020.119981

Zhang Y, Wang H, Yang R, Zhang Y, Chen Y, Jiang C, Li X (2024) Synergistic therapeutic effects of D-Mannitol–Cerium–Quercetin (Rutin) coordination polymer nanoparticles on acute lung injury. Molecules 29. https://doi.org/10.3390/molecules29122819

Zhen AX, Piao MJ, Kang KA, Fernando PDSM, Kang HK, Koh YS, Yi JM, Hyun JW (2019) Niacinamide protects skin cells from oxidative stress induced by particulate matter. Biomol Ther (Seoul) 27. https://doi.org/10.4062/biomolther.2019.061

Zhu Q, Xiao Y, Jiang M, Liu X, Cui Y, Hao H, Flaker GC, Liu Q, Zhou S, Liu Z (2022) N-acetylcysteine attenuates atherosclerosis progression in aging LDL receptor deficient mice with preserved M2 macrophages and increased CD146. Atherosclerosis 357. https://doi.org/10.1016/j.atherosclerosis.2022.08.008

Zoi V, Galani V, Lianos GD, Voulgaris S, Kyritsis AP, Alexiou GA (2021) The role of curcumin in cancer treatment. Biomedicine 9:1086

Chapter 14
Structural and Clinical Perspectives of Antioxidants in Oxidative Stress-Mediated Ailments

Gul Naz Fatima, Hera Fatma, and Shailendra K. Saraf

Abbreviations

AO	Antioxidant
CAT	Catalase
FRs	Free radicals
GPx	Glutathione peroxidase
GST	Glutathione S-transferase
N-AOs	Natural antioxidants
NO	Nitric oxide
PDB	Protein Data Bank
RNS	Reactive nitrogen species
ROM	Reactive oxygen metabolite
ROS	Reactive oxygen species
RSS	Reactive sulphur species
S-AOs	Synthetic antioxidants
SAR/QSAR	Structure-activity relationship/Quantitative structure-activity relationship
SOD	Superoxide dismutase
SVR	Support vector regression
TrxR	Thioredoxin reductase

G. N. Fatima · H. Fatma · S. K. Saraf (✉)
Division of Pharmaceutical Chemistry, Faculty of Pharmacy, Babu Banarasi Das Northern India Institute of Technology, Lucknow, UP, India

N. Srivastava, S. A. Saraf (eds.), *Pharmacological Strategies for Prevention and Treatment of Oxidative Stress*, Oxidative Stress in Applied Basic Research and Clinical Practice, https://doi.org/10.1007/978-3-032-19837-2_14

14.1 Introduction

Antioxidants delay or prevent oxidation when present in modest amounts relative to the substrate (Rice-Evans and Burdon 1994). They prevent cell damage by inhibiting molecular oxidation. Oxidative reactions produce free radicals (FRs), which are extremely reactive entities with one or more unpaired electrons (Mamta et al. 2014). External sources like ozone, air pollution, X-rays, smoking, and industrial chemicals (Preiser 2012), the peroxisomes, the phagocytosis, the mitochondria, enzymes including xanthine oxidase, and conditions like ischemia generate FR (Lobo et al. 2010). The ROS, reactive oxygen species, the RNS, reactive nitrogen species, and the RSS, reactive sulphur species behave as FRs (Atta et al. 2017). The ROS include FR as well as non-FR oxygen-containing species such as H_2O_2, OH, O^{2-}, and ½ O_2 (Poljsak et al. 2013). They rapidly interact with the membrane, causing cell degeneration and cell death (Sindhi et al. 2013). The polyunsaturated fatty acids are among the preferred oxidation targets for ROS (Rahal et al. 2014). An imbalance in ROS/RNS develops oxidative stress within the cells (Pisoschi and Pop 2015). Gene expression changes help cells tolerate mild oxidative stress by increasing antioxidant defense systems. However, excessive oxidative stress can damage cells (Willcox et al. 2004) by destruction of proteins, DNA, and lipids (Halliwell 1996). FR can be scavenged either by enzymatic or non-enzymatic processes. Nitric oxide (NO) is rapidly oxidized by oxyhemoglobin to generate nitrate (NO_3), the body's main stable NO oxidation product (Fang et al. 2002). Enzymes such as superoxide dismutase-SOD, glutathione peroxidase-GP_X, and catalase-CAT form the enzyme-antioxidant system (Salehi et al. 2018). The non-enzymatic system includes lipophilic and hydrophilic antioxidants (Sies 1991) which block the FR chain reactions by interacting with the ROS and the RON (Liguori et al. 2018). The antioxidant system includes both the endogenous, and dietary sources that protect the body from FRs (Lachance et al. 2001).

The oxidative stress is capable of causing a number of diseases thereby affecting a large population, directly or indirectly. Thus, the measurement of oxidant levels is of prime importance. A new reactive oxygen metabolite (ROMs) kit has been developed to measure the plasma and other body fluid oxidant levels (Rajendran et al. 2014). Various antioxidants, of either natural or synthetic origin, have come in to clinical use for the prevention as well as treatment of oxidative stress-mediated disorders and diseases. These agents are highly effective and the search for more refined versions and new candidates continues.

14.2 Classification of Antioxidants

The antioxidants can be classified into diverse classes based on their origin. The following is a list of the classification (Flieger et al. 2021).

I. Natural antioxidants:

   (a) Exogeneous-

      (i) Vitamins- Ascorbic acid, ascorbate, tocopherols, tocotrienols, retinol.
      (ii) Trace elements- Zinc, selenium, copper, iron, manganese.
      (iii) Carotenoids- Zeaxanthin, lutein, lycopene, α and β-carotenes.
      (iv) Polyphenols- Flavones, flavanol, flavanones.

   (b) Endogenous-

      (i) Enzymatic

         - Primary defense system- Catalase, superoxide dismutase, glutathione peroxidase.
         - Secondary defense system- glutathione reductase, G-6-PD—Glucose-6-phosphate dehydrogenase.

      (ii) Non-enzymatic

         - Low molecular weight.

            – Lipophilic- Ubiquinone, lipoic acid, plasmalogen.
            – Hydrophilic- Glutathione, melatonin, conjugate bilirubin, uric acid, amino acids.

         - Metal binding protein- Ferritin, albumin, ceruloplasmin.

II. Synthetic antioxidants:

   (a) Phenolic- Tert-butylhydroquinone-TBHQ, butylated hydroxyanisole-BHA, butylated hydroxytoluene-BHT.
   (b) Nano-antioxidants- Metallic nanoparticles, oxides, antioxidant functionalized NPs.

There are anti-oxidants, that can be classified based on the nuclei present in the structure, as given in Table 14.1. In addition to the nuclei as given in the table, the antioxidants may also contain pyran, imidazole, pyrazine, benzothiazole, pyrazole, pyrimidine, piperazine, purine, pyridine, indole, quinoline, as found in kojic acid, ergothioneine, cyclo (his-pro), pramipexole, edaravone, allopurinol, tirilazad, pentoxifylline, niacinamide, melatonin, rebamipide, respectively.

14.3 Measurement of the Antioxidant Activity: Total Antioxidant Capacity (TAC) (Sun et al. 2018)

Chemical-based antioxidant assay: -

1. Hydrogen atom transfer- HAT-based assay: Oxygen radical absorbance capacity assays (ORAC) assay.

Table 14.1 List of few antioxidants classified based on the nucleus present in their structure

Nucleus	Compounds	References
Cyclohexene	Zeaxanthin, Lutein, Idebenone, Ubiquinone	Gueven et al. (2021), Mrowicka et al. (2022)
Chromen-4-one	Baicalein, Quercetin, Fisetin, Silibinin, Tricetin	Gao et al. (1999)
Chromen-6-yl	D-α-Tocopherol acetate, D- α-Tocopherol succinate	Cheng et al. (2016)
Phenol	Apocynin, Hydroxytyrosol, Bisphenol A, Butylated hydroxyanisole	Amjad et al. (2020)
Sulfanyl	Tetradecylthioacetic acid, Acetylcysteine zinc Acetylcysteine, Probucol	Ezeriņa et al. (2018)
Furan	Sodium ascorbate, Ascorbic acid Magnesium ascorbate, Ferrous ascorbate, Sodium ascorbate	Clark et al. (2010), Jakubek et al. (2023)
Azanium	Ergothioneine, Acetylcanitine	Cheah and Halliwell (2012)
Phenanthrene	Carnosic acid, Ginsenosides Tirilazad, Sitosterol	Loussouarn et al. (2017), Morshed et al. (2023)

2. Single electron transfer- SET based assay:
   (a) Spectroscopic methods- These include total radical trapping antioxidant parameter assay, the TRAP assay, total oxyradical scavenging capacity assay, the TOSC assay, ferric reducing antioxidant power assay, the FRAP assay, and crocin bleaching assay, the CBA.
   (b) Electrochemical methods- The methods include the ferricyanide-prussian blue assay, cyclic voltammetry-based assay, differential pulse-based assay, and the DPV assay.
   (c) Nanotechnological methods- These include silver nanoparticle-based assay, the SNPAC assay, gold nanoparticle-based assay, and the AuNPs assay.
3. Mixed mode assay- The assay methods include 2,2′-azinobis (3-ethylbenzothiazoline-6-sulfonic acid) assay, the ABTS assay, troloxequivalent antioxidant capacity assay, the TEAC assay, and 2,2-diphenyl-1-picrylhydrazyl radical scavenging assay, the DPPH assay.

In-vivo antioxidant assay:

1. Cellular antioxidant activity-CAA assay: The assay employs a fluorescent probe known as dichlorofluorescin-DCF to determine the efficacy of antioxidants (Kellett et al. 2018).
2. Biomarkers of oxidative stress and the related methods: The assay methods include ROS, RNS and superoxide anion radical scavenging assays, the luminol-based chemiluminescence assay, and the nitroblue tetrazolium-based chemiluminescence assay.

14.4 Mechanism of Action

The antioxidants can reduce oxidative damage by either of the following mechanisms (Lü et al. 2010).

(a) *FR scavenging*: The $ABTS^+$, the DPPS, and the N, N-dimethyl-*p*-phenylenediamine dihydrochloride cation radical, $DMPD^+$ assays are based on the use of superoxide anion radical, $O_2^{-\bullet}$. These are the common methods to determine antioxidant capacity of vegetable extract, food or beverages (Ak and Gülçin 2008; Lü et al. 2010).
(b) *Inhibition of FR generating enzymes*: The NADPH oxidase is a group of enzymes that catalyzes transfer one electron from the donor NADPH to a molecule of oxygen, resulting in generating $O_2^{-\bullet}$ (Lü et al. 2010). Xanthine oxidase (XO), is such an example, which catalyzes the biosynthesis of uric acid, oxidation of hypoxanthine and xanthine to uric acid, thereby, generating superoxide radical and H_2O_2, resulting in elevating the oxidation level (Berry and Hare 2004). Treatment with an XO inhibitor significantly prevents atherosclerosis and the onset of endothelial dysfunction in mice (Schröder et al. 2006). The activity of NADPH oxidase can be quantified by measuring the absorbance at a wavelength of 340 nm in a buffer system of Tris-EDTA (Lü et al. 2010). The superoxide radical production can also be determined from the reaction mixture by adding cytochrome C (Fang et al. 2005).
(c) *Activation of internal antioxidant enzyme*: The intracellular antioxidant enzyme systems are key protectors against the ROS. They are synthesized within the cell and play a vital role in defense against FRs. The key enzymes in the defense system include CAT, SOD, TrxR (thioredoxin reductase), GPX, and glutathione S-transferase (GST). The enzyme SOD converts two superoxide radicals into one molecule of hydrogen peroxide and a molecule of oxygen. The GPX and GST are responsible for removing peroxides generated during metabolism. The TrxR serves as electron donor to ribonucleotide reductase, and is the key enzyme for DNA synthesis (Lü et al. 2010).

14.5 Natural Antioxidants (N-AOs) Versus Synthetic Antioxidants (S-AOs)

There are numerous differences between natural and synthetically derived antioxidants. Other than the source, the antioxidants differ in safety and potential, utility, cost-effectiveness and impact on the environment.

The primary antioxidants found in diverse plant species include phenolic compounds, phenolic acids, terpenoids, flavonoids, coumarins, and sesquiterpene lactones (Augustyniak et al. 2010), while the synthetic ones include BHA, BHT, and TBHQ (Xu et al. 2021). The S-AOs employed are intended to prevent rancidity and deterioration. Their extensive use is due to their chemical and thermal stability, enabling large-scale production, as well as their significant antioxidant properties (Gonçalves-Filho and De Souza 2022). However, they possess the drawback that their extended use may adversely affect living cells. For example, prolonged use of BHA has demonstrated an onset of stomach cancers, DNA damage, and several other detrimental effects in animal feeding trials (Xu et al. 2021). Because of the safety issues of the S-AOs, the FDA has set a limit on its content in food items, which should not exceed 0.02% of the total oil or fat content of the food (Galal et al. 2024). The thermal stability of synthetic and natural antioxidants has been investigated using gravimetric thermoanalytical techniques, employing both dynamic and isothermal analysis at 110 °C. The synthetic antioxidants exhibited resistance in the order: PG>TBHA>BHA>BHT, and the natural antioxidants in the order: α-tocopherol>caffeic acid>ferulic acid>gallic acid (Viana Da Silva et al. 2022). In comparing consumer demand, N-AOs are favored over S-AOs. This is especially evident in the food industry (Caleja et al. 2017) as there is minimal variation in antioxidant potential while ensuring safety. Also, no significant changes in flavor and color of the items are observed due to them (Jayathilakan et al. 2007). The cosmetic and pharmaceutical industry is observed to follow the same trend. The N-AOs are predominantly utilized in formulations intended for anti-ageing and skin protection (Choi et al. 2024). The antioxidants also present emerging environmental concerns, particularly the synthetic phenolic antioxidants-SPAs, amine antioxidants-AAs, and organophosphite antioxidants-OPAs. Their extensive utilization renders them substantial indoor pollutants, raising considerable concern for individuals who predominantly inhabit indoor settings (Ji et al. 2023). The N-AOs may incur higher expenses attributed to sourcing and processing costs. For instance, extracting antioxidants from rare plants or fruits can incur significant costs. The S-AOs are generally more cost-effective to manufacture (Lourenço et al. 2019).

14.6 Inorganic Antioxidants

Inorganic small molecules, particularly the organo-selenium and the organosulfur compounds, have been found to possess significant antioxidant potential. Their advantages include improved bioavailability and increased integration into amino acids and proteins. Inorganic selenium compounds such as selenite, selenate, selenium dioxide, and sodium selenide have been clinically established in the treatment of cancer. Studies suggest that they may act as pro-oxidants, leading to DNA damage and cellular apoptosis, a mechanism crucial in cancer therapy. Sodium selenite is widely utilized in oncological therapy. This underscores the potential of inorganic small molecules as beneficial dietary supplements that can suppress tumour cell proliferation (Ramoutar and Brumaghim 2010).

14.7 Antioxidants in Traditional Indian (Ayurvedic) Medicine

Amritotharanam Kashyam, an ayurvedic decoction of herbs, is a potent antioxidant (Raman et al. 2023) with numerous health benefits, and is valuable as hepatoprotective and in lowering serum cholesterol and sugar levels (Dhane et al. 2023). Ashwagandha churna is also a powerful antioxidant that offers protection against cellular damage caused by FR (Singh et al. 2011). Ashwagandha is used in the treatment of several health conditions such as Alzheimer's disease, Huntington's disease, Parkinson's disease, alcohol withdrawal syndrome, and obsessive-compulsive disorder. It has effects like anti-inflammatory, antibacterial, infertility, anticancer, cardioprotective, and many more (Mikulska et al. 2023). It is also useful in conditions of stress, anxiety, and sleep disorders (Health 2023). Amalaki rasayana is an Indian traditional drug used as a rejuvenating medicine. It has been reported to exhibit promising clinical use in reversing ventricular hypertrophy and age-related cardiac dysfunction in rat models (Kumar et al. 2017) and as an anti-aging agent (Samarakoon et al. 2011). Arjuna Ksheera Paka is an ayurvedic formulation known for its cardioprotective properties. The extract of *T. arjuna* exhibits antioxidant activity by scavenging DPPH FRs in a linear fashion. Polyphenols like phenolic acids (gallic acid) and hydrophilic flavonoid (quercetin) possess FR scavenging properties (Dube et al. 2017). Ksheera Paka formulations primarily utilize cow milk as the base ingredient (Gadgil and Nene 2013). Bhasmas are ayurvedic metallic or mineral products prepared using herbal juice or decoction (Chavare et al. 2017). They have analgesic and antioxidant activity, and display stimulatory effects on peritoneal macrophages (Pal et al. 2014). *Tamra bhasma, lauha bhasma, mandura bhasma, abhraka bhasma* etc. are the examples of bhasmas used as antioxidants. Similarly, *triphala* is an ayurvedic triherbal formulation composed of an equiproportional mixture of *amalaka, haritaki* and *bibhitaki* fruits. It is rich in antioxidants and is used as a food supplement (Parveen et al. 2018). It possesses numerous clinical benefits because of its gastrointestinal, anti-obesogenic, and antidiabetic potential, along with stress-reducing capabilities, cardiovascular benefits, radioprotective effects, antineoplastic, antimicrobial, and anti-inflammatory effects, among various others (Peterson et al. 2017).

14.8 Food as Antioxidants

The colored fruit drinks from grapes, various berries, and pomegranate are popular because of their capability to scavenge FRs. The commercially available pomegranate drinks have three times the antioxidant activity of red wine and green tea. The main antioxidants in pomegranate juice include hydrolysable tannins, anthocyanins, and ellagic acid derivatives (Gil et al. 2000). Of the Concord, purple, green, and red grapes, the green and red grapes have lower TAC than Concord and purple grapes, which contain anthocyanins and flavanols (Callaghan et al. 2013). Blueberries, due to their high polyphenolic content, are considered superfruits with strong in-vitro antioxidant potential (Huang et al. 2012; Kalt et al. 2020). Blackberries, due to their high phenolic content, are among the most antioxidant-rich fruits. Strawberry fruit's bioactive phenolic compounds, such as hydroxycinnamic acids, ellagic acids, ellagitannins, flavanols, and anthocyanins, have the property to scavenge chemically produced radicals, thus preventing LDL oxidation (Huang et al. 2012). Olives are rich in lipophilic and hydrophilic phenolics. Olive polyphenols reduce oxidative damage that causes wrinkles, skin dryness, and hyperproliferation (Ghanbari et al. 2012). Fig fruit has high antioxidant capacity. Dried figs have more antioxidant activity than fresh ones (Arvaniti et al. 2019). Since the peel is responsible for figs' antioxidant properties, dark-skinned figs contain more polyphenols, flavonoids, and anthocyanins, making them stronger antioxidants (Harzallah et al. 2016). Cherries display antioxidant actions due to the presence of anthocyanins and phenolic acids (Ferretti et al. 2010). Pears have high levels of total phenolics, flavonoids, and triterpenes, indicating their antioxidant capabilities, both in the peels and the meat (Li et al. 2014). Watermelon is rich in lycopene, vitamin A, vitamin C, and antioxidants. It lowers the levels of triglycerides, LDL, thiobarbituric acid reactive compounds, and helps boost antioxidant capacity in obese individuals (Manivannan et al. 2020). Guava leaf polysaccharides (GLP) are efficient dietary antioxidants found in guava. The sulfated GLP has been found to effectively fight DPPH, hydroxyl, and alkyl radicals (Kumar et al. 2021). Oranges are rich in vitamin C and polyphenols, such as hydroxycinnamic acids (Franke et al. 2005; Park et al. 2014).

Bioactive substances including polyphenols and tocopherols are found abundantly in cocoa beans (Oracz and Żyżelewicz 2020). Apricot contains abundant bioactive phytochemicals including carotenes, lycopene, β-cryptoxanthin, phytoene, phytofluene, lutein, flavonoids, and phenolic compounds (Alajil et al. 2021). Tomatoes contains vitamin E, ascorbic acid, flavonoids, carotenoids, and phenolic acids (Bianchi et al. 2023). Mango, papaya, and green leafy vegetables are among others containing phytochemicals that act as antioxidants. Due to the presence of a large range of constituents with antioxidant capacity, vegetables and fruits are the natural source of antioxidants useful in the prevention of oxidative-stress mediated disorders (Table 14.2).

14.9 Quantitative Structure Activity Relationship

Although numerous experimental techniques exist to assess the antioxidant activity of molecules, these methods tend to be both time consuming and expensive. Alternatively, employing the quantitative structure-activity relationship (QSAR) methods allows the screening of compounds to assess their potency (Deeb and Goodarzi 2017). The QSAR aims to establish a relationship between the physico-chemical/structural descriptors of related compounds, and their biological and

Table 14.2 Some of the diseases where antioxidants are clinically useful

Disease	Description	Reference
Cardiovascular diseases	The LDL is a major contributor to coronary heart disease (CHD) and atherosclerosis. The antioxidants that inhibit LDL oxidation may protect against CHD. By improving lipid profiles and preventing lipid peroxidation, the antioxidants can lower the risk of cardiovascular diseases.	Młynarska et al. (2024)
Alzheimer's disease	Palmatine, an antioxidant, demonstrates ROS and RNS scavenging activity, along with anti-Alzheimer's effect via the cholinesterase (ChEs) and amyloid-beta (Aβ) pathways. The natural antioxidant can diminish amyloid precursor protein and neutralize FRs, hence inhibiting Aβ aggregation.	Feng et al. (2012)
Cancer	A promising anticancer strategy involves modulating ROS levels which may help mitigate ROS-induced carcinogenesis and progression of cancer.	Luo et al. (2022)
Parkinson's disease	The mechanisms of protection include interactions with metal complexes, scavenging of FRs, alterations in enzyme activity, effects on intracellular signaling pathways, and modulation of gene expression.	Mittal et al. (2023)
Hypertension	Antioxidants inhibit NADPH oxidase, a principal generator of ROS in the vascular wall. They help in enhancing the activity of endothelial nitric oxide synthase to reduce the oxidative stress, and thus lower the blood pressure.	Amponsah-Offeh et al. (2023)
Huntington's disease	Enhancement in expression of enzymes through modulation of the Nrf-2/ARE pathway or by elevating PGC-1α (peroxisome proliferator activated receptor γ-coactivator 1α) expression holds significance in the development of therapies against Huntington's disease. The antioxidant vitamins safeguard cell membranes from oxidation and function through the glutathione peroxidase pathway.	Johri and Beal (2012)
Wound healing	The regulation of redox equilibrium via modulation of reactive oxygen species and antioxidant concentrations is a focal point in the process of injured tissue healing.	Comino-Sanz et al. (2021)
Diabetes	In the etiology of type 2 diabetes, oxidative stress plays an important role, and exacerbates the pathology and consequences by disrupting the regulatory mechanisms associated with insulin resistance and β-cell malfunction.	Fatima et al. (2023)
Obesity	Antioxidants can regulate lipid metabolism in obesity. This may help prevent accumulation of excessive lipid in the adipose tissue, and ectopic lipid in the non-adipose tissue, thereby, helpful against obesity associated metabolic disturbances.	Almoraie and Shatwan (2024)
Antimicrobial	Antioxidants with hydroxyl groups interact with microbial cell membrane, affecting transport, adhesion, enzyme activity, and enterotoxin generation.	Suriyaprom et al. (2022)
Arthritis	Antioxidants can destroy the reactive species in joints, eliminate FRs, and reduce the oxidative stress linked to the advancement of arthritis.	Mobasheri et al. (2012)
Skin disorder	Antioxidants scavenge the ROS, reducing oxidative damage to cells and tissues.	De Simoni et al. (2024)

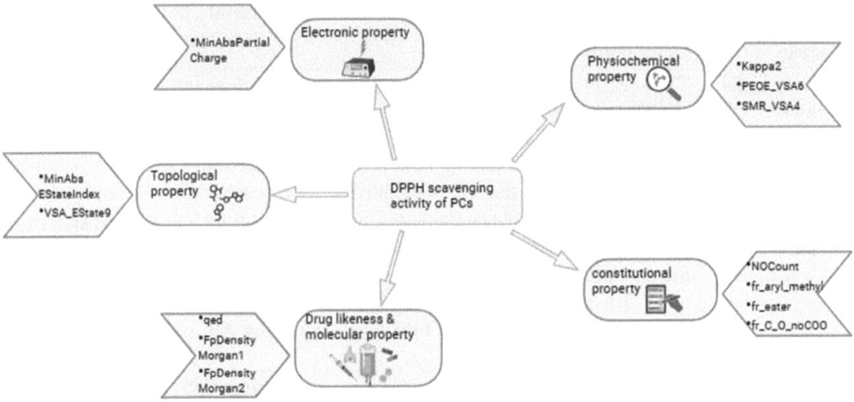

Fig. 14.1 Interrelationship between molecular descriptors and anti-oxidant activity of the phenolic compounds (Lu et al. 2022)

toxicological profile, as well as their physical properties. The molecular descriptors typically encompass parameters that consider electronic properties, topology, hydrophobicity, and steric effects (Fig. 14.1). An essential element in the QSAR process involves identifying descriptors that are rich in information for a molecule/a fragment. Upon development, QSARs yield predictive models of biological activity and may reveal the underlying mechanisms of action (Amić et al. 2003). In order to develop QSAR models certain software such as SPARTAN, DRAGON, QSARINS, SPPS, CQSAR, QSAR-Co, MOE, Discovery Studio, ADMEWORKS ModelBuilder, Schrodinger Suite, LigandScout, Biovia Material Studio, OPEN3DQSAR, LQTAgrid, Python, YASARA, PyRx, MARVIN can be used (Erzincan et al. 2015). For example, Lien et al. developed a model and conducted validation to assess the antioxidative potential of flavonoids, associated phenolic compounds, and vitamin E derivatives through the calculation of physiochemical parameters. The model utilized several parameters to determine the redox potential of the molecules, including the heat of formation of phenoxyl radicals (Hfr), the heat of formation of the parent phenol (Hfp), the energy of the highest occupied molecular orbital of the parent compound (Ehomo), the energy of the lowest unoccupied molecular orbital of the radicals (Elumo-r), the Brown constant ($\Sigma\sigma+$), and the number of hydroxy groups (OH), the half peak oxidation potential (Ep/2). The physicochemical parameters were also correlated with TEAC. The research facilitated the development of a model that may be valuable in forecasting the antioxidant potential of molecules that vary from the parent compound's structure solely by the substituent on the ring.

Flavonoids, chemically classified as benzo-γ-pyrones, are naturally occurring compounds recognized for their antioxidant properties. The family includes flavones, isoflavones, flavanones, biflavanones, flavanols, catechins, and anthocyanidins. The fundamental structure comprises fifteen carbon atoms organized into three distinct rings. The structural differences arise from the number of hydroxyl groups, and their substitution pattern, as well as the degree of glycosylation of these groups (Parshad et al. 2016). To establish the QSAR relations for flavonoids, one may apply stepwise regression (Gonzales et al. 2015), multiple linear regression (Glevitzky et al. 2019; Zapadka et al. 2019), and partial least squares analysis (Roy et al. 2008). Duchowicz et al. developed a simpler method for establishing QSAR and designed novel molecules with targeted antioxidant activity, focusing solely on constitutional and electro-topological parameters. Utilizing the model developed, derivatives of anthocyanin could be obtained, with significant antioxidative potential comparable to alpha-tocopherol or quercetin (Duchowicz et al. 2019). The QSAR model utilizing molecular descriptors to establish a relationship between the structure of benzenesulphonamide and the antioxidant activity was also studied. Three descriptors, namely electrophilicity, self-consistent field (SCF), and molecular refractivity have shown a significant impact on the antioxidant capacity (Abdulrahman et al. 2022). Similarly, a study focusing on the development of QSAR models for tripeptides has been carried out. It established the presence of C-terminal cysteine and tyrosine as the key players in the antioxidant activity of tripeptides. Seven feature selection strategies and pair-wise correlation, along with fourteen machine learning methods, were utilized to generate the models to screen the indices for activity (Du et al. 2022). The developed QSAR models were expected to accurately predict the TEAC for the potential of molecules as antioxidants (Jeličić et al. 2022).

Phenolic compounds are widely present in nature; however, their complex structures make it challenging to establish a direct relationship with their antioxidant activity. This renders basic linear models insufficient for laying out the effects of structural factors on antioxidant efficacy (Fernández et al. 2005; Katritzky et al. 2010; Shi 2021). Therefore, it is essential to implement machine learning algorithms, such as artificial neural networks (ANNs) and multiple linear regression (MLR), to improve the predictability of QSAR (Niu et al. 2007; Žuvela et al. 2018). A study conducted by Shi et al. presented a support vector regression (SVR)-QSAR model for the phenolic compounds. The Pearson correlation coefficient method was utilized for the selection of parameters, ensuring robust predictive stability (Shi 2021).

14.10 Structure Activity Relationship

Gallic acid, chemically 3,4,5-trihydroxybenzoic acid, is a secondary metabolite, polyphenolic in nature, and widely distributed in plants. Researchers in this field suggest that the known derivatives of gallic acid may have a probable antioxidant action. n-Dodecyl gallate, n-Octyl gallate, n-Propyl gallate, Methyl gallate, 3,4,5-Trimethoxybenzoic acid, 5-Hydroxyl-3,4-dimethoxybenzoic acid, and syringic acid were compared in a study by Lu et al, concluding a structure-activity relationship between them. It was found that when compared to gallic acid, the esters of gallic acid, namely, propyl gallate, methyl gallate, octyl gallate and dodecyl gallate, showed potential antioxidant action in an increasing order when tested against

6-hydroxydopamine-induced dopaminergic cell apoptosis. However, the methylated gallic acids, viz. 3,4,5-Trimethoxybenzoic acid, 5-Hydroxyl-3,4-dimethoxybenzoic acid, and Syringic acid, exhibited lower antioxidant activity when compared to one another. This may be associated with the hydrophobicity factor, which facilitates the entry of compounds into the cellular system, thereby enabling their action. Thus, it may be concluded that neuroprotective action of a compound against damaged brought in by the oxidative stress is contingent not only upon its ability to neutralizing FRs, but also relies on its hydrophobic characteristic enabling it to traverse cell membranes to gain entry to the targets (Hara et al. 2003; Kumar et al. 1995; Lu et al. 2006) making the molecules clinically useful.

Studies show that the introduction of another -OH group at ortho/para position in polyphenols results in enhanced antioxidant activity, as seen in catechol and hydroquinone. However, when positioned at the meta-level, it demonstrates minimal activity. The observed result can be attributed to the strong electron-donating ability of the hydroxyl groups located either at the ortho or the para positions, whereas the hydroxyl group at the meta-position exhibits electron-withdrawing characteristics. It is well recognised that electron-donating groups improve the stability of the resulting phenoxy radicals through inductive or resonance effects. A reduction in O-H bond energy is associated with an increase in radical scavenging activity. Conversely, electron-withdrawing groups enhance the stabilization of phenols while simultaneously destabilizing the resulting radicals (Ali et al. 2013; Foti 2007). Additionally, hydrogen bonding may occur between the unpaired electron of the phenoxyl group and the adjacent hydroxyl group in catechols, thereby increasing the stability of the resulting radicals compared to the parent diols. Similarly, o- and p-aminophenols demonstrate notable activity. The presence of alkyl groups in any position (ortho, meta, or para) enhances the stabilization of phenoxy radicals through the inductive effect, consequently improving their antioxidative potential. The effects of thymol and carvacrol illustrate this concept well. The antioxidative activity of the compounds can be determined using molecular docking software. A study conducted by Oliveira et al. utilized the N-methyl-D-aspartate-glycine binding site from Rattus norvegicus, with PDB ID 4KFQ. The Fig. 14.2 shows interaction pattern of carvacrol in the binding pocket of the target protein. The presence of three alkyl groups in BHT leads to high activity, and indicates that BHT exhibits a unique antiradical mechanism (Bondet et al. 1997). A possible substitution identified in polyphenols is the methoxy group. A study conducted by Ali et al. found that the presence of only a methoxy group, as observed in guaiacol, or in combination with a carboxyl group, as seen in vanillic acid, exhibited moderate activity levels of 28.3% and 25.1%, respectively. The presence of two methoxy groups, as observed in syringic acid, resulted in a notably higher activity of 90.4%. The findings validated the role of alkoxy groups as efficient electron-donating entities that contribute to increased electron density and stabilize phenoxyl radicals. It is important to highlight that the methoxy group, unlike alkyl groups, must be positioned at either the ortho or the para position to function effectively as a good electron-donating group. It was supported by the Brown parameter, where $\sigma p+$ was -0.78, and $\sigma o+$ was -0.51. Conversely, the meta-positioned methoxy group typically functions as an electron-

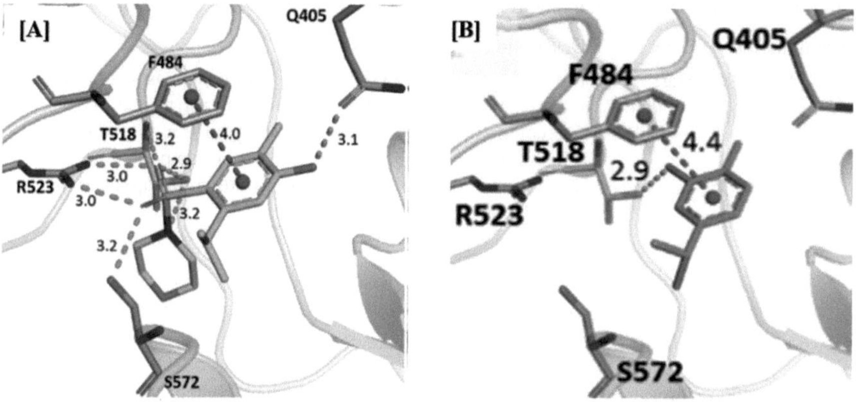

Fig. 14.2 The interaction diagrams of the redocked ligand (**a**) and carvacrol (**b**) into the binding site of the target protein (PDB ID 4FFQ)

withdrawing group, as evidenced by the sigma parameter, with σm equal to 0.12. Nonetheless, the significant electron-withdrawing effect of an oxygen phenoxy radical leads to the m-methoxy group displaying limited electron-donating characteristics, with σm + measured at −0.14. The findings suggest that the alkoxy groups, when strategically positioned, appear to enhance activity in an additive manner, which may be applied and explored in a QSAR study (Ali et al. 2013).

14.11 Patents

Recent times have witnessed aggressive research in establishing the utility of antioxidants in the oxidative stress-mediated diseases and disorders. Many patents have been subsequently filed for the new findings. A few of the patents pertaining to the antioxidants are given in Table 14.3.

14.12 Conclusion

The chapter discusses the characteristics of antioxidants, concerning their origin and the chemical moiety they possess. It also examines the various assay methods used to determine antioxidant capability. In clinical settings, antioxidants have demonstrated a significant role as therapeutic agents in the treatment of stress-mediated diseases and disorders. The chapter describes the mechanisms by which antioxidants may exhibit their action. This has been illustrated through examples, highlighting the relevance of antioxidants in various cases. A quantitative structure-activity relationship employing model sets to assess the efficacy of antioxidants in combating

Table 14.3 Some patents pertaining to antioxidants

Patent number	Date	Description
US 11,919,838 B2	Mar. 5, 2024	It provides new compounds having antioxidant as well as anti-inflammatory property along with manufacturing method and use. The compounds which were covered by generic formula are referred to as oleanolic acid derivatives or compounds of the invention or compounds of the disclosure in this application (US11919838B2 2024). 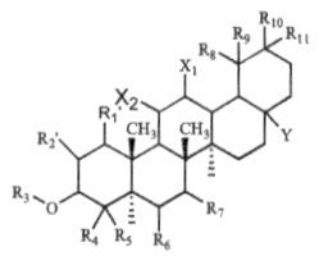
US 12,011,496 B2	Jun. 18, 2024	It provides a topical formulation which contains the extract of macqui berry having anthocyanins, which have a very high oxygen radical absorbing capacity. It is for the prevention or treatment of damaged skin, particularly resulting from chronoaging and photo-aging (US12011496B2 2024).
US2023/0310700 A1	Oct. 5, 2023	This study involved a novel dermal filler, which is used for treating skin conditions. It contains biocompatible polymer like hyaluronic acid which includes a crosslinking component, and a vitamin, like vitamin C. More specifically, it relates to injectable dermal filler composition including antioxidants (US2023/0310700A1 2023).
US 11,122,830 B2	Sep. 21, 2021	This invention provided an antioxidant producing bacterium. The process for producing antioxidant-enriched fruit extract was given, wherein the fruit extract is fermented with a bacterial strain. It is used in the preparation of cosmetics, pharmaceutical compositions as well as nutraceuticals, beverages, dietary, and functional food supplements (US11122830B2 2021).
US 10,683,455 B2	Jun. 16, 2020	It describes highly effective macromolecules which incorporate at least two antioxidant moieties with different reactivity. It describes their superior antioxidant activity as compared to the commercial antioxidants (US10683455B2 2020). 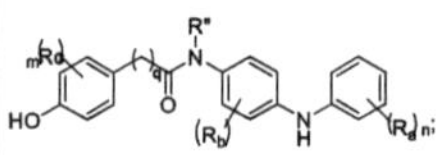
US 10,858,381 B2	Dec. 8, 2020	The invention presents compounds that include novel amine-functionalized porphyrins, along with their salts and esters. They provide antioxidant, anti-inflammatory, and anti-microbial properties. Such substances also exhibit anti-fouling properties (US10858381B2 2020). 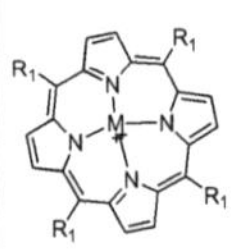
US 2019/0285657 A1	Sep. 19, 2019	This invention describes a method for determining the antioxidant activity of topical formulations. It also describes the testing to check the ability for inhibiting UV-radiation-induced lipid peroxidation on the skin and UV radiation induced ROS formation throughout the dermis (US2019/0285657A1 2019).

(continued)

Table 14.3 (continued)

Patent number	Date	Description
US 10, 045, 923 B2	Aug. 14, 2018	The patent describes novel compounds that demonstrate skin-whitening characteristics, as well as PPAR inhibition and antioxidant activity. The skin whitening mechanism entails the inhibition of tyrosinase. The antioxidant abilities assist in the prevention and cure of skin ageing. The compound is useful in prevention and management of obesity, cardiovascular disease, and metabolic disorders by utilizing PPARα and PPARγ receptors (US10045923 2018). 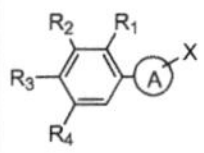
US 9,585,822 B2	Mar. 7, 2017	The study describes a topical cosmetic or dermatological composition, which contains an extract blend of botanical antioxidants in effective amount. It protects from the damaging oxidation process in the skin (US9585822B2 2017).
US 9,707,417 B2	Jul.18, 2017	The study describes the use of benzotropolone and its derivatives as antioxidants and UV absorbers, and its use in cosmetic preparations. It will prevent FR initiated damage and also serve as a defense line by absorbing UV light (US9707417B2 2017). 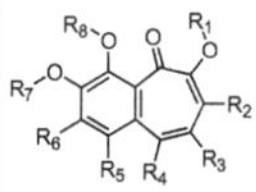
US 9,713,604 B2	Jul. 25, 2017	The invention presents a topical formulation comprising of vitamin E, vitamin C, and a polyphenolic compound as antioxidants. This serves to cure or prevent skin damage associated with free nitrogen, oxygen, and radicals (US9713604B2 2017).
US 8,710,266 B2	Apr. 29, 2014	The patent pertains to compounds that possess the dual functionalities of aromatic amines and phenols. It involves a very potent antioxidant macromolecule that comprises of two antioxidant molecules with distinct reactivities. One distinctive aspect of this antioxidant is its enhanced solubility in several commercially available oils and lubricants, in contrast to the conventional antioxidants (US8710266B2 2014). 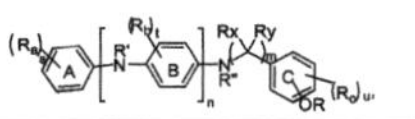
US 8,435,574 B2	May 7, 2013	This invention pertains to natural remedies and the formulation of compositions designed to enhance antioxidant capacity by stimulating the subject's innate cellular defenses. The SOD, CAT, and GP_X constitute the primary cellular defenses against ROS-induced cellular damage (US8435574B2 2013).

illnesses has also been included. Furthermore, data from various experimental sources have been utilized to deduce the structure-activity relationship of the antioxidants. In conclusion, the chapter highlights antioxidants as pharmacological agents, delineates methodologies to develop their QSAR models, establishes their SAR, discusses their clinical applications in oxidative stress-mediated disease, and provides a summary of relevant patents.

References

Abdulrahman SH, Al-Khayat RZ, Ali WKJ (2022) QSAR of antioxidant activity of some novel sulfonamide derivatives. Egypt J Chem 65:489–497

Ak T, Gülçin I (2008) Antioxidant and radical scavenging properties of curcumin. Chem Biol Interact 174:27–37

Alajil O, Sagar VR, Kaur C, Rudra SG, Sharma R, Kaushik R, Verma MK, Tomar M, Kumar M, Mekhemar M (2021) Nutritional and phytochemical traits of apricots (Prunus armeniaca L.) for application in nutraceutical and health industry. Foods 10:1344

Ali HM, Abo-Shady A, Sharaf Eldeen HA, Soror HA, Shousha WG, Abdel-Barry OA, Saleh AM (2013) Structural features, kinetics and SAR study of radical scavenging and antioxidant activities of phenolic and anilinic compounds. Chem Cent J 7:1–9

Almoraie NM, Shatwan IM (2024) The potential effects of dietary antioxidants in obesity: a comprehensive review of the literature. Healthcare 12:416. MDPI

Amić D, Davidović-Amić D, Bešlo D, Trinajstić NJ (2003) Structure-radical scavenging activity relationships of flavonoids. Croat Chem Acta 76:55–61

Amjad S, Rahman MS, Pang M-G (2020) Role of antioxidants in alleviating bisphenol a toxicity. Biomolecules 10:1105

Amponsah-Offeh M, Diaba-Nuhoho P, Speier S, Morawietz H (2023) Oxidative stress, antioxidants and hypertension. Antioxidants 12:281

Arvaniti OS, Samaras Y, Gatidou G, Thomaidis NS, Stasinakis AS (2019) Review on fresh and dried figs: chemical analysis and occurrence of phytochemical compounds, antioxidant capacity and health effects. Food Res Int 119:244–267

Atta EM, Mohamed NH, Abdelgawad AAJ (2017) Antioxidants: an overview on the natural and synthetic types. Eur Chem Bull 6:365–375

Augustyniak A, Bartosz G, Čipak A, Duburs G, Horáková LU, Łuczaj W, Majekova M, Odysseos AD, Rackova L, Skrzydlewska E (2010) Natural and synthetic antioxidants: an updated overview. Free Radic Res 44:1216–1262

Berry CE, Hare JM (2004) Xanthine oxidoreductase and cardiovascular disease: molecular mechanisms and pathophysiological implications. J Physiol 555:589–606

Bianchi A, Vitale E, Guerretti V, Palumbo G, De Clemente I, Vitale L, Arena C, De Maio A (2023) Antioxidant characterization of six tomato cultivars and derived products destined for human consumption. Antioxidants 12:761

Bondet V, Brand-Williams W, Berset C (1997) Kinetics and mechanisms of antioxidant activity using the DPPH. Free radical method. LWT-Food Sci Technol 30:609–615

Caleja C, Barros L, Antonio AL, Oliveira MBP, Ferreira ICFR (2017) A comparative study between natural and synthetic antioxidants: evaluation of their performance after incorporation into biscuits. Food Chem 216:342–346

Callaghan CM, Leggett RE, Levin RM (2013) A comparison of total antioxidant capacities of concord, purple, red, and green grapes using the CUPRAC assay. Antioxidants 2:257–264

Chavare A, Chowdari P, Ghosh S, Pawar P, Patkar M, Dakave S, Gota V, Chaudhari P, Chiplunkar S, Kane S (2017) Safety and bioactivity studies of Jasad Bhasma and its in-process intermediate in Swiss mice. J Ethnopharmacol 197:73–86

Cheah IK, Halliwell B (2012) Ergothioneine; antioxidant potential, physiological function and role in disease. Biochim Biophys Acta (BBA) 1822:784–793

Cheng K, Niu Y, Zheng X, Zhang H, Chen Y, Zhang M, Huang X, Zhang L, Zhou Y, Wang T (2016) A comparison of natural (D-α-tocopherol) and synthetic (DL-α-tocopherol acetate) vitamin E supplementation on the growth performance, meat quality and oxidative status of broilers. Asian Australas J Anim Sci 29:681

Choi HY, Lee YJ, Kim CM, Lee YM (2024) Revolutionizing cosmetic ingredients: harnessing the power of antioxidants, probiotics, plant extracts, and peptides in personal and skin care products. Cosmetics 11:157

Clark AC, Vestner J, Barril C, Maury C, Prenzler PD, Scollary GR (2010) The influence of stereochemistry of antioxidants and flavanols on oxidation processes in a model wine system: ascorbic acid, erythorbic acid,(+)-catechin and (−)-epicatechin. J Agric Food Chem 58:1004–1011

Comino-Sanz IM, López-Franco MD, Castro B, Pancorbo-Hidalgo PL (2021) The role of antioxidants on wound healing: a review of the current evidence. J Clin Med 10:3558

De Simoni E, Candelora M, Belleggia S, Rizzetto G, Molinelli E, Capodaglio I, Ferretti G, Bacchetti T, Offidani A, Simonetti O (2024) Role of antioxidants supplementation in the treatment of atopic dermatitis: a critical narrative review. Front Nutr 11:1393673

Deeb O, Goodarzi M (2017) QSAR of antioxidants. In: Oncology: breakthroughs in research and practice. IGI Global

Dhane K, Gupta MK, Hyam S, Patil AJIJPER (2023) Preparation and in-vitro and in-vivo evaluation of Ayurvedic formulation "Amruthotharam" formulated by classical and modern technique. Indian J Pharm Educ Res 57:S126–S134

Du Z, Wang D, Li Y (2022) Comprehensive evaluation and comparison of machine learning methods in QSAR modeling of antioxidant tripeptides. ACS Omega 7:25760–25771

Dube N, Nimgulkar C, Bharatraj DK (2017) Validation of therapeutic anti-inflammatory potential of Arjuna Ksheera Paka–a traditional Ayurvedic formulation of Terminalia arjuna. J Tradit Complement Med 7:414–420

Duchowicz PR, Szewczuk NA, Pomilio AB (2019) QSAR studies of the antioxidant activity of anthocyanins. J Food Sci Technol 56:5518–5530

Erzincan P, Saçan MT, Yüce-Dursun B, Danış Ö, Demir S, Erdem S, Ogan A (2015) Qsar models for antioxidant activity of new Coumarin derivatives. SAR QSAR Environ Res 26:721–737

Ezeriņa D, Takano Y, Hanaoka K, Urano Y, Dick TPJCCB (2018) N-acetyl cysteine functions as a fast-acting antioxidant by triggering intracellular H2S and sulfane sulfur production. Cell Chem Biol 25:447–459. e4

Fang Y-Z, Yang S, Wu G (2002) Free radicals, antioxidants, and nutrition. Nutrition 18:872–879

Fang J, Lu J, Holmgren A (2005) Thioredoxin reductase is irreversibly modified by curcumin: a novel molecular mechanism for its anticancer activity. J Biol Chem 280:25284–25290

Fatima MT, Bhat AA, Nisar S, Fakhro KA, Akil ASAS (2023) The role of dietary antioxidants in type 2 diabetes and neurodegenerative disorders: an assessment of the benefit profile. Heliyon 9: 1

Feng Y, Wang XJOM, Longevity C (2012) Antioxidant therapies for Alzheimer's disease. Oxidative Med Cell Longev 2012:472932

Fernández M, Caballero J, Helguera AM, Castro EA, González MP (2005) Quantitative structure–activity relationship to predict differential inhibition of aldose reductase by flavonoid compounds. Bioorg Med Chem 13:3269–3277

Ferretti G, Bacchetti T, Belleggia A, Neri D (2010) Cherry antioxidants: from farm to table. Molecules 15:6993–7005

Flieger J, Flieger W, Baj J, Maciejewski R (2021) Antioxidants: classification, natural sources, activity/capacity measurements, and usefulness for the synthesis of nanoparticles. Materials 14: 4135

Foti MC (2007) Antioxidant properties of phenols. J Pharm Pharmacol 59:1673–1685

Franke AA, Cooney RV, Henning SM, Custer LJ (2005) Bioavailability and antioxidant effects of orange juice components in humans. J Agric Food Chem 53:5170–5178

Gadgil S, Nene P (2013) Revisiting the potential of ksheerapaka formulation and its applications. Asian J Pharma Biol Res 3:2

Galal SAB, Elzanfaly ES, Hussien EM, Amer EAH, Zaazaa HE (2024) Spectrofluorimetric determination of butylated hydroxytoluene and butylated hydroxyanisole in their combined formulation: application to butylated hydroxyanisole residual analysis in milk and butter. Sci Rep 14:4498

Gao Z, Huang K, Yang X, Xu H (1999) Free radical scavenging and antioxidant activities of flavonoids extracted from the radix of Scutellaria baicalensis Georgi. Biochim Biophys Acta Gen Subj 1472:643–650

Ghanbari R, Anwar F, Alkharfy KM, Gilani A-H, Saari N (2012) Valuable nutrients and functional bioactives in different parts of olive (Olea europaea L.)—a review. Int J Mol Sci 13:3291–3340

Gil MI, Tomás-Barberán FA, Hess-Pierce B, Holcroft DM, Kader AA (2000) Antioxidant activity of pomegranate juice and its relationship with phenolic composition and processing. J Agric Food Chem 48:4581–4589

Glevitzky I, Dumitrel GA, Glevitzky M, Pasca B, Otrisal P, Bungau S, Cioca G, Pantis C, Popa M (2019) Statistical analysis of the relationship between antioxidant activity and the structure of flavonoid compounds. Rev Chim 70:3103–3107

Gonçalves-Filho D, De Souza D (2022) Detection of synthetic antioxidants: what factors affect the efficiency in the chromatographic analysis and in the electrochemical analysis? Molecules 27: 7137

Gonzales GB, Van Camp J, Zotti M, Kobayashi V, Grootaert C, Raes K, Smagghe G (2015) Two-and three-dimensional quantitative structure–permeability relationship of flavonoids in Caco-2 cells using stepwise multiple linear regression (SMLR), partial least squares regression (PLSR), and pharmacophore (GALAHAD)-based comparative molecular similarity index analysis (COMSIA). Med Chem Res 24:1696–1706

Gueven N, Ravishankar P, Eri R, Rybalka E (2021) Idebenone: when an antioxidant is not an antioxidant. Redox Biol 38:101812

Halliwell BJAIP (1996) Antioxidants: the basics-what they are and how to evaluate them. Adv Pharmacol 38:3–20

Hara H, Ohta M, Ohta K, Kuno S, Adachi T (2003) Increase of antioxidative potential by tert-butylhydroquinone protects against cell death associated with 6-hydroxydopamine-induced oxidative stress in neuroblastoma SH-SY5Y cells. Mol Brain Res 119:125–131

Harzallah A, Bhouri AM, Amri Z, Soltana H, Hammami M (2016) Phytochemical content and antioxidant activity of different fruit parts juices of three figs (Ficus carica L.) varieties grown in Tunisia. Ind Crop Prod 83:255–267

Health NIO (2023) Ashwagandha: is it helpful for stress, anxiety, or sleep [Online]. [Accessed 24]

Huang W-Y, Zhang H-C, Liu W-X, Li C (2012) Survey of antioxidant capacity and phenolic composition of blueberry, blackberry, and strawberry in Nanjing. J Zhejiang Univ Sci B 13:94–102

Jakubek P, Suliborska K, Kuczyńska M, Asaduzzaman M, Parchem K, Koss-Mikołajczyk I, Kusznierewicz B, Chrzanowski W, Namieśnik J, Bartoszek A (2023) The comparison of antioxidant properties and nutrigenomic redox-related activities of vitamin C, C-vitamers, and other common ascorbic acid derivatives. Free Radic Biol Med 209:239–251

Jayathilakan K, Sharma G, Radhakrishna K, Bawa A (2007) Antioxidant potential of synthetic and natural antioxidants and its effect on warmed-over-flavour in different species of meat. Food Chem 105:908–916

Jeličić M-L, Kovačić J, Cvetnić M, Mornar A, Amidžić Klarić D (2022) Antioxidant activity of pharmaceuticals: predictive QSAR modeling for potential therapeutic strategy. Pharmaceuticals 15:791

Ji X, Liang J, Wang Y, Liu X, Li Y, Liu Q, Liu RJ (2023) Synthetic antioxidants as contaminants of emerging concern in indoor environments: knowns and unknowns. Environ Sci Technol 57: 21550–21557

Johri A, Beal MF (2012) Antioxidants in Huntington's disease. Biochim Biophys Acta (BBA) 1822:664–674

Kalt W, Cassidy A, Howard LR, Krikorian R, Stull AJ, Tremblay F, Zamora-Ros R (2020) Recent research on the health benefits of blueberries and their anthocyanins. Adv Nutr 11:224–236

Katritzky AR, Kuanar M, Slavov S, Hall CD, Karelson M, Kahn I, Dobchev DA (2010) Quantitative correlation of physical and chemical properties with chemical structure: utility for prediction. Chem Rev 110:5714–5789

Kellett ME, Greenspan P, Pegg RB (2018) Modification of the cellular antioxidant activity (CAA) assay to study phenolic antioxidants in a Caco-2 cell line. Food Chem 244:359–363

Kumar R, Agarwal AK, Seth PK (1995) Free radical-generated neurotoxicity of 6-hydroxydopamine. J Neurochem 64:1703–1707

Kumar V, Kshemada K, Ajith KG, Binil RS, Deora N, Sanjay G, Jaleel A, Muraleedharan T, Anandan E, Mony R (2017) Amalaki rasayana, a traditional Indian drug enhances cardiac mitochondrial and contractile functions and improves cardiac function in rats with hypertrophy. Sci Rep 7:1–17

Kumar M, Tomar M, Amarowicz R, Saurabh V, Nair MS, Maheshwari C, Sasi M, Prajapati U, Hasan M, Singh S (2021) Guava (Psidium guajava L.) leaves: nutritional composition, phytochemical profile, and health-promoting bioactivities. Foods 10:752

Lachance PA, Nakat Z, Jeong WS (2001) Antioxidants: an integrative approach. Nutrition 17:835–838

Li X, Wang T, Zhou B, Gao W, Cao J, Huang L (2014) Chemical composition and antioxidant and anti-inflammatory potential of peels and flesh from 10 different pear varieties (Pyrus spp.). Int J Mol Sci 152:531–538

Liguori I, Russo G, Curcio F, Bulli G, Aran L, Della-Morte D, Gargiulo G, Testa G, Cacciatore F, Bonaduce D (2018) Oxidative stress, aging, and diseases. Clin Interv Aging 13:757–772

Lobo V, Patil A, Phatak A, Chandra N (2010) Free radicals, antioxidants and functional foods: impact on human health. Pharmacogn Rev 4:118

Lourenço SC, Moldão-Martins M, Alves VD (2019) Antioxidants of natural plant origins: from sources to food industry applications. Molecules 24:4132

Loussouarn M, Krieger-Liszkay A, Svilar L, Bily A, Birtić S, Havaux M (2017) Carnosic acid and carnosol, two major antioxidants of rosemary, act through different mechanisms. Plant Physiol 175:1381–1394

Lu Z, Nie G, Belton PS, Tang H, Zhao B (2006) Structure–activity relationship analysis of antioxidant ability and neuroprotective effect of gallic acid derivatives. Neurochem Int 48: 263–274

Lü JM, Lin PH, Yao Q, Chen C (2010) Chemical and molecular mechanisms of antioxidants: experimental approaches and model systems. J Cell Mol Med 14:840–860

Lu A, Yuan S-M, Xiao H, Yang D-S, Ai Z-Q, Li Q-Y, Zhao Y, Chen Z-Z, Wu XM (2022) QSAR study of phenolic compounds and their anti-DPPH radical activity by discriminant analysis. Sci Rep 12:7860

Luo M, Zhou L, Huang Z, Li B, Nice EC, Xu J, Huang C (2022) Antioxidant therapy in cancer: rationale and progress. Antioxidants 11:1128

Mamta, Misra K, Dhillon GS, Brar SK, Verma M (2014) Antioxidants. In: Biotransformation of waste biomass into high value biochemicals, pp 117–138

Manivannan A, Lee E-S, Han K, Lee H-E, Kim DS (2020) Versatile nutraceutical potentials of watermelon—a modest fruit loaded with pharmaceutically valuable phytochemicals. Molecules 25:5258

Mikulska P, Malinowska M, Ignacyk M, Szustowski P, Nowak J, Pesta K, Szeląg M, Szklanny D, Judasz E, Kaczmarek G (2023) Ashwagandha (Withania somnifera) - current research on the health-promoting activities: a narrative review. Pharmaceutics 15:1057

Mittal P, Dhankhar S, Chauhan S, Garg N, Bhattacharya T, Ali M, Chaudhary AA, Rudayni HA, Al-Zharani M, Ahmad W (2023) A review on natural antioxidants for their role in the treatment of Parkinson's disease. Pharmaceuticals 16:908

Młynarska E, Hajdys J, Czarnik W, Fularski P, Leszto K, Majchrowicz G, Lisińska W, Rysz J, Franczyk B (2024) The role of antioxidants in the therapy of cardiovascular diseases - a literature review. Nutrients 16:2587

Mobasheri A, Shakibaei M, Biesalski HK, Henrotin Y (2012) Antioxidants in the treatment of osteoarthritis and bone mineral loss. In: Studies on arthritis and joint disorders. Springer

Morshed MN, Ahn JC, Mathiyalagan R, Rupa EJ, Akter R, Karim MR, Jung DH, Yang DU, Yang DC, Jung SK (2023) Antioxidant activity of Panax ginseng to regulate ROS in various chronic diseases. Appl Sci 13:2893

Mrowicka M, Mrowicki J, Kucharska E, Majsterek I (2022) Lutein and zeaxanthin and their roles in age-related macular degeneration - neurodegenerative disease. Nutrients 14:827

Niu B, Lu WC, Yang SS, Cai YD, Li GZ (2007) Support vector machine for SAR/QSAR of phenethyl-amines. Acta Pharmacol Sin 28:1075–1086

Oracz J, Żyżelewicz DJA (2020) Antioxidants in cocoa. Antioxidants 9(12):1230

Pal D, Sahu CK, Haldar AH (2014) Bhasma: the ancient Indian nanomedicine. J Adv Pharma Technol Res 5:4–12

Park J-H, Lee M, Park E (2014) Antioxidant activity of orange flesh and peel extracted with various solvents. Prev Nutr Food Sci 19:291

Parshad B, Duraisamy A, Saini S, Yadav P, Vats P, Sunil K (2016) Synthesis and SAR study of antioxidant potential of polyhydroxy coumarin derivatives. Med Chem 6:506–514

Parveen R, Shamsi TN, Singh G, Athar T, Fatima S (2018) Phytochemical analysis and in-vitro biochemical characterization of aqueous and methanolic extract of Triphala, a conventional herbal remedy. Biotechnol Rep 17:126–136

Peterson CT, Denniston K, Chopra D (2017) Therapeutic uses of triphala in ayurvedic medicine. J Altern Complement Med 23:607–614

Pisoschi AM, Pop A (2015) The role of antioxidants in the chemistry of oxidative stress: a review. Eur J Med Chem 97:55–74

Poljsak B, Šuput D, Milisav I (2013) Achieving the balance between ROS and antioxidants: when to use the synthetic antioxidants. Oxid Med Cell Longev 2013:956792

Preiser JC (2012) Oxidative stress. J Parenter Enter Nutr 36:147–154

Rahal A, Kumar A, Singh V, Yadav B, Tiwari R, Chakraborty S, Dhama K (2014) Oxidative stress, prooxidants, and antioxidants: the interplay. Biomed Res Int 2014:761264

Rajendran P, Nandakumar N, Rengarajan T, Palaniswami R, Gnanadhas EN, Lakshminarasaiah U, Gopas J, Nishigaki I (2014) Antioxidants and human diseases. Clin Chim Acta 436:332–347

Raman LS, Shanmuganathan A, Chandrashekar S, Kaliyaperumal P, Perumal E, Mudiganti RKR, Nachammai K, Kulanthaivel L, Subbaraj GK, Chandrasekaran K (2023) Antioxidant, anti-inflammatory, and anticarcinogenic efficacy of an ayurvedic formulation: Amritotharanam Kashyam. Adv Pharmacol Pharma Sci 2023:3387261

Ramoutar RR, Brumaghim JL (2010) Antioxidant and anticancer properties and mechanisms of inorganic selenium, oxo-sulfur, and oxo-selenium compounds. Cell Biochem Biophys 58:1–23

Rice-Evans CA, Burdon RH (eds) (1994) Free radical damage and its control. Elsevier

Roy K, Roy PPJCB, Design D (2008) Comparative QSAR studies of CYP1A2 inhibitor flavonoids using 2D and 3D descriptors. Chem Biol Drug Des 72:370–382

Salehi B, Martorell M, Arbiser JL, Sureda A, Martins N, Maurya PK, Sharifi-Rad M, Kumar P, Sharifi-Rad J (2018) Antioxidants: positive or negative actors? Biomolecules 8:124

Samarakoon S, Chandola H, Shukla V (2011) Evaluation of antioxidant potential of Amalakayas Rasayana: a polyherbal Ayurvedic formulation. Int J Ayurveda Res 2:23

Schröder K, Vecchione C, Jung O, Schreiber JG, Shiri-Sverdlov R, Van Gorp PJ, Busse R, Brandes RP (2006) Xanthine oxidase inhibitor tungsten prevents the development of atherosclerosis in ApoE knockout mice fed a Western-type diet. Free Radical Biol Med 41:1353–1360
Shi Y (2021) Support vector regression-based QSAR models for prediction of antioxidant activity of phenolic compounds. Sci Rep 11:8806
Sies H (1991) Oxidative stress: from basic research to clinical application. Am J Med 91:S31–S38
Sindhi V, Gupta V, Sharma K, Bhatnagar S, Kumari R, Dhaka NJ (2013) Potential applications of antioxidants–a review. J Pharm Res 7:828–835
Singh N, Bhalla M, De Jager P, Gilca M (2011) An overview on ashwagandha: a Rasayana (rejuvenator) of Ayurveda. Afr J Tradit Complement Altern Med 8
Sun Y, Yang C, Tsao R (2018) Nomenclature and general classification of antioxidant activity/capacity assays. In: Measurement of antioxidant activity & capacity: recent trends and applications, pp 1–19
Suriyaprom S, Mosoni P, Leroy S, Kaewkod T, Desvaux M, Tragoolpua Y (2022) Antioxidants of fruit extracts as antimicrobial agents against pathogenic bacteria. Antioxidants 11:602
US10045923 (2018) Compound having skin-whitening, anti-oxidizing and PPAR activities and medical use thereof. US patent application
US10683455B2 (2020) Macromolecular antioxidants based on dual type moiety per molecule: structures, methods of making and using the same. US patent application
US10858381B2 (2020) Antioxidant compounds and their use. US patent application
US11122830B2 (2021) Antioxidant producing bacterium and uses thereof. US patent application
US11919838B2 (2024) Antioxidant inflammation modulators: oleanolic acid derivatives with amino and other modifications at C-17. US patent application
US12011496B2 (2024) Topical macqui berry formulation. US patent application
US2019/0285657A1 (2019) Method of selecting antioxidants for use in topically applied compositions. US patent application
US2023/0310700A1 (2023) Dermal filler compositions including antioxidants. US patent application
US8435574B2 (2013) Compositions for alleviating inflammation and oxidative stress in a mammal. US patent application
US8710266B2 (2014) Nitrogen and hindered phenol containing dual functional macromolecular antioxidants: synthesis, performances and applications. US patent application
US9585822B2 (2017) Methods of preparing and using botanical antioxidant compositions. US patent application
US9707417B2 (2017) Use of benzotropolone derivatives as UV absorbers and antioxidants and their use in sunscreens and/or cosmetic compositions. US patent application
US9713604B2 (2017) Antioxidant compositions and methods of using the same. US patent application
Viana Da Silva M, Santos MRC, Alves Silva IR, Macedo Viana EB, Dos Anjos DA, Santos IA, Barbosa De Lima NG, Wobeto C, Jorge N, Lannes SCDS (2022) Synthetic and natural antioxidants used in the oxidative stability of edible oils: an overview. Food Rev Intl 38:349–372
Willcox JK, Ash SL, Catignani GL (2004) Antioxidants and prevention of chronic disease. Crit Rev Food Sci Nutr 44:275–295
Xu X, Liu A, Hu S, Ares I, Martínez-Larrañaga M-R, Wang X, Martínez M, Anadón A, Martínez MA (2021) Synthetic phenolic antioxidants: metabolism, hazards and mechanism of action. Food Chem 353:129488
Zapadka M, Kaczmarek M, Kupcewicz B, Dekowski P, Walkowiak A, Kokotkiewicz A, Łuczkiewicz M, Buciński A (2019) An application of Qsrr approach and multiple linear regression method for lipophilicity assessment of flavonoids. J Pharm Biomed Anal 164:681–689
Žuvela P, David J, Wong MW (2018) Interpretation of ANN-based QSAR models for prediction of antioxidant activity of flavonoids. J Comput Chem 39:953–963

Chapter 15
Pharmacological Insights into Antioxidants for Oxidative Stress Management

Prajakta C. Ghume, Divyansh Khatri, and Sapana Kushwaha

Abbreviations

ATP	Adenosine-5-triphosphate
CAD	Coronary artery disease
CAT	Catalase
ETC	Electron transport chain
ETFQOR	Electron transferring flavoprotein Q oxidoreductase
GPDH	Glycerol-3-phosphate dehydrogenase
GPx	Glutathione peroxidase
GRx	Glutathione reductase
H_2O	Water
H_2O_2	Hydrogen peroxide
Mn-SOD	Manganese superoxide dismutase
NADPH	Nicotinamide adenine dinucleotide phosphate
O_2	Molecular oxygen
O_2^-	Superoxide radicals

Prajakta C. Ghume and Divyansh Khatri contributed equally with all other contributors.

P. C. Ghume · D. Khatri
Department of Regulatory Toxicology, National Institute of Pharmaceutical Education and Research, Raebareli (NIPER-R), Transit Campus, Lucknow, Uttar Pradesh, India

S. Kushwaha (✉)
Department of Pharmacology and Toxicology, National Institute of Pharmaceutical Education and Research, Raebareli (NIPER-R), Transit Campus, Lucknow, Uttar Pradesh, India
e-mail: sapana.kushwaha@niperrbl.ac.in

N. Srivastava, S. A. Saraf (eds.), *Pharmacological Strategies for Prevention and Treatment of Oxidative Stress*, Oxidative Stress in Applied Basic Research and Clinical Practice, https://doi.org/10.1007/978-3-032-19837-2_15

PKC	Protein kinase C
ROS	Reactive oxygen species
SERCA 2A	Sarco/endoplasmic reticulum Ca^{2+}-ATPase
SOD	Superoxide dismutase

15.1 Introduction

15.1.1 Definition and Overview

Oxygen shows dual effects, both beneficial as well as potentially damaging side effects, so known as Janus gas. Due to its reactivity, oxygen participates in the electron transport chain and via oxidative phosphorylation generation of adenosine-5-triphosphate (ATP). It is necessary for the continuous development of multicellular organisms, but it also tends to harm the organism if its balance gets disturbed. The reactive oxygen species (ROS) are continuously being developed in the organism and have biological functions in normal conditions. Still, under stress, the amount of these ROS started to increase which further caused an imbalance in the oxidant–antioxidant ratio, hence causing oxidative stress. The oxidative stress can be defined as, an imbalance in oxidants and antioxidants in favour of the oxidants, which leads to a disruption of redox signalling and control and molecular damage (Sies et al. 2017).

According to its intensity oxidative stress can be classified into, physiological oxidative stress also known as "oxidative eustress" which has a physiological role in the organism, and the second one is toxic oxidative stress known as "oxidative burden" which causes damage to the organism (Sies 2020) (Fig. 15.1).

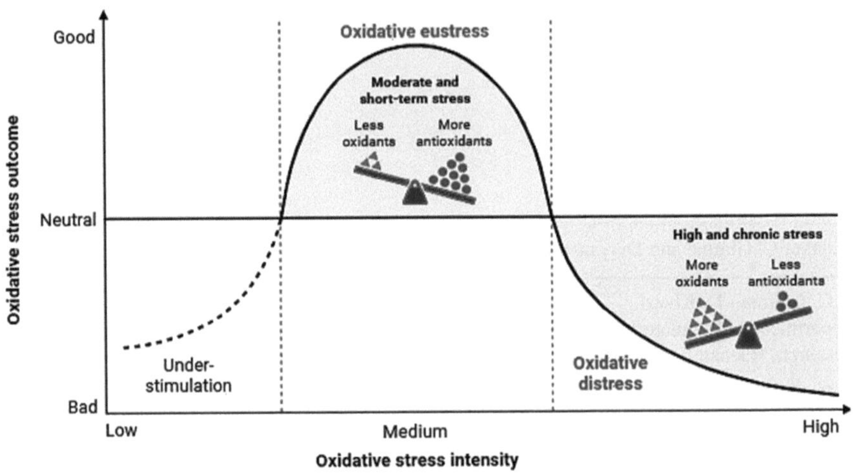

Fig. 15.1 Effect of oxidant and antioxidant imbalance on oxidative stress outcome. (Figures created in Biorender.com)

15.1.2 Historical Perspective of Oxidative Stress

The fundamental concept of stress and its physiological responses was first introduced by Selye in 1936 (Selye 1936). In the book, oxidative stress is introduced much later, it is reported to have discussed the idea in 1985 (Sies 1985). In the year 1970, this term "oxidative stress" was first utilized for the red blood cells as it describes adding hydrogen peroxide effects, marking new toxicology areas of research. Early studies focused on the harmful effects of drugs that caused "redox cycling" and lipid damage. Many anticancer drugs work based on the concept of redox cycling to produce their toxic effects (Sies et al. 2017).

15.1.3 The Concept of Redox Balance

Redox balance is the condition in which the rate of oxidation, essentially the loss of electrons, is equal to the rate of reduction, which involves gaining electrons. This balance maintains reactive species at nontoxic levels. Within a biological framework, redox balance refers to the ongoing interplay between oxidizing agents, which are sometimes associated with oxidative stress, and reducing agents that help neutralize their effects. It is also essential for healthy cellular function (Le Gal et al. 2021).

15.1.3.1 Importance of Redox Reactions

Most living organisms carry out biological processes whose redox mechanism is a pivotal feature in it. Examples are metabolism, particularly cellular respiration where ATP syntheses is boosted through electron transport chains; signal transduction since their signaling pathways also govern other features such as cellular growth, differentiation, and programmed cell death among others. Further, they carry out antioxidant defenses; this takes place since antioxidant substances will detoxify excess generated reactive oxygen species; an instance being derived from either biological mechanism of metabolic events or exposure by the body environment (Dröge 2002).

15.1.3.2 Consequences of Imbalance

When redox balance is perturbed favouring oxidation cells may suffer from oxidative stress. Oxidative stress may cause, cell damage, oxidation stress damages lipids, proteins and DNA causes aging and disease progression, dysregulation of redox homeostasis is implicated in various diseases such as, neurodegenerative diseases, cancer, metabolic disorders, and cardiovascular diseases (Chaudhary et al. 2023).

15.2 Mechanisms of ROS Generation

The reactive oxygen species (ROS) can be produced through either from endogenous or exogenous processes. The primary endogenous source for the generation of ROS is the mitochondria, primarily due to their involvement in ATP production through oxidative phosphorylation. During this process, molecular oxygen (O_2) is reduced to water (H_2O) in the electron transport chain (ETC), leading to the generation of superoxide radicals ($O_2^- O_2^-$) (Enríquez 2016). Key sites for superoxide production in mammalian mitochondria include complex I (site IQ), ubiquinone binding site, complex III (site IIIQo), glycerol 3-phosphate dehydrogenase (GPDH), Flavin in complex I (site IF), electron transferring flavoprotein Q oxidoreductase (ETFQOR), pyruvate and 2-oxoglutarate dehydrogenases (Brand 2010), except complex III site and GPDH, all the other sites release O_2^- in the mitochondrial matrix (Hajam et al. 2022).

These complexes predominantly release superoxide into the mitochondrial matrix, where it is converted into hydrogen peroxide (H_2O_2) by manganese superoxide dismutase (Mn-SOD). The subsequent conversion of H_2O_2 to hydroxyl radicals can occur via Fenton reactions (Okado-Matsumoto and Fridovich 2001).

The other notable ROS source is the cytochrome P450 enzymatic activity involved in the metabolism of numerous organic chemicals such as lipids and steroids (Hajam et al. 2022). The superoxide, hydrogen peroxide, and other reactive metabolites formed during this process add to the overall oxidative stress at the cellular level (Yasui et al. 2005).

Mitochondrial sources are not the only source of ROS generation, there are other cellular processes involved, NADPH oxidases, these enzymes generate superoxide during the oxidative burst in immune cells (Hernansanz-Agustín and Enríquez 2021), xanthine oxidase, involvement of purine metabolism, this enzyme is responsible for generating superoxide and hydrogen peroxide, lipoxygenases and cyclooxygenases, these enzymes that are involved in lipid metabolism can also lead to the generation of ROS as byproducts, monoamine oxidases, these enzymes degrade neurotransmitters, producing hydrogen peroxide as a metabolic byproduct (Hajam et al. 2022; Hernansanz-Agustín and Enríquez 2021).

Conditions leading to excessive ROS production can occur under various pathological conditions, including, hyperoxia, elevated oxygen levels can lead to increased superoxide formation, inflammation, immune responses often involve the activation of NADPH oxidases, resulting in an oxidative burst, metabolic dysregulation, conditions like diabetes can lead to increased glucose autoxidation, generating free radicals and depleting antioxidant defences (Huchzermeyer et al. 2022; Hajam et al. 2022) (Fig. 15.2).

The figure shows various mechanisms for the production of reactive oxygen species, i.e. radiation, NADPH oxidase, through the uptake of microorganisms, xenobiotics, oxidative phosphorylation in mitochondria, stress in the endoplasmic reticulum, by various enzymatic reactions, and metabolic reactions.

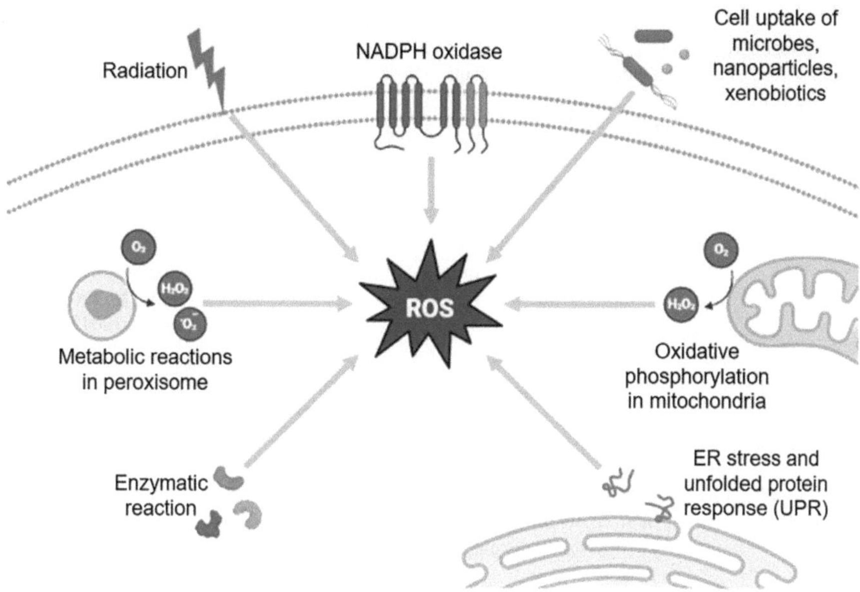

Fig. 15.2 Mechanisms for the production of ROS. (Figures created in Biorender.com)

15.3 Oxidative Stress and Human Diseases

The oxidative stress is one of the main causes of various diseases that occur in humans, there are many diseases which either caused by oxidative stress or aided by it, like various neurodegenerative diseases, such as Alzheimer's, Parkinson's, multiple sclerosis, Amyotrophic lateral sclerosis, Duchenne muscular atrophy, etc. (Rekatsina et al. 2020), are associated with oxidative stress, as well as many cardiovascular diseases (Daiber and Chlopicki 2020), cancer, diabetes (Ghasemi-Dehnoo et al. 2020) etc. are also associated with oxidative stress.

15.3.1 Neurodegenerative Diseases and Oxidative Stress

Several interrelated mechanisms account for the contribution of ROS to neurodegeneration. Firstly, oxidative stress, the imbalance between ROS production and the neutralization capability of the body which leads to neuronal injury. One of the mechanisms is mitochondrial dysfunction, which leads to cellular damage and excessive ROS production caused by mitochondrial dysfunction enhances this vicious cycle of injury. Moreover, misfolded proteins accumulate in neurodegenerative diseases and induce oxidative stress and inflammation causing further neuronal damage (Rekatsina et al. 2020). Table 15.1 shows the various neurodegenerative diseases and their underlying pathologies associated with oxidative stress:

Table 15.1 Pathologies of neurodegenerative diseases associated with oxidative stress

Neurodegenerative Disease	Underlying mechanisms	Pathology Associated with Oxidative Stress	Reference
Alzheimer's Disease	Aggregation of amyloid-beta (Aβ) plaques and neurofibrillary tangles formed by tau protein	Oxidative stress is implicated in the formation of these plaques and tangles.	Awasthi et al. (2005), Giraldo et al. (2014)
Parkinson's Disease	Loss of dopaminergic neurons in the substantia nigra and accumulation of alpha-synuclein	Oxidative stress leads to mitochondrial dysfunction and neuron damage.	Solleiro-Villavicencio and Rivas-Arancibia (2018), Puspita et al. (2017)
Multiple Sclerosis	Demyelination and axonal damage	Oxidative stress is linked with inflammation and the recruitment of reactive oxygen species-producing lesions.	Haider et al. (2011), Fischer et al. (2013)
Amyotrophic Lateral Sclerosis	Degeneration of motor neurons with mutations in superoxide dismutase 1 (SOD1)	Oxidative stress results from mitochondrial dysfunction and contributes to motor neuron injury.	Malaspina et al. (2015), D'Amico et al. (2013)
Duchenne Muscular Dystrophy	Muscle wasting with chronic inflammation	Oxidative stress exacerbates the inflammatory response, contributing to muscle degradation.	Petrillo et al. (2017)

15.3.2 Cardiovascular Diseases in Association with Oxidative Stress

In cardiovascular system, ROS, under physiological conditions, produced in lesser amount and play role in detoxification and cellular function and signaling but under conditions such as atherosclerosis, hypertension, coronary artery disease (CAD), heart failure, stroke, etc. ROS causes oxidative stress and more harm to the cardiovascular system. One of the example of such is, the oxidation of sarco/endoplasmic reticulum Ca^{2+}-ATPase (SERCA 2A) (Lancel et al. 2010) and contractile proteins like tropomyosin and actin, ROS could modulate contractility leading to contractile dysfunction (Dubois-Deruy et al. 2020). Table 15.2 comprises of various disease pathologies and underlying mechanisms associated with oxidative stress:

15.3.3 Carcinogenesis Associated with Oxidative Stress

Cancer is marked by excessive oxidative stress, mainly because of the overproduction of ROS from metabolic derangements and interruption of the signaling pathways. The increased levels of ROS can cause the carcinogenesis and progression of tumors by producing genetic mutations and triggering pro-oncogenic pathways (Luo et al. 2022).

Table 15.2 Pathologies of cardiovascular diseases associated with oxidative stress

Cardiovascular Disease	Underlying mechanisms	Pathology Associated with Oxidative Stress	Reference
Atherosclerosis	Endothelial dysfunction and plaque formation	Imbalance in reactive oxygen and nitrogen species (RONS)	Batty et al. (2022), Daiber and Chlopicki (2020)
Hypertension	Reduced nitric oxide (NO) availability and impaired vasodilation	Increased oxidative stress damages endothelial cells	Griendling et al. (2021)
Coronary Artery Disease (CAD)	Arterial blockage and inflammation	Elevated oxidative stress contributes to lipid peroxidation and inflammation	Vichova and Motovska (2013), Daiber and Chlopicki (2020)
Heart Failure	Cardiac remodelling and impaired energy metabolism	Mitochondrial oxidative stress promotes pathological changes	Tsutsui et al. (2011)
Stroke	Neuronal injury and vascular damage	Oxidative damage exacerbates ischemic and reperfusion injury	Shirley et al. (2014)

Excessive levels of ROS, originating from disturbed redox homeostasis, are critical for converting normal cells to malignant ones and sustaining malignancy via two major pathways. Firstly, ROS directly oxidize macromolecules (nucleic acids, proteins, lipids, and glucose) and cause genetic mutations, leading to DNA damage and abnormal inflammation (Zhong and Yin 2015). Another way is by oxidative stress-induced dysregulation of redox signaling that promotes cancer development (Luo et al. 2022).

Hydrogen peroxide (H_2O_2)-induced oxidative stress can be mitigated by the interaction between STUB1 and BMAL1, which subsequently reduces cell senescence (Li et al. 2024) (Fig. 15.3). Table 15.3 shows carcinogenic mechanisms associated with oxidative stress on different organs:

15.3.4 Diabetes and Oxidative Stress

Diabetes mellitus is a metabolic disorder with severe complications including retinopathy, neuropathy, nephropathy, and cardiovascular diseases. The pathogenesis mechanism in diabetes with oxidative stress encompasses an important role of hyperglycaemia. Oxidative stress has clinical associations and genetic associations with diabetes–related complications. Oxidative stress-driven pathogenesis mechanisms proliferate in diabetes (Ghasemi-Dehnoo et al. 2020; Asmat et al. 2016).

Oxidative stress is closely associated with diabetes mellitus via different pathways. Polyol pathway is activated by hyperglycaemia, depriving tissues of NADPH and glutathione, increases tissue susceptibility to oxidative damage (Kilanczyk et al.

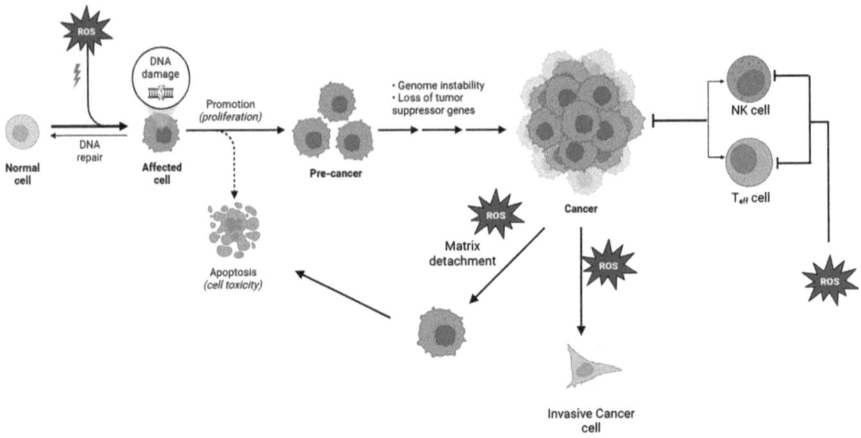

Fig. 15.3 ROS induced carcinogenesis

Table 15.3 Mechanisms of carcinogenesis associated with oxidative stress

Organ	Carcinogenesis Mechanisms	Association with oxidative stress	Reference
Lung	K-RAS activation promoting tumorigenesis via the AGT/ANG II/AT1R-STAT3 pathway.	Oxidative stress induces senescence through mitochondrial dysfunction and ROS production.	Volonte et al. (2021)
Colon	Oxidative stress-induced fibroblasts releasing SASP containing GDF15, promoting a pro-carcinogenic tumor environment.	APC mutations delaying senescence onset, enhancing oxidative DNA damage and tumorigenesis.	Tortora et al. (2022)
Liver	Gestational arsenite exposure activates TGF-β and oxidative stress, causing liver carcinogenesis.	ROS generation regulating senescence via the P53/P21 pathway and suppressing tumor growth.	Chibaya et al. (2021)
Breast	Senescent MCF-7 cells promote metabolomic reprogramming and tumor proliferation.	Oxidative stress-induced senescence inhibiting tumor progression via mitochondrial dysfunction.	Kwon et al. (2023)
Prostate	Stromal cell downregulation of TRPC3 increasing ROS and SASP secretion, enhancing cancer cell proliferation.	Senescence-associated oxidative stress promoting stromal cells to support cancer growth.	Farfariello et al. (2022)
Head and Neck	Gas plasma treatments inducing oxidative stress and cellular senescence, reducing cancer cell proliferation.	Enhanced ROS levels through EGFR signaling promote senescence and tumor progression.	Berner et al. (2023)
Stomach	Cystathionine-β-synthase essential for AKT-induced senescence; its knockout reduces oxidative stress and senescence, enhancing cell growth.	AKT/mTOR pathway regulating ROS-induced cellular changes influencing senescence and cancer development.	Zhu et al. (2022)

Table 15.4 Mechanisms involved in diabetes associated with oxidative stress

Mechanism	Description	Implications ROS for Diabetes Mellitus	Reference
Polyol Pathway Flux	Excess glucose is converted to sorbitol and then fructose, consuming NADPH and reducing glutathione production.	Increased cellular sensitivity to oxidative stress due to decreased antioxidant capacity.	Kilanczyk et al. (2016)
AGE Precursor Formation	Hyperglycaemia leads to advanced glycation end product (AGE) precursor production, altering protein function and signaling pathways.	Contributes to vascular damage and chronic inflammation.	Ighodaro (2018)
Protein Kinase C Activation	Intracellular hyperglycaemia activates PKC isoforms, altering gene expression and reducing nitric oxide production.	Impairs vascular function and increases production of harmful vasoconstrictors.	Dowling et al. (2017)
Hexosamine Pathway Activation	Excess fructose-6-phosphate enters the hexosamine pathway, modifying transcription factors through glucosamine derivatives.	Leads to pathological changes in gene expression and protein function, impacting cellular metabolism and structure.	Asmat et al. (2016)
Mitochondrial Superoxide Production	Hyperglycaemia drives excess ROS generation in mitochondria, disrupting cellular respiration and energy production.	Directly damages DNA, proteins, and lipids, exacerbating complications like nephropathy and neuropathy.	Asmat et al. (2016)

2016). It also promotes the development of advanced glycation end products, which alter proteins and signaling pathways, stimulating inflammation and tissue damage. High glucose also stimulates PKC isoforms, resulting in altered gene expression, decreased nitric oxide production, and impaired endothelial function. Second, the hexosamine pathway results in transcription factor modifications, changes in cellular metabolism (Ghasemi-Dehnoo et al. 2020) (Table 15.4).

15.4 Body's Defence Against Oxidative Stress

There are multiple ways for the body to fight oxidative stress by creating antioxidants, whether endogenously produced in situ (endogenous antioxidants) or externally supplied, via food (exogenous antioxidants). The function of antioxidants are to neutralize the excess of free radicals, and shield the cells from their toxic effects and to aid disease prevention (Pham-Huy et al. 2008).

The antioxidant defence of the body generally works by two mechanisms, i.e. chain breaking and/or prevention. In the former, when an electron is released by a radical, a second radical is generated, same action is exerted by the last one on another molecule and until a chain breaking antioxidant stabilises the free radical formed, the process continues or a disintegrated in offensive product is formed.

Lipid peroxidation is one of the classical examples of chain reaction. Antioxidant enzymes such as superoxide dismutase, catalase, and glutathione peroxidase, by reducing the chain initiation, can prevent the oxidation, by stabilization of transition metal ions or by scavenging the free radicals (Pham-Huy et al. 2008; Young and Woodside 2001).

15.5 Classification of Antioxidants

The endogenous antioxidants, classified into enzymatic and non-enzymatic antioxidants. Majorly involved enzymes in the neutralisation of oxidative stress are, superoxide dismutase (SOD), catalase (CAT), glutathione reductase (GRx) and peroxidase (GPx) (Willcox et al. 2004). By the catalysis of dismutation of superoxide anion radical ($O_2^{-\bullet}$) into hydrogen peroxide (H_2O_2) through reduction, SOD becomes the first line of defence against free radicals. The CAT or GPx converts the H_2O_2 into water (H_2O) and oxygen (O_2) (Dröge 2002; Pham-Huy et al. 2008).

The non-enzymatic antioxidants further subdivided into metabolic and nutrient antioxidants. The metabolic antioxidants are endogenous and generated through metabolism, e.g. lipoid acid glutathione, L-arginine, coenzyme Q10, melatonin, uric acid, etc. (Dröge 2002). The antioxidants are exogenous and can't be generated by the body and should be taken via food, e.g. vitamins such as vitamin E & vitamin C, carotenoids, trace metals, flavonoids, etc. (Pham-Huy et al. 2008) (Fig. 15.4).

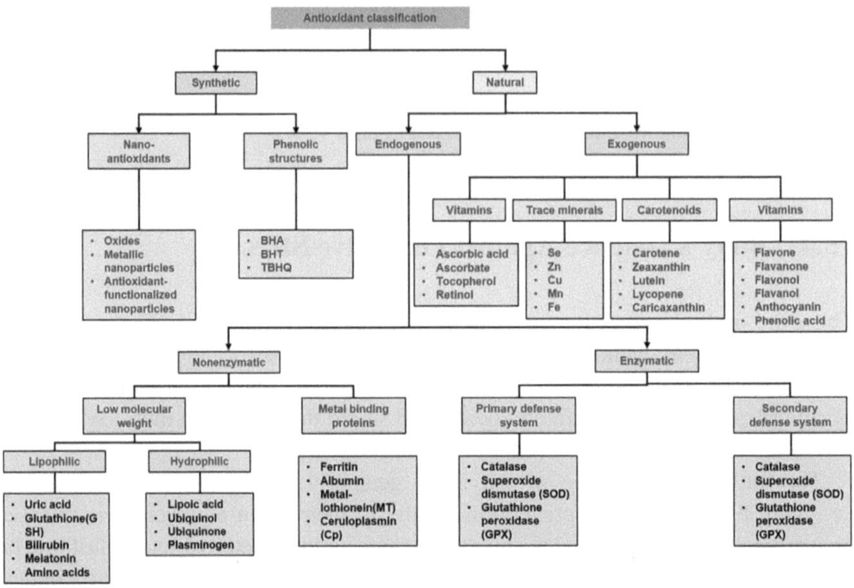

Fig. 15.4 Classification of antioxidants. (Figures created in Biorender.com)

15.6 Pharmacological Insights into Antioxidants

Antioxidants have a critical role in the body's defence system against oxidative stress, an imbalance between antioxidants and free radicals. This leads to damage the cells and is related with various diseases as discussed above in this chapter. To understand the pharmacological insights into antioxidants involves their mechanism of action sources and implications for health (Flieger et al. 2021).

Antioxidants mitigate oxidative stress through several mechanisms, i.e., free radical scavenging, regeneration of other antioxidants, enzymatic actions, modulation of signaling pathways, stimulation of endogenous antioxidant production, etc. (Huchzermeyer et al. 2022).

15.6.1 Free Radical Scavenging

In free radical scavenging, antioxidants neutralise free radicals by donating electrons so the chain reaction (Fig. 15.5), which could lead to cellular damage, to be stopped, this protects the cellular components such as lipids, proteins and DNA from oxidative stress (Flieger et al. 2021). Examples of free radical scavengers are, vitamin C (ascorbic acid), vitamin E (α-tocopherol), etc. (Chaudhary et al. 2023).

Vitamin C can donate electrons to free radicals for scavenge their activity, ascorbic acid can form dehydroascorbate (A) and semidehydroascorbate radical (A^{-}), by reacting with singlet oxygen ($O_2^{\bullet-}$), peroxyl radical ($ROO^{\bullet}$), and

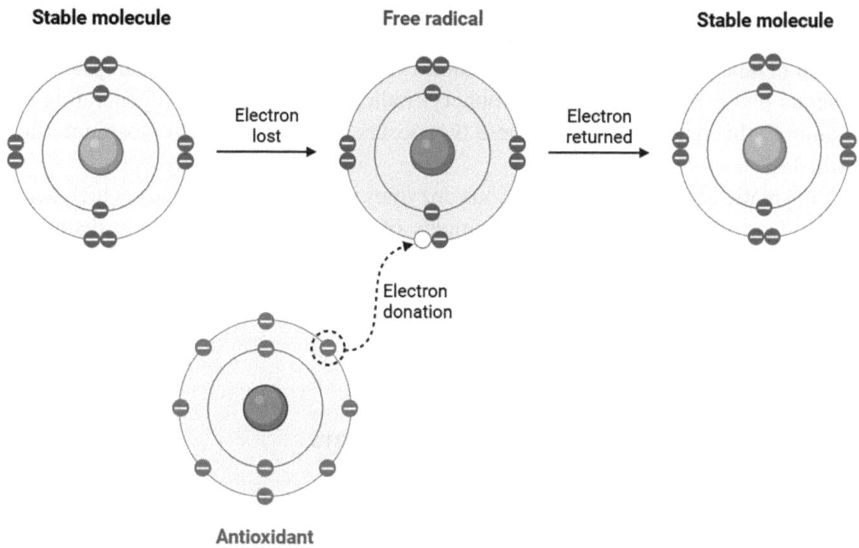

Fig. 15.5 Antioxidant mechanism of free radical scavenging

hydrogen peroxide (H_2O_2) and hydroxyl radical (OH) (Phaniendra et al. 2015). As shown in the following equations:

$$AH^- + \bullet OH \underset{\text{Vit C}}{\rightarrow} H2O + A^{\bullet -}$$

$$AH^- + 1O2 + H^+ \underset{\text{Vit C}}{\rightarrow} H2O2 + A^{\bullet -}$$

$$AH^- + ROO\bullet \underset{\text{Vit C}}{\rightarrow} RH + A^{\bullet -}$$

$$AH^- + H_2O_2 + H^+ \underset{\text{Vit C}}{\rightarrow} 2H_2O + A$$

Vitamin E (α-tocopherol) is antioxidant, interrupts the lipid peroxidation and shown major role in modulation of, ROS dependent, cell signaling pathways they act as free radical scavengers by scavenging •OH, O2•-, and $^1O_2$. It disrupts the lipid peroxidation by donating the hydrogen atom to fatty peroxyl radical.

$$\alpha - \text{tocopherol} + LOO- \rightarrow \alpha - \text{tocopherol.} + LOOH$$

15.6.2 Regeneration of Another Antioxidant

Another mechanism is the regeneration of other antioxidants, some antioxidants are there which can regenerate other antioxidants, for example, vitamin C regenerates vitamin E from its oxidised form, enhancing its protective effects (Ryan et al. 2010). Vitamin E is a lipid-soluble vitamin which acts as an antioxidant within biological membranes and neutralizes free radicals, in order to do that vitamin E itself gets oxidized forming vitamin E radical (tocopheryl radical) by reacting with peroxyl radical (Traber and Stevens 2011). Vitamin C is a water-soluble vitamin present in aqueous phase that reacts with tocopheryl radical and reduces it, regenerating the vitamin E, this reaction between these two occurs at the membrane-cytosol interface (Niki 1987).

Enzymatic actions, there are various enzymes produced by the body which act as antioxidant and neutralize ROS, examples of enzymes are superoxide dismutase (SOD), which converts superoxide radicals into H_2O_2, and catalase, breaks down H_2O_2 into water and oxygen, glutathione peroxide, reduces H_2O_2 using glutathione as a substrate (Huchzermeyer et al. 2022).

15.6.3 Modulation of Signaling Pathways

Modulation of signaling pathways, many antioxidants can influence signaling pathways related to inflammation and cell survival, one the examples is compounds like quercetin, which inhibits the NLRP3 inflammasome, and leads to reduced

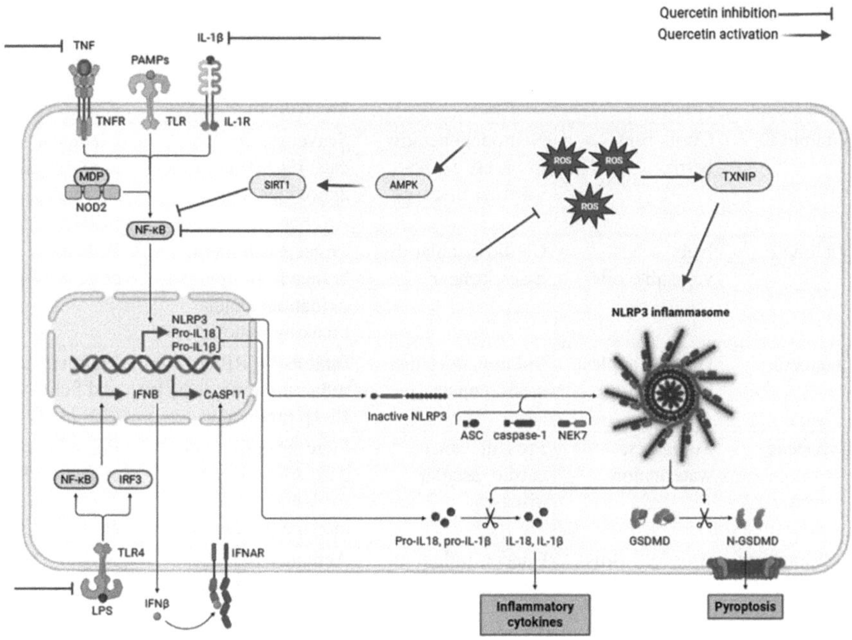

Fig. 15.6 Inhibition of NLRP3 inflammasome by quercetin. (Image created by biorender)

production of inflammatory cytokines (Pham-Huy et al. 2008). Quercetin inhibits the activation of NLRP3 by inhibiting both initiation and activation phase. Initiation phase, quercetin inhibits NFκB from entering the nucleus and inhibits the expression of NLRP3 inflammasome genes, it promotes the AMPK/SIRT1 signaling pathway, quercetin inhibits ROS levels by promoting mitochondrial autophagy and regulating the related oxidative stress, quercetin inhibits expression of thioredoxin-inter acting protein (TXNIP), an activator of the NLRP3 inflammasome as shown in (Fig. 15.6) (Wu et al. 2024).

15.6.4 *Therapeutic Potential of Antioxidants*

Antioxidants are potent agents in the fight against oxidative stress-related diseases. They protect the body by reducing the free radicals. When free radicals are present, they contribute to an excessive event of inflammation, deplete antioxidants, and even affect signaling pathways that can result in cell apoptosis or necrosis (Chaudhary et al. 2023). A diet that is high in antioxidants produced naturally from whole foods–vegetables, fruits, nuts, and seeds to mention a few, is often a direct remedy for life-threatening conditions. It does so using preventing or controlling diseases, such as neurodegenerative disorders, cancer, diabetes, and heart disease (Hajam et al. 2022).

Table 15.5 Showing antioxidants, their origin, role and mechanisms

Antioxidant	Origin	Therapeutic Potential in Diseases	Mechanism of Action	Reference
Vitamin C	Citrus fruits, berries	Neurodegenerative diseases, cardiovascular diseases	Scavenges free radicals; regenerates other antioxidants	Chaudhary et al. (2023), Sil et al. (2016)
Vitamin E	Nuts, seeds, vegetable oils	Cardiovascular diseases, cancer	Protects cell membranes from lipid peroxidation; chain-breaking antioxidant	Willcox et al. (2004)
Quercetin	Onions, apples, berries	Inflammatory diseases, cancer	Inhibits NLRP3 inflammasome; reduces IL-1β production	Talirevic and Sehovic (2012)
Lycopene	Tomatoes, watermelon	Prostate cancer, cardiovascular diseases	Scavenges free radicals; stimulates Nrf2-mediated antioxidant enzyme production	Pham-Huy et al. (2008), Adewoyin et al. (2017)
Sulforaphane	Broccoli, Brussels sprouts	Cancer, inflammatory diseases	Activates Nrf2 pathway; upregulates endogenous antioxidants	Ho et al. (2009)
Glutathione	Produced in the body (from cysteine, glutamine, glycine)	Liver diseases, neurodegenerative disorders	Reduces oxidative stress by detoxifying hydrogen peroxide and free radicals	Chaudhary et al. (2023)
Curcumin	Turmeric	Cancer, arthritis	Inhibits NF-κB and other pro-inflammatory pathways	Khanna et al. (2017)
Resveratrol	Grapes, red wine	Cardiovascular diseases, neurodegenerative diseases	Activates SIRT1 pathway; exhibits anti-inflammatory effects	Arús et al. (2017), Chaudhary et al. (2023)
Selenium	Brazil nuts, seafood	Cancer, cardiovascular diseases	Acts as a cofactor for glutathione peroxidase	Pham-Huy et al. (2008)

Evidence supporting that the synergistic effects in whole foods are frequently greater than those of single antioxidant supplements, thus, justifies the claim that the dietary sources are the best path to health (Willcox et al. 2004). Tables 15.5 and 15.6 show various natural and synthetic antioxidants, their role, and mechanisms in diseases.

15.6.5 Microalgae as Natural Antioxidant Sources

Microalgae have gained exposure among the research community as the case of a natural source of antioxidants with potential value due to their capacity to produce a various bioactive compound such as carotenoids, polyphenols, and tocopherols

Table 15.6 Synthetic antioxidants, their role, origin, and mechanisms

Synthetic Antioxidant	Origin	Therapeutic Potential in Diseases	Mechanism of Action	Reference
Butylated Hydroxyanisole (BHA)	Petroleum derivatives	Food preservation, potential cancer prevention	Prevents lipid oxidation; scavenges free radicals	Chaudhary et al. (2023)
Butylated Hydroxytoluene (BHT)	Petroleum derivatives	Food preservation, potential cardiovascular protection	Scavenges free radicals; stabilizes food products	Ansar et al. (2013)
Propyl Gallate	Derived from gallic acid	Food preservation, potential anti-cancer effects	Inhibits lipid peroxidation; scavenges free radicals	Flieger et al. (2021)
Tocopherol Acetate	Vitamin E derivative	Skin protection, anti-aging, potential cancer prevention	Acts as a chain-breaking antioxidant; protects cell membranes	Santos et al. (2021)
Sodium Ascorbate	Vitamin C derivative	Immune support, potential cancer therapy	Scavenges free radicals; regenerates other antioxidants	Bedhiafi et al. (2022)

(Vignaud et al. 2023). At the same time, these bioactive compounds have a high degree of antioxidant activity. Therefore, they can be readily utilized to impede the action of ROS and to enhance the cellular lifespan and functional capacity of the organs for repairing of damaged tissues secondary to prolonged oxidative stress that is a result of aging. Sustainable growth of microalgae offers a good opportunity for the increased production of these antioxidants, thus, supporting the goals of environmental sustainability (Eze et al. 2023). Prospective studies will be dedicated to the modification of specialized plant technology (SPT) techniques for acquiring better extraction and purification technology together with the calculation of the bioavailability and potency of microalgae-originated nutraceuticals and pharmaceuticals (Pereira et al. 2024).

15.6.6 Development of Nanoantioxidants

The advent of nanotechnology has led to the creation of the so-called "nanoantioxidants" which are supposed to boost the targeted delivery and efficiency of traditional antioxidants. By attaching natural antioxidants to nanoparticles, scientists are able to improve their stability and biologically effectiveness (Hu et al. 2017). As an example, it has been demonstrated that silica nanoparticles with antioxidants functionalized have higher radical scavenging abilities if compared to the non-nanoparticle-form (Deligiannakis et al. 2012). The upcoming studies would possibly include different nanomaterials in antioxidant delivery systems in order to make the treatment of oxidative stress-related conditions more effective (Shah et al. 2022).

Table 15.6 shows various synthetic antioxidants, their role, and mechanisms in diseases:

15.7 Limitations and Challenges of Antioxidant Therapy

Antioxidant therapy is a promising treatment method for oxidative stress-related diseases that have many significant drawbacks. Thus, it has a low bioavailability and is not sufficiently transported to the main point of its action, such as the CNS, which, indeed, lessens its therapeutic utility. In conjunction with this, high doses can be consumptive and toxic while they can induce reductive stress and hence defeating the intended purposes. That is not all, with a surplus of antioxidants, the whole pattern of exercise may outstrip the features of its own and lead cancer patients to get sick (cancer) more often as the cancer treatments may become less effective. On the one hand, the clinical results are frequently unsystematic being numerous trials that did not confirm great advantages, on the other hand, the information gives reason to believe that antioxidants obtained from whole foods are superior to isolated supplements. Together, these aspects draw attention to the necessity for cognition and additional studies in the area of antioxidant therapy (Al-Madhagi and Masoud 2024; Forman and Zhang 2021).

15.8 Conclusions

This chapter reviews the effects of the oxidative stress to the human body and reveals the potential of these dysfunctions in various diseases such as neurodegenerative disorders, cardiovascular diseases, and cancer. The chapter introduces the oxidative and antioxidant balance in the living world by uncovering the mechanisms of ROS generation and the dual roles they play in the organs or tissues of the biological system. On top of that, the chapter analyses the basic pharmacological and clinical aspects of antioxidants. As we move on in antioxidant research, the coming together of alternative means, for example, personalized therapies and the mining of natural sources, brings the expectations for the better prevention and treatment of diseases.

15.9 Future Perspectives in Antioxidant Research

The future of antioxidant research is poised for significant advancements, driven by innovative approaches and a deeper understanding of oxidative stress mechanisms. It will be extended substantially in terms of the use of microalgae (Pereira et al. 2024), deeply utilizing nanotechnology (Shah et al. 2022). These, changes not only a new path to health for humans but also the hope to cure all the diseases related to oxidative stress through their effective fight.

References

Adewoyin M, Ibrahim M, Roszaman R, Md Isa ML, Mat Alewi NA, Abdul Rafa AA, Anuar MNN (2017) Male infertility: the effect of natural antioxidants and phytocompounds on seminal oxidative stress. Diseases 5:9

Al-Madhagi H, Masoud A (2024) Limitations and challenges of antioxidant therapy. Phytother Res 38(12):5549–5566

Ansar S, Tabassum H, Jameil NA (2013) Protective effect of butylated hydroxytoluene on ferric nitrilotriacetate induced hepatotoxicity and oxidative stress in mice. Hum Exp Toxicol 32:513–521

Arús BA, Souza DG, Bellaver B, Souza DO, Goncalves C-A, Quincozes-Santos A, Bobermin LD (2017) Resveratrol modulates GSH system in C6 astroglial cells through heme oxygenase 1 pathway. Mol Cell Biochem 428:67–77

Asmat U, Abad K, Ismail K (2016) Diabetes mellitus and oxidative stress—A concise review. Saudi Pharm J 24:547–553

Awasthi A, Matsunaga Y, Yamada T (2005) Amyloid-beta causes apoptosis of neuronal cells via caspase cascade, which can be prevented by amyloid-beta-derived short peptides. Exp Neurol 196:282–289

Batty M, Bennett MR, Yu E (2022) The role of oxidative stress in atherosclerosis. Cells 11:3843

Bedhiafi T, Inchakalody VP, Fernandes Q, Mestiri S, Billa N, Uddin S, Merhi M, Dermime S (2022) The potential role of vitamin C in empowering cancer immunotherapy. Biomed Pharmacother 146:112553

Berner J, Miebach L, Kordt M, Seebauer C, Schmidt A, Lalk M, Vollmar B, Metelmann H-R, Bekeschus S (2023) Chronic oxidative stress adaptation in head and neck cancer cells generates slow-cyclers with decreased tumour growth in vivo. Br J Cancer 129:869–883

Brand MD (2010) The sites and topology of mitochondrial superoxide production. Exp Gerontol 45:466–472

Chaudhary P, Janmeda P, Docea AO, Yeskaliyeva B, Abdull Razis AF, Modu B, Calina D, Sharifi-Rad J (2023) Oxidative stress, free radicals and antioxidants: potential crosstalk in the pathophysiology of human diseases. Front Chem 11:1158198

Chibaya L, Karim B, Zhang H, Jones SN (2021) Mdm2 phosphorylation by Akt regulates the p53 response to oxidative stress to promote cell proliferation and tumorigenesis. Proc Natl Acad Sci 118:e2003193118

D'amico E, Factor-Litvak P, Santella RM, Mitsumoto H (2013) Clinical perspective on oxidative stress in sporadic amyotrophic lateral sclerosis. Free Radic Biol Med 65:509–527

Daiber A, Chlopicki S (2020) Revisiting pharmacology of oxidative stress and endothelial dysfunction in cardiovascular disease: evidence for redox-based therapies. Free Radic Biol Med 157:15–37

Deligiannakis Y, Sotiriou GA, Pratsinis SE (2012) Antioxidant and antiradical SiO2 nanoparticles covalently functionalized with gallic acid. ACS Appl Mater Interfaces 4:6609–6617

Dowling CM, Hayes SL, Phelan JJ, Cathcart MC, Finn SP, Mehigan B, Mccormick P, Coffey JC, O'Sullivan J, Kiely PA (2017) Expression of protein kinase C gamma promotes cell migration in colon cancer. Oncotarget 8:72096

Dröge W (2002) Free radicals in the physiological control of cell function. Physiol Rev 82(1):47–95

Dubois-Deruy E, Peugnet V, Turkieh A, Pinet F (2020) Oxidative stress in cardiovascular diseases. Antioxidants 9:864

Enríquez JA (2016) Supramolecular organization of respiratory complexes. Annu Rev Physiol 78: 533–561

Eze CN, Onyejiaka CK, Ihim SA, Ayoka TO, Aduba CC, K Ndukwe J, Nwaiwu O, Onyeaka H (2023) Bioactive compounds by microalgae and potentials for the management of some human disease conditions. AIMS Microbiol 9:55

Farfariello V, Gordienko DV, Mesilmany L, Touil Y, Germain E, Fliniaux I, Desruelles E, Gkika D, Roudbaraki M, Shapovalov G (2022) TRPC3 shapes the ER-mitochondria Ca2+ transfer characterizing tumour-promoting senescence. Nat Commun 13:956

Fischer MT, Wimmer I, Höftberger R, Gerlach S, Haider L, Zrzavy T, Hametner S, Mahad D, Binder CJ, Krumbholz M (2013) Disease-specific molecular events in cortical multiple sclerosis lesions. Brain 136:1799–1815

Flieger J, Flieger W, Baj J, Maciejewski R (2021) Antioxidants: classification, natural sources, activity/capacity measurements, and usefulness for the synthesis of nanoparticles. Materials 14: 4135

Forman HJ, Zhang H (2021) Targeting oxidative stress in disease: promise and limitations of antioxidant therapy. Nat Rev Drug Discov 20:689–709

Ghasemi-Dehnoo M, Amini-Khoei H, Lorigooini Z, Rafieian-Kopaei M (2020) Oxidative stress and antioxidants in diabetes mellitus. Asian Pac J Trop Med 13:431–438

Giraldo E, Lloret A, Fuchsberger T, Viña J (2014) Aβ and tau toxicities in Alzheimer's are linked via oxidative stress-induced p38 activation: protective role of vitamin E. Redox Biol 2:873–877

Griendling KK, Camargo LL, Rios FJ, Alves-Lopes R, Montezano AC, Touyz RM (2021) Oxidative stress and hypertension. Circ Res 128:993–1020

Haider L, Fischer MT, Frischer JM, Bauer J, Höftberger R, Botond G, Esterbauer H, Binder CJ, Witztum JL, Lassmann H (2011) Oxidative damage in multiple sclerosis lesions. Brain 134: 1914–1924

Hajam YA, Rani R, Ganie SY, Sheikh TA, Javaid D, Qadri SS, Pramodh S, Alsulimani A, Alkhanani MF, Harakeh S (2022) Oxidative stress in human pathology and aging: molecular mechanisms and perspectives. Cells 11:552

Hernansanz-Agustín P, Enríquez JA (2021) Generation of reactive oxygen species by mitochondria. Antioxidants 10:415

Ho E, Clarke JD, Dashwood RH (2009) Dietary sulforaphane, a histone deacetylase inhibitor for cancer prevention. J Nutr 139:2393–2396

Hu B, Liu X, Zhang C, Zeng X (2017) Food macromolecule based nanodelivery systems for enhancing the bioavailability of polyphenols. J Food Drug Anal 25:3–15

Huchzermeyer B, Menghani E, Khardia P, Shilu A (2022) Metabolic pathway of natural antioxidants, antioxidant enzymes and ROS providence. Antioxidants 11:761

Ighodaro OM (2018) Molecular pathways associated with oxidative stress in diabetes mellitus. Biomed Pharmacother 108:656–662

Khanna S, Jaiswal KS, Gupta B (2017) Managing rheumatoid arthritis with dietary interventions. Front Nutr 4:301603

Kilanczyk E, Saraswat Ohri S, Whittemore SR, Hetman M (2016) Antioxidant protection of NADPH-depleted oligodendrocyte precursor cells is dependent on supply of reduced glutathione. ASN Neuro 8:1759091416660404

Kwon Y, Lee H, Park H, Lee B, Kwon T-U, Kwon Y-J, Chun Y-J (2023) YPEL3 expression induces cellular senescence via the Hippo signaling pathway in human breast cancer cells. Toxicol Res 39:711–719

Lancel S, Qin F, Lennon SL, Zhang J, Tong X, Mazzini MJ, Kang YJ, Siwik DA, Cohen RA, Colucci WS (2010) Oxidative posttranslational modifications mediate decreased SERCA activity and myocyte dysfunction in Gαq-overexpressing mice. Circ Res 107:228–232

Le Gal K, Schmidt EE, Sayin VI (2021) Cellular redox homeostasis. Antioxidants 10:1377

Li D, Yu Q, Wu R, Tuo Z, Wang J, Ye L, Shao F, Chaipanichkul P, Yoo KH, Wei W (2024) Interactions between oxidative stress and senescence in cancer: mechanisms, therapeutic implications, and future perspectives. Redox Biol 73:103208

Luo M, Zhou L, Huang Z, Li B, Nice EC, Xu J, Huang C (2022) Antioxidant therapy in cancer: rationale and progress. Antioxidants 11:1128

Malaspina A, Puentes F, Amor S (2015) Disease origin and progression in amyotrophic lateral sclerosis: an immunology perspective. Int Immunol 27:117–129

Niki E (1987) Interaction of ascorbate and alpha-tocopherol. Ann N Y Acad Sci 498:186–199

Okado-Matsumoto A, Fridovich I (2001) Subcellular distribution of superoxide dismutases (SOD) in rat liver: Cu, Zn-SOD in mitochondria. J Biol Chem 276:38388–38393

Pereira L, Cotas J, Valado A (2024) Antioxidants from microalgae and their potential impact on human well-being. Explor Drug Sci 2:292–321

Petrillo S, Pelosi L, Piemonte F, Travaglini L, Forcina L, Catteruccia M, Petrini S, Verardo M, D'Amico A, Musarò A (2017) Oxidative stress in Duchenne muscular dystrophy: focus on the NRF2 redox pathway. Hum Mol Genet 26:2781–2790

Pham-Huy LA, He H, Pham-Huy C (2008) Free radicals, antioxidants in disease and health. Int J Biomed Sci 4:89

Phaniendra A, Jestadi DB, Periyasamy L (2015) Free radicals: properties, sources, targets, and their implication in various diseases. Indian J Clin Biochem 30:11–26

Puspita L, Chung SY, Shim J-W (2017) Oxidative stress and cellular pathologies in Parkinson's disease. Mol Brain 10:1–12

Rekatsina M, Paladini A, Piroli A, Zis P, Pergolizzi JV, Varrassi G (2020) Pathophysiology and therapeutic perspectives of oxidative stress and neurodegenerative diseases: a narrative review. Adv Ther 37:113–139

Ryan MJ, Dudash HJ, Docherty M, Geronilla KB, Baker BA, Haff GG, Cutlip RG, Alway SE (2010) Vitamin E and C supplementation reduces oxidative stress, improves antioxidant enzymes and positive muscle work in chronically loaded muscles of aged rats. Exp Gerontol 45:882–895

Santos JS, Tavares GD, Barradas TN (2021) Vitamin E and derivatives in skin health promotion. In: Vitamin E in health and disease-interactions, diseases and health aspects

Selye H (1936) A syndrome produced by diverse nocuous agents. Nature 138:32–32

Shah ST, Chowdhury ZZ, Simarani K, Basirun WJ, Badruddin IA, Hussien M, Alrobei H, Kamangar S (2022) Nanoantioxidants: the fourth generation of antioxidants—recent research roadmap and future perspectives. Coatings 12:1568

Shirley R, Ord EN, Work LM (2014) Oxidative stress and the use of antioxidants in stroke. Antioxidants 3:472–501

Sies H (1985) What is oxidative stress? Oxidative stress and vascular disease. Springer

Sies H (2020) Findings in redox biology: from H2O2 to oxidative stress. J Biol Chem 295:13458–13473

Sies H, Berndt C, Jones DP (2017) Oxidative stress. Annu Rev Biochem 86:715–748

Sil S, Ghosh T, Gupta P, Ghosh R, Kabir SN, Roy A (2016) Dual role of vitamin C on the neuroinflammation mediated neurodegeneration and memory impairments in colchicine induced rat model of Alzheimer disease. J Mol Neurosci 60:421–435

Solleiro-Villavicencio H, Rivas-Arancibia S (2018) Effect of chronic oxidative stress on neuroinflammatory response mediated by CD4+ T cells in neurodegenerative diseases. Front Cell Neurosci 12:114

Talirevic E, Sehovic J (2012) Quercetin in the treatment of dyslipidemia. Med Arch 66:87

Tortora K, Margheri F, Luceri C, Mocali A, Ristori S, Magnelli L, Caderni G, Giovannelli L (2022) Colon fibroblasts from Pirc rats (F344/NTac-Apc am1137) exhibit a proliferative and inflammatory phenotype that could support early stages of colon carcinogenesis. Int J Cancer 150: 362–373

Traber MG, Stevens JF (2011) Vitamins C and E: beneficial effects from a mechanistic perspective. Free Radic Biol Med 51:1000–1013

Tsutsui H, Kinugawa S, Matsushima S (2011) Oxidative stress and heart failure. Am J Phys Heart Circ Phys 301:H2181–H2190

Vichova T, Motovska Z (2013) Oxidative stress: predictive marker for coronary artery disease. Exp Clin Cardiol 18:e88

Vignaud J, Loiseau C, Hérault J, Mayer C, Côme M, Martin I, Ulmann L (2023) Microalgae produce antioxidant molecules with potential preventive effects on mitochondrial functions and skeletal muscular oxidative stress. Antioxidants 12:1050

Volonte D, Sedorovitz M, Cespedes VE, Beecher ML, Galbiati F (2021) Cell autonomous angiotensin II signaling controls the pleiotropic functions of oncogenic K-Ras. J Biol Chem 296: 100242

Willcox JK, Ash SL, Catignani GL (2004) Antioxidants and prevention of chronic disease. Crit Rev Food Sci Nutr 44:275–295

Wu J, Lv T, Liu Y, Liu Y, Han Y, Liu X, Peng X, Tang F, Cai J (2024) The role of quercetin in NLRP3-associated inflammation. Inflammopharmacology 32:3585–3610

Yasui H, Hayashi S, Sakurai H (2005) Possible involvement of singlet oxygen species as multiple oxidants in p450 catalytic reactions. Drug Metab Pharmacokinet 20:1–13

Young I, Woodside J (2001) Antioxidants in health and disease. J Clin Pathol 54:176–186

Zhong H, Yin H (2015) Role of lipid peroxidation derived 4-hydroxynonenal (4-HNE) in cancer: focusing on mitochondria. Redox Biol 4:193–199

Zhu H, Chan KT, Huang X, Cerra C, Blake S, Trigos AS, Anderson D, Creek DJ, De Souza DP, Wang X (2022) Cystathionine-β-synthase is essential for AKT-induced senescence and suppresses the development of gastric cancers with PI3K/AKT activation. elife 11:e71929

Index

N. Srivastava, S. A. Saraf (eds.), *Pharmacological Strategies for Prevention and Treatment of Oxidative Stress*, Oxidative Stress in Applied Basic Research and Clinical Practice, https://doi.org/10.1007/978-3-032-19837-2

MIX
Papier aus verantwortungsvollen Quellen
Paper from responsible sources
FSC® C105338

If you have any concerns about our products,
you can contact us on
ProductSafety@springernature.com

In case Publisher is established outside the EU,
the EU authorized representative is:
Springer Nature Customer Service Center GmbH
Europaplatz 3, 69115 Heidelberg, Germany

Printed by Libri Plureos GmbH
in Hamburg, Germany